STUDY SMARTER

CHAPTER
Test Prep
VIDEO CD

Step-by-step solutions on video for all chapter test exercises from the text

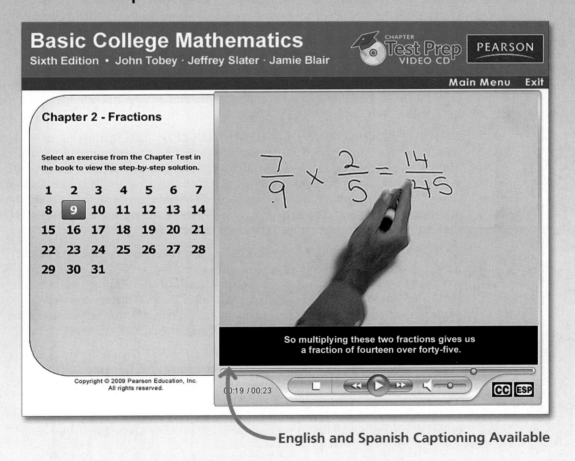

Basic College Mathematics
Sixth Edition · John Tobey · Jeffrey Slater · Jamie Blair

CHAPTER Test Prep VIDEO CD   PEARSON

Main Menu   Exit

Chapter 2 - Fractions

Select an exercise from the Chapter Test in the book to view the step-by-step solution.

1	2	3	4	5	6	7
8	**9**	10	11	12	13	14
15	16	17	18	19	20	21
22	23	24	25	26	27	28
29	30	31

$$\frac{7}{9} \times \frac{2}{5} = \frac{14}{45}$$

So multiplying these two fractions gives us a fraction of fourteen over forty-five.

00:19 / 00:23

CC ESP

English and Spanish Captioning Available

INCLUDED WITH EVERY NEW COPY OF THIS TEXTBOOK!

D1515284

With this edition of the Tobey/Slater/Blair Developmental Mathematics series, we are committed to helping you get the most out of your learning experience by showing you how important math can be in your daily lives. For example, the new *Use Math to Save Money* feature presents practical examples of how math can help you save money, cut costs, and spend less. One of the best ways you can save money is to pass your course the first time you take it. This text and its features and MyMathLab can help you do that.

The best place to start is the ***How Am I Doing? Guide to Math Success.*** This clear path for you to follow is based upon how our successful students have utilized the textbook in the past. Here is how it works:

EXAMPLES and PRACTICE PROBLEMS: When you study an Example, you should immediately do the Practice Problem that follows to make sure you understand each step in solving a particular problem. The worked-out solution to every Practice Problem can be found in the back of the text, starting at page SP-1, so you can check your work and receive immediate guidance in case you need to review.

EXERCISE SETS—Practice, Practice, Practice: You learn math by *doing* math. The best way to learn math is to *practice, practice, practice*. Be sure that you complete every exercise your instructor assigns as homework. In addition, check your answers to the odd-numbered exercises in the back of the text to see whether you have correctly solved each problem.

QUICK QUIZ: After every exercise set, be sure to do the problems in the Quick Quiz. This will tell you immediately if you have understood the key points of the homework exercises.

CONCEPT CHECK: At the end of the Quick Quiz is a concept check. This will test your understanding of the key concept of the section. It will ask you to explain in your own words how a procedure works. It will help you clarify your knowledge of the methods of the section.

HOW AM I DOING? MID-CHAPTER REVIEW: This feature allows you to check if you understand the important concepts covered to that point in a particular chapter. Many students find that halfway through a chapter is a crucial point for review because so many different types of problems have been covered. This review covers each of the types of problems from the first half of the chapter. Do these problems and check your answers at the back of the text. If you need to review any of these problems, simply refer back to the section and objective indicated next to the answer.

HOW AM I DOING? CHAPTER TEST: This test (found at the end of every chapter) provides you with an excellent opportunity to both practice and review for any test you will take in class. Take this test to see how much of the chapter you have mastered. By checking your answers, you can once again refer back to the section and the objective of any exercise you want to review further. This allows you to see at once what has been learned and what still needs more study as you prepare for your test or exam.

HOW AM I DOING? CHAPTER TEST PREP VIDEO CD: If you need to review any of the exercises from the *How Am I Doing? Chapter Test,* this video CD found at the front of the text provides a clear explanation of how to do each step of every problem on the test. Simply insert the CD into a computer and watch a math instructor solve each of the Chapter Test exercises in detail. By reviewing these problems, you can study through any points of difficulty and better prepare yourself for your upcoming test or exam.

These steps provide a clear path you can follow in order to successfully complete your math course. More importantly, the ***How Am I Doing? Guide to Math Success*** is a tool to help you achieve an understanding of mathematics. We encourage you to take advantage of this new feature.

John Tobey and Jeffrey Slater
North Shore Community College

Jamie Blair
Orange Coast College

Essentials of Basic College Mathematics

Annotated Instructor's Edition

2nd Edition

Essentials of Basic College Mathematics

John Tobey

North Shore Community College
Danvers, Massachusetts

Jeffrey Slater

North Shore Community College
Danvers, Massachusetts

Jamie Blair

Orange Coast College
Costa Mesa, California

With Contributions from Jennifer Crawford

Prentice Hall
is an imprint of

Upper Saddle River, NJ 07458

Editorial Director, Mathematics: *Christine Hoag*
Editor in Chief: *Paul Murphy*
Executive Project Manager: *Kari Heen*
Senior Project Editor: *Lauren Morse*
Assistant Editors: *Georgina Brown and Christine Whitlock*
Production Management: *Elm Street Publishing Services*
Senior Managing Editor: *Linda Mihatov Behrens*
Operations Specialist: *Ilene Kahn*
Senior Operations Supervisor: *Diane Peirano*
Marketing Manager: *Marlana Voerster*
Marketing Assistant: *Nathaniel Koven*
Art Director: *Heather Scott*
Interior/Cover Designer: *Tamara Newnam*
AV Project Manager: *Thomas Benfatti*
Executive Manager, Course Production: *Peter Silvia*
Media Producer: *Audra J. Walsh*
Associate Producer: *Emilia Yeh*
Manager, Content Development: *Rebecca Williams*
QA Manager: *Marty Wright*
Senior Content Developer: *Mary Durnwald*
Photo Research Development Manager: *Elaine Soares*
Image Permission Coordinator: *Kathy Gavilanes*
Photo Researcher: *Stephen Forsling*
Manager, Cover Visual Research and Permissions: *Karen Sanatar*
Cover Image: *Ryan McVay/Photodisc/Getty Images, Inc.*
Compositor: *Macmillan Publishing Solutions*
Art Studios: *Scientific Illustrators and Laserwords*

Photo credits appear on page P-1, which constitutes a continuation of the copyright page.

Prentice Hall
is an imprint of

PEARSON

© 2009, 2006 by Pearson Education, Inc.
Pearson Prentice Hall
Pearson Education, Inc.
Upper Saddle River, New Jersey 07458

Printed in the United States of America

10 9 8 7 6 5 4 3 2 1

ISBN 10: 0-321-57066-9 (Annotated Instructor's Edition)
ISBN 13: 978-0-321-57066-6 (Annotated Instructor's Edition)

ISBN 10: 0-321-57065-0 (Student Edition)
ISBN 13: 978-0-321-57065-9 (Student Edition)

Pearson Education Ltd., London
Pearson Education Singapore, Pte. Ltd.
Pearson Education Canada, Inc.
Pearson Education, Japan
Pearson Education Australia PTY, Limited

Pearson Education North Asia, Ltd., Hong Kong
Pearson Educación de Mexico, S.A. de C.V.
Pearson Education Malaysia, Pte. Ltd.
Pearson Education Upper Saddle River,
    New Jersey

This book is dedicated to Nancy Tobey
A loving wife for forty-one years,
An outstanding mother of three children,
A joyful and thankful grandmother,
A true friend

Contents

CHAPTER 1

Whole Numbers   1

CHAPTER 2

Fractions   105

CHAPTER 3

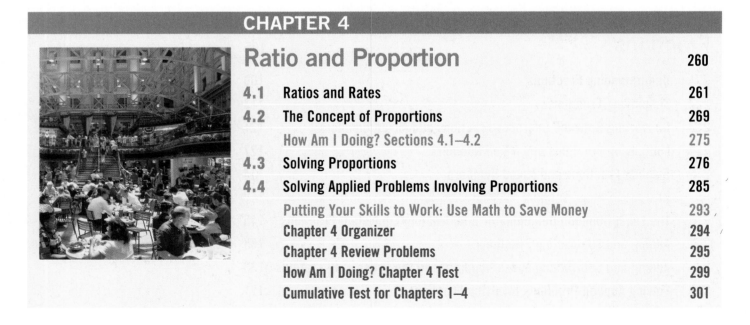

Decimals   194

CHAPTER 4

Ratio and Proportion   260

CHAPTER 5

Percent   302

Preface

TO THE INSTRUCTOR

One of the hallmark characteristics of *Essentials of Basic College Mathematics* that makes the text easy to learn and teach from is the building-block organization. Each section is written to stand on its own, and every homework set is completely self-testing. Exercises are paired and graded and are of varying levels and types to ensure that all skills and concepts are covered. As a result, the text offers students an effective and proven learning program suitable for a variety of course formats—including lecture-based classes; discussion-oriented classes; distance learning classes; modular, self-paced courses; mathematics laboratories; and computer-supported centers. The book has been written to be especially helpful in online courses. The authors usually teach at least one course online each semester.

   Essentials of Basic College Mathematics is part of a series that includes the following:

   Tobey/Slater/Blair, *Basic College Mathematics,* Sixth Edition
   Tobey/Slater/Blair, *Essentials of Basic College Mathematics,* Second Edition
   Blair/Tobey/Slater, *Prealgebra,* Fourth Edition
   Tobey/Slater, *Beginning Algebra,* Seventh Edition
   Tobey/Slater/Blair, *Beginning Algebra: Early Graphing,* Second Edition
   Tobey/Slater, *Intermediate Algebra,* Sixth Edition
   Tobey/Slater/Blair, *Beginning and Intermediate Algebra,* Third Edition

We have visited and listened to teachers across the country and have incorporated a number of suggestions into this edition to help you with the particular learning delivery system at your school. The following pages describe the key continuing features and changes in the second edition.

NEW! FEATURES IN THE SECOND EDITION

Quick Quiz

At the end of each problem section, there is a **Quick Quiz** for the student. The quiz contains three problems that cover all of the essential content of that entire section of the book. If a student can do those three problems, the student has mastered the mathematics skills of that section. If a student cannot do those three problems, the student is made aware that further study is needed to obtain mastery. At the end of the Quick Quiz is a **Concept Check** question. This question stresses a mastery of the concepts of each section of the book. The question asks the student to explain how and why a solution method actually works. It forces the student to analyze problems and reflect on the mathematical concepts that have been learned. The student is asked to explain in his or her own words the mathematical procedures that have been practiced in a given section.

Classroom Quiz

Adjacent to each Quick Quiz for the student, the Annotated Instructor's Edition of the book contains a **Classroom Quiz.** This quiz allows the instructor to give a short quiz in class that covers the essential content of each section of the book. Immediately, an instructor can find out if the students have mastered the material in this section or not. No more will an instructor have to rush to pick out the right balance of problems that test a student's knowledge of a section. The Classroom Quiz is instantly available to assist the instructor in assessing student knowledge.

Putting Your Skills to Work: Use Math to Save Money

Each chapter of the book presents a simple, down-to-earth, practical example of how to save money. Students are given a straightforward, realistic way to cut costs and spend less. They are shown logical ways to get out of debt. They are given motivating examples of how other students saved money. Students are very motivated to read these articles and soon begin to see how the course will actually help them in everyday life. Many of these activities were contributed by professors, and our thanks go out to Mary Pearce, Suellen Robinson, Mike Yarbrough, Betty Ludlum, Connie Buller, Armando Perez, and Maria Luisa Mendez for their insightful contributions.

Teaching Examples

Having in-class practice problems readily available is extremely helpful to both new and experienced instructors. These instructor examples, called **Teaching Examples,** are included in the margins of the Annotated Instructor's Edition for practice in class.

KEY FEATURES IN THE SECOND EDITION

The **How Am I Doing? Guide to Math Success** shows students how they can effectively use this textbook to succeed in their mathematics course. This clear path for them to follow is based upon how students have successfully utilized this textbook in the past.

The text has been designed so the **Examples and Practice Problems** are clearly connected in a cohesive unit. This encourages students to try the Practice Problem associated with each Example to ensure they understand each step in solving a particular problem. The worked-out solution to every Practice Problem can be found in the back of the text so students can check their work. **New!** to this edition, the answers are now included next to each Practice Problem in the Annotated Instructor's Edition.

Each **exercise set** progresses from easy to medium to challenging problems, with appropriate quantities of each, and has paired even and odd problems. All concepts are fully represented with every Example from the section covered by a group of exercises. Exercise sets include **Mixed Practice** problems, which require students to identify the type of problem and the best method they should use to solve it, as well as **Verbal and Writing Skills** exercises, which allow students to explain new concepts fully in their own words. Throughout the text the application exercises have been updated. These **Applications** relate to everyday life, global issues beyond the borders of the United States, and other academic disciplines. Roughly 25 percent of the applications have been contributed by actual students based on scenarios they have encountered in their home or work lives.

Many students find that halfway through a chapter is a crucial point for review because so many different types of problems have been covered. The **How Am I Doing? Mid-Chapter Review** covers each of the types of problems from the first half of the chapter and allows instructors to check if students understand the important concepts covered to that point. Specific section and objective references are provided with each answer to indicate where a student should look for further review.

Developing Problem-Solving Abilities

As authors, we are committed to producing a textbook that emphasizes mathematical reasoning and problem-solving techniques as recommended by AMATYC, NCTM, AMS, NADE, MAA, and other bodies. To this end, the problem sets are built on a wealth of real-life and real-data applications. Unique problems have been developed and incorporated into the exercise sets that help train students in data interpretation, mental mathematics, estimation, geometry and graphing, number sense, critical thinking, and decision making.

The successful **Mathematics Blueprint for Problem Solving** strengthens problem-solving skills by providing a consistent and interactive outline to help students organize their approach to problem solving. Once students fill in the blueprint, they can refer back to their plan as they do what is needed to solve the problem. Because of its flexibility, this feature can be used with single-step problems, multi-step problems, applications, and nonroutine problems that require problem-solving strategies.

The **Developing Your Study Skills** boxes are integrated throughout the text to provide students with techniques for improving their study skills and succeeding in math courses.

Integration and Emphasis on Geometry Due to the emphasis on geometry on many statewide exams, geometry problems are integrated throughout the text. Examples and exercises that incorporate a principle of geometry are marked with a triangle icon for easy identification.

When students encounter mathematics in real-world publications, they often encounter data represented in a **graph, chart, or table** and are asked to make a reasonable conclusion based on the data presented. This emphasis on graphical interpretation is a continuing trend with today's expanding technology. In this text, students are asked to make simple interpretations, to solve medium-level problems, and to investigate challenging applied problems based on the data shown in a chart, graph, or table.

Mastering Mathematical Concepts

Text features that develop the mastery of concepts include the following:

Concise **Learning Objectives** listed at the beginning of each section allow students to preview the goals of that section.

To Think About questions extend the concept being taught, providing the opportunity for all students to stretch their minds, to look for patterns, and to make conclusions based on their previous experience. These critical-thinking questions may follow Examples in the text and appear in the exercise sets.

Almost every exercise set concludes with a section of **Cumulative Review** problems. These problems review topics previously covered, and are designed to assist students in retaining the material.

Calculator boxes are placed in the margin of the text to alert students to a scientific calculator application. In the exercise section a scientific calculator icon is used to indicate problems that are designed for solving with a calculator. There is also instruction on how to use a scientific calculator in an appendix.

Reviewing Mathematical Concepts

At the end of each chapter, we have included problems and tests to provide your students with several different formats to help them review and reinforce the ideas that they have learned. This assists them not only with that specific chapter, but reviews previously covered topics as well.

The concepts and mathematical procedures covered are reviewed at the end of each chapter in a unique **Chapter Organizer.** It lists concepts and methods, and provides a completely worked-out example for each type of problem.

Chapter Review Problems are grouped by section as a quick refresher at the end of the chapter. They can also be used by the student as a quiz of the chapter material.

Found at the end of the chapter, the **How Am I Doing? Chapter Test** is a representative review of the material from that particular chapter that simulates an actual testing format. This provides the students with a gauge of their preparedness for the actual examination.

At the end of each chapter is a **Cumulative Test.** One-half of the content of each cumulative test is based on the math skills learned in previous chapters. By completing these tests for each chapter, the students build confidence that they have mastered not only the contents of the chapter but those of previous chapters as well.

RESOURCES FOR THE STUDENT

Student Solutions Manual
(ISBNs: 0-321-56851-6, 978-0-321-56851-9)

- Solutions to all odd-numbered section exercises
- Solutions to every exercise (even and odd) in the Quick Quiz, mid-chapter reviews, chapter reviews, chapter tests, and cumulative reviews

Worksheets for Classroom or Lab Practice
(ISBNs: 0-321-57775-2, 978-0-321-57775-7)

- Extra practice exercises for every section of the text with ample space for students to show their work

Chapter Test Prep Video CD
Provides step-by-step video solutions to each problem in each How Am I Doing? Chapter Test in the textbook. Automatically included with every new copy of the text, inside the front cover.

Lecture Series on DVD
(ISBNs: 0-321-57795-7, 978-0-321-57795-5)

- Organized by section, contain problem-solving techniques and examples from the textbook
- Step-by-step solutions to selected exercises from each textbook section

MathXL® Tutorials on CD
(ISBNs: 0-321-57777-9, 978-0-321-57777-1)
This interactive tutorial CD-ROM provides:

- Algorithmically generated practice exercises correlated at the objective level
- Practice exercises accompanied by an example and a guided solution
- Tutorial video clips within the exercise to help students visualize concepts
- Easy-to-use tracking of student activity and scores and printed summaries of students' progress

RESOURCES FOR THE INSTRUCTOR

Annotated Instructor's Edition
(ISBNs: 0-321-57066-9, 978-0-321-57066-6)

- Complete student text with answers to all practice problems, section exercises, mid-chapter reviews, chapter reviews, chapter tests, cumulative tests, and practice final exam.

- Teaching Tips placed in the margin at key points where students historically need extra help
- **New!** Teaching Examples provide in-class practice problems and are placed in the margins accompanying each example.

Instructor's Solutions Manual
(ISBNs: 0-321-56853-2, 978-0-321-56853-3)

- Detailed step-by-step solutions to the even-numbered section exercises
- Solutions to every exercise (odd and even) in the Classroom Quiz, mid-chapter reviews, chapter reviews, chapter tests, cumulative tests, and practice final

Instructor's Resource Manual with Tests and Mini-Lectures
(ISBNs: 0-321-56854-0, 978-0-321-56854-0)

- For each section there is one Mini-Lecture with key learning objectives, classroom examples, and teaching notes.
- Two short group activities per chapter are provided in a convenient ready-to-use handout format.
- Three forms of additional practice exercises that help instructors support students of different ability and skill levels.
- Answers are included for all items.
- Alternate test forms with answers:
  - Two Chapter Pretests per chapter (1 free response, 1 multiple choice)
  - Six Chapter Tests per chapter (3 free response, 3 multiple choice)
  - Two Cumulative Tests per even-numbered chapter (1 free response, 1 multiple choice)
  - Two Final Exams (1 free response, 1 multiple choice)

TestGen®

- Enables instructors to build, edit, print, and administer tests.
- Features a computerized bank of questions developed to cover all text objectives.
- Creates multiple but equivalent versions of the same question or test with the click of a button.
- Instructors can modify questions or add new questions.
- Tests can be printed or administered online.

The software and testbank are available for download from Pearson Education's online catalog.

Pearson Adjunct Support Center

The Pearson Adjunct Support Center is staffed by qualified mathematics instructors with more than 50 years of combined experience at both the community college and university level. Assistance is provided for faculty in the following areas:

- Suggested syllabus consultation
- Tips on using materials packed with your book
- Book-specific content assistance
- Teaching suggestions including advice on classroom strategies

MEDIA RESOURCES

MathXL® www.mathxl.com

MathXL is a powerful online homework, tutorial, and assessment system that accompanies Pearson Education textbooks in mathematics and statistics. With MathXL, instructors can create, edit, and assign online homework and tests using algorithmically generated exercises correlated at the objective level to the textbook. They can also create and assign their own online exercises and import TestGen tests for added flexibility. All student work is tracked in MathXL's online gradebook. Students can take chapter tests in MathXL and receive personalized study plans based on their test results. The study plan diagnoses weaknesses and links students directly to tutorial exercises for the objectives they need to study and retest. Students can also access supplemental animations and video clips directly from selected exercises. MathXL is available to qualified adopters. For more information, visit our Web site at www.mathxl.com, or contact your sales representative.

MyMathLab® www.mymathlab.com

MyMathLab is a series of text-specific, easily customizable online courses for Pearson Education textbooks in mathematics and statistics. Powered by CourseCompass™ (our online teaching and learning environment) and MathXL® (our online homework, tutorial, and assessment system), MyMathLab gives instructors the tools they need to deliver all or a portion of their course online, whether students are in a lab or working at home or elsewhere. MyMathLab provides a rich and flexible set of course materials, featuring free-response exercises that are algorithmically generated for unlimited practice and mastery. Students can also use online tools, such as video lectures, animations, and a multimedia textbook, to independently improve their understanding and performance. Instructors can use MyMathLab's homework and test managers to select and assign online exercises correlated directly to the textbook, and they can create and assign their own online exercises and import TestGen tests for added flexibility.

MyMathLab's online gradebook—designed specifically for mathematics and statistics—automatically tracks students' homework and test results and gives the instructor control over how to calculate final grades. Instructors can also add offline (paper-and-pencil) grades to the gradebook. MyMathLab also includes access to the Pearson Tutor Center, which provides students with tutoring via toll-free phone, fax, e-mail, and interactive Web sessions. MyMathLab is available to qualified adopters. For more information, visit our Web site at www.mymathlab.com, or contact your sales representative.

ACKNOWLEDGMENTS

This book is the product of many years of work and many contributions from faculty and students across the country. We would like to thank the many reviewers and participants in focus groups and special meetings with the authors in preparation of previous editions. Our deep appreciation to each of the following:

John Akutagawa, *Heald Business College*

George J. Apostolopoulos, *DeVry Institute of Technology*

Sohrab Bakhtyari, *St. Petersburg Junior College—Clearwater*

Katherine Barringer, *Central Virginia Community College*

Christine R. Bauman, *Clark College at Larch*

Rita Beaver, *Valencia Community College*

Gopa Bhowmick, *Mississippi Gulf Coast College*

Jamie Blair, *Orange Coast College*

Jon Blakely, *College of the Sequoias*

Larry Blevins, *Tyler Junior College*

Matt Bourez, *College of the Sequoias*

Vernon Bridges, *Durham Technical Community College*

Jared Burch, *College of the Sequoias*

Connie Buller, *Metropolitan Community College*

Oscar Caballero III, *Laredo Community College*

Brenda Callis, *Rappahannock Community College*

Joan P. Capps, *Raritan Valley Community College*

Robert Christie, *Miami-Dade Community College*

Nelson Collins, *Joliet Junior College*

Mike Contino, *California State University at Heyward*

Yen-Phi (Faye) Dang, *Joliet Junior College*

Callie Jo Daniels, *St. Charles County Community College*

Ky Davis, *Muskingum Area Technical College*

Judy Dechene, *Fitchburg State University*

Floyd L. Downs, *Arizona State University*

Barbara Edwards, *Portland State University*

Disa Enegren, *Rose State College*

Janice F. Gahan-Rech, *University of Nebraska at Omaha*

Naomi Gibbs, *Pitt Community College*

Colin Godfrey, *University of Massachusetts, Boston*

Nancy Graham, *Rose State College*

Mary Beth Headlee, *Manatee Community College*

Laura Huerta, *Laredo Community College*

Joe Karnowski, *Norwalk Community College*

Kay Kriewald, *Laredo Community College*

Douglas Lewis, *Yakima Valley Community College*

Sharon Louvier, *Lee College*

Luanne Lundberg, *Clark College*

Doug Mace, *Baker College*

Carl Mancuso, *William Paterson College*

James A. Matovina, *Community College of Southern Nevada*

Janet McLaughlin, *Montclair State College*

Beverly Meyers, *Jefferson College*

Wayne L. Miller, *Lee College*

Gloria Mills, *Tarrant County Junior College*

Norman Mittman, *Northeastern Illinois University*

Jody E. Murphy, *Lee College*

Katrina Nichols, *Delta College*

Henri Onuigbo, *Wayne County Community College*

Leticia M. Oropesa, *University of Miami*

Sandra Orr, *West Virginia State Community College*

Jim Osborn, *Baker College*

Linda Padilla, *Joliet Junior College*

Catherine Panik, *Manatee Community College—South Campus*

Gary Phillips, *Clark College*

Elizabeth A. Polen, *County College of Morris*

Joel Rappaport, *Miami Dade Community College*

Jack Roberts, *Ivy Tech State, Sellersburg*

Ronald Ruemmler, *Middlesex County College*

Dennis Runde, *Manatee Community College*

Cindy Satriano, *Albuquerque Technical Vocational Institute*

Sally Search, *Tallahassee Community College*

Jeffrey Simmons, *Ivy Tech State, Ft. Wayne*

Richard Sturgeon, *University of Southern Maine*

Ara B. Sullenberger, *Tarrant County Community College*

Brad Sullivan, *Community College of Denver*

Margie Thrall, *Manatee Community College*

Michael Trappuzanno, *Arizona State University*

Bettie Truitt, *Black Hawk College*

Cora S. West, *Florida Community College at Jacksonville*

Jacquelyne Wing, *Angelina College*

Jerry Wisnieski, *Des Moines Community College*

In addition, we want to thank the following individuals for providing splendid insight and suggestions for this new edition:

Suzanne Battista, *St. Petersburg Junior College—Clearwater*

Karen Bingham, *Clarion College*

Nadine Branco, *Western Nevada Community College*

Connie Buller, *Metropolitan Community College*

John Close, *Salt Lake Community College*

Patricia Donovan, *San Joaquin Delta College*

Colin Godfrey, *UMass Boston*

Shanna Goff, *Grand Rapids Community College*

Edna Greenwood, *Tarrant County College, Northwest*

Peter Kaslik, *Pierce College*

Joyce Keenan, *Horry-Georgetown Technical College*

Carolyn Krause, *Delaware Technical and Community College*

Nam Lee, *Griffin Technical Institute*

Tanya Lee, *Career Technical College*

Betty Ludlum, *Austin Community College*

Mary Marlin, *Western Virginia Northern Community College*

Carolyn T. McIntyre, *Horry-Georgetown Technical College*

Maria Luisa Mendez, *Laredo Community College*

Steven J. Meyer, *Erie Community College*

Marcia Mollé, *Metropolitan Community College*

Jay L. Novello, *Horry-Georgetown Technical College*

Sandra Lee Orr, *West Virginia State University*

Mary Pearce, *Wake Technical Community College*

Armando Perez, *Laredo Community College*

Regina Pierce, *Davenport University*

Anne Praderas, *Austin Community College*

Suellen Robinson, *North Shore Community College*

Kathy Ruggieri, *Lansdale School of Business*

Randy Smith, *Des Moines Area Community College*

Dina Spain, *Horry-Georgetown Technical College*

Lori Welder, *PACE Institute*

Kimberly Williams-Brito, *Cosumnes River College*

Michael Yarbrough, *Cosumnes River College*

We have been greatly helped by a supportive group of colleagues who not only teach at North Shore Community College but have also provided a number of ideas as well as extensive help on all of our mathematics books. Our special best wishes to our colleague Bob Campbell, who recently retired. He has given us a friendly smile and encouraging ideas for 35 years! Also, a special word of thanks to Wally Hersey, Judy Carter, Rick Ponticelli, Lora Connelly, Sharyn Sharaf, Donna Stefano, Nancy Tufo, Elizabeth Lucas, Anne O'Shea, Marsha Pease, Walter Stone, Evangeline Cornwall, Rumiya Masagutova, Charles Peterson, and Neha Jain.

Jenny Crawford provided major contributions to this revision. She provided new problems, new ideas, and great mental energy. She greatly assisted us during the production process. She made helpful decisions. Her excellent help was much appreciated. She has become an essential part of our team as we work to provide the best possible textbook.

A special word of thanks goes to Cindy Trimble and Associates for their excellent work in accuracy checking manuscript and page proofs, as well as Twin Prime Editorial for their assistance reviewing page proofs.

Each textbook is a combination of ideas, writing, and revisions from the authors and wise editorial direction and assistance from the editors. We especially want to thank our editor at Pearson Education—Paul Murphy. He has a true vision of how authors and editors can work together as partners, and it has been a rewarding experience to create and revise textbooks together with him. We especially want to thank our Project Manager Lauren Morse for patiently answering questions and solving many daily problems. We also want to thank our entire team at Pearson Education—Marlana Voerster, Nathaniel Koven, Christine Whitlock, Georgina Brown, Linda Behrens, Tom Benfatti, Heather Scott, Ilene Kahn, Audra Walsh, and MiMi Yeh—as well as Allison Campbell and Karin Kipp at Elm Street Publishing Services for their assistance during the production process.

Nancy Tobey served as our administrative assistant. Daily she was involved with mailing, photocopying, collating, and taping. A special thanks goes to Nancy. We could not have finished the book without you.

Book writing is impossible for us without the loyal support of our families. Our deepest thanks and love to Nancy, Johnny, Melissa, Marcia, Shelley, Rusty, and Abby. Your understanding, your love and help, and your patience have been a source of great encouragement. Finally, we thank God for the strength and energy to write and the opportunity to help others through this textbook.

We have spent more than 37 years teaching mathematics. Each teaching day, we find that our greatest joy is helping students learn. We take a personal interest in ensuring that each student has a good learning experience in taking this course. If you have some personal comments, suggestions, or ideas for future editions of this textbook, please write to us at:

Prof. John Tobey, Prof. Jeffrey Slater, and Prof. Jamie Blair
Pearson Education
Office of the College Mathematics Editor
75 Arlington Street, Suite 300
Boston, MA 02116

or e-mail us at

jtobey@northshore.edu

We wish you success in this course and in your future life!

John Tobey
Jeffrey Slater
Jamie Blair

Diagnostic Pretest: Essentials of Basic College Mathematics

Chapter 1

1. Add. $3846 + 527$

2. Divide. $58\overline{)1508}$

3. Subtract.
$$\begin{array}{r} 12{,}807 \\ -11{,}679 \end{array}$$

4. The highway department used 115 truckloads of sand. Each truck held 8 tons of sand. How many tons of sand were used?

Chapter 2

5. Add. $\dfrac{3}{7} + \dfrac{2}{5}$

6. Multiply and simplify. $3\dfrac{3}{4} \times 2\dfrac{1}{5}$

7. Subtract. $2\dfrac{1}{6} - 1\dfrac{1}{3}$

8. Mike's car traveled 237 miles on $7\dfrac{9}{10}$ gallons of gas. How many miles per gallon did he achieve?

Chapter 3

9. Multiply.
$$\begin{array}{r} 51.06 \\ \times\, 0.307 \end{array}$$

10. Divide. $0.026\overline{)0.0884}$

11. The copper pipe was 24.375 centimeters long. Paula had to shorten it by cutting off 1.75 centimeters. How long will the copper pipe be when it is shortened?

12. Russ bicycled 20.5 miles on Monday, 5.8 miles on Tuesday, and 14.9 miles on Wednesday. How many miles did he bicycle on those three days?

1.	4373
2.	26
3.	1128
4.	920 tons of sand
5.	$\dfrac{29}{35}$
6.	$8\dfrac{1}{4}$
7.	$\dfrac{5}{6}$
8.	30 miles per gallon
9.	15.67542
10.	3.4
11.	22.625 centimeters
12.	41.2 miles

13. ___n = 10.3___

14. ___n = 352___

15. ___$1080___

16. ___225 miles___

17. ___37.5%___

18. ___7728___

19. ___3900 students___

20. ___0.3% are defective___

Chapter 4

Solve each proportion problem. Round to the nearest tenth if necessary.

13. $\dfrac{3}{7} = \dfrac{n}{24}$

14. $\dfrac{0.5}{0.8} = \dfrac{220}{n}$

15. Wally's Landscape earned $600 for mowing lawns at 25 houses last week. At that rate, how much would he earn for doing 45 houses?

16. Two cities that are actually 300 miles apart appear to be 8 inches apart on the road map. How many miles apart are two cities that appear to be 6 inches apart on the map?

Chapter 5

Round to the nearest tenth if necessary.

17. Change to a percent: $\dfrac{3}{8}$

18. 138% of 5600 is what number?

19. At Mountainview College 53% of the students are women. There are 2067 women at the college. How many students are at the college?

20. At a manufacturing plant it was discovered that 9 out of every 3000 parts made were defective. What percent of the parts are defective?

Essentials of Basic College Mathematics

CHAPTER 1

For many years there were many more drivers in the United States than there were passenger cars. Over the years, that trend has changed. Now there are more passenger cars than there are drivers in the United States. When did that change occur? How many more cars are there than drivers? The mathematics you learn in this chapter will help you to answer these kinds of questions.

Whole Numbers

Student Learning Objectives

After studying this section, you will be able to:

 1 Write numbers in expanded form.

2 Write whole numbers in standard notation.

3 Write a word name for a number and write a number for a word name.

4 Read numbers in tables.

1 Writing Numbers in Expanded Form

To count a number of objects or to answer the question "How many?" we use a set of numbers called **whole numbers.** These whole numbers are as follows.

$$0, 1, 2, 3, 4, 5, 6, 7, 8, 9, 10, 11, 12, 13, 14, 15, \ldots$$

There is no largest whole number. The three dots . . . indicate that the set of whole numbers goes on indefinitely. Our number system is based on tens and ones and is called the **decimal system** (or the **base 10 system**). The numbers 0, 1, 2, 3, 4, 5, 6, 7, 8, 9 are called **digits.** The position, or placement, of the digits in the number tells the value of the digits. For example, in the number 521, the "5" means 5 hundreds (500). In the number 54, the "5" means 5 tens (50).

For this reason, our number system is called a **place-value system.**

Consider the number 5643. We will use a place-value chart to illustrate the value of each digit in the number 5643.

Place-value Chart

Millions			Thousands			Ones		
					5	6	4	3
Hundred millions	Ten millions	Millions	Hundred thousands	Ten thousands	Thousands	Hundreds	Tens	Ones

The value of the number is 5 thousands, 6 hundreds, 4 tens, 3 ones.

The place-value chart shows the value of each place, from ones on the right to hundred millions on the left. When we write very large numbers, we place a comma after every group of three digits called a **period,** moving from right to left. This makes the number easier to read. It is usually agreed that a four-digit number does not have a comma, but that numbers with five or more digits do. So 32,000 would be written with a comma but 7000 would not.

To show the value of each digit in a number, we sometimes write the number in expanded notation. For example, 56,327 is 5 ten thousands, 6 thousands, 3 hundreds, 2 tens, and 7 ones. In **expanded notation,** this is

$$50,000 + 6000 + 300 + 20 + 7.$$

EXAMPLE 1 Write each number in expanded notation.

(a) 2378   **(b)** 538,271   **(c)** 980,340,654

Solution

(a) Sometimes it helps to say the number to yourself.

$$
\begin{array}{ccccccccc}
& \text{two thousand} & & \text{three hundred} & & \text{seventy} & & \text{eight} \\
2378 = & 2000 & + & 300 & + & 70 & + & 8
\end{array}
$$

(b)
Expanded notation
$$538{,}271 = 500{,}000 + 30{,}000 + 8000 + 200 + 70 + 1$$

(c) When 0 is used as a placeholder, you do not include it in the expanded form.

Expanded notation
$$980{,}340{,}654 = 900{,}000{,}000 + 80{,}000{,}000 + 300{,}000 + 40{,}000 + 600 + 50 + 4$$

Practice Problem 1 Write each number in expanded notation.

(a) 3182   **(b)** 520,890   **(c)** 709,680,059

(a) $3000 + 100 + 80 + 2$ (b) $500{,}000 + 20{,}000 + 800 + 90$
(c) $700{,}000{,}000 + 9{,}000{,}000 + 600{,}000 + 80{,}000 + 50 + 9$

NOTE TO STUDENT: *Fully worked-out solutions to all of the Practice Problems can be found at the back of the text starting at page SP-1*

2 Writing Whole Numbers in Standard Notation

The way that you usually see numbers written is called **standard notation.**
980,340,654 is the standard notation for the number nine hundred eighty million, three hundred forty thousand, six hundred fifty-four.

EXAMPLE 2 Write each number in standard notation.

(a) $500 + 30 + 8$
(b) $300{,}000 + 7000 + 40 + 7$

Solution

(a) 538

(b) Be careful to keep track of the place value of each digit. You may need to use 0 as a placeholder.

$$
\begin{array}{c}
\text{3 hundred thousand} \\
300{,}000 + 7000 + 40 + 7 = 307{,}047 \\
\text{7 thousand}
\end{array}
$$

We needed to use 0 in the ten thousands place and in the hundreds place.

Practice Problem 2 Write each number in standard notation.

(a) $400 + 90 + 2$ 492   **(b)** $80{,}000 + 400 + 20 + 7$ 80,427

Teaching Example 1 Write each number in expanded notation

(a) 3549 **(b)** 146,285 **(c)** 403,621,017

Ans:
(a) three thousand five hundred forty nine
$$3000 + 500 + 40 + 9$$
(b) $100{,}000 + 40{,}000 + 6000 + 200 + 80 + 5$
(c) $400{,}000{,}000 + 3{,}000{,}000 + 600{,}000 + 20{,}000 + 1000 + 10 + 7$

Teaching Example 2 Write each number in standard notation

(a) $600 + 10 + 3$
(b) $500{,}000 + 80{,}000 + 200 + 4$

Ans: (a) 613   **(b)** 580,204

Teaching Tip You can remind students that a few ancient cultures actually avoided the use of zero by never writing numbers like 40. Instead they would make the number larger by writing 41 or smaller by writing 39. Needless to say, such a culture had a hard time developing effective records for business transactions.

Teaching Example 3 The population of Butler County is 3,057,489. In the number 3,057,489

(a) How many thousands are there?

(b) What is the value of the digit 5?

(c) How many millions are there?

(d) In what place is the digit 0?

Ans: (a) 7   **(b)** 50,000   **(c)** 3

(d) hundred thousands place

NOTE TO STUDENT: *Fully worked-out solutions to all of the Practice Problems can be found at the back of the text starting at page SP-1*

EXAMPLE 3   Last year the population of Central City was 1,509,637. In the number 1,509,637

(a) How many ten thousands are there?   **(b)** How many tens are there?

(c) What is the value of the digit 5?   **(d)** In what place is the digit 6?

Solution   A place-value chart will help you identify the value of each place.

(a) Look at the digit in the ten thousands place. There are 0 ten thousands.

(b) Look at the digit in the tens place. There are 3 tens.

(c) The digit 5 is in the hundred thousands place. The value of the digit is 5 hundred thousand or 500,000.

(d) The digit 6 is in the hundreds place.

Practice Problem 3   The campus library has 904,759 books.

(a) What digit tells the number of hundreds? 7

(b) What digit tells the number of hundred thousands? 9

(c) What is the value of the digit 4? 4000

(d) What is the value of the digit 9? Why does this question have two answers? 900,000 for the first 9; 9 for the last 9

③ Writing Word Names for Numbers and Numbers for Word Names

A number has the same *value* no matter how we write it. For example, "a million dollars" means the same as "$1,000,000." In fact, any number in our number system can be written in several ways or forms:

• Standard notation	521
• Expanded notation	500 + 20 + 1
• Word name	five hundred twenty-one

You may want to write a number in any of these ways. To write a check, you need to use both standard notation and words.

Teaching Tip Additional coverage on balancing a checkbook can be found in the Consumer Finance Appendix.

To write a word name, start from the left. Name the number in each period, followed by the name of the period, and a comma. The last period name, "ones," is not used.

EXAMPLE 4 Write a word name for 364,128,957.

Solution

Place-value Chart

Billions			Millions			Thousands			Ones		
			3	6	4	1	2	8	9	5	7
Hundreds	Tens	Ones	Hundreds	Tens	Ones	Hundreds	Tens	Ones	Hundreds	Tens	Ones

We want to write a word name for 364, 128, 957.

　　three hundred sixty-four million,⎤

　　　one hundred twenty-eight thousand,⎦

　　　　nine hundred fifty-seven

　　The answer is three hundred sixty-four million, one hundred twenty-eight thousand, nine hundred fifty-seven.

Practice Problem 4 Write a word name for 267,358,981.

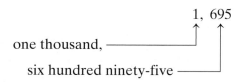

two hundred sixty-seven million, three hundred fifty-eight thousand, nine hundred eighty-one

EXAMPLE 5 Write the word name for each number.

(a) 1695　　　　**(b)** 200,470　　　　**(c)** 7,003,038

Solution Look at the place-value chart if you need help identifying the place for each digit.

(a) To help us, we will put in the optional comma: 1,695.

　　　　　　　　　　　　　1, 695

　　　one thousand, ────────┘

　　　　six hundred ninety-five ──────┘

　　The word name is one thousand, six hundred ninety-five.

(b)

　　　　　　　　　　　　　200, 470

　　　two hundred thousand, ──────┘

　　　　four hundred seventy ──────┘

　　The word name is two hundred thousand, four hundred seventy.

(c)

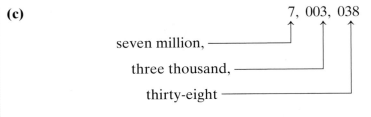

The word name is seven million, three thousand, thirty-eight.

Practice Problem 5 Write the word name for each number.

(a) 2736 **(b)** 980,306 **(c)** 12,000,021

(a) two thousand, seven hundred thirty-six **(b)** nine hundred eighty thousand, three hundred six **(c)** twelve million, twenty-one

CAUTION: DO NOT USE THE WORD <u>AND</u> FOR WHOLE NUMBERS. Many people use the word *and* when giving the word name for a whole number. For example, you might hear someone say the number 34,507 as "thirty-four thousand, five hundred *and* seven." However, this is not technically correct. In mathematics we do NOT use the word *and* when writing word names for whole numbers. In Chapter 3 we will use the word *and* to represent the decimal point. For example, 59.76 will have the word name "fifty-nine *and* seventy-six hundredths."

Very large numbers are used to measure quantities in some disciplines, such as distance in astronomy and the national debt in macroeconomics. We can extend the place-value chart to include these large numbers.

The national debt for the United States as of November 22, 2003, was $6,923,886,720,833. This number is indicated in the following place-value chart.

Place-value Chart

Trillions			Billions			Millions			Thousands			Ones		
		6	9	2	3	8	8	6	7	2	0	8	3	3

EXAMPLE 6 Write the number for the national debt for the United States as of November 22, 2003, in the amount of $6,923,886,720,833 using a word name.

Solution The national debt on November 22, 2003, was six trillion, nine hundred twenty-three billion, eight hundred eighty-six million, seven hundred twenty thousand, eight hundred thirty-three dollars.

Practice Problem 6 As of January 1, 2004, the estimated population of the world was 6,393,646,525. Write this world population using a word name.

The world population on January 1, 2004, was six billion, three hundred ninety-three million, six hundred forty-six thousand, five hundred twenty-five.

Occasionally you may want to write a word name as a number.

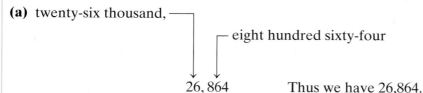 **Write each number in standard notation.**

(a) twenty-six thousand, eight hundred sixty-four

(b) two billion, three hundred eighty-six million, five hundred forty-seven thousand, one hundred ninety

Solution

(a) twenty-six thousand, ⌐ eight hundred sixty-four

26, 864   Thus we have 26,864.

(b) two billion, ⌐ three hundred eighty-six million,

⌐ five hundred forty-seven thousand,

⌐ one hundred ninety

2, 386, 547, 190   Thus we have 2,386,547,190.

Practice Problem 7 Write in standard notation.

(a) eight hundred three   803

(b) thirty thousand, two hundred twenty-nine   30,229

Teaching Example 7 Write each number in standard notation.

(a) six thousand, eighty-seven

(b) one hundred six million, two hundred fifty-three thousand, four hundred thirty-five

Ans: (a) 6087   **(b)** 106,253,435

 Reading Numbers in Tables

Sometimes numbers found in charts and tables are abbreviated. Look at the chart below from the U.S. Bureau of the Census. Notice that the second line tells us the numbers represent thousands. To understand what these numbers mean, think "thousands." If the number 23 appears across from 1740 for New Hampshire, the 23 represents 23 thousand. 23 thousand is 23,000. Note that census figures for some colonies are not available for certain years.

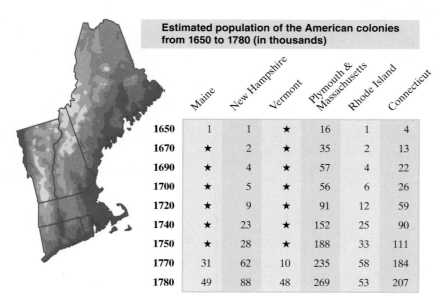

Estimated population of the American colonies from 1650 to 1780 (in thousands)

	Maine	New Hampshire	Vermont	Plymouth & Massachusetts	Rhode Island	Connecticut
1650	1	1	★	16	1	4
1670	★	2	★	35	2	13
1690	★	4	★	57	4	22
1700	★	5	★	56	6	26
1720	★	9	★	91	12	59
1740	★	23	★	152	25	90
1750	★	28	★	188	33	111
1770	31	62	10	235	58	184
1780	49	88	48	269	53	207

Teaching Example 8 Refer to the chart on the previous page to answer the following questions. Write each number in standard notation.

(a) What was the estimated population of Rhode Island in 1650?

(b) What was the estimated population of New Hampshire in 1740?

(c) What was the estimated population of Connecticut in 1750?

Ans: (a) 1000   **(b)** 23,000   **(c)** 111,000

NOTE TO STUDENT: *Fully worked-out solutions to all of the Practice Problems can be found at the back of the text starting at page SP-1*

EXAMPLE 8   Refer to the chart on the previous page to answer the following questions. Write each number in standard notation.

(a) What was the estimated population of Maine in 1780?

(b) What was the estimated population of Plymouth and Massachusetts in 1720?

(c) What was the estimated population of Rhode Island in 1700?

Solution

(a) To read the chart, first look for Maine along the top. Read down to the row for 1780. The number is 49. In this chart 49 means 49 thousands.

$$49 \text{ thousands} \Rightarrow 49{,}000$$

(b) Read the column of the chart for Plymouth and Massachusetts. The number for Plymouth and Massachusetts in the row for 1720 is 91. This means 91 thousands. We will write this as 91,000.

(c) Read the column of the chart for Rhode Island. The number for Rhode Island in the row for 1700 is 6. This means 6 thousands. We will write this as 6000.

TO THINK ABOUT: Interpreting Data in a Table Why do you think Plymouth and Massachusetts had the largest population for the years shown in the table?

Practice Problem 8   Refer to the chart on the previous page to answer the following questions. Write each number in standard notation.

(a) What was the estimated population of Connecticut in 1670? 13,000

(b) What was the estimated population of New Hampshire in 1780? 88,000

(c) What was the estimated population of Vermont in 1770? 10,000

Developing Your Study Skills

Class Participation

People learn mathematics through active participation, not through observation from the sidelines. If you want to do well in this course, get involved in all course activities. If you are in a traditional mathematics class, sit near the front where you can see and hear well, where your focus is on the material being covered in class. Ask questions, be ready to contribute toward solutions, and take part in all classroom activities. Your contributions are valuable to the class and to yourself. Class participation requires an investment of yourself in the learning process, which you will find pays huge dividends.

If you are in an online class or nontraditional class, be sure to e-mail the teacher or talk to the tutor on duty. Ask questions. Think about the concepts. Make your mind interact with the textbook. Be mentally involved. This active mental interaction is the key to your success.

Write each number in expanded notation.

1. 6731 6000 + 700 + 30 + 1

2. 9519 9000 + 500 + 10 + 9

3. 108,276 100,000 + 8000 + 200 + 70 + 6

4. 701,285 700,000 + 1000 + 200 + 80 + 5

5. 23,761,345 20,000,000 + 3,000,000 + 700,000 + 60,000 + 1000 + 300 + 40 + 5

6. 46,198,253 40,000,000 + 6,000,000 + 100,000 + 90,000 + 8000 + 200 + 50 + 3

7. 103,260,768 100,000,000 + 3,000,000 + 200,000 + 60,000 + 700 + 60 + 8

8. 820,310,574 800,000,000 + 20,000,000 + 300,000 + 10,000 + 500 + 70 + 4

Write each number in standard notation.

9. 600 + 70 + 1 671

10. 500 + 90 + 6 596

11. 9000 + 800 + 60 + 3 9863

12. 7000 + 600 + 50 + 2 7652

13. 40,000 + 800 + 80 + 5 40,885

14. 60,000 + 7000 + 200 + 4 67,204

15. 700,000 + 6000 + 200 706,200

16. 300,000 + 40,000 + 800 340,800

Verbal and Writing Skills

17. In the number 437,521
(a) What digit tells the number of thousands? 7
(b) What is the value of the digit 3? 30,000

18. In the number 805,712 **(a)** 0
(a) What digit tells the number of ten thousands?
(b) What is the value of the digit 8? 800,000

19. In the number 1,214,847
(a) What digit tells the number of hundred thousands? 2
(b) What is the value of the digit? 200,000

20. In the number 6,789,345
(a) What digit tells the number of thousands? 9
(b) What is the value of the digit? 9000

Write a word name for each number.

21. 142
one hundred forty-two

22. 376
three hundred seventy-six

23. 9304
nine thousand, three hundred four

24. 7606
seven thousand, six hundred six

25. 36,118 thirty-six thousand, one hundred eighteen

26. 55,742 fifty-five thousand, seven hundred forty-two

27. 105,261 one hundred five thousand, two hundred sixty-one

28. 370,258 three hundred seventy thousand, two hundred fifty-eight

29. 14,203,326 fourteen million, two hundred three thousand, three hundred twenty-six

30. 68,089,213 sixty-eight million, eighty-nine thousand, two hundred thirteen

31. 4,302,156,200 four billion, three hundred two million, one hundred fifty-six thousand, two hundred

32. 7,436,210,400 seven billion, four hundred thirty-six million, two hundred ten thousand, four hundred

Write each number in standard notation.

33. one thousand, five hundred sixty-one 1561

34. three thousand, one hundred eighty-nine 3189

35. thirty-three thousand, eight hundred nine 33,809

36. two hundred three thousand, three hundred seventy-four 203,374

37. one hundred million, seventy-nine thousand, eight hundred twenty-six 100,079,826

38. four hundred fifty million, three hundred thousand, two hundred forty-nine 450,300,249

Applications *When writing a check, a person must write the word name for the dollar amount of the check.*

39. *Personal Finance* Alex bought new equipment for his laboratory for $1965. What word name should he write on the check?
one thousand, nine hundred sixty-five

40. *Personal Finance* Alex later bought a new personal computer for $6383. What word name should he write on the check?
six thousand, three hundred eighty-three

In exercises 41–44, use the following chart prepared with data from the U.S. Bureau of the Census. Notice that the second line tells us that the numbers represent millions. These values are only approximate values representing numbers written to the nearest million. They are not exact census figures.

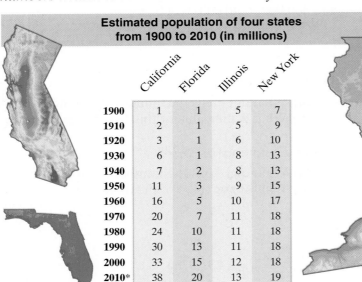

Estimated population of four states from 1900 to 2010 (in millions)

	California	Florida	Illinois	New York
1900	1	1	5	7
1910	2	1	5	9
1920	3	1	6	10
1930	6	1	8	13
1940	7	2	8	13
1950	11	3	9	15
1960	16	5	10	17
1970	20	7	11	18
1980	24	10	11	18
1990	30	13	11	18
2000	33	15	12	18
2010*	38	20	13	19

*estimated

Source: U.S. Bureau of the Census

41. *Historical Analysis* What was the estimated population of New York in 1910?
9 million or 9,000,000

42. *Historical Analysis* What was the estimated population of Florida in 1970?
7 million or 7,000,000

43. *Historical Analysis* What is the estimated population of California in 2010?
38 million or 38,000,000

44. *Historical Analysis* What was the estimated population of Illinois in 1940?
8 million or 8,000,000

In exercises 45–48, use the following chart:

Number of Flights and Passengers for Selected Airlines in 1999, 2000, and 2005 (in thousands)

Airline	1999		2000		2005	
	Flights*	Passengers	Flights*	Passengers	Flights	Passengers
American	740	72,567	791	77,185	780	75,300
Continental	428	40,059	423	40,989	401	39,520
Delta	930	101,843	922	101,809	900	98,360
Northwest	552	50,441	565	52,566	448	49,690

*Includes passenger and freight flights
Source: Bureau of Transportation Statistics

45. *Airline Travel* How many flights did Delta have in 1999? 930,000

46. *Airline Travel* How many passengers flew on American flights in 2000? 77,185,000

47. *Airline Travel* How many passengers flew on Northwest flights in 2000? 52,566,000

48. *Airline Travel* How many flights did Continental have in 2005? 401,000

49. *Physics* The speed of light is approximately 29,979,250,000 centimeters per second. **(a)** 5
   (a) What digit tells the number of ten thousands?
   (b) What digit tells the number of ten billions? 2

▲**50.** *Earth Science* The circumference of Earth at the equator is 131,480,184 feet. **(a)** 3
   (a) What digit tells the number of ten millions?
   (b) What digit tells the number of hundred thousands? 4

51. *Blood Vessels* There are about 316,820,000 feet of blood vessels in an adult human body **(a)** 2
   (a) What digit tells the number of ten thousands?
   (b) What digit tells the number of ten millions? 1

52. *Historical Analysis* The world's population is expected to reach 7,900,000,000 by the year 2020, according to the U.S. Bureau of the Census.
   (a) Which digit tells the number of hundred millions? 9
   (b) Which digit tells the number of billions? 7

53. Write in standard notation: six hundred thirteen trillion, one billion, thirty-three million, two hundred eight thousand, three. 613,001,033,208,003

54. Write in standard notation: nine hundred fourteen trillion, two billion, fifty-two million, four hundred nine thousand, six. 914,002,052,409,006

To Think About

55. Write a word name for 3,682,968,009,931,960,747. (*Hint:* The digit 1 followed by 18 zeros represents the number *1 quintillion*. 1 followed by 15 zeros represents the number *1 quadrillion*.)
three quintillion, six hundred eighty-two quadrillion, nine hundred sixty-eight trillion, nine billion, nine hundred thirty-one million, nine hundred sixty thousand, seven hundred forty-seven

56. The number 50,000,000,000,000,000,000 is represented on some scientific calculators as 5 E 19. We will cover this in more detail in a later chapter. However, for the present we can see that this is a convenient notation that allows us to record very large whole numbers. Note that this number (50 quintillion) is a 5 followed by 19 zeros. Write in standard form the number that would be represented on a calculator as 6 E 22.
60,000,000,000,000,000,000,000

57. Think about the discussion in exercise 56. If the number 4 E 20 represented on a scientific calculator was divided by 2, what number would be the result? Write your answer in standard form.
You would obtain 2 E 20. This is 200,000,000,000,000,000,000 in standard form.

58. Consider all the whole numbers between 200 and 800 that contain the digit 6. How many such numbers are there?
195

Quick Quiz 1.1

1. Write in expanded notation. 73,952
70,000 + 3000 + 900 + 50 + 2

2. Write a word name. 8,932,475
eight million, nine hundred thirty-two thousand, four hundred seventy-five

3. Write in standard notation.
Nine hundred sixty-four thousand, two hundred fifty-seven 964,257

4. **Concept Check** Explain why the zeros are needed when writing the following number in standard notation: three hundred sixty-eight million, five hundred twenty-two. Answers may vary

Classroom Quiz 1.1 You may use these problems to quiz your students' mastery of Section 1.1.

1. Write in expanded notation. 41,127
Ans: 40,000 + 1000 + 100 + 20 + 7

2. Write a word name. 5,327,896
Ans: five million, three hundred twenty-seven thousand, eight hundred ninety-six

3. Write in standard notation.
Four hundred twenty-two thousand, nine hundred eighty-five **Ans:** 422,985

Student Learning Objectives

After studying this section, you will be able to:

1. Master basic addition facts.

2. Add several single-digit numbers.

3. Add several-digit numbers when carrying is not needed.

4. Add several-digit numbers when carrying is needed.

5. Review the properties of addition.

6. Apply addition to real-life situations.

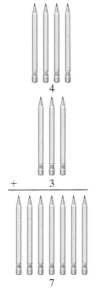

4

+ 3

7

Teaching Example 1 Add.

(a) $4 + 6$ **(b)** $7 + 7$ **(c)** $0 + 8$

Ans: **(a)** 10   **(b)** 14   **(c)** 8

1 Mastering Basic Addition Facts

We see the addition process time and time again. Carpenters add to find the amount of lumber they need for a job. Auto mechanics add to make sure they have enough parts in the inventory. Bank tellers add to get cash totals.

What is addition? We do addition when we put sets of objects together.

■■■■■   ■■■■■■■   ■■■■■■■■■■■■

5 objects   +   7 objects   =   12 objects

$$5 + 7 = 12$$

Usually when we add numbers, we put one number under the other in a column. The numbers being added are called **addends.** The result is called the **sum.**

Suppose that we have four pencils in the car and we bring three more pencils from home. How many pencils do we have with us now? We add 4 and 3 to obtain a value of 7. In this case, the numbers 4 and 3 are the addends and the answer 7 is the sum.

$$\begin{array}{r} 4 \\ + 3 \\ \hline 7 \end{array}$$ addend / addend / sum

Think about what we do when we add 0 to another number. We are not making a change, so whenever we add zero to another number, that number will be the sum. Since this is always true, this is called a *property*. Since the sum is identical to the number added to zero, this is called the **identity property of zero.**

> **EXAMPLE 1** Add.
>
> **(a)** $8 + 5$ **(b)** $3 + 7$ **(c)** $9 + 0$
>
> **Solution**
>
> **(a)** $\begin{array}{r} 8 \\ + 5 \\ \hline 13 \end{array}$ **(b)** $\begin{array}{r} 3 \\ + 7 \\ \hline 10 \end{array}$ **(c)** $\begin{array}{r} 9 \\ + 0 \\ \hline 9 \end{array}$ ◄——— *Note:* When we add zero to any other number, that number is the sum.

> **Practice Problem 1** Add.
>
> **(a)** $\begin{array}{r} 7 \\ + 5 \\ \hline 12 \end{array}$ **(b)** $\begin{array}{r} 9 \\ + 4 \\ \hline 13 \end{array}$ **(c)** $\begin{array}{r} 3 \\ + 0 \\ \hline 3 \end{array}$

The following table shows the basic addition facts. You should know these facts. If any of the answers don't come to you quickly, now is the time to learn them. To check your knowledge try Exercises 1.2, exercises 3 and 4.

Basic Addition Facts

+	0	1	2	3	4	5	6	7	8	9
0	0	1	2	3	4	5	6	7	8	9
1	1	2	3	4	5	6	7	8	9	10
2	2	3	4	5	6	7	8	9	10	11
3	3	4	5	6	7	8	9	10	11	12
4	4	5	6	7	8	9	10	11	12	13
5	5	6	7	8	9	10	11	12	13	14
6	6	7	8	9	10	11	12	13	14	15
7	7	8	9	10	11	12	13	14	15	16
8	8	9	10	11	12	13	14	15	16	17
9	9	10	11	12	13	14	15	16	17	18

To use the table to find the sum $4 + 7$, read across the top of the table to the 4 column, and then read down the left to the 7 row. The box where the 4 and 7 meet is 11, which means that $4 + 7 = 11$. Now read across the top to the 7 column and down the left to the 4 row. The box where these numbers meet is also 11. We can see that the order in which we add the numbers does not change the sum. $4 + 7 = 11$, and $7 + 4 = 11$. We call this the **commutative property of addition.**

This property does not hold true for everything in our lives. When you put on your socks and then your shoes, the result is not the same as if you put on your shoes first and then your socks! Can you think of any other examples where changing the order in which you add things would change the result?

 Adding Several Single-Digit Numbers

If more than two numbers are to be added, we usually add from the first number to the next number and mentally note the sum. Then we add that sum to the next number, and so on.

EXAMPLE 2   Add. $3 + 4 + 8 + 2 + 5$

Solution   We rewrite the addition problem in a column format.

$$
\begin{array}{r}
3 \\
4 \\
8 \\
2 \\
+\,5 \\
\hline
22
\end{array}
$$

$\left.\begin{array}{l}3 \\ 4\end{array}\right\}\,3 + 4 = 7$ Mentally, we do these steps.

$7 + 8 = 15$

$15 + 2 = 17$

$17 + 5 = 22$

Practice Problem 2   Add. $7 + 6 + 5 + 8 + 2$ 28

Because the order in which we add numbers doesn't matter, we can choose to add from the top down, from the bottom up, or in any other way. One shortcut is to add first any numbers that will give a sum of 10, or 20, or 30, and so on.

EXAMPLE 3 Add.   3
4
8
2
+ 6

Solution   We mentally group the numbers into tens.

The sum is 10 + 10 + 3 or 23.

Practice Problem 3 Add. 1 + 7 + 2 + 9 + 3   22

3 Adding Several-Digit Numbers When Carrying Is Not Needed

Of course, many numbers that we need to add have more than one digit. In such cases, we must be careful to first add the digits in the ones column, then the digits in the tens column, then those in the hundreds column, and so on. Notice that we move from *right to left.*

EXAMPLE 4 Add. 4304 + 5163

Solution   4 3 0 4
+ 5 1 6 3
9 4 6 7

sum of 4 ones + 3 ones = 7 ones

sum of 0 tens + 6 tens = 6 tens

sum of 3 hundreds + 1 hundred = 4 hundreds

sum of 4 thousands + 5 thousands = 9 thousands

Practice Problem 4 Add.

8246
+ 1702
9948

4 Adding Several-Digit Numbers When Carrying Is Needed

When you add several whole numbers, often the sum in a column is greater than 9. However, we can only use *one* digit in any one place. What do we do with a two-digit sum? Look at the following example.

EXAMPLE 5 Add. 45 + 37

Teaching Example 5 Add. 52 + 19

Ans: 71

Solution

$$\begin{array}{r} {}^{1} \\ 4\ 5 \\ +3\ 7 \\ \hline 2 \end{array}$$

5 ones and 7 ones = 12.
We rename 12 in expanded notation: 1 ten + 2 ones.
We place the 2 ones in the ones column.
We carry the 1 ten over to the tens column.

Note: Placing the 1 in the next column is often called "carrying the one."

$$\begin{array}{r} {}^{1} \\ 4\ 5 \\ +3\ 7 \\ \hline 8\ 2 \end{array}$$

Now we can add the digits in the tens column.

Thus, 45 + 37 = 82.

Practice Problem 5 Add.

$$\begin{array}{r} 56 \\ +\ 36 \\ \hline 92 \end{array}$$

NOTE TO STUDENT: *Fully worked-out solutions to all of the Practice Problems can be found at the back of the text starting at page SP-1*

Often you must use carrying several times by bringing the left digit into the next column to the left.

EXAMPLE 6 Add. 257 + 688 + 94

Teaching Example 6 Add. 392 + 57 + 726

Ans: 1175

Solution

Thousands Column | Hundreds Column | Tens Column | Ones Column

$$\begin{array}{r} {}^{2}\ {}^{1} \\ 2\ 5\ 7 \\ 6\ 8\ 8 \\ +\quad 9\ 4 \\ \hline 1\ 0\ 3\ 9 \end{array}$$

In the ones column we add 7 + 8 + 4 = 19. Because 19 is 1 ten and 9 ones, we place 9 in the ones column and carry 1 to the top of the tens column.

In the tens column we add 1 + 5 + 8 + 9 = 23. Because 23 tens is 2 hundreds and 3 tens, we place the 3 in the tens column and carry 2 to the top of the hundreds column.

In the hundreds column we add 2 + 2 + 6 = 10 hundreds. Because 10 hundreds is 1 thousand and 0 hundreds, we place the 0 in the hundreds column and place the 1 in the thousands column.

Teaching Tip Remind students that when they carry a digit such as in Example 6, they may write down the digit they are carrying. Some students were probably criticized in elementary school for showing the carrying step. In college, students should feel free to write down the carrying step if it is needed. Of course, if students can do that part in their heads, there is no need to write down the carrying digit.

Practice Problem 6 Add. 789 + 63 + 297 1149

We can add numbers in more than one way. To add $5 + 3 + 7$ we can first add the 5 and 3. We do this by using parentheses to show the first operation to be done. This shows us that $5 + 3$ is to be grouped together.

$$5 + 3 + 7 = (5 + 3) + 7 = 15$$
$$= \quad 8 \quad + 7 = 15$$

We could add the 3 and 7 first. We use parentheses to show that we group $3 + 7$ together and that we will add these two numbers first.

$$5 + 3 + 7 = 5 + (3 + 7) = 15$$
$$= 5 + \quad 10 \quad = 15$$

The way we group numbers to be added does not change the sum. This property is called the **associative property of addition.**

⑤ Reviewing the Properties of Addition

Look again at the three properties of addition we have discussed in this section.

1. Associative Property of Addition When we add three numbers, we can group them in any way.	$(8 + 2) + 6 = 8 + (2 + 6)$ $10 + 6 = 8 + 8$ $16 = 16$
2. Commutative Property of Addition Two numbers can be added in either order with the same result.	$5 + 12 = 12 + 5$ $17 = 17$
3. Identity Property of Zero When zero is added to a number, the sum is that number.	$8 + 0 = 8$ $0 + 5 = 5$

Because of the commutative and associative properties of addition, we can check our addition by adding the numbers in the opposite order.

Teaching Example 7

(a) Add the numbers. $6037 + 928 + 65$

(b) Check by reversing the order. $65 + 928 + 6037$

Ans: (a) 7030   **(b)** 7030

Teaching Tip Some students lack confidence that they will be able to find their own errors. As a classroom activity, have students add $258 + 167 + 879$. Then have them add $879 + 167 + 258$. The sum is 1304. If you ask students how many of them made an error and detected it by adding the numbers in the opposite order and getting a different answer, there will usually be several students in the class who raise their hands.

EXAMPLE 7   (a) Add the numbers. $39 + 7284 + 3132$

(b) Check by reversing the order of addition.

Solution

(a)
$$\begin{array}{r} \overset{1\,1}{39} \\ 7284 \\ +\,3132 \\ \hline 10{,}455 \end{array}$$
Addition

(b)
$$\begin{array}{r} \overset{1\,1}{3132} \\ 7284 \\ +\quad 39 \\ \hline 10{,}455 \end{array}$$
Check by reversing the order.

The sum is the same in each case.

Practice Problem 7

(a) Add.
$$\begin{array}{r} 127 \\ 9876 \\ +\ 342 \end{array}$$ 10,345

(b) Check by reversing the order.
$$\begin{array}{r} 342 \\ 9876 \\ +\ 127 \end{array}$$
same; 10,345

Applying Addition to Real-Life Situations

We use addition in all kinds of situations. There are several key words in word problems that imply addition. For example, it may be stated that there are 12 math books, 9 chemistry books, and 8 biology books on a book shelf. To find the *total* number of books implies that we add the numbers 12 + 9 + 8. Other key words are *how much, how many,* and *all*.

Sometimes a problem will have more information than you will need to answer the question. If you have too much information, to solve the problem you will need to separate out the facts that are not important. The following three steps are involved in the problem-solving process.

Step 1 Understand the problem.
Step 2 Calculate and state the answer.
Step 3 Check.

We may not write all of these steps down, but they are the steps we use to solve all problems.

EXAMPLE 8 The bookkeeper for Smithville Trucking was examining the following data for the company checking account.

Monday:	$23,416 was deposited and $17,389 was debited.
Tuesday:	$44,823 was deposited and $34,089 was debited.
Wednesday:	$16,213 was deposited and $20,057 was debited.

What was the total of all deposits during this period?

Solution

Step 1 *Understand the problem.*
Total implies that we will use addition. Since we don't need to know about the debits to answer this question, we use only the *deposit* amounts.

Step 2 *Calculate and state the answer.*

		$\overset{1\;1\;\;\;1}{23{,}416}$
Monday:	$23,416 was deposited.	
Tuesday:	$44,823 was deposited.	44,823
Wednesday:	$16,213 was deposited.	$+\;16{,}213$
		84,452

A total of $84,452 was deposited on those three days.

Step 3 *Check.*
You may add the numbers in reverse order to check. We leave the check up to you.

Practice Problem 8 North University has 23,413 men and 18,316 women. South University has 19,316 men and 24,789 women. East University has 20,078 men and 22,965 women. What is the total enrollment of *women* at the three universities?   66,070 total women

Teaching Example 8 Last summer, the Wild River Amusement Park had the following attendance numbers.

In June, 13,458 adults attended and 14,986 children attended.
In July, 22,473 adults attended and 27,396 children attended.
In August, 20,721 adults attended and 23,058 children attended.

How many children attended the amusement park during those three months?

Ans: 65,440 children

NOTE TO STUDENT: Fully worked-out solutions to all of the Practice Problems can be found at the back of the text starting at page SP-1

EXAMPLE 9 Mr. Ortiz has a rectangular field whose length is 400 feet and whose width is 200 feet. What is the total number of feet of fence that would be required to fence in the field?

Solution

1. *Understand the problem.*
   To help us to get a picture of what the field looks like, we will draw a diagram.

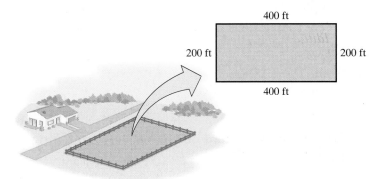

Note that ft is the abbreviation for feet. ft means feet.

2. *Calculate and state the answer.*
   Since the fence will be along each side of the field, we add the lengths all around the field.

$$\begin{array}{r} 200 \\ 400 \\ 200 \\ + \ 400 \\ \hline 1200 \end{array}$$

The amount of fence that would be required is 1200 feet.

3. *Check.*
   Regroup the addends and add.

$$\begin{array}{r} 200 \\ 200 \\ 400 \\ + \ 400 \\ \hline 1200 \end{array} \checkmark$$

Practice Problem 9 In Vermont, Gretchen fenced the rectangular field on which her sheep graze. The length of the field is 2000 feet and the width of the field is 1000 feet. What is the perimeter of the field? (*Hint:* The "distance around" an object [such as a field] is called the *perimeter.*) 6000 ft

Developing Your Study Skills

Getting Organized for an Exam

Studying adequately for an exam requires careful preparation. Begin early so that you will be able to spread your review over several days. Even though you may still be learning new material at this time, you can be reviewing concepts previously learned in the chapter. Giving yourself plenty of time for review will take the pressure off. You need this time to process what you have learned and to tie concepts together.

Adequate preparation enables you to feel confident and to think clearly with less tension and anxiety.

Verbal and Writing Skills

1. Explain in your own words. Answers may vary. Samples are below.

  (a) the commutative property of addition You can change the order of the addends without changing the sum.
  (b) the associative property of addition You can group the addends in any way without changing the sum.

2. When zero is added to any number, it does not change that number. Why do you think this is called the identity property of zero? When zero is added to any number, the sum is identical to that number.

Complete the addition facts for each table. Strive for total accuracy, but work quickly. Allow a maximum of five minutes for each table.

3.

+	3	5	4	8	0	6	7	2	9	1
2	5	7	6	10	2	8	9	4	11	3
7	10	12	11	15	7	13	14	9	16	8
5	8	10	9	13	5	11	12	7	14	6
3	6	8	7	11	3	9	10	5	12	4
0	3	5	4	8	0	6	7	2	9	1
4	7	9	8	12	4	10	11	6	13	5
1	4	6	5	9	1	7	8	3	10	2
8	11	13	12	16	8	14	15	10	17	9
6	9	11	10	14	6	12	13	8	15	7
9	12	14	13	17	9	15	16	11	18	10

4.

+	1	6	5	3	0	9	4	7	2	8
3	4	9	8	6	3	12	7	10	5	11
9	10	15	14	12	9	18	13	16	11	17
4	5	10	9	7	4	13	8	11	6	12
0	1	6	5	3	0	9	4	7	2	8
2	3	8	7	5	2	11	6	9	4	10
7	8	13	12	10	7	16	11	14	9	15
8	9	14	13	11	8	17	12	15	10	16
1	2	7	6	4	1	10	5	8	3	9
6	7	12	11	9	6	15	10	13	8	14
5	6	11	10	8	5	14	9	12	7	13

Add.

5.
$$\begin{array}{r} 4 \\ 2 \\ 8 \\ +\,9 \\ \hline 23 \end{array}$$

6.
$$\begin{array}{r} 4 \\ 6 \\ 2 \\ +\,7 \\ \hline 19 \end{array}$$

7.
$$\begin{array}{r} 2 \\ 6 \\ 7 \\ 8 \\ +\,3 \\ \hline 26 \end{array}$$

8.
$$\begin{array}{r} 1 \\ 5 \\ 5 \\ 9 \\ +\,9 \\ \hline 29 \end{array}$$

9.
$$\begin{array}{r} 18 \\ 36 \\ +\,3 \\ \hline 57 \end{array}$$

10.
$$\begin{array}{r} 63 \\ 11 \\ +\,6 \\ \hline 80 \end{array}$$

11.
$$\begin{array}{r} 63 \\ 24 \\ +\,12 \\ \hline 99 \end{array}$$

12.
$$\begin{array}{r} 54 \\ 21 \\ +\,23 \\ \hline 98 \end{array}$$

13.
$$\begin{array}{r} 3315 \\ 726 \\ +\,84 \\ \hline 4125 \end{array}$$

14.
$$\begin{array}{r} 5773 \\ 425 \\ +\,67 \\ \hline 6265 \end{array}$$

15.
$$\begin{array}{r} 5631 \\ 2344 \\ +\,2019 \\ \hline 9994 \end{array}$$

16.
$$\begin{array}{r} 5017 \\ 2984 \\ +\,1328 \\ \hline 9329 \end{array}$$

17.
$$\begin{array}{r} 8235 \\ +\,5626 \\ \hline 13{,}861 \end{array}$$

18.
$$\begin{array}{r} 6753 \\ +\,3265 \\ \hline 10{,}018 \end{array}$$

19.
$$\begin{array}{r} 62{,}504 \\ +\,54{,}736 \\ \hline 117{,}240 \end{array}$$

20.
$$\begin{array}{r} 83{,}596 \\ +\,56{,}384 \\ \hline 139{,}980 \end{array}$$

Add from the top. Then check by adding in the reverse order.

21.
```
  36
  41
  25
   6
+ 13
```
121

22.
```
  24
  39
  16
  14
+  9
```
102

23.
```
  207
   15
    3
   57
+ 861
```
1143

24.
```
  426
   39
    6
   52
+ 802
```
1325

Add.

25.
```
    85
   256
    55
+ 9734
```
10,130

26.
```
   582
  1674
   336
+ 8458
```
11,050

27.
```
  1,362,214
  7,002,316
+ 3,214,896
```
11,579,426

28.
```
  4,002,983
  2,134,702
+ 3,592,001
```
9,729,686

29.
```
  837,241,000
+ 298,039,240
```
1,135,280,240

30.
```
  982,306,000
+ 583,215,320
```
1,565,521,320

31.
```
    516,208
     24,317
+ 1,763,295
```
2,303,820

32.
```
    32,500
   763,420
+ 2,837,667
```
3,633,587

33. 25 + 130 + 70 + 75 300

34. 125 + 60 + 140 + 75 400

35. 102 + 50 + 98 + 35 + 50 335

36. 20 + 205 + 95 + 42 + 80 442

Applications

37. *Consumer Mathematics* Vanessa took her children shopping for the new school year. She spent $455 on clothes, $186 on shoes, and $82 on supplies. What was the total amount of money Vanessa spent? $723

38. *Consumer Mathematics* Richy has a part-time job as a dog walker. He saves all of the money he earns for a vacation. He earned $235 in June, $198 in July, and $282 in August. What is the total amount of money Richy saved? $715

39. *Personal Finance* Sheila owns a studio where she teaches music classes to children. Two months ago she made a profit of $1875. Last month she made $1930 and this month she earned $1744. What is the total amount for the three months? $5549

40. *Consumer Mathematics* Terrell flies to several cities each month for his job. During the past three months he has spent $2230, $2655, and $2570 on airline tickets. What is the total amount for the three months? $7455

▲ **41. *Geometry*** Nate wants to put a fence around his backyard. The sketch below indicates the length of each side of the yard. What is the total number of feet of fence he needs for his backyard? 468 feet

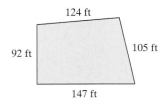

▲ **42. *Geometry*** Jessica has a field with the length of each side as labeled on the sketch. What is the total number of feet of fence that would be required to fence in the field? (Find the perimeter of the field.) 2335 feet

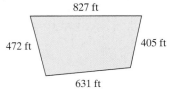

▲ **43.** *Geography* The Pacific Ocean, the world's largest, has an area of 64,000,000 square miles. The Atlantic Ocean has an area of 31,800,000 square miles. The Indian Ocean has an area of 25,300,000 square miles. What is the total area for these oceans? 121,100,000 square miles

▲ **44.** *Geography* The Arctic Ocean has an area of 5,400,000 square miles. The Mediterranean Sea has an area of 1,100,000 square miles. The Caribbean Sea has an area of 1,000,000 square miles. What is the total area for these bodies of water? 7,500,000 square miles

45. *Geography* The Nile River is Africa's longest river, measuring 7,272,320 yards. The second and third longest rivers in Africa are the Congo River, measuring 5,104,000 yards, and the Niger River, which measures 4,558,400 yards. What is the total length of these rivers? 16,934,720 yards

▲ **46.** *Geography* The world's three largest lakes are the Caspian Sea at 152,239 square miles, Lake Superior at 31,820 square miles, and Lake Victoria at 26,828 square miles. What is the total area of these three lakes? 210,887 square miles

In exercises 47–48, be sure you understand the problem and then choose the numbers you need in order to answer each question. Then solve the problem.

47. *Education* The admissions department of a competitive university is reviewing applications to see whether students are *eligible* or *ineligible* for student aid. On Monday, 415 were found eligible and 27 ineligible. On Tuesday, 364 were found eligible and 68 ineligible. On Wednesday, 159 were found eligible and 102 ineligible. On Thursday, 196 were found eligible and 61 ineligible.

(a) How many students were eligible for student aid over the four days? 1134 students

(b) How many students were considered in all?
1392 students

48. *Manufacturing* The quality control division of a motorcycle company classifies the final assembled bike as *passing* or *failing* final inspection. In January, 14,311 vehicles passed whereas 56 failed. In February, 11,077 passed and 158 failed. In March, 12,580 passed and 97 failed.

(a) How many motorcycles passed the inspection during the three months? 37,968 motorcycles

(b) How many motorcycles were assembled during the three months in all? 38,279 motorcycles

Use the following facts to solve exercises 49 and 50. It is 87 miles from Springfield to Weston. It is 17 miles from Weston to Boston. Driving directly, it is 98 miles from Springfield to Boston. It is 21 miles from Boston to Hamilton.

49. *Geography* If Melissa drives from Springfield to Weston, then from Weston to Boston, and finally directly home to Springfield, how many miles does she drive? 202 miles

50. *Geography* If Marcia drives from Hamilton to Boston, then from Boston to Weston, and then from Weston to Springfield, how many miles does she drive? 125 miles

▲ **51.** *Geometry* Walter Swensen is examining the fences of a farm in Caribou, Maine. One field is in the shape of a four-sided figure with no sides equal. The field is enclosed with 2387 feet of wooden rail fence. The first side is 568 feet long, while the second side is 682 feet long. The third side is 703 feet long. How long is the fourth side?
434 feet

▲ **52.** *Geometry* Carlos Sontera is walking to examine the fences of a ranch in El Paso, Texas. The field he is examining is in the shape of a rectangle. The perimeter of the rectangle is 3456 feet. One side of the rectangle is 930 feet long. How long are the other sides? (*Hint:* The opposite sides of a rectangle are equal.) Two sides are 930 feet long and two sides are 798 feet long.

53. *Personal Finance* Answer using the information in the following Western University expense chart for the current academic year.

Western University Yearly Expenses	In-State Student, U.S. Citizen	Out-of-State Student, U.S. Citizen	Foreign Student
Tuition	$3640	$5276	$8352
Room	1926	2437	2855
Board	1753	1840	1840

How much is the total cost for tuition, room, and board for

(a) an out-of-state U.S. citizen? $9553
(b) an in-state U.S. citizen? $7319
(c) a foreign student? $13,047

To Think About *In exercises 54–55, add.*

54. 2,368,521,788 + 5,721,368,701 + 4,027,399,206 12,117,289,695

55. 89 + 166 + 23 + 45 + 72 + 190 + 203 + 77 + 18 + 93 + 46 + 73 + 66 1161

56. What would happen if addition were not commutative? Answers may vary. A sample is: You could not add the addends in reverse order to check the addition.

57. What would happen if addition were not associative? Answers may vary. A sample is: You could not group the addends in groups that sum to 10s to make column addition easier.

Cumulative Review *Write the word name for each number.*

58. **[1.1.3]** 76,208,941 seventy-six million, two hundred eight thousand, nine hundred forty-one

59. **[1.1.3]** 121,000,374 one hundred twenty-one million, three hundred seventy-four

Write each number in standard notation.

60. **[1.1.3]** eight million, seven hundred twenty-four thousand, three hundred ninety-six 8,724,396

61. **[1.1.3]** nine million, fifty-one thousand, seven hundred nineteen 9,051,719

62. **[1.1.3]** twenty-eight million, three hundred eighty-seven thousand, eighteen 28,387,018

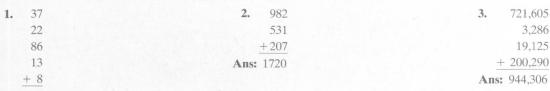

Quick Quiz 1.2 Add.

1.
```
   56
   38
   92
   17
 +  9
  212
```

2.
```
  831
  276
+ 508
 1615
```

3.
```
  681,302
    5,126
   18,371
+ 300,012
1,004,811
```

4. Concept Check Explain how you would use carrying when performing the calculation 4567 + 3189 + 895.

Answers may vary

Classroom Quiz 1.2 You may use these problems to quiz your students' mastery of Section 1.2.

Add.

1.
```
   37
   22
   86
   13
 +  8
```
Ans: 166

2.
```
  982
  531
+ 207
```
Ans: 1720

3.
```
  721,605
    3,286
   19,125
+ 200,290
```
Ans: 944,306

① Mastering Basic Subtraction Facts

Subtraction is used day after day in the business world. The owner of a bakery placed an ad for his cakes in a local newspaper to see if this might increase his profits. To learn how many cakes had been sold, at closing time he subtracted the number of cakes remaining from the number of cakes the bakery had when it opened. To figure his profits, he subtracted his costs (including the cost of the ad) from his sales. Finally, to see if the ad paid off, he subtracted the profits he usually made in that period from the profits after advertising. He needed subtraction to see whether it paid to advertise.

What is subtraction? We do subtraction when we take objects away from a group. If you have 12 objects and take away 3 of them, 9 objects remain.

Student Learning Objectives

After studying this section, you will be able to:

① **Master basic subtraction facts.**

② **Subtract whole numbers when borrowing is not necessary.**

③ **Subtract whole numbers when borrowing is necessary.**

④ **Check the answer to a subtraction problem.**

⑤ **Apply subtraction to real-life situations.**

■■■■■■■■ (■■■)   ■■■■■■■■■

12 objects − 3 objects   =   9 objects

$$12 - 3 = 9$$

If you earn \$400 per month, but have \$100 taken out for taxes, how much do you have left?

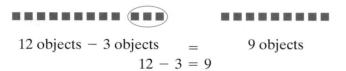

\$400   −   \$100   =   \$300

| salary | subtraction symbol | amount withheld | amount left |

We can use addition to help with a subtraction problem.

To subtract: $200 - 196 =$ what number

We can think: $196 +$ what number $= 200$

Usually when we subtract numbers, we put one number under the other in a column. When we subtract one number from another, the answer is called the **difference.**

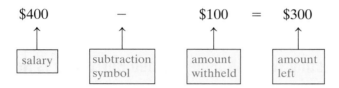

$$\begin{array}{cccc} 9 & 8 & 12 & 17 \\ -2 & -3 & -6 & -9 \\ \hline 7 & 5 & 6 & 8 \end{array}$$

Each of these is called the difference of the two numbers.

The other two parts of a subtraction problem have labels, although you will not often come across them. The number being subtracted is called the **subtrahend.** The number being subtracted from is called the **minuend.**

$$\begin{array}{rl} 17 & \text{minuend} \\ -\ 9 & \text{subtrahend} \\ \hline 8 & \text{difference} \end{array}$$

In this case, the number 17 is called the *minuend.* The number 9 is called the *subtrahend.* The number 8 is called the *difference.*

QUICK RECALL OF SUBTRACTION FACTS It is helpful if you can subtract quickly. See if you can do Example 1 correctly in 15 seconds or less. Repeat again with Practice Problem 1. Strive to obtain all answers correctly in 15 seconds or less.

Teaching Example 1 Subtract.
(a) $7 - 3$ (b) $11 - 5$ (c) $14 - 6$
(d) $19 - 0$ (e) $17 - 9$
Ans: (a) 4   (b) 6   (c) 8   (d) 19   (e) 8

EXAMPLE 1   Subtract.

(a) $8 - 2$ (b) $13 - 5$ (c) $12 - 4$
(d) $15 - 8$ (e) $16 - 0$

Solution

(a) $\begin{array}{r} 8 \\ -2 \\ \hline 6 \end{array}$ (b) $\begin{array}{r} 13 \\ -5 \\ \hline 8 \end{array}$ (c) $\begin{array}{r} 12 \\ -4 \\ \hline 8 \end{array}$

(d) $\begin{array}{r} 15 \\ -8 \\ \hline 7 \end{array}$ (e) $\begin{array}{r} 16 \\ -0 \\ \hline 16 \end{array}$

NOTE TO STUDENT: *Fully worked-out solutions to all of the Practice Problems can be found at the back of the text starting at page SP-1*

Practice Problem 1   Subtract.

(a) $\begin{array}{r} 9 \\ -6 \\ \hline 3 \end{array}$ (b) $\begin{array}{r} 12 \\ -5 \\ \hline 7 \end{array}$ (c) $\begin{array}{r} 17 \\ -8 \\ \hline 9 \end{array}$ (d) $\begin{array}{r} 14 \\ -0 \\ \hline 14 \end{array}$ (e) $\begin{array}{r} 18 \\ -9 \\ \hline 9 \end{array}$

 Subtracting Whole Numbers When Borrowing Is Not Necessary

When we subtract numbers with more than two digits, in order to keep track of our work, we line up the ones column, the tens column, the hundreds column, and so on. Note that we begin with the ones column, and move from right to left.

Teaching Example 2 Subtract. $6857 - 4326$
Ans: 2531

EXAMPLE 2   Subtract. $9867 - 3725$

Solution

$$\begin{array}{r} 9\ 8\ 6\ 7 \\ -3\ 7\ 2\ 5 \\ \hline 6\ 1\ 4\ 2 \end{array}$$

7 ones $-$ 5 ones $=$ 2 ones
6 tens $-$ 2 tens $=$ 4 tens
8 hundreds $-$ 7 hundreds $=$ 1 hundred
9 thousands $-$ 3 thousands $=$ 6 thousands

Practice Problem 2   Subtract. $7695 - 3481$ 4214

Subtracting Whole Numbers When Borrowing Is Necessary

In the subtraction that we have looked at so far, each digit in the upper number (the minuend) has been greater than the digit in the lower number (the subtrahend) for each place value. Many times, however, a digit in the lower number is greater than the digit in the upper number for that place value.

$$
\begin{array}{r}
42 \\
- 28 \\
\end{array}
$$

The digit in the ones place in the lower number, the 8 of 28, is greater than the number in the ones place in the upper number, the 2 of 42. To subtract, we must *rename* 42, using place values. This is called **borrowing.**

EXAMPLE 3   Subtract. 42 − 28

Teaching Example 3 Subtract. 53 − 36

Ans: 17

Solution

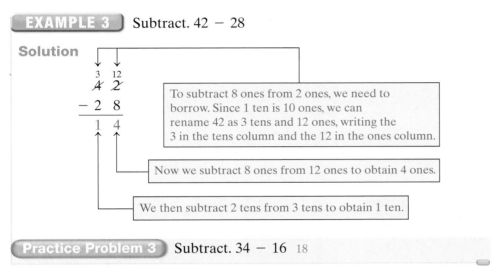

Practice Problem 3   Subtract. 34 − 16   18

NOTE TO STUDENT: *Fully worked-out solutions to all of the Practice Problems can be found at the back of the text starting at page SP-1*

EXAMPLE 4   Subtract. 864 − 548

Teaching Example 4 Subtract. 762 − 235

Ans: 527

Solution

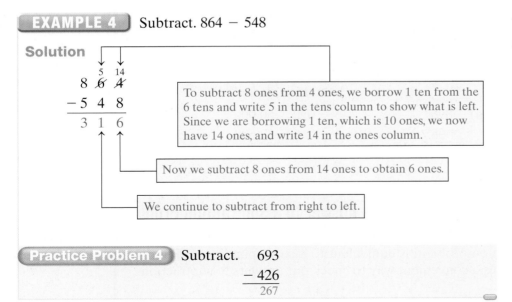

Practice Problem 4   Subtract.
$$
\begin{array}{r}
693 \\
- 426 \\
\hline
267 \\
\end{array}
$$

EXAMPLE 5 Subtract. 8040 − 6375

Solution

To subtract 5 from 0, we borrow 1 ten from the 4 tens to make 3 tens and 10 ones. 10 − 5 = 5

$$
\begin{array}{r}
\overset{\overset{9}{\cancel{10}}\ \overset{13}{}}{\overset{7\ \ \cancel{10}\ \ \cancel{3}\ \ 10}{8\ \ \cancel{0}\ \ \cancel{4}\ \ \cancel{0}}} \\
-\ 6\ \ 3\ \ 7\ \ 5 \\
\hline
1\ \ 6\ \ 6\ \ 5
\end{array}
$$

To subtract 7 tens from the 3 tens, we need to borrow 1 hundred to make 10 tens. Since we find a 0 in the hundreds column, first we borrow 1 thousand to make 10 hundreds. We show the number of thousands that are left, and write the 10 in the hundreds column. Now we borrow 1 hundred, show the number of hundreds that are left, and add the 10 tens to the 3 tens. We now do the subtraction.
13 tens − 7 tens = 6 tens

9 hundreds − 3 hundreds = 6 hundreds

7 thousands − 6 thousands = 1 thousand

Practice Problem 5 Subtract. 9070 − 5886 3184

EXAMPLE 6 Subtract.

(a) 9521 − 943   **(b)** 40,000 − 29,056

Solution

(a)
$$
\begin{array}{r}
\overset{\overset{14}{}\ \overset{11}{}}{\overset{8\ \ \cancel{4}\ \ \cancel{1}\ \ 11}{\cancel{9}\ \ \cancel{5}\ \ \cancel{2}\ \ \cancel{1}}} \\
-\ \ \ 9\ \ 4\ \ 3 \\
\hline
8\ \ 5\ \ 7\ \ 8
\end{array}
$$

(b)
$$
\begin{array}{r}
\overset{3\ \ 9\ \ 9\ \ 9\ \ 10}{\cancel{4}\ \ \cancel{0},\ \cancel{0}\ \ \cancel{0}\ \ \cancel{0}} \\
-\ 2\ \ 9,\ 0\ \ 5\ \ 6 \\
\hline
1\ \ 0,\ 9\ \ 4\ \ 4
\end{array}
$$

Practice Problem 6 Subtract.

(a)
$$
\begin{array}{r}
8964 \\
-\ \ 985 \\
\hline
7979
\end{array}
$$

(b)
$$
\begin{array}{r}
50,000 \\
-\ 32,508 \\
\hline
17,492
\end{array}
$$

 Checking the Answer to a Subtraction Problem

We observe that when 9 − 7 = 2 it follows that 7 + 2 = 9. Each subtraction problem is equivalent to a corresponding addition problem. This gives us a convenient way to check our answers to subtraction.

EXAMPLE 7 Check this subtraction problem.

$$5829 - 3647 = 2182$$

Teaching Example 7 Check this subtraction problem.

$$7396 - 2849 = 4547$$

Ans: $2849 + 4547 = 7396$

Solution

```
   5 8 2 9 ←─────────── The sum should equal 5829, which it does.
 − 3 6 4 7                We have checked our work, and it is correct.
   2 1 8 2    then    3 6 4 7
                    + 2 1 8 2
                      5 8 2 9 ←
```

Practice Problem 7 Check this subtraction problem.

$$9763 - 5732 = 4031 \quad 5732 + 4031 = 9763$$

NOTE TO STUDENT: Fully worked-out solutions to all of the Practice Problems can be found at the back of the text starting at page SP-1

EXAMPLE 8 Subtract and check your answers.

(a) $156{,}000 - 29{,}326$ **(b)** $1{,}264{,}308 - 1{,}057{,}612$

Teaching Example 8 Subtract and check your answers.

(a) $347{,}000 - 52{,}183$

(b) $5{,}283{,}175 - 2{,}734{,}093$

Ans: **(a)** $294{,}817$ **(b)** $2{,}549{,}082$

Solution

(a)
```
    156,000 ←──────────┌It checks.┐
  −  29,326      29,326      │
    126,674   + 126,674      │
                156,000 ←────┘
```

(b)
```
   1,264,308 ←──────────┌It checks.┐
 − 1,057,612    1,057,612      │
     206,696  +   206,696      │
              1,264,308 ←──────┘
```

Practice Problem 8 Subtract and check your answers.

(a)
```
    284,000
 −   96,327
    187,673
```

(b)
```
   8,526,024
 − 6,397,518
   2,128,506
```

Subtraction can be used to solve word problems. Some problems can be expressed (and solved) with an **equation.** An equation is a number sentence with an equal sign, such as

$$10 = 4 + x$$

Here we use the letter x to represent a number we do not know. When we write $10 = 4 + x$, we are stating that 10 is equal to 4 added to some other number. Since $10 - 4 = 6$, we would assume that the number is 6. If we substitute 6 for x in the equation, we have two values that are the same.

$$10 = 4 + x$$
$$10 = 4 + 6 \quad \text{Substitute 6 for } x.$$
$$10 = 10 \quad\quad \text{Both sides of the equation are the same.}$$

Teaching Tip Taking the time to emphasize the idea of a variable in very simple terms in such problems as $10 = 4 + x$ will make the use of variables in later chapters much easier for the students to learn.

We can write an equation when one of the addends is not known, then use subtraction to solve for the unknown.

EXAMPLE 9   The librarian knows that he has eight world atlases and that five of them are in full color. How many are not in full color?

Solution   We represent the number that we don't know as x and write an equation, or mathematical sentence.

$$8 = 5 + x$$

To solve an equation means to find those values that will make the equation true. We solve this equation by reasoning and by a knowledge of the relationship between addition and subtraction.

$$8 = 5 + x \text{ is equivalent to } 8 - 5 = x$$

We know that $8 - 5 = 3$. Then $x = 3$. We can check the answer by substituting 3 for x in the original equation.

$$8 = 5 + x$$
$$8 = 5 + 3 \quad \text{True} \checkmark$$

We see that $x = 3$ checks, so our answer is correct. There are three atlases not in full color.

Practice Problem 9   Form an equation for each of the following problems. Solve the equation in order to answer the question.

(a) The Salem Harbormaster's daily log noted that seventeen fishing vessels left the harbor yesterday during daylight hours. Walter was at the harbor all morning and saw twelve fishing vessels leave in the morning. How many vessels left in the afternoon? (Assume that sunset was at 6 P.M.) 5 vessels

(b) The Appalachian Mountain Club noted that twenty-two hikers left to climb Mount Washington during the morning. By 4 P.M., ten of them had returned. How many of the hikers were still on the mountain? 12 hikers

⑤ Applying Subtraction to Real-Life Situations

We use subtraction in all kinds of situations. There are several key words in word problems that imply subtraction. Words that involve comparison, such as *how much more, how much greater,* or how much a quantity *increased* or *decreased,* all imply subtraction. The *difference* between two numbers implies subtraction.

EXAMPLE 10 Look at the following population table.

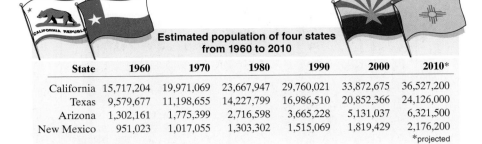

State	1960	1970	1980	1990	2000	2010*
California	15,717,204	19,971,069	23,667,947	29,760,021	33,872,675	36,527,200
Texas	9,579,677	11,198,655	14,227,799	16,986,510	20,852,366	24,126,000
Arizona	1,302,161	1,775,399	2,716,598	3,665,228	5,131,037	6,321,500
New Mexico	951,023	1,017,055	1,303,302	1,515,069	1,819,429	2,176,200

Estimated population of four states from 1960 to 2010

*projected

Source: U.S. Bureau of the Census

Teaching Example 10 Refer to the population table.

(a) In 1960, how many more people lived in Arizona than in New Mexico?

(b) How much did the population of Texas increase from 1980 to 2000?

(c) In 2010, the projected population of Texas will be how much greater than the projected population of Arizona and New Mexico combined?

Ans: (a) 351,138 **(b)** 6,624,567

(c) 15,628,300

(a) In 1980, how much greater was the population of Texas than that of Arizona?

(b) How much did the population of California increase from 1960 to 2000?

(c) How much greater was the population of California in 1990 than that of the other three states combined?

Solution

(a)

14,227,799	1980 population of Texas
− 2,716,598	1980 population of Arizona
11,511,201	difference

The population of Texas was greater by 11,511,201.

(b)

33,872,675	2000 population of California
− 15,717,204	1960 population of California
18,155,471	difference

The population of California increased by 18,155,471 in those 40 years.

(c) First we need to find the total population in 1990 of Texas, Arizona, and New Mexico.

16,986,510	1990 population of Texas
3,665,228	1990 population of Arizona
+ 1,515,069	1990 population of New Mexico
22,166,807	

We use subtraction to compare this total with the population of California.

29,760,021	1990 population of California
− 22,166,807	
7,593,214	

The population of California in 1990 was 7,593,214 more than the population of the other three states combined.

Practice Problem 10

(a) In 1980, how much greater was the population of California than the population of Texas? 9,440,148

(b) How much did the population of Texas increase from 1960 to 1970?

1,618,978

NOTE TO STUDENT: Fully worked-out solutions to all of the Practice Problems can be found at the back of the text starting at page SP-1

EXAMPLE 11   The number of real estate transfers in several towns during the years 2007 to 2009 is given in the following bar graph.

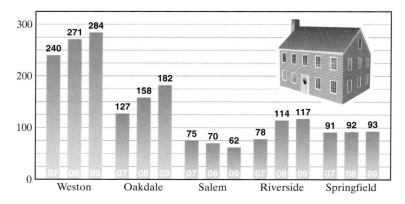

(a) What was the increase in homes sold in Weston from 2008 to 2009?

(b) What was the decrease in homes sold in Salem from 2007 to 2009?

(c) Between what two years did Oakdale have the greatest increase in sales?

Solution

(a) From the labels on the bar graph we see that 284 homes were sold in 2009 in Weston and 271 homes were sold in 2008. Thus the increase can be found by subtracting 284 − 271 = 13. There was an increase of 13 homes sold in Weston from 2008 to 2009.

(b) In 2007, 75 homes were sold in Salem. In 2009, 62 homes were sold in Salem. The decrease in the number of homes sold is 75 − 62 = 13. There was a decrease of 13 homes sold in Salem from 2007 to 2009.

(c) Here we will need to make two calculations in order to decide where the greatest increase occurs.

$$
\begin{array}{ll}
158 & \text{2008 sales} \\
-127 & \text{2007 sales} \\
\hline
31 & \text{Sales increase} \\
 & \text{from 2007 to 2008}
\end{array}
\qquad
\begin{array}{ll}
182 & \text{2009 sales} \\
-158 & \text{2008 sales} \\
\hline
24 & \text{Sales increase} \\
 & \text{from 2008 to 2009}
\end{array}
$$

The greatest increase in sales in Oakdale occurred from 2007 to 2008.

Practice Problem 11   Based on the preceding bar graph, answer the following questions.

(a) What was the increase in homes sold in Riverside from 2007 to 2008? 36

(b) How many more homes were sold in Springfield in 2007 than in Riverside in 2007? 13 more homes

(c) Between what two years did Weston have the greatest increase in sales? between 2007 and 2008

Verbal and Writing Skills

1. Explain how you can check a subtraction problem.

In subtraction the minuend minus the subtrahend equals the difference. To check the problem we add the subtrahend and the difference to see if we get the minuend. If we do, the answer is correct.

2. Explain how you use borrowing to calculate 107 − 88.

Since there are not enough ones to subtract 8 ones from 7 ones, we borrow. This means that we change the 1 hundred to an equivalent 10 tens. From the 10 tens we borrow one, making it 9 tens and 10 ones. Now we have 7 ones and 10 ones or 17 ones. 17 ones subtract 8 ones is 9 ones, and 9 tens subtract 8 tens is 1 ten. Thus 107 − 88 = 19.

3. Explain what number should be used to replace the question mark in the subtraction equation 32?5 − 1683 = 1592.

We know that 1683 + 1592 = 32?5. Therefore if we add 8 tens and 9 tens we get 17 tens, which is 1 hundred and 7 tens. Thus the ? should be replaced by 7.

4. Explain what steps need to be done to calculate 7 feet − 11 inches.

In subtraction we can subtract only numbers representing the same unit. Thus we need to change 7 feet to a number that measures inches. Since 1 foot equals 12 inches, 7 feet equals 84 inches. Now we subtract: 84 inches − 11 inches = 73 inches.

Try to do exercises 5–20 in one minute or less with no errors.

Subtract.

5.　8
　　− 3
　　‾‾5

6.　17
　　− 8
　　‾‾9

7.　15
　　− 9
　　‾‾6

8.　14
　　− 5
　　‾‾9

9.　16
　　− 0
　　‾16

10.　17
　　− 9
　　‾‾8

11.　18
　　− 9
　　‾‾9

12.　12
　　− 7
　　‾‾5

13.　11
　　− 4
　　‾‾7

14.　15
　　− 8
　　‾‾7

15.　13
　　− 7
　　‾‾6

16.　16
　　− 9
　　‾‾7

17.　11
　　− 8
　　‾‾3

18.　10
　　− 7
　　‾‾3

19.　15
　　− 6
　　‾‾9

20.　12
　　− 5
　　‾‾7

Subtract. Check your answers by adding.

21.　47　　26
　　− 26　+ 21
　　‾21　　47

22.　96　　51
　　− 51　+ 45
　　‾45　　96

23.　85　　73
　　− 73　+ 12
　　‾12　　85

24.　77　　36
　　− 36　+ 41
　　‾41　　77

25.　379　　36
　　− 36　+ 343
　　‾343　　379

26.　189　　65
　　− 65　+ 124
　　‾124　　189

27.　869　　548
　　− 548　+ 321
　　‾321　　869

28.　659　　247
　　− 247　+ 412
　　‾412　　659

29.　4799　　596
　　− 596　+ 4203
　　‾4203　　4799

30.　5780　　530
　　− 530　+ 5250
　　‾5250　　5780

31.　155,835　　12,600
　　− 12,600　+ 143,235
　　‾143,235　　155,835

32.　243,951　　12,400
　　− 12,400　+ 231,551
　　‾231,551　　243,951

33.　986,302　　433,201
　　− 433,201　+ 553,101
　　‾553,101　　986,302

34.　807,965　　304,214
　　− 304,214　+ 503,751
　　‾503,751　　807,965

Check each subtraction. If the problem has not been done correctly, find the correct answer.

35.
$$\begin{array}{r} 129 \\ -\ 19 \\ \hline 110 \end{array}$$
$$\begin{array}{r} 19 \\ +110 \\ \hline 129 \end{array}$$
Correct

36.
$$\begin{array}{r} 186 \\ -\ 45 \\ \hline 141 \end{array}$$
$$\begin{array}{r} 45 \\ +141 \\ \hline 186 \end{array}$$
Correct

37.
$$\begin{array}{r} 8596 \\ -3215 \\ \hline 5781 \end{array}$$
$$\begin{array}{r} 3215 \\ +5781 \\ \hline 8996 \end{array}$$
Incorrect
Correct answer: 5381

38.
$$\begin{array}{r} 9956 \\ -7254 \\ \hline 2702 \end{array}$$
$$\begin{array}{r} 7254 \\ +2702 \\ \hline 9956 \end{array}$$
Correct

39.
$$\begin{array}{r} 6030 \\ -5020 \\ \hline 1020 \end{array}$$
$$\begin{array}{r} 5020 \\ +1020 \\ \hline 6040 \end{array}$$
Incorrect
Correct answer: 1010

40.
$$\begin{array}{r} 7890 \\ -3200 \\ \hline 7670 \end{array}$$
$$\begin{array}{r} 3200 \\ +7670 \\ \hline 10,876 \end{array}$$
Incorrect
Correct answer: 4690

41.
$$\begin{array}{r} 47,869 \\ -33,846 \\ \hline 13,023 \end{array}$$
$$\begin{array}{r} 33,846 \\ +13,023 \\ \hline 46,869 \end{array}$$
Incorrect
Correct answer: 14,023

42.
$$\begin{array}{r} 99,583 \\ -41,181 \\ \hline 58,402 \end{array}$$
$$\begin{array}{r} 41,181 \\ +58,402 \\ \hline 99,583 \end{array}$$
Correct

Subtract. Use borrowing if necessary.

43.
$$\begin{array}{r} 98 \\ -52 \\ \hline 46 \end{array}$$

44.
$$\begin{array}{r} 86 \\ -33 \\ \hline 53 \end{array}$$

45.
$$\begin{array}{r} 174 \\ -\ 82 \\ \hline 92 \end{array}$$

46.
$$\begin{array}{r} 136 \\ -\ 95 \\ \hline 41 \end{array}$$

47.
$$\begin{array}{r} 647 \\ -263 \\ \hline 384 \end{array}$$

48.
$$\begin{array}{r} 706 \\ -435 \\ \hline 271 \end{array}$$

49.
$$\begin{array}{r} 955 \\ -237 \\ \hline 718 \end{array}$$

50.
$$\begin{array}{r} 861 \\ -345 \\ \hline 516 \end{array}$$

51.
$$\begin{array}{r} 20,000 \\ -\ 9285 \\ \hline 10,715 \end{array}$$

52.
$$\begin{array}{r} 50,000 \\ -\ 7338 \\ \hline 42,662 \end{array}$$

53.
$$\begin{array}{r} 152,000 \\ -117,908 \\ \hline 34,092 \end{array}$$

54.
$$\begin{array}{r} 361,000 \\ -121,520 \\ \hline 239,480 \end{array}$$

55.
$$\begin{array}{r} 45,312 \\ -37,865 \\ \hline 7447 \end{array}$$

56.
$$\begin{array}{r} 64,381 \\ -29,997 \\ \hline 34,384 \end{array}$$

57.
$$\begin{array}{r} 2,378,862 \\ -1,469,932 \\ \hline 908,930 \end{array}$$

58.
$$\begin{array}{r} 3,554,830 \\ -1,710,913 \\ \hline 1,843,917 \end{array}$$

Solve.

59. $x + 14 = 19$
$x = 5$

60. $x + 35 = 50$
$x = 15$

61. $28 = x + 20$
$x = 8$

62. $25 = x + 18$
$x = 7$

63. $100 + x = 127$
$x = 27$

64. $140 + x = 200$
$x = 60$

Applications

65. *Current Events* In one of the 2006 district races in Texas for U.S. Senate, a total of 161,160 votes were cast for two candidates. Republican Ralph Hall received 106,268 votes to beat Democrat Glenn Melancon. How many votes did Melancon receive? 54,892 votes

66. *Current Events* In one of the 2006 district races in California for U.S. Senate, a total of 138,203 votes were cast for two candidates. Democrat Jane Harman received 91,951 votes to beat Republican Brian Gibson. How many votes did Gibson receive? 46,252 votes

67. *Population Trends* In 2006, the population of Ireland was approximately 4,062,235. In the same year, the population of Portugal was approximately 10,605,870. How much less than the population of Portugal was the population of Ireland in 2006? 6,543,635

68. *Geography* The Nile River, the longest river in the world, is approximately 22,070,400 feet long. The Yangtze Kiang River, which is the longest river in China, is approximately 19,018,560 feet long. How much longer is the Nile River than the Yangtze Kiang River? 3,051,840 feet

69. *Personal Finance* Michaela's gross pay on her last paycheck was $1280. Her deductions totaled $318 and she deposited $200 into her savings account. She put the remaining amount into her checking account to pay bills. How much did Michaela put into her checking account? $762

70. *Personal Finance* Adam earned $3450 last summer at his construction job. He owed his brother $375 and saved $2300 to pay for his college tuition. He used the remaining amount as a down payment for a car. How much did Adam have for the down payment? $775

Population Trends *In answering exercises 71–78, consider the following population table.*

	1960	1970	1980	1990	2000	2010*
Illinois	10,081,158	11,110,285	11,427,409	11,430,602	12,051,683	13,216,340
Michigan	7,823,194	8,881,826	9,262,044	9,295,297	9,679,052	9,769,131
Indiana	4,662,498	5,195,392	5,490,212	5,544,159	6,045,521	6,178,300
Minnesota	3,413,864	3,806,103	4,075,970	4,375,099	4,830,784	5,263,820

Source: U.S. Census Bureau   *estimated

71. How much did the population of Minnesota increase from 1960 to 2000? 1,416,920 people

72. How much did the population of Michigan increase from 1960 to 2000? 1,855,858 people

73. In 1960, how much greater was the population of Illinois than the populations of Indiana and Minnesota combined? 2,004,796 people

74. In 2000, how much greater was the population of Illinois than the populations of Indiana and Minnesota combined? 1,175,378 people

75. How much did the population of Illinois increase from 1970 to 1990? 320,317 people

76. How much did the population of Michigan increase from 1970 to 1990? 413,471 people

77. Compare your answers to exercises 75 and 76. How much greater was the population increase of Michigan than the population increase of Illinois from 1970 to 1990? 93,154 people

78. In 2010, what will be the difference in population between the state with the highest population and the state with the lowest population? 7,952,520 people

Real Estate *The number of real estate transfers in several towns during the years 2005 to 2007 is given in the following bar graph. Use the bar graph to answer exercises 79–86. The figures in the bar graph reflect sales of single-family detached homes only.*

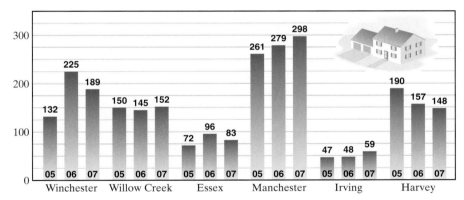

79. What was the increase in the number of homes sold in Winchester from 2005 to 2006? 93 homes

80. What was the increase in the number of homes sold in Irving from 2006 to 2007? 11 homes

81. What was the decrease in the number of homes sold in Essex from 2006 to 2007? 13 homes

82. What was the decrease in the number of homes sold in Harvey from 2005 to 2006? 33 homes

83. Between what two years did the greatest change occur in the number of homes sold in Willow Creek? between 2006 and 2007

84. Between what two years did the greatest change occur in the number of homes sold in Manchester? between 2006 and 2007

85. A real estate agent was trying to determine which two towns were closest to having the same number of sales in 2007. Which two towns should she select? Willow Creek and Harvey

86. A real estate agent was trying to determine which two towns were closest to having the same number of sales in 2005. Which two towns should he select? Winchester and Willow Creek

To Think About ·

87. In general, subtraction is not commutative. If a and b are whole numbers, $a - b \neq b - a$. For what types of numbers would it be true that $a - b = b - a$?

It is true if a and b represent the same number, for example, if $a = 10$ and $b = 10$.

88. In general, subtraction is not associative. For example, $8 - (4 - 3) \neq (8 - 4) - 3$. In general, $a - (b - c) \neq (a - b) - c$. Can you find some numbers a, b, c for which $a - (b - c) = (a - b) - c$? (Remember, do operations inside the parentheses first.)

It is true for all a and b if $c = 0$, for example, if $a = 5$, $b = 3$, and $c = 0$.

89. *Consumer Mathematics* Walter Swensen wants to replace some of the fences on a farm in Caribou, Maine. The wooden rail fence costs about $60 for wood and $50 for labor to install a fence that is 12 feet long. His son estimated he would need 276 feet of new fence. However, when he measured it he realized he would only need 216 new feet of fence. What is the difference in cost of his son's estimate versus his estimate with regard to how many feet of fence are needed? $550

90. *Consumer Mathematics* Carlos Sontera is replacing some barbed-wire fence on a ranch in El Paso, Texas. The barbed wire and poles for 12 feet of fence cost about $80. The labor cost to install 12 feet of fence is about $40. A ranch hand reported that 300 new feet of fence were needed. However, when Carlos actually rode out there and measured it, he found that only 228 new feet of fence were needed. What is the difference in cost of the ranch hand's estimate versus Carlos's estimate of how many feet of fence are needed? $720

Cumulative Review

91. **[1.1.3]** Write in standard notation: eight million, four hundred sixty-six thousand, eighty-four

8,466,084

92. **[1.1.3]** Write a word name for 296,308.

two hundred ninety-six thousand, three hundred eight

93. **[1.2.4]** Add. $25 + 75 + 80 + 20 + 18$ 218

94. **[1.2.4]** Add.

$$\begin{array}{r} 278{,}563 \\ + 896{,}187 \\ \hline 1{,}174{,}750 \end{array}$$

Quick Quiz 1.3 Subtract.

1.
$$\begin{array}{r} 5392 \\ -\ 938 \\ \hline 4454 \end{array}$$

2.
$$\begin{array}{r} 609{,}240 \\ -\ 386{,}307 \\ \hline 222{,}933 \end{array}$$

3.
$$\begin{array}{r} 17{,}200{,}300 \\ -\ 11{,}562{,}178 \\ \hline 5{,}638{,}122 \end{array}$$

4. Concept Check Explain how you would use borrowing when performing the calculation $12{,}345 - 11{,}976$.

Answers may vary

Classroom Quiz 1.3 You may use these problems to quiz your students' mastery of Section 1.3.

Subtract.

1.
$$\begin{array}{r} 7631 \\ -892 \\ \hline \end{array}$$
Ans: 6739

2.
$$\begin{array}{r} 706{,}350 \\ -\ 287{,}809 \\ \hline \end{array}$$
Ans: 418,541

3.
$$\begin{array}{r} 26{,}300{,}500 \\ -\ 18{,}279{,}156 \\ \hline \end{array}$$
Ans: 8,021,344

① Mastering Basic Multiplication Facts

Like subtraction, multiplication is related to addition. Suppose that the pastry chef at the Gourmet Restaurant bakes croissants on a sheet that holds four croissants across, with room for three rows. How many croissants does the sheet hold?

We can add $4 + 4 + 4$ to get the total, or we can use a shortcut: three rows of four is the same as 3 times 4, which equals 12. This is **multiplication,** a shortcut for repeated addition.

The numbers that we multiply are called **factors.** The answer is called the **product.** For now, we will use $\times$ to show multiplication. 3×4 is read "three times four."

$$\underbrace{3}_{\text{factor}} \quad \times \quad \underbrace{4}_{\text{factor}} \quad = \quad \underbrace{12}_{\text{product}} \qquad \begin{array}{r} 3 \\ \times\ 4 \\ \hline 12 \end{array} \begin{array}{l} \text{factor} \\ \text{factor} \\ \text{product} \end{array}$$

Your skill in multiplication depends on how well you know the basic multiplication facts. Look at the table on page 36. You should learn these facts well enough to quickly and correctly give the products of any two factors in the table. To check your knowledge, try Exercises 1.4, exercises 3 and 4.

Study the table to see if you can discover any properties of multiplication. What do you see as results when you multiply zero by any number? When you multiply any number times zero, the result is zero. That is the **multiplication property of zero.**

$$2 \times 0 = 0 \qquad 5 \times 0 = 0 \qquad 0 \times 6 = 0 \qquad 0 \times 0 = 0$$

You may recall that zero plays a special role in addition. Zero is the *identity element* for addition. When we add any number to zero, that number does not change. Is there an identity element for multiplication? Look at the table. What is the identity element for multiplication? Do you see that it is 1? The **identity element for multiplication** is 1.

$$5 \times 1 = 5 \qquad 1 \times 5 = 5$$

What other properties of addition hold for multiplication? Is multiplication commutative? Does the order in which you multiply two numbers change the results? Find the product of 3×4. Then find the product of 4×3.

$$3 \times 4 = 12$$
$$4 \times 3 = 12$$

The **commutative property of multiplication** tells us that when we multiply two numbers, changing the order of the numbers gives the same result.

Student Learning Objectives

After studying this section, you will be able to:

① **Master basic multiplication facts.**

② **Multiply a single-digit number by a several-digit number.**

③ **Multiply a whole number by a power of 10.**

④ **Multiply a several-digit number by a several-digit number.**

⑤ **Use the properties of multiplication to perform calculations.**

⑥ **Apply multiplication to real-life situations.**

Teaching Tip Remind students that they may need to practice or relearn facts that they cannot instantly recall from the multiplication table. As with the addition table, some students will need to make and use multiplication flash cards in order to master the basic multiplication facts.

Basic Multiplication Facts

×	0	1	2	3	4	5	6	7	8	9	10	11	12
0	0	0	0	0	0	0	0	0	0	0	0	0	0
1	0	1	2	3	4	5	6	7	8	9	10	11	12
2	0	2	4	6	8	10	12	14	16	18	20	22	24
3	0	3	6	9	12	15	18	21	24	27	30	33	36
4	0	4	8	12	16	20	24	28	32	36	40	44	48
5	0	5	10	15	20	25	30	35	40	45	50	55	60
6	0	6	12	18	24	30	36	42	48	54	60	66	72
7	0	7	14	21	28	35	42	49	56	63	70	77	84
8	0	8	16	24	32	40	48	56	64	72	80	88	96
9	0	9	18	27	36	45	54	63	72	81	90	99	108
10	0	10	20	30	40	50	60	70	80	90	100	110	120
11	0	11	22	33	44	55	66	77	88	99	110	121	132
12	0	12	24	36	48	60	72	84	96	108	120	132	144

QUICK RECALL OF MULTIPLICATION FACTS It is helpful if you can multiply quickly. See if you can do Example 1 correctly in 15 seconds or less. Repeat again with Practice Problem 1. Strive to obtain all answers correctly in 15 seconds or less.

Teaching Example 1 Multiply.

(a) 4×5 **(b)** 6×9 **(c)** 8×3

(d) 7×7 **(e)** 8×4

Ans: **(a)** 20 **(b)** 54 **(c)** 24
      (d) 49 **(e)** 32

EXAMPLE 1 Multiply.

(a) 5×7 **(b)** 8×9 **(c)** 6×8

(d) 9×3 **(e)** 7×8

Solution

(a) $\begin{array}{r} 5 \\ \times\ 7 \\ \hline 35 \end{array}$ **(b)** $\begin{array}{r} 8 \\ \times\ 9 \\ \hline 72 \end{array}$ **(c)** $\begin{array}{r} 6 \\ \times\ 8 \\ \hline 48 \end{array}$

(d) $\begin{array}{r} 9 \\ \times\ 3 \\ \hline 27 \end{array}$ **(e)** $\begin{array}{r} 7 \\ \times\ 8 \\ \hline 56 \end{array}$

NOTE TO STUDENT: *Fully worked-out solutions to all of the Practice Problems can be found at the back of the text starting at page SP-1*

Practice Problem 1 Multiply.

(a) $\begin{array}{r} 8 \\ \times\ 8 \\ \hline 64 \end{array}$ **(b)** $\begin{array}{r} 7 \\ \times\ 6 \\ \hline 42 \end{array}$ **(c)** $\begin{array}{r} 5 \\ \times\ 8 \\ \hline 40 \end{array}$

(d) $\begin{array}{r} 9 \\ \times\ 7 \\ \hline 63 \end{array}$ **(e)** $\begin{array}{r} 9 \\ \times\ 9 \\ \hline 81 \end{array}$

Multiplying a Single-Digit Number by a Several-Digit Number

EXAMPLE 2   Multiply. 4312×2

Solution   We first multiply the ones column, then the tens column, and so on, moving right to left.

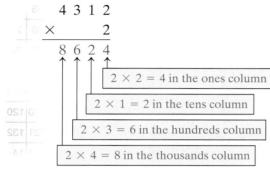

$$
\begin{array}{r}
4\ 3\ 1\ 2 \\
\times \quad\quad 2 \\
\hline
8\ 6\ 2\ 4
\end{array}
$$

$2 \times 2 = 4$ in the ones column

$2 \times 1 = 2$ in the tens column

$2 \times 3 = 6$ in the hundreds column

$2 \times 4 = 8$ in the thousands column

Practice Problem 2   Multiply. 3021×3 9063

Teaching Example 2 Multiply. 2013×3

Ans: 6039

NOTE TO STUDENT: *Fully worked-out solutions to all of the Practice Problems can be found at the back of the text starting at page SP-1*

Usually, we will have to carry one digit of the result of some of the multiplication into the next left-hand column.

EXAMPLE 3   Multiply. 36×7

Solution

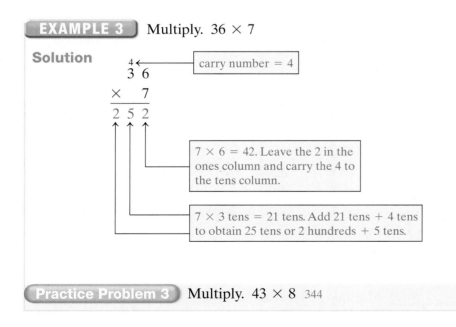

$$
\begin{array}{r}
4 \\
3\ 6 \\
\times \quad 7 \\
\hline
2\ 5\ 2
\end{array}
$$

carry number = 4

$7 \times 6 = 42$. Leave the 2 in the ones column and carry the 4 to the tens column.

7×3 tens $= 21$ tens. Add 21 tens $+ 4$ tens to obtain 25 tens or 2 hundreds $+ 5$ tens.

Practice Problem 3   Multiply. 43×8 344

Teaching Example 3 Multiply. 54×6

Ans: 324

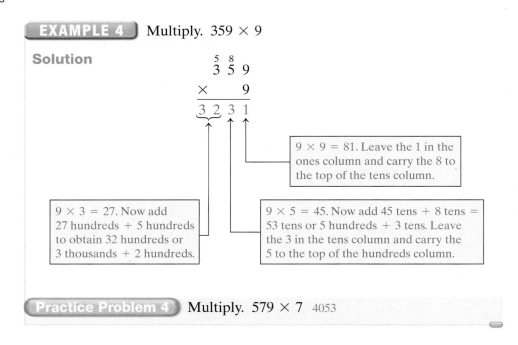

EXAMPLE 4   Multiply. 359 × 9

Solution

$$
\begin{array}{r}
\overset{5}{}\overset{8}{} \\
3\;5\;9 \\
\times\qquad 9 \\
\hline
3\;2\;3\;1
\end{array}
$$

$9 \times 9 = 81$. Leave the 1 in the ones column and carry the 8 to the top of the tens column.

$9 \times 3 = 27$. Now add 27 hundreds + 5 hundreds to obtain 32 hundreds or 3 thousands + 2 hundreds.

$9 \times 5 = 45$. Now add 45 tens + 8 tens = 53 tens or 5 hundreds + 3 tens. Leave the 3 in the tens column and carry the 5 to the top of the hundreds column.

Practice Problem 4   Multiply. 579 × 7   4053

3 Multiplying a Whole Number by a Power of 10

Observe what happens when a number is multiplied by 10, 100, 1000, 10,000, and so on.

$$56 \times 10 \quad = \quad 560 \qquad (\text{one zero})$$
$$56 \times 100 \quad = \quad 5600 \qquad (\text{two zeros})$$
$$56 \times 1000 \quad = \quad 56{,}000 \qquad (\text{three zeros})$$
$$56 \times 10{,}000 \quad = \quad 560{,}000 \qquad (\text{four zeros})$$

A **power of 10** is a whole number that begins with 1 and ends in one or more zeros. The numbers 10, 100, 1000, 10,000, and so on are powers of 10.

> To multiply a whole number by a power of 10:
>
> 1. Count the number of zeros in the power of 10.
>
> 2. Attach that number of zeros to the right side of the other whole number to obtain the answer.

EXAMPLE 5   Multiply 358 by each number.

(a) 10   **(b)** 100   **(c)** 1000   **(d)** 100,000

Solution

(a) $358 \times 10 = 3580$ (one zero)   **(b)** $358 \times 100 = 35{,}800$ (two zeros)

(c) $358 \times 1000 = 358{,}000$ (three zeros)

(d) $358 \times 100{,}000 = 35{,}800{,}000$ (five zeros)

Practice Problem 5 Multiply 1267 by each number.

(a) 10   **(b)** 1000   **(c)** 10,000   **(d)** 1,000,000

  12,670   1,267,000   12,670,000   1,267,000,000

How can we handle zeros in multiplication involving a number that is not 10, 100, 1000, or any other power of 10? Consider 32×400. We can rewrite 400 as 4×100, which gives us $32 \times 4 \times 100$. We can simply multiply 32×4 and then attach two zeros for the factor 100. We find that $32 \times 4 = 128$. Attaching two zeros gives us 12,800, or $32 \times 400 = 12,800$.

Teaching Tip You may need to do several problems of the type 345×30 and 2300×50 until students grasp the shortcut of counting the number of end zeros and then adding them at the end. Show them that to multiply 2300×50, you merely multiply $23 \times 5 = 115$ and then add the three zeros to obtain the final answer $2300 \times 50 = 115,000$.

EXAMPLE 6 Multiply.

(a) 12×3000 **(b)** 25×600 **(c)** 430×260

Solution

(a) $12 \times 3000 = 12 \times 3 \times 1000 = 36 \times 1000 = 36,000$
(b) $25 \times 600 = 25 \times 6 \times 100 = 150 \times 100 = 15,000$
(c) $430 \times 260 = 43 \times 26 \times 10 \times 10 = 1118 \times 100 = 111,800$

Teaching Example 6 Multiply.
(a) 23×2000 **(b)** 34×400
(c) 210×370

Ans: (a) 46,000   **(b)** 13,600   **(c)** 77,700

Practice Problem 6 Multiply.

(a) $9 \times 60,000$ **(b)** 15×400 **(c)** 270×800

  540,000   6000   216,000

 4 ## Multiplying a Several-Digit Number by a Several-Digit Number

EXAMPLE 7 Multiply. 234×21

Teaching Example 7 Multiply. 413×13

Ans: 5369

Solution   We can consider 21 as 2 tens (20) and 1 one (1). First we multiply 234 by 1.

We also multiply 234×20. This gives us two **partial products.**

$$\begin{array}{r} 234 \\ \times\ \ 1 \\ \hline 234 \end{array} \qquad \begin{array}{r} 234 \\ \times\ 20 \\ \hline 4680 \end{array}$$

Now we combine these two operations together by adding the two partial products to reach the final product, which is the solution.

$$\begin{array}{r} 2\ 3\ 4 \\ \times\ \ 2\ 1 \\ \hline 2\ 3\ 4 \\ 4\ 6\ 8\ 0 \\ \hline 4\ 9\ 1\ 4 \end{array}$$

⟵ Multiply 234×1.
⟵ Multiply 234×20.
⟵ Add the two partial products.

Practice Problem 7 Multiply. 323×32 10,336

EXAMPLE 8 Multiply. 671×35

Solution

$$
\begin{array}{r}
6\ 7\ 1 \\
\times\ \ \ 3\ 5 \\
\hline
3\ 3\ 5\ 5 \\
2\ 0\ 1\ 3\ 0 \\
\hline
2\ 3\ 4\ 8\ 5
\end{array}
$$

←———— First multiply 671×5.

←———— Now multiply 671×30.

←———— Now add the two partial products.

Note: We could omit zero on this line and leave the ones place blank.

Practice Problem 8 Multiply. 385×69 26,565

EXAMPLE 9 Multiply. 14×20

Solution

$$
\begin{array}{r}
1\ 4 \\
\times\ 2\ 0 \\
\hline
0 \\
2\ 8\ 0 \\
\hline
2\ 8\ 0
\end{array}
$$

←———— Multiply 14 by 0.

←———— Multiply 14 by 2 tens.

Now add the → partial products.

Place 28 with the 8 in the tens column. To line up the digits for adding, we can insert a 0 in the ones column.

Notice that you will also get this result if you multiply $14 \times 2 = 28$ and then attach a zero to multiply it by 10: 280.

Practice Problem 9 Multiply. 34×20 680

EXAMPLE 10 Multiply. 120×40

Solution

$$
\begin{array}{r}
1\ 2\ 0 \\
\times\ \ \ 4\ 0 \\
\hline
0 \\
4\ 8\ 0\ 0 \\
\hline
4\ 8\ 0\ 0
\end{array}
$$

←———— Multiply 120×0.

←———— Multiply 120 by 4 tens.

Now add the → partial products.

The answer is 480 tens. We place the 0 of the 480 in the tens column. To line up the digits for adding, we can insert a 0 in the ones column.

Notice that this result is the same as $12 \times 4 = 48$ with two zeros attached: 4800.

Practice Problem 10 Multiply. 130×50 6500

EXAMPLE 11 Multiply. 684×763

Solution

$$
\begin{array}{r}
6\ 8\ 4 \\
\times 7\ 6\ 3 \\
\hline
2\ 0\ 5\ 2 \\
4\ 1\ 0\ 4 \\
4\ 7\ 8\ 8 \\
\hline
5\ 2\ 1\ 8\ 9\ 2
\end{array}
$$

← Multiply 684×3.

← Multiply 684×60. Note that we omit the final zero.

← Multiply 684×700. Note that we omit the final two zeros.

Practice Problem 11 Multiply. 923×675 623,025

Teaching Example 11 Multiply. 437×258

Ans: 112,746

NOTE TO STUDENT: Fully worked-out solutions to all of the Practice Problems can be found at the back of the text starting at page SP-1

⑤ Using the Properties of Multiplication to Perform Calculations

When we add three numbers, we use the associative property. Recall that the associative property allows us to group the three numbers in different ways. Thus to add $9 + 7 + 3$, we can group the numbers as $9 + (7 + 3)$ because it is easier to find the sum. $9 + (7 + 3) = 9 + 10 = 19$. We can demonstrate that multiplication is also associative.

$$\text{Is this true?} \quad 2 \times (5 \times 3) = (2 \times 5) \times 3$$
$$2 \times (15) = (10) \times 3$$
$$30 = 30$$

The final product is the same in both cases.

The way we group numbers to be multiplied does not change the product. This property is called the **associative property of multiplication.**

EXAMPLE 12 Multiply. $14 \times 2 \times 5$

Solution Since we can group any two numbers together, let's take advantage of the ease of multiplying by 10.

$$14 \times 2 \times 5 = 14 \times (2 \times 5) = 14 \times 10 = 140$$

Practice Problem 12 Multiply. $25 \times 4 \times 17$ 1700

Teaching Example 12 Multiply. $20 \times 5 \times 35$

Ans: 3500

For convenience, we list the properties of multiplication that we have discussed in this section.

1. Associative Property of Multiplication. When we multiply three numbers, the multiplication can be grouped in any way. $\quad (7 \times 3) \times 2 = 7 \times (3 \times 2)$ $\quad 21 \times 2 = 7 \times 6$ $\quad 42 = 42$

2. **Commutative Property of Multiplication.** Two numbers can be multiplied in either order with the same result.

$9 \times 8 = 8 \times 9$
$72 = 72$

3. **Identity Property of One.** When one is multiplied by a number, the result is that number.

$7 \times 1 = 7$
$1 \times 15 = 15$

4. **Multiplication Property of Zero.** The product of any number and zero yields zero as a result.

$0 \times 14 = 0$
$2 \times 0 = 0$

Sometimes you can use several properties in one problem to make the calculation easier.

Teaching Example 13 Multiply.
$25 \times 9 \times 7 \times 4$

Ans: 6300

EXAMPLE 13 Multiply. $7 \times 20 \times 5 \times 6$

Solution
$$
\begin{aligned}
7 \times 20 \times 5 \times 6 &= 7 \times (20 \times 5) \times 6 \quad \text{Associative property} \\
&= 7 \times 6 \times (20 \times 5) \quad \text{Commutative property} \\
&= 42 \times 100 \\
&= 4200
\end{aligned}
$$

Practice Problem 13 Multiply. $8 \times 4 \times 3 \times 25$ 2400

Thus far we have discussed the properties of addition and the properties of multiplication. There is one more property that links both operations.

Before we discuss that property, we will illustrate several different ways of showing multiplication. The following are all the ways to show "3 times 4."

3×4	$(3)(4)$	$3(4)$ $(3)4$	$3 \cdot 4$	$3 * 4$
with an $\times$	with two sets of parentheses	with a single set of parentheses	with a dot	with a star

We will use parentheses to mean multiplication when we use the **distributive property.**

SIDELIGHT: The Distributive Property

Why does our method of multiplying several-digit numbers work? Why can we say that 234×21 is the same as $234 \times 1 + 234 \times 20$?

The *distributive property of multiplication over addition* allows us to distribute the multiplication and then add the results. To illustrate, 234×21 can be written as $234(20 + 1)$. By the distributive property

$$
\begin{aligned}
234(20 + 1) &= (234 \times 20) + (234 \times 1) \\
&= \quad 4680 \quad + \quad 234 \\
&= \quad 4914
\end{aligned}
$$

This is what we actually do when we multiply.

$$
\begin{array}{r}
234 \\
\times \quad 21 \\
\hline
234 \\
4680 \\
\hline
4914 \\
\end{array}
$$

DISTRIBUTIVE PROPERTY OF MULTIPLICATION OVER ADDITION

Multiplication can be distributed over addition without changing the result.

$$5 \times (10 + 2) = (5 \times 10) + (5 \times 2)$$

 ## Applying Multiplication to Real-Life Situations

To use multiplication in word problems, the number of items or the value of each item must be the same. Recall that multiplication is a quick way to do repeated addition where *each addend is the same.* In the beginning of the section we showed three rows of four croissants to illustrate 3×4. The number of croissants in each row was the same, 4. Look at another example. If we had six nickels, we could use multiplication to find the total value of the coins because the value of each nickel is the same, 5¢. Since $6 \times 5 = 30$, six nickels are worth 30¢.

In the following example the word *average* is used. The word *average* has several different meanings. In this example, we are told that the *average annual salary* of an employee at Software Associates is $42,132. This means that we can calculate the total payroll as if each employee made $42,132 even though we know that the president probably makes more than any other employee.

EXAMPLE 14 The average annual salary of an employee at Software Associates is $42,132. There are 38 employees. What is the annual payroll?

Solution

$$
\begin{array}{r}
\$42{,}132 \\
\times \qquad 38 \\
\hline
337\ 056 \\
1263\ 96 \\
\hline
1{,}601{,}016 \\
\end{array}
$$

The total annual payroll is $1,601,016.

Practice Problem 14 The average cost of a new car sold last year at Westover Chevrolet was $17,348. The dealership sold 378 cars. What were the total sales of cars at the dealership last year? $6,557,544

Another useful application of multiplication is area. The following example involves the area of a rectangle.

Teaching Example 14 An average of 3152 cars cross over the West River Bridge each weekday. There are 260 weekdays each year. How many cars crossed the bridge last year?

Ans: 819,520 cars

NOTE TO STUDENT: Fully worked-out solutions to all of the Practice Problems can be found at the back of the text starting at page SP-1

▲ **EXAMPLE 15** What is the area of a rectangular hallway that measures 4 feet wide and 8 feet long?

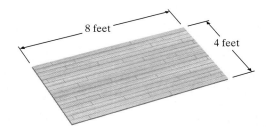

Solution The **area** of a rectangle is the product of the length times the width. Thus for this hallway

$$\text{Area} = 8 \text{ feet} \times 4 \text{ feet} = 32 \text{ square feet.}$$

The area of the hallway is 32 square feet.

Note: All measurements for area are given in square units such as square feet, square meters, square yards, and so on.

▲ **Practice Problem 15** What is the area of a rectangular rug that measures 5 yards by 7 yards? 35 square yards

Developing Your Study Skills

Why Is Homework Necessary?

Mathematics involves mastering a set of skills that you learn by practicing, not by watching someone else do it. Your instructor may make solving a mathematics problem look very easy, but for you to learn the necessary skills, you must practice them over and over again, just as your instructor once had to. There is no other way. Learning mathematics is like learning to play a musical instrument, to type, or to play a sport. No matter how much you watch someone else do it, no matter how many books you read on "how to" do it, no matter how easy it seems to be, the key to success is practice on a regular basis.

Homework provides this practice. The amount of practice needed varies for each individual, but usually students need to do most or all of the exercises provided at the end of each section in the text. The more exercises you do, the better you get. Some exercises in a set are more difficult than others, and some stress different concepts. Only by working all the exercises will you cover the full range of skills.

Verbal and Writing Skills

1. Explain in your own words.

Answers may vary. Samples follow.

(a) the commutative property of multiplication

You can change the order of the factors without changing the product.

(b) the associative property of multiplication

You can group the factors in any way without changing the product.

2. How does the distributive property of multiplication over addition help us to multiply 4×13?

You can write 13 as $10 + 3$ and distribute 4 over the addition. $4 \times (10 + 3) = (4 \times 10) + (4 \times 3)$

Complete the multiplication facts for each table. Strive for total accuracy, but work quickly. (Allow a maximum of six minutes for each table.)

3.

×	6	2	3	8	0	5	7	9	12	4
5	30	10	15	40	0	25	35	45	60	20
7	42	14	21	56	0	35	49	63	84	28
1	6	2	3	8	0	5	7	9	12	4
0	0	0	0	0	0	0	0	0	0	0
6	36	12	18	48	0	30	42	54	72	24
2	12	4	6	16	0	10	14	18	24	8
3	18	6	9	24	0	15	21	27	36	12
8	48	16	24	64	0	40	56	72	96	32
4	24	8	12	32	0	20	28	36	48	16
9	54	18	27	72	0	45	63	81	108	36

4.

×	2	7	0	5	3	4	8	12	6	9
1	2	7	0	5	3	4	8	12	6	9
6	12	42	0	30	18	24	48	72	36	54
5	10	35	0	25	15	20	40	60	30	45
3	6	21	0	15	9	12	24	36	18	27
0	0	0	0	0	0	0	0	0	0	0
9	18	63	0	45	27	36	72	108	54	81
4	8	28	0	20	12	16	32	48	24	36
7	14	49	0	35	21	28	56	84	42	63
2	4	14	0	10	6	8	16	24	12	18
8	16	56	0	40	24	32	64	96	48	72

Multiply.

5.   32
    × 3
    96

6.   21
    × 4
    84

7.   14
    × 5
    70

8.   15
    × 6
    90

9.   87
    × 6
    522

10.   95
    × 7
    665

11.   231
    × 3
    693

12.   313
    × 3
    939

13.   276
    × 7
    1932

14.   538
    × 8
    4304

15.   6102
    × 3
    18,306

16.   5203
    × 2
    10,406

17.   12,203
    × 3
    36,609

18.   31,206
    × 3
    93,618

19.   5218
    × 6
    31,308

20.   3215
    × 6
    19,290

21.   12,526
    × 8
    100,208

22.   48,761
    × 7
    341,327

23.   344,601
    × 9
    3,101,409

24.   257,021
    × 9
    2,313,189

Multiply by powers of 10.

25.   156
    × 10
    1560

26.   278
    × 10
    2780

27.   27,158
    × 100
    2,715,800

28.   89,361
    × 100
    8,936,100

29.   482
    ×1000
    482,000

30.   579
    ×1000
    579,000

31.   37,256
    × 10,000
    372,560,000

32.   614,260
    × 10,000
    6,142,600,000

Multiply by multiples of 10.

33.   423
        × 20
        8460

34.   332
        × 30
        9960

35.   2120
        × 30
        63,600

36.   4230
        × 20
        84,600

37.   14,000
        × 4000
     56,000,000

38.   62,000
        × 3000
     186,000,000

Multiply.

39.   514
        × 12
        1028
         514
        6168

40.   432
        × 13
        1296
         432
        5616

41.   146
        × 54
         584
         730
        7884

42.   163
        × 35
         815
         489
        5705

43.   89
        ×64
         356
         534
        5696

44.   68
        ×49
         612
         272
        3332

45.   607
        × 25
        3 035
        12 14
       15,175

46.   780
        × 24
        3 120
        15 60
       18,720

47.   544
        × 38
        4 352
        16 32
       20,672

48.   652
        × 92
        1 304
        58 68
       59,984

49.   912
        × 76
        5 472
        63 84
       69,312

50.   498
        × 39
        4 482
        14 94
       19,422

51.   5123
        × 29
        46 107
       102 46
       148,567

52.   1268
        × 38
        10 144
        38 04
       48,184

53.   9053
        × 91
         9 053
       814 77
       823,823

54.   3078
        × 72
         6 156
       215 46
       221,616

55.   4326
        × 435
        21 630
       129 78
      1 730 4
      1,881,810

56.   3725
        × 546
        22 350
       149 00
      1 862 5
      2,033,850

57.   678
        ×132
        1 356
        20 34
        67 8
       89,496

58.   392
        ×187
        2 744
        31 36
        39 2
       73,304

Mixed Practice

59.   2076
        × 105
        10 380
        00 00
       207 6
       217,980

60.   5092
        × 302
        10 184
        00 00
      1 527 6
      1,537,784

61.   1324
        × 2004
        5 296
        2 648
       2,653,296

62.   2074
        × 1003
        6 222
        2 074
       2,080,222

63.   12,000
        × 60
       720,000

64.   15,200
        × 30
       456,000

65.   250
        × 40
       10,000

66.   302
        × 30
       9060

67.   302
        × 300
       90,600

68.   3000
        × 302
       906,000

69. $7 \cdot 2 \cdot 5$
70

70. $8 \cdot 3 \cdot 2$
48

71. $11 \cdot 7 \cdot 4$
308

72. $15 \cdot 4 \cdot 4$
240

73. 412×33
13,596

74. 526×21
11,046

75. $5 \cdot 8 \cdot 4 \cdot 10$
1600

76. $5 \cdot 10 \cdot 18 \cdot 2$
1800

77. What is x if
$x = 8 \cdot 7 \cdot 6 \cdot 0?$
$x = 0$

78. What is x if
$x = 3 \cdot 12 \cdot 0 \cdot 5?$
$x = 0$

Applications

▲ **79. Geometry** Find the area of a patio that is 16 feet wide and 24 feet long.
384 square feet

▲ **80. Geometry** Find the area of a calculator screen that is 15 millimeters wide and 60 millimeters long.
900 square millimeters

▲ **81. Consumer Mathematics** Don Williams and his wife want to put down new carpet in the living room and the hallway of their house. The living room measures 12 feet by 14 feet. The hallway measures 9 feet by 3 feet. If the living room and the hallway are rectangular in shape, how many square feet of new carpet do Don and his wife need?
195 square feet

▲ **82. Wildlife Management** Robert Tobey in Copper Center, Alaska, wants to put a field under helicopter surveillance because of a roving pack of wolves that are destroying other wildlife in the area. The field consists of two rectangular regions. The first one is 4 miles by 5 miles. The second one is 12 miles by 8 miles. How many square miles does he want to place under surveillance?
116 square miles

83. Business Decisions The student commons food supply needs to purchase espresso coffee. Find the cost of purchasing 240 pounds of espresso coffee beans at $5 per pound.
$1200

84. Business Decisions The music department of Wheaton College wishes to purchase 345 sets of headphones at the music supply store at a cost of $8 each. What will be the total amount of the purchase?
$2760

85. Personal Finance Helen pays $266 per month for her car payment on her new Honda Civic. What is her automobile payment cost for a one-year period?
$3192

86. Personal Finance A company rents a Ford Escort for a salesman at $276 per month for eight months. What is the cost for the car rental during this time?
$2208

87. Environmental Studies Marcos has a Toyota Corolla that gets 34 miles per gallon during highway driving. Approximately how far can he travel if he has 18 gallons of gas in the tank?
612 miles

88. Environmental Studies Cheryl has a subcompact car that gets 48 miles per gallon on highway driving. Approximately how far can she travel if she has 12 gallons of gas in the tank?
576 miles

89. Personal Finance Sylvia worked as a camp counselor for 12 weeks during the summer. She earned $420 per week. What is the total amount Sylvia earned during the summer?
$5040

90. Personal Finance Each time Jorge receives a paycheck, $125 is put into his IRA retirement savings account. If he gets paid twice a month, how much does Jorge contribute to his IRA in one year? $3000

91. International Relations The country of Haiti has an average per capita (per person) income of $1070. If the approximate population of Haiti is 6,890,000, what is the approximate total yearly income of the entire country?
$7,372,300,000

92. International Relations The country of the Netherlands (Holland) has an approximate population of 15,800,000. The average per capita (per person) income is $22,000. What is the approximate total yearly income of the entire country?
$347,600,000,000

To Think About *Use the following information to answer exercises 93–96. There are 98 puppies in a room, with an assortment of black and white ears and paws. 18 puppies have totally black ears and 2 white paws; 26 puppies have 1 black ear and 4 white paws; and 54 puppies have no black ears and 1 white paw.*

93. How many black paws are in the room? 198

94. How many white paws are in the room? 194

95. How many black ears are in the room? 62

96. How many white ears are in the room? 134

In exercises 97–100, find the value of x in each equation.

97. $5(x) = 40$ $x = 8$

98. $7(x) = 56$ $x = 8$

99. $72 = 8(x)$ $x = 9$

100. $63 = 9(x)$ $x = 7$

101. Would the distributive property of multiplication be true for Roman numerals such as $(XII) \times (IV)$? Why or why not?
No, it would not always be true. In our number system $62 = 60 + 2$. But in Roman numerals IV ≠ I + V. The digit system in Roman numerals involves subtraction. Thus $(XII) \times (IV) \neq (XII \times I) + (XII \times V)$.

102. We saw that multiplication is distributive over addition. Is it distributive over subtraction? Why or why not? Give examples.
yes, $5 \times (8 - 3) = 5 \times 8 - 5 \times 3$,
$a \times (b - c) = (a \times b) - (a \times c)$

Cumulative Review

103. **[1.3.3]** Subtract.
$$\begin{array}{r} 34,084 \\ - \ 27,328 \\ \hline 6,756 \end{array}$$

104. **[1.2.4]** Add.
$$\begin{array}{r} 263 \\ 27 \\ 891 \\ 5 \\ + \ 63 \\ \hline 1249 \end{array}$$

105. **[1.3.5]** *Personal Finance* Adam Goulet has $1278 in his checking account. After writing checks for $345 and $128, how much is left in his account? $805

106. **[1.3.5]** *Personal Finance* Petra Mayer was earning $1672 each month at her job. She received a raise and now her paychecks are $1758. What was the increase? $86

107. **[1.3.5]** *Population Studies* The population of Paynesville in 1995 was 34,988. In 2007, the population had increased to 37,125. By how many people did the net population increase?
2137 people

108. **[1.3.5]** *International Relations* In 2000, the gross domestic product of Spain was $720,800,000,000. In 2005, it had grown to $1,113,539,000,000. How much did the gross domestic product increase from 2000 to 2005?
$392,739,000,000

Quick Quiz 1.4 Multiply.

1.
$$\begin{array}{r} 34,986 \\ \times \qquad 5 \\ \hline 174,930 \end{array}$$

2.
$$\begin{array}{r} 79 \\ \times 64 \\ \hline 5056 \end{array}$$

3.
$$\begin{array}{r} 698 \\ \times 297 \\ \hline 207,306 \end{array}$$

4. **Concept Check** Explain what you do with the zeros when you multiply 3457×2008. Answers may vary

Classroom Quiz 1.4 You may use these problems to quiz your students' mastery of Section 1.4.

Multiply.

1.
$$\begin{array}{r} 26,523 \\ \times \qquad 8 \\ \hline \end{array}$$
Ans: 212,184

2.
$$\begin{array}{r} 83 \\ \times 57 \\ \hline \end{array}$$
Ans: 4731

3.
$$\begin{array}{r} 782 \\ \times 345 \\ \hline \end{array}$$
Ans: 269,790

1 Mastering Basic Division Facts

Suppose that we have eight quarters and want to divide them into two equal piles. We would discover that each pile contains four quarters.

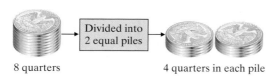

8 quarters   4 quarters in each pile

In mathematics we would express this thought by saying that

$$8 \div 2 = 4.$$

We know that this answer is right because two piles of four quarters is the same dollar amount as eight quarters. In other words, we know that $8 \div 2 = 4$ because $2 \times 4 = 8$. These two mathematical sentences are called **related sentences.** The division sentence $8 \div 2 = 4$ is related to the multiplication sentence $2 \times 4 = 8$.

In fact, in mathematics we usually define **division** in terms of multiplication. The answer to the division problem $12 \div 3$ is that number which when multiplied by 3 yields 12. Thus

$$12 \div 3 = 4 \quad \text{because } 3 \times 4 = 12.$$

Suppose that a surplus of \$30 in the French Club budget at the end of the year is to be equally divided among the five club members. We would want to divide the \$30 into five equal parts. We would write $30 \div 5 = 6$ because $5 \times 6 = 30$. Thus each of the five people would get \$6 in this situation.

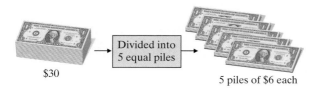

\$30   5 piles of \$6 each

As a mathematical sentence, $30 \div 5 = 6$.
The division problem $30 \div 5 = 6$ could also be written $\frac{30}{5} = 6$ or $5 \overline{)30}$.

When referring to division, we sometimes use the words **divisor, dividend,** and **quotient** to identify the three parts.

$$\text{divisor} \overline{)\text{dividend}}^{\text{quotient}}$$

With $30 \div 5 = 6$, 30 is the dividend, 5 is the divisor, and 6 is the quotient.

$$\text{divisor} \rightarrow 5 \overline{)30}^{\; 6 \; \leftarrow \text{quotient}} \leftarrow \text{dividend}$$

So the quotient is the answer to a division problem. It is important that you be able to do short problems involving basic division facts quickly.

Student Learning Objectives

After studying this section, you will be able to:

1 Master basic division facts.

2 Perform division by a one-digit number.

3 Perform division by a two- or three-digit number.

4 Apply division to real-life situations.

Teaching Tip Throughout all mathematics courses from basic mathematics to calculus, the words *divisor, dividend,* and *quotient* are used extensively. Stress to students that in any given division problem they need to be able to recognize which number is the divisor, which is the dividend, and which is the quotient.

EXAMPLE 1 Divide.

(a) $12 \div 4$ **(b)** $81 \div 9$ **(c)** $56 \div 8$ **(d)** $54 \div 6$

Solution

(a) $4\overline{)12}$ gives 3 **(b)** $9\overline{)81}$ gives 9 **(c)** $8\overline{)56}$ gives 7 **(d)** $6\overline{)54}$ gives 9

Practice Problem 1 Divide.

(a) $36 \div 4$ 9   **(b)** $25 \div 5$ 5   **(c)** $72 \div 9$ 8   **(d)** $30 \div 6$ 5

Zero can be divided by any nonzero number, but division by zero is not possible. Why is this?

Suppose that we could divide by zero. Then $7 \div 0 =$ some number. Let us represent "some number" by the letter a.

$$\text{If } 7 \div 0 = a, \text{ then } 7 = 0 \times a,$$

because every division problem has a related multiplication problem. But zero times any number is zero, $0 \times a = 0$. Thus

$$7 = 0 \times a = 0.$$

That is, $7 = 0$, which we know is not true. Therefore, our assumption that $7 \div 0 = a$ is wrong. Thus we conclude that we cannot divide by zero. Mathematicians state this by saying, "Division by zero is **undefined**."

It is helpful to remember the following basic concepts:

> **DIVISION PROBLEMS INVOLVING THE NUMBER 1 AND THE NUMBER 0**
>
> **1.** Any nonzero number divided by itself is $1(7 \div 7 = 1)$.
>
> **2.** Any number divided by 1 remains unchanged ($29 \div 1 = 29$).
>
> **3.** Zero may be divided by any nonzero number; the result is always zero ($0 \div 4 = 0$).
>
> **4.** Zero can never be the divisor in a division problem ($3 \div 0$ is undefined).

EXAMPLE 2 Divide, if possible. If it is not possible, state why.

(a) $8 \div 8$ **(b)** $9 \div 1$ **(c)** $0 \div 6$ **(d)** $20 \div 0$

Solution

(a) $\dfrac{8}{8} = 1$ Any number divided by itself is 1.

(b) $\dfrac{9}{1} = 9$ Any number divided by 1 remains unchanged.

(c) $\dfrac{0}{6} = 0$ Zero divided by any nonzero number is zero.

(d) $\dfrac{20}{0}$ cannot be done   Division by zero is undefined.

Practice Problem 2 Divide, if possible.

(a) $7 \div 1$　7　　**(b)** $\dfrac{9}{9}$　1　　**(c)** $\dfrac{0}{5}$　0　　**(d)** $12 \div 0$　cannot be done

② Performing Division by a One-Digit Number

Our accuracy with division is improved if we have a checking procedure. For each division fact, there is a related multiplication fact.

$$\text{If } 20 \div 4 = 5, \quad \text{then } 20 = 4 \times 5.$$
$$\text{If } 36 \div 9 = 4, \quad \text{then } 36 = 9 \times 4.$$

We will often use multiplication to check our answers.

When two numbers do not divide exactly, a number called the **remainder** is left over. For example, 13 cannot be divided exactly by 2. The number 1 is left over. We call this 1 the *remainder*.

$$
\begin{array}{r}
6 \\
2\overline{)13} \\
\underline{12} \\
1 \leftarrow \text{remainder}
\end{array}
$$

Thus $13 \div 2 = 6$ with a remainder of 1. We can abbreviate this answer as

6 R 1.

To check this division, we multiply $2 \times 6 = 12$ and add the remainder: $12 + 1 = 13$. That is, $(2 \times 6) + 1 = 13$. The result will be the dividend if the division was done correctly. The following box shows you how to check a division that has a remainder.

$$(\text{divisor} \times \text{quotient}) + \text{remainder} = \text{dividend}$$

EXAMPLE 3 Divide. $33 \div 4$. Check your answer.

Solution
$$
\begin{array}{r}
8 \rightarrow \text{How many times can 4 be divided into 33?}\quad 8. \\
4\overline{)33} \\
\underline{32} \leftarrow \text{What is } 8 \times 4?\quad 32. \\
1 \leftarrow \text{32 subtracted from 33 is 1.}
\end{array}
$$

The answer is 8 with a remainder of 1. We abbreviate this as 8 R 1.

CHECK.
$$
\begin{array}{r}
8 \\
\underline{\times\ 4} \\
32 \quad \text{Multiply. } 8 \times 4 = 32. \\
\underline{+\ 1} \quad \text{Add the remainder. } 32 + 1 = 33. \\
33 \quad \text{Because the dividend is 33, the answer is correct.}
\end{array}
$$

Practice Problem 3 Divide. $45 \div 6$. Check your answer.　7 R 3

Teaching Tip It is critical that all students master division with zero. They need to know that zero can be divided by any nonzero number, but that division by zero is never allowed. A simple example usually helps them to see the logic of the situation. A student club with profits of $400 and 10 members could distribute 400 divided by 10 = $40 to each member. A student club with profits of $0 and 10 members would only be able to distribute $0 divided by 10 = $0 to each member. A student club with profits of $400 and 0 members could not distribute any money to each member. 400 divided by zero cannot be done. Therefore division by zero is **undefined.**

Teaching Tip Be sure to go over in detail how to check a division problem if there is a remainder. A few students will find this idea totally new and will need to see a few examples worked out before they understand the concept.

Teaching Example 3 Divide. $42 \div 8$. Check your answer.

Ans: 5 R 2

EXAMPLE 4 Divide. 158 ÷ 5. Check your answer.

Solution

$$
\begin{array}{r}
31 \\
5\overline{)158} \\
\end{array}
$$

5 divided into 15? 3.

15 ← What is 3 × 5? 15.

08 ← 15 subtract 15? 0. Bring down 8.

5 ← 5 divided into 8? 1. What is 1 × 5? 5.

3 ← 8 subtract 5? 3.

The answer is 31 R 3.

CHECK.

$$
\begin{array}{r}
31 \\
\times\ 5 \\
\hline
155 \\
\end{array}
$$

Multiply. 31 × 5 = 155.

+ 3 Add the remainder 3.

158 Because the dividend is 158, the answer is correct.

Practice Problem 4 Divide. 129 ÷ 6. Check your answer. 21 R 3

EXAMPLE 5 Divide. 3672 ÷ 7

Solution

$$
\begin{array}{r}
524 \\
7\overline{)3672} \\
\end{array}
$$

How many times can 7 be divided into 36? 5.

35 ← What is 5 × 7? 35.

17 ← 36 subtract 35? 1. Bring down 7.

14 ← 7 divided into 17? 2. What is 2 × 7? 14.

32 ← 17 subtract 14? 3. Bring down 2.

28 ← 7 divided into 32? 4. What is 4 × 7? 28.

4 ← 32 subtract 28? 4.

The answer is 524 R 4.

Practice Problem 5 Divide. 4237 ÷ 8 529 R 5

③ Performing Division by a Two- or Three-Digit Number

When the divisor has more than one digit, an estimation technique may help. Figure how many times the first digit of the divisor goes into the first two digits of the dividend. Try this answer as the first number in the quotient.

EXAMPLE 6 Divide. 283 ÷ 41

Solution

First guess:

$$
\begin{array}{r}
7 \\
41\overline{)283} \\
287 \\
\end{array}
$$

too large

How many times can the first digit of the divisor (4) be divided into the first two digits of the dividend (28)? 7. We try the answer 7 as the first number of the quotient. We multiply 7 × 41 = 287. We see that 287 is larger than 283.

Second guess:

$$\begin{array}{r} 6 \\ 41\overline{)283} \\ \underline{246} \\ 37 \end{array}$$

Because 7 is slightly too large, we try 6.

$246 \leftarrow$ 6×41? 246.

$37 \leftarrow$ 283 subtract 246? 37.

The answer is 6 **R** 37. (Note that the remainder must always be less than the divisor.)

Practice Problem 6   Divide. $243 \div 32$ 7 R 19

EXAMPLE 7   Divide. $33,897 \div 56$

Teaching Example 7 Divide. $36,161 \div 45$.

Ans: 803 R 26

Solution

First guess:

$$\begin{array}{r} 60 \\ 56\overline{)33897} \\ \underline{336} \\ 29 \end{array}$$

How many times can 33 be divided by 5? 6.

$336 \leftarrow$ What is 6×56? 336.

338 subtract 336? 2. Bring down 9.

56 cannot be divided into 29. Write 0 in quotient.

Second set of steps:

$$\begin{array}{r} 605 \\ 56\overline{)33897} \\ \underline{336} \\ 297 \\ \underline{280} \\ 17 \end{array}$$

Bring down 7.

How many times can 5 be divided into 29? 5.

$280 \leftarrow$ What is 5×56? 280. Subtract $297 - 280$.

Remainder is 17.

The answer is 605 **R** 17.

Practice Problem 7   Divide. $42,183 \div 33$ 1278 R 9

EXAMPLE 8   Divide. $5629 \div 134$

Teaching Example 8 Divide. $6513 \div 118$. Check your answer.

Ans: 55 R 23

Solution

$$\begin{array}{r} 42 \\ 134\overline{)5629} \\ \underline{536} \\ 269 \\ \underline{268} \\ 1 \end{array}$$

How many times does 134 divide into 562?
We guess by saying that 1 divides into 5 five times, but this is too large. ($5 \times 134 = 670$!)
So we try 4. What is 4×134? 536.
Subtract $562 - 536$. We obtain 26. Bring down 9.
How many times does 134 divide into 269?
We guess by saying that 1 divided into 2 goes 2 times.
What is 2×134? 268. Subtract $269 - 268$.
The remainder is 1.

The answer is 42 **R** 1.

Practice Problem 8   Divide. $3227 \div 128$ 25 R 27

Applying Division to Real-Life Situations

When you solve a word problem that requires division, you will be given the total number and asked to calculate the number of items in each group or to calculate the number of groups. In the beginning of this section we showed eight quarters (the total number) and we divided them into two equal piles (the number of groups). Division was used to find how many quarters were in each pile (the number in each group). That is, 8 ÷ 2 = 4. There were four quarters in each pile.

Let's look at another example. Suppose that $30 is to be divided equally among the members of a group. If each person receives $6, how many people are in the group? We use division, 30 ÷ 6 = 5, to find that there are five people in the group.

You will find many real world examples where you know the total cost of several identical items, and you need to find the cost per item. You will encounter this situation in the following example.

EXAMPLE 9 City Service Realty just purchased nine identical computers for the real estate agents in the office. The total cost for the nine computers was $25,848. What was the cost of one computer? Check your answer.

Solution   To find the cost of one computer, we need to divide the total cost by 9. Thus we will calculate 25,848 ÷ 9.

$$
\begin{array}{r}
2872 \\
9\overline{)25848} \\
\underline{18} \\
78 \\
\underline{72} \\
64 \\
\underline{63} \\
18 \\
\underline{18} \\
0
\end{array}
$$

Therefore, the cost of one computer is $2872. In order to check our work we will need to see if nine computers each costing $2872 will in fact result in a total of $25,848. We use multiplication to check division.

$$
\begin{array}{r}
2872 \\
\times\quad 9 \\
\hline
25848 \quad ✓
\end{array}
$$

We did obtain 25,848. Our answer is correct.

Practice Problem 9 The Dallas police department purchased seven identical used police cars at a total cost of $117,964. Find the cost of one used car. Check your answer. $16,852

In the following example you will see the word *average* used as it applies to division. The problem states that a car traveled 1144 miles in 22 hours. The problem asks you to find the average speed in miles per hour. This means that we will treat the problem as if the speed of the car were the same during each hour of the trip. We will use division to solve.

EXAMPLE 10 A car traveled from California to Texas, a distance of 1144 miles, in 22 hours. What was the average speed in miles per hour?

Solution When doing distance problems, it is helpful to remember that distance ÷ time = rate. We need to divide 1144 miles by 22 hours to obtain the rate or speed in miles per hour.

$$
\begin{array}{r}
52 \\
22\overline{)1144} \\
\underline{110} \\
44 \\
\underline{44} \\
0
\end{array}
$$

The car traveled an average of 52 miles per hour.

Practice Problem 10 An airplane traveled 5138 miles in 14 hours. What was the average speed in miles per hour? 367 mph

Teaching Example 10 A bird that is migrating south for the winter flew 255 miles in 15 days. What was the average number of miles it flew each day?

Ans: The bird flew an average of 17 miles each day.

Developing Your Study Skills

Taking Notes in Class

An important part of mathematics studying is taking notes. In order to take meaningful notes, you must be an active listener. Keep your mind on what the instructor is saying, and be ready with questions whenever you do not understand something.

If you have previewed the lesson material, you will be prepared to take good notes. The important concepts will seem somewhat familiar. You will have a better idea of what needs to be written down. If you frantically try to write all that the instructor says or copy all the examples done in class, you may find your notes nearly worthless when you look at them at home. You may find that you are unable to make sense of what you have written.

Write down *important* ideas and examples as the instructor lectures, making sure that you are listening and following the logic. Include any helpful hints or suggestions that your instructor gives you or refers to in your text. You will be amazed at how easily you will forget these if you do not write them down. Try to review your notes the *same day* sometime after class. You will find the material in your notes easier to understand if you have attended class within the last few hours.

Successful note taking requires active listening and processing. Stay alert in class. You will realize the advantages of taking your own notes over copying those of someone else.

Verbal and Writing Skills

1. Explain in your own words what happens when you

(a) divide a nonzero number by itself.
When you divide a nonzero number by itself, the result is 1.

(b) divide a number by 1.
When you divide a number by 1, the result is that number.

(c) divide zero by a nonzero number.
When you divide zero by a nonzero number, the result is zero.

(d) divide a nonzero number by 0.
You cannot divide a number by 0. Division by zero is undefined.

Divide. See if you can work exercises 2–30 in three minutes or less.

2. $5\overline{)35}$ (7)

3. $6\overline{)42}$ (7)

4. $4\overline{)32}$ (8)

5. $8\overline{)24}$ (3)

6. $9\overline{)27}$ (3)

7. $5\overline{)25}$ (5)

8. $7\overline{)49}$ (7)

9. $9\overline{)36}$ (4)

10. $4\overline{)16}$ (4)

11. $7\overline{)21}$ (3)

12. $9\overline{)81}$ (9)

13. $5\overline{)30}$ (6)

14. $6\overline{)54}$ (9)

15. $7\overline{)63}$ (9)

16. $4\overline{)28}$ (7)

17. $8\overline{)72}$ (9)

18. $8\overline{)64}$ (8)

19. $6\overline{)36}$ (6)

20. $9\overline{)72}$ (8)

21. $1\overline{)9}$ (9)

22. $1\overline{)8}$ (8)

23. $10\overline{)0}$ (0)

24. $7\overline{)0}$ (0)

25. $9 \div 0$ undefined

26. $12 \div 0$ undefined

27. $\dfrac{0}{8}$ 0

28. $\dfrac{0}{7}$ 0

29. $6 \div 6$ 1

30. $5 \div 5$ 1

Divide. In exercises 31–42, check your answer.

31. $29 \div 6$

$$6\overline{)29} \quad \text{4 R 5}$$
$$\underline{24}$$
$$5$$

Check
$$\begin{array}{r}4\\ \times\ 6\\ \hline 24\\ +\ 5\\ \hline 29\end{array}$$

32. $42 \div 8$

$$8\overline{)42} \quad \text{5 R 2}$$
$$\underline{40}$$
$$2$$

Check
$$\begin{array}{r}5\\ \times\ 8\\ \hline 40\\ +\ 2\\ \hline 42\end{array}$$

33. $76 \div 8$

$$8\overline{)76} \quad \text{9 R 4}$$
$$\underline{72}$$
$$4$$

Check
$$\begin{array}{r}9\\ \times\ 8\\ \hline 72\\ +\ 4\\ \hline 76\end{array}$$

34. $75 \div 9$

$$9\overline{)75} \quad \text{8 R 3}$$
$$\underline{72}$$
$$3$$

Check
$$\begin{array}{r}8\\ \times\ 9\\ \hline 72\\ +\ 3\\ \hline 75\end{array}$$

35. $128 \div 5$

$$5\overline{)128} \quad \text{25 R 3}$$
$$\underline{10}$$
$$28$$
$$\underline{25}$$
$$3$$

Check
$$\begin{array}{r}25\\ \times\ 5\\ \hline 125\\ +\ 3\\ \hline 128\end{array}$$

36. $6\overline{)103} \quad \text{17 R 1}$

$$\underline{6}$$
$$43$$
$$\underline{42}$$
$$1$$

Check
$$\begin{array}{r}17\\ \times\ 6\\ \hline 102\\ +\ 1\\ \hline 103\end{array}$$

37. $9\overline{)196} \quad \text{21 R 7}$

$$\underline{18}$$
$$16$$
$$\underline{9}$$
$$7$$

Check
$$\begin{array}{r}21\\ \times\ 9\\ \hline 189\\ +\ 7\\ \hline 196\end{array}$$

38. $8\overline{)427} \quad \text{53 R 3}$

$$\underline{40}$$
$$27$$
$$\underline{24}$$
$$3$$

Check
$$\begin{array}{r}53\\ \times\ 8\\ \hline 424\\ +\ 3\\ \hline 427\end{array}$$

39. $9\overline{)288} \quad \text{32}$

$$\underline{27}$$
$$18$$
$$\underline{18}$$
$$0$$

Check
$$\begin{array}{r}32\\ \times\ 9\\ \hline 288\end{array}$$

40. $7\overline{)294} \quad \text{42}$

$$\underline{28}$$
$$14$$
$$\underline{14}$$
$$0$$

Check
$$\begin{array}{r}42\\ \times\ 7\\ \hline 294\end{array}$$

41. $5\overline{)185} \quad \text{37}$

$$\underline{15}$$
$$35$$
$$\underline{35}$$
$$0$$

Check
$$\begin{array}{r}37\\ \times\ 5\\ \hline 185\end{array}$$

42. $8\overline{)224} \quad \text{28}$

$$\underline{16}$$
$$64$$
$$\underline{64}$$
$$0$$

Check
$$\begin{array}{r}28\\ \times\ 8\\ \hline 224\end{array}$$

43. $4\overline{)1289} \quad \text{322 R 1}$

$$\underline{12}$$
$$8$$
$$\underline{8}$$
$$9$$
$$\underline{8}$$
$$1$$

44. $3\overline{)758} \quad \text{252 R 2}$

$$\underline{6}$$
$$15$$
$$\underline{15}$$
$$8$$
$$\underline{6}$$
$$2$$

45. $6\overline{)763} \quad \text{127 R 1}$

$$\underline{6}$$
$$16$$
$$\underline{12}$$
$$43$$
$$\underline{42}$$
$$1$$

46.
$$\begin{array}{r} 57\ R\ 4 \\ 7\overline{)403} \\ \underline{35} \\ 53 \\ \underline{49} \\ 4 \end{array}$$

47.
$$\begin{array}{r} 563 \\ 8\overline{)4504} \\ \underline{40} \\ 50 \\ \underline{48} \\ 24 \\ \underline{24} \\ 0 \end{array}$$

48.
$$\begin{array}{r} 455 \\ 9\overline{)4095} \\ \underline{36} \\ 49 \\ \underline{45} \\ 45 \\ \underline{45} \\ 0 \end{array}$$

49.
$$\begin{array}{r} 1122\ R\ 1 \\ 3\overline{)3367} \\ \underline{3} \\ 3 \\ \underline{3} \\ 6 \\ \underline{6} \\ 7 \\ \underline{6} \\ 1 \end{array}$$

50.
$$\begin{array}{r} 1347\ R\ 4 \\ 6\overline{)8086} \\ \underline{6} \\ 20 \\ \underline{18} \\ 28 \\ \underline{24} \\ 46 \\ \underline{42} \\ 4 \end{array}$$

51.
$$\begin{array}{r} 2056\ R\ 2 \\ 8\overline{)16,450} \\ \underline{16} \\ 45 \\ \underline{40} \\ 50 \\ \underline{48} \\ 2 \end{array}$$

52.
$$\begin{array}{r} 3021\ R\ 1 \\ 6\overline{)18,127} \\ \underline{18} \\ 12 \\ \underline{12} \\ 7 \\ \underline{6} \\ 1 \end{array}$$

53.
$$\begin{array}{r} 2562\ R\ 3 \\ 5\overline{)12,813} \\ \underline{10} \\ 28 \\ \underline{25} \\ 31 \\ \underline{30} \\ 13 \\ \underline{10} \\ 3 \end{array}$$

54.
$$\begin{array}{r} 4027\ R\ 7 \\ 8\overline{)32,223} \\ \underline{32} \\ 22 \\ \underline{16} \\ 63 \\ \underline{56} \\ 7 \end{array}$$

55. $185 \div 6$
$$\begin{array}{r} 30\ R\ 5 \\ 6\overline{)185} \\ \underline{18} \\ 5 \\ \underline{0} \\ 5 \end{array}$$

56. $202 \div 5$
$$\begin{array}{r} 40\ R\ 2 \\ 5\overline{)202} \\ \underline{20} \\ 2 \\ \underline{0} \\ 2 \end{array}$$

57. $267 \div 52$
$$\begin{array}{r} 5\ R\ 7 \\ 52\overline{)267} \\ \underline{260} \\ 7 \end{array}$$

58. $324 \div 36$
$$\begin{array}{r} 9 \\ 36\overline{)324} \\ \underline{324} \\ 0 \end{array}$$

59. $427 \div 61$
$$\begin{array}{r} 7 \\ 61\overline{)427} \\ \underline{427} \\ 0 \end{array}$$

Mixed Practice

60.
$$\begin{array}{r} 6 \\ 72\overline{)432} \\ \underline{432} \\ 0 \end{array}$$

61.
$$\begin{array}{r} 418\ R\ 8 \\ 12\overline{)5024} \\ \underline{48} \\ 22 \\ \underline{12} \\ 104 \\ \underline{96} \\ 8 \end{array}$$

62.
$$\begin{array}{r} 523\ R\ 11 \\ 13\overline{)6810} \\ \underline{65} \\ 31 \\ \underline{26} \\ 50 \\ \underline{39} \\ 11 \end{array}$$

63.
$$\begin{array}{r} 48\ R\ 12 \\ 30\overline{)1452} \\ \underline{120} \\ 252 \\ \underline{240} \\ 12 \end{array}$$

64.
$$\begin{array}{r} 28\ R\ 5 \\ 40\overline{)1125} \\ \underline{80} \\ 325 \\ \underline{320} \\ 5 \end{array}$$

65.
$$\begin{array}{r} 845 \\ 7\overline{)5915} \\ \underline{56} \\ 31 \\ \underline{28} \\ 35 \\ \underline{35} \\ 0 \end{array}$$

66.
$$\begin{array}{r} 768 \\ 8\overline{)6144} \\ \underline{56} \\ 54 \\ \underline{48} \\ 64 \\ \underline{64} \\ 0 \end{array}$$

67.
$$\begin{array}{r} 210\ R\ 8 \\ 36\overline{)7568} \\ \underline{72} \\ 36 \\ \underline{36} \\ 8 \\ \underline{0} \\ 8 \end{array}$$

68.
$$\begin{array}{r} 110\ R\ 7 \\ 32\overline{)3527} \\ \underline{32} \\ 32 \\ \underline{32} \\ 7 \\ \underline{0} \\ 7 \end{array}$$

69.
$$\begin{array}{r} 14\ R\ 2 \\ 182\overline{)2550} \\ \underline{182} \\ 730 \\ \underline{728} \\ 2 \end{array}$$

70.
$$\begin{array}{r} 104\ R\ 6 \\ 19\overline{)1982} \\ \underline{19} \\ 82 \\ \underline{76} \\ 6 \end{array}$$

71.
$$\begin{array}{r} 4\ R\ 4 \\ 174\overline{)700} \\ \underline{696} \\ 4 \end{array}$$

72.
$$\begin{array}{r} 7 \\ 128\overline{)896} \\ \underline{896} \\ 0 \end{array}$$

73.
$$\begin{array}{r} 125 \\ 224\overline{)28,000} \\ \underline{224} \\ 560 \\ \underline{448} \\ 1120 \\ \underline{1120} \\ 0 \end{array}$$

74.
$$\begin{array}{r} 134 \\ 235\overline{)31,490} \\ \underline{235} \\ 799 \\ \underline{705} \\ 940 \\ \underline{940} \\ 0 \end{array}$$

Solve.

75. $518 \div 14 = x$. What is the value of x? $x = 37$

76. $1572 \div 131 = x$. What is the value of x? $x = 12$

Applications

77. *Sports* A *run* in skiing is going from the top of the ski lift to the bottom. If over seven days, 431,851 runs were made, what was the average number of ski runs per day? 61,693 runs

78. *Farming* Western Saddle Stable uses 21,900 pounds of feed per year to feed its 30 horses. How much does each horse eat per year? 730 pounds

79. *Sports* Coach Deno Johnson purchased 9 pairs of cross-country skis for his team. He spent a total of $2592. How much did each pair of skis cost? $288

80. *Business Finances* During the 2006–2007 winter season, Indiana's Department of Transportation spent $9,120,000 on 76 new snowplows. How much did each snowplow cost? $120,000

81. *Business Finances* A horse and carriage company in New York City bought seven new carriages at exactly the same price each. The total bill was $147,371. How much did each carriage cost? $21,053

82. *Real Estate* A group of eight friends invested the same amount each in a beach property that sold for $369,432. How much did each friend pay? $46,179

83. *Business Finances* Appleton Community College spent $10,290 to equip their math center with 42 new flat-panel monitors. How much did each monitor cost? $245

84. *Business Finances* Metropolitan College spent $13,020 on new bookcases for their faculty offices. If 70 faculty members received new bookcases, how much did each bookcase cost? $186

85. *Business Planning* The 2nd Avenue Delicatessen is making bagel sandwiches for a New York City Marathon party. The sandwich maker has 360 bagel halves, 340 slices of turkey, and 330 slices of Swiss cheese. If he needs to make sandwiches each consisting of two bagel halves, two slices of turkey, and two slices of Swiss cheese, what is the greatest number of sandwiches he can make? 165 sandwiches

▲ **86.** *Geometry* Ace Landscaping is mowing a rectangular lawn that has an area of 2652 square feet. The company keeps a record of all lawn mowed in terms of length, width, square feet, and number of minutes it takes to mow the lawn. The width of the lawn is 34 feet. However, the page that lists the length of the lawn is soiled and the number cannot be read. Determine the length of the lawn. 78 feet

87. *Business Management* Dick Wightman is managing a company that is manufacturing and shipping modular homes in Canada. He has a truck that has made the trip from Toronto, Ontario, to Halifax, Nova Scotia, 12 times and has made the return run from Halifax to Toronto 12 times. The distance from Toronto to Halifax is 1742 kilometers.

 (a) How many kilometers has the truck traveled on these 12 trips from Toronto to Halifax and back? 41,808 kilometers

 (b) If Dick wants to limit the truck to a total of 50,000 kilometers driven this year, how many more kilometers can the truck be driven? 8192 kilometers

88. *Space Travel* The space shuttle has recently gone through a number of repairs and improvements. NASA recently approved the use of a shuttle control panel that has an area of 3526 square centimeters. The control panel is rectangular. The width of the panel is 43 centimeters. What is the length of the panel? 82 centimeters

To Think About

89. Division is not commutative. For example, $12 \div 4 \neq 4 \div 12$. If $a \div b = b \div a$, what must be true of the numbers a and b besides the fact that $b \neq 0$ and $a \neq 0$?

a and b must represent the same number. For example, if $a = 12$, then $b = 12$.

90. You can think of division as repeated subtraction. Show how $874 \div 138$ is related to repeated subtraction.

$$
\begin{array}{ll}
874 & \\
-138 & \text{first time} \\
\overline{736} & \\
-138 & \text{second time} \\
\overline{598} & \\
-138 & \text{third time} \\
\overline{460} & \\
-138 & \text{fourth time} \\
\overline{322} & \\
-138 & \text{fifth time} \\
\overline{184} & \\
-138 & \text{sixth time} \\
\overline{46} &
\end{array}
$$

$$
\begin{array}{r}
6 \text{ R } 46 \\
138\overline{)874} \\
828 \\
\hline
46
\end{array}
$$

Cumulative Review

Solve.

91. [1.4.4]
$$
\begin{array}{r}
108 \\
\times\ 50 \\
\hline
5400
\end{array}
$$

92. [1.4.4]
$$
\begin{array}{r}
7162 \\
\times\ 145 \\
\hline
35\ 810 \\
286\ 48 \\
716\ 2 \\
\hline
1{,}038{,}490
\end{array}
$$

93. [1.2.4] $316{,}214 + 89{,}981$
$$
\begin{array}{r}
316{,}214 \\
+\ 89{,}981 \\
\hline
406{,}195
\end{array}
$$

94. [1.3.3] $1{,}360{,}000 - 1{,}293{,}156$
$$
\begin{array}{r}
1{,}360{,}000 \\
-\ 1{,}293{,}156 \\
\hline
66{,}844
\end{array}
$$

Quick Quiz 1.5 Divide. If there is a remainder, be sure to state it as part of your answer.

1. $9\overline{)4203}$ 467

2. $8\overline{)26{,}299}$ 3287 R 3

3. $76\overline{)24{,}928}$ 328

4. Concept Check When performing the division problem $2956 \div 43$, you need to decide how many times 43 goes into 295. Explain how you would decide this. Answers may vary

Classroom Quiz 1.5 You may use these problems to quiz your students' mastery of Section 1.5.

Divide. If there is a remainder, be sure to state it as part of your answer.

1. $8\overline{)2944}$ **Ans:** 368

2. $7\overline{)25{,}869}$ **Ans:** 3695 R 4

3. $56\overline{)13{,}272}$ **Ans:** 237

1.	Seventy-eight million, three hundred ten thousand, four hundred thirty-six
2.	30,000 + 8000 + 200 + 40 + 7
3.	5,064,122
4.	2,747,000
5.	2,802,000
6.	244
7.	50,570
8.	1,351,461
9.	3993
10.	76,311
11.	1,981,652
12.	108
13.	100,000
14.	18,606
15.	3740
16.	331,420
17.	10,605
18.	7376 R 1
19.	26 R 8
20.	139

How are you doing with your homework assignments in Sections 1.1 to 1.5? Do you feel you have mastered the material so far? Do you understand the concepts you have covered? Before you go further in the textbook, take some time to do each of the following problems.

1.1

1. Write in words. 78,310,436

2. Write in expanded notation. 38,247

3. Write in standard notation. five million, sixty-four thousand, one hundred twenty-two

Use the following table to answer questions 4 and 5.

Public High School Graduates (in thousands)

1980	2,747
1995	2,273
2000	2,583
2005	2,641
2010*	2,802

*estimated

Source: U.S. Department of Education

4. How many public school graduates were there in 1980?

5. How many public school graduates are expected in 2010?

1.2 *Add.*

6.
13
31
88
43
+ 69

7.
28,318
5,039
+ 17,213

8.
833,576
+ 517,885

1.3 *Subtract.*

9.
5728
−1735

10.
100,450
− 24,139

11.
45,861,413
− 43,879,761

1.4 *Multiply.*

12. $9 \times 6 \times 1 \times 2$

13. $50 \times 10 \times 200$

14.
2658
× 7

15.
68
× 55

16.
365
× 908

1.5 *Divide. If there is a remainder, be sure to state it as part of the answer.*

17. $8\overline{)84,840}$ **18.** $7\overline{)51,633}$ **19.** $76\overline{)1984}$ **20.** $42\overline{)5838}$

Now turn to page SA-2 for the answer to each of these problems. Each answer also includes a reference to the objective in which the problem is first taught. If you missed any of these problems, you should stop and review the Examples and Practice Problems in the referenced objective. A little review now will help you master the material in the upcoming sections of the text.

Evaluating Expressions with Whole-Number Exponents

Sometimes a simple math idea comes "disguised" in technical language. For example, an **exponent** is just a "shorthand" number that saves writing multiplication of the same numbers.

$$10^3 \quad \text{The exponent 3 means} \quad 10 \times 10 \times 10$$
$$\text{(which takes longer to write).}$$

The product 5×5 can be written as 5^2. The small number 2 is called the *exponent*. The exponent tells us how many factors are in the multiplication. The number 5 is called the **base.** The base is the number that is multiplied.

$$3 \times 3 \times 3 \times 3 = 3^4 \longleftarrow \text{exponent}$$
$$\underset{\text{base}}{\uparrow}$$

In 3^4 the base is 3 and the exponent is 4. (The 4 is sometimes called the *superscript*.) 3^4 is read as "three to the fourth power."

EXAMPLE 1 Write each product in exponent form.

(a) $15 \times 15 \times 15$ **(b)** $7 \times 7 \times 7 \times 7 \times 7$

Solution

(a) $15 \times 15 \times 15 = 15^3$ **(b)** $7 \times 7 \times 7 \times 7 \times 7 = 7^5$

Practice Problem 1 Write each product in exponent form.

(a) $12 \times 12 \times 12 \times 12$ 12^4 **(b)** $2 \times 2 \times 2 \times 2 \times 2 \times 2$ 2^6

EXAMPLE 2 Find the value of each expression.

(a) 3^3 **(b)** 7^2 **(c)** 2^5 **(d)** 1^8

Solution

(a) To find the value of 3^3, multiply the base 3 by itself 3 times.
$$3^3 = 3 \times 3 \times 3 = 27$$

(b) To find the value of 7^2, multiply the base 7 by itself 2 times.
$$7^2 = 7 \times 7 = 49$$

(c) $2^5 = 2 \times 2 \times 2 \times 2 \times 2 = 32$

(d) $1^8 = 1 \times 1 \times 1 \times 1 \times 1 \times 1 \times 1 \times 1 = 1$

Practice Problem 2 Find the value of each expression.

(a) 12^2 144 **(b)** 6^3 216 **(c)** 2^6 64 **(d)** 1^{10} 1

Student Learning Objectives

After studying this section, you will be able to:

 Evaluate expressions with whole-number exponents.

 Perform several arithmetic operations in the proper order.

Teaching Tip Students who have never used exponents before often write the exponent with the same-size digit as the base. Remind them that exponents are written with a smaller digit than the base.

Teaching Example 1 Write each product in exponent form.
(a) 27×27 **(b)** $5 \times 5 \times 5 \times 5$
Ans: (a) 27^2 **(b)** 5^4

NOTE TO STUDENT: Fully worked-out solutions to all of the Practice Problems can be found at the back of the text starting at page SP-1

Teaching Example 2 Find the value of each expression.
(a) 9^2 **(b)** 4^3 **(c)** 1^9 **(d)** 2^4
Ans: (a) 81 **(b)** 64 **(c)** 1 **(d)** 16

If a whole number does not have a visible exponent, the exponent is understood to be 1. Thus

$$3 = 3^1 \quad \text{and} \quad 10 = 10^1.$$

Large numbers are often expressed as a power of 10.

$10^1 = 10 = 1 \text{ ten}$ $10^4 = 10{,}000 = 1 \text{ ten thousand}$
$10^2 = 100 = 1 \text{ hundred}$ $10^5 = 100{,}000 = 1 \text{ hundred thousand}$
$10^3 = 1000 = 1 \text{ thousand}$ $10^6 = 1{,}000{,}000 = 1 \text{ million}$

What does it mean to have an exponent of zero? What is 10^0? Any whole number that is not zero can be raised to the zero power. The result is 1. Thus $10^0 = 1, 3^0 = 1, 5^0 = 1$, and so on. Why is this? Let's reexamine the powers of 10. As we go down one line at a time, notice the pattern that occurs.

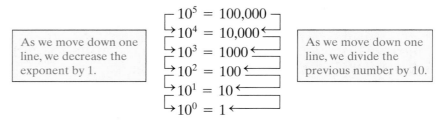

| As we move down one line, we decrease the exponent by 1. | $10^5 = 100{,}000$
 $10^4 = 10{,}000$
 $10^3 = 1000$
 $10^2 = 100$
 $10^1 = 10$
 $10^0 = 1$ | As we move down one line, we divide the previous number by 10. |

Therefore, we present the following definition.

> For any whole number a other than zero, $a^0 = 1$.

If numbers with exponents are added to other numbers, it is first necessary to **evaluate**, or find the value of, the number that is raised to a power. Then we may combine the results with another number.

EXAMPLE 3 Find the value of each expression.

(a) $3^4 + 2^3$ **(b)** $5^3 + 7^0$ **(c)** $6^3 + 6$

Solution

(a) $3^4 + 2^3 = (3)(3)(3)(3) + (2)(2)(2) = 81 + 8 = 89$
(b) $5^3 + 7^0 = (5)(5)(5) + 1 = 125 + 1 = 126$
(c) $6^3 + 6 = (6)(6)(6) + 6 = 216 + 6 = 222$

NOTE TO STUDENT: Fully worked-out solutions to all of the Practice Problems can be found at the back of the text starting at page SP-1

Practice Problem 3 Find the value of each expression.

(a) $7^3 + 8^2$ 407   **(b)** $9^2 + 6^0$ 82   **(c)** $5^4 + 5$ 630

② Performing Several Arithmetic Operations in the Proper Order

Sometimes the order in which we do things is not important. The order in which chefs hang up their pots and pans probably does not matter. The order in which they add and mix the elements in preparing food, however, makes all the difference in the world! If various cooks follow a recipe, though, they will get similar results. The recipe assures that the results will be consistent. It shows the **order of operations.**

In mathematics the order of operations is a list of priorities for working with the numbers in computational problems. This mathematical "recipe" tells how to handle certain indefinite computations. For example, how does a person find the value of $5 + 3 \times 2$?

A problem such as $5 + 3 \times 2$ sometimes causes students difficulty. Some people think $(5 + 3) \times 2 = 8 \times 2 = 16$. Some people think $5 + (3 \times 2) = 5 + 6 = 11$. Only one answer is right, 11. To obtain the right answer, follow the steps outlined in the following box.

ORDER OF OPERATIONS

In the absence of grouping symbols:

Do first	**1.** Simplify any expressions with exponents.
↓	**2.** Multiply or divide from left to right.
Do last	**3.** Add or subtract from left to right.

EXAMPLE 4 Evaluate. $3^2 + 5 - 4 \times 2$

Solution

$$\begin{aligned}
3^2 + 5 - 4 \times 2 &= 9 + 5 - 4 \times 2 \quad \text{Evaluate the expression with exponents.}\\
&= 9 + 5 - 8 \quad \text{Multiply from left to right.}\\
&= 14 - 8 \quad \text{Add from left to right.}\\
&= 6 \quad \text{Subtract.}
\end{aligned}$$

Practice Problem 4 Evaluate. $7 + 4^3 \times 3$ 199

EXAMPLE 5 Evaluate. $5 + 12 \div 2 - 4 + 3 \times 6$

Solution There are no numbers to raise to a power, so we first do any multiplication or division in order from *left to right*.

$$\begin{aligned}
5 + 12 \div 2 - 4 + 3 \times 6 & \quad \text{Multiply or divide from left to right.}\\
= 5 + 6 - 4 + 3 \times 6 & \quad \text{Divide.}\\
= 5 + 6 - 4 + 18 & \quad \text{Multiply. Add or subtract from left to right.}\\
= 11 - 4 + 18 & \quad \text{Add.}\\
= 7 + 18 & \quad \text{Subtract.}\\
= 25 & \quad \text{Add.}
\end{aligned}$$

Practice Problem 5 Evaluate. $37 - 20 \div 5 + 2 - 3 \times 4$ 23

EXAMPLE 6 Evaluate. $2^3 + 3^2 - 7 \times 2$

Solution

$$\begin{aligned}
2^3 + 3^2 - 7 \times 2 &= 8 + 9 - 7 \times 2 \quad \text{Evaluate exponent expressions } 2^3 = 8 \text{ and } 3^2 = 9.\\
&= 8 + 9 - 14 \quad \text{Multiply.}\\
&= 17 - 14 \quad \text{Add.}\\
&= 3 \quad \text{Subtract.}
\end{aligned}$$

Practice Problem 6 Evaluate. $4^3 - 2 + 3^2$ 71

You can change the order in which you compute by using grouping symbols. Place the numbers you want to calculate first within parentheses. This tells you to do those calculations first.

> **ORDER OF OPERATIONS**
>
> With grouping symbols:
>
> Do first   **1.** Perform operations inside parentheses.
>
> $\downarrow$ **2.** Simplify any expressions with exponents.
>
>   **3.** Multiply or divide from left to right.
>
> Do last   **4.** Add or subtract from left to right.

Teaching Example 7 Evaluate.
$8 + 36 \div (3 \times 4) - 5$

Ans: 6

EXAMPLE 7   Evaluate. $2 \times (7 + 5) \div 4 + 3 - 6$

Solution   First, we combine numbers inside the parentheses by adding the 7 to the 5. Next, because multiplication and division have equal priority, we work from left to right doing whichever of these operations comes first.

$$
\begin{aligned}
2 \times (7 + 5) &\div 4 + 3 - 6 \\
&= 2 \times 12 \div 4 + 3 - 6 \quad \text{Parentheses.} \\
&= 24 \div 4 + 3 - 6 \quad \text{Multiply.} \\
&= 6 + 3 - 6 \quad \text{Divide.} \\
&= 9 - 6 \quad \text{Add.} \\
&= 3 \quad \text{Subtract.}
\end{aligned}
$$

NOTE TO STUDENT: *Fully worked-out solutions to all of the Practice Problems can be found at the back of the text starting at page SP-1*

Practice Problem 7   Evaluate. $(17 + 7) \div 6 \times 2 + 7 \times 3 - 4$ 25

Teaching Example 8 Evaluate.
$20 - 14 \div 2 + 3^2 - 2 \times (9 - 6)$

Ans: 16

EXAMPLE 8   Evaluate. $4^3 + 18 \div 3 - 2^4 - 3 \times (8 - 6)$

Solution

$$
\begin{aligned}
4^3 + 18 &\div 3 - 2^4 - 3 \times (8 - 6) \\
&= 4^3 + 18 \div 3 - 2^4 - 3 \times 2 \quad \text{Work inside the parentheses.} \\
&= 64 + 18 \div 3 - 16 - 3 \times 2 \quad \text{Evaluate exponents.} \\
&= 64 + 6 - 16 - 3 \times 2 \quad \text{Divide.} \\
&= 64 + 6 - 16 - 6 \quad \text{Multiply.} \\
&= 70 - 16 - 6 \quad \text{Add.} \\
&= 54 - 6 \quad \text{Subtract.} \\
&= 48 \quad \text{Subtract.}
\end{aligned}
$$

Practice Problem 8   Evaluate. $5^2 - 6 \div 2 + 3^4 + 7 \times (12 - 10)$ 117

Verbal and Writing Skills

1. Explain what the expression 5^3 means. Evaluate 5^3. 5^3 means $5 \times 5 \times 5$. $5^3 = 125$.

2. In exponent notation, the __exponent__ tells how many times to multiply the base.

3. In exponent notation, the __base__ is the number that is multiplied.

4. 10^5 is read as __10 to the 5th power__.

5. Explain the order in which we perform mathematical operations to ensure consistency.
To ensure consistency we
1. perform operations inside parentheses
2. simplify any expressions with exponents
3. multiply or divide from left to right
4. add or subtract from left to right

6. Use the order of operations to solve $12 \times 5 + 3 \times 5 + 7 \times 5$. Is this the same as $5(12 + 3 + 7)$? Why or why not?
110   yes; because of the distributive property.

Write each number in exponent form.

7. $6 \times 6 \times 6 \times 6$ 6^4

8. $2 \times 2 \times 2 \times 2 \times 2$ 2^5

9. $5 \times 5 \times 5 \times 5 \times 5 \times 5$ 5^6

10. $3 \times 3 \times 3 \times 3 \times 3 \times 3$ 3^6

11. $9 \times 9 \times 9 \times 9$ 9^4

12. $1 \times 1 \times 1 \times 1 \times 1 \times 1 \times 1$ 1^7

13. 9 9^1

14. 27 27^1

Find the value of each expression.

15. 2^4 16

16. 3^3 27

17. 4^3 64

18. 5^2 25

19. 6^2 36

20. 10^3 1000

21. 10^4 10,000

22. 1^{20} 1

23. 1^{17} 1

24. 2^5 32

25. 2^6 64

26. 4^2 16

27. 3^5 243

28. 12^2 144

29. 15^2 225

30. 3^4 81

31. 7^3 343

32. 5^4 625

33. 4^4 256

34. 7^2 49

35. 9^0 1

36. 8^0 1

37. 25^2 625

38. 20^3 8000

39. 10^6 1,000,000

40. 8^1 8

41. 13^2 169

42. 11^2 121

43. 9^1 9

44. 14^2 196

45. 8^2 64

46. 5^3 125

47. $3^2 + 1^2$
$9 + 1 = 10$

48. $7^0 + 4^3$
$1 + 64 = 65$

49. $2^3 + 10^2$
$8 + 100 = 108$

50. $7^3 + 4^2$
$343 + 16 = 359$

51. $8^3 + 8$
$512 + 8 = 520$

52. $9^2 + 9$
$81 + 9 = 90$

Work each exercise, using the correct order of operations.

53. $9 \times 10 - 35$
$90 - 35 = 55$

54. $9 \times 7 + 42$
$63 + 42 = 105$

55. $3 \times 9 - 10 \div 2$
$27 - 5 = 22$

56. $4 \times 6 - 24 \div 4$
$24 - 6 = 18$

57. $48 \div 2^3 + 4$
$48 \div 8 + 4$
$= 6 + 4 = 10$

58. $4^3 \div 4 - 11$
$64 \div 4 - 11$
$= 16 - 11 = 5$

59. $3 \times 6^2 - 50$
$3 \times 36 - 50$
$= 108 - 50 = 58$

60. $2 \times 12^2 - 80$
$2 \times 144 - 80$
$= 288 - 80 = 208$

61. $10^2 + 3 \times (8 - 3)$
$100 + 3 \times 5$
$= 100 + 15 = 115$

62. $4^3 - 5 \times (9 + 1)$
$64 - 5 \times 10$
$= 64 - 50 = 14$

63. $(400 \div 20) \div 20$
$20 \div 20 = 1$

64. $(600 \div 30) \div 20$
$20 \div 20 = 1$

65. $950 \div (25 \div 5)$
$950 \div 5 = 190$

66. $875 \div (35 \div 7)$
$875 \div 5 = 175$

67. $(12)(5) - (12 + 5)$
$60 - 17 = 43$

68. $(3)(60) - (60 + 3)$
$180 - 63 = 117$

69. $3^2 + 4^2 \div 2^2$
$9 + 16 \div 4 = 9 + 4 = 13$

70. $7^2 + 9^2 \div 3^2$
$49 + 81 \div 9 = 49 + 9 = 58$

71. $(6)(7) - (12 - 8) \div 4$
$42 - 4 \div 4 = 42 - 1 = 41$

72. $(8)(9) - (15 - 5) \div 5$
$72 - 10 \div 5 = 72 - 2 = 70$

73. $100 - 3^2 \times 4$
$100 - 9 \times 4 = 100 - 36 = 64$

74. $130 - 4^2 \times 5$
$130 - 16 \times 5 = 130 - 80 = 50$

75. $5^2 + 2^2 + 3^3$
$25 + 4 + 27 = 56$

76. $2^3 + 3^2 + 4^3$
$8 + 9 + 64 = 81$

77. $72 \div 9 \times 3 \times 1 \div 2$
$8 \times 3 \times 1 \div 2$
$= 24 \div 2 = 12$

78. $120 \div 30 \times 2 \times 5 \div 8$
$4 \times 2 \times 5 \div 8$
$= 40 \div 8 = 5$

79. $12^2 - 2 \times 0 \times 5 \times 6$
$144 - 0 = 144$

80. $8^2 - 4 \times 3 \times 0 \times 7$
$64 - 0 = 64$

Mixed Practice *Work each exercise, using the correct order of operations.*

81. $4^2 \times 6 \div 3$
$16 \times 6 \div 3$
$= 96 \div 3 = 32$

82. $7^2 \times 3 \div 3$
$49 \times 3 \div 3$
$= 147 \div 3 = 49$

83. $60 - 2 \times 4 \times 5 + 10$
$60 - 40 + 10$
$= 20 + 10 = 30$

84. $75 - 3 \times 5 \times 2 + 15$
$75 - 30 + 15$
$= 45 + 15 = 60$

85. $3 + 3^2 \times 6 + 4$
$3 + 9 \times 6 + 4$
$= 3 + 54 + 4 = 61$

86. $5 + 4^3 \times 2 + 7$
$5 + 64 \times 2 + 7$
$= 5 + 128 + 7 = 140$

87. $32 \div 2 \times (3 - 1)^4$
$32 \div 2 \times (2)^4$
$= 32 \div 2 \times 16 = 16 \times 16 = 256$

88. $24 \div 3 \times (5 - 3)^2$
$24 \div 3 \times (2)^2$
$= 24 \div 3 \times 4 = 8 \times 4 = 32$

89. $3^2 \times 6 \div 9 + 4 \times 3$
$9 \times 6 \div 9 + 4 \times 3$
$= 6 + 12 = 18$

90. $5^2 \times 3 \div 25 + 7 \times 6$
$25 \times 3 \div 25 + 7 \times 6$
$= 3 + 42 = 45$

91. $6^2 + 5^0 + 2^3$
$36 + 1 + 8 = 45$

92. $8^0 + 7^2 + 3^3$
$1 + 49 + 27 = 77$

93. $1200 - 2^3(3) \div 6$
$1200 - 8(3) \div 6 = 1200 - 4 = 1196$

94. $2150 - 3^4(2) \div 9$
$2150 - 81(2) \div 9 = 2150 - 18 = 2132$

95. $120 \div (30 + 10) - 1$
$120 \div 40 - 1 = 3 - 1 = 2$

96. $100 - 48 \div (2 \times 3)$
$100 - 48 \div 6 = 100 - 8 = 92$

97. $120 \div 30 + 10 - 1$
$4 + 10 - 1 = 13$

98. $100 - 48 \div 2 \times 3$
$100 - 24 \times 3 = 100 - 72 = 28$

99. $5 \times 2 + (7 - 4)^3 + 2^0$
$5 \times 2 + (3)^3 + 2^0 = 10 + 27 + 1 = 38$

100. $9 \times 8 + 5^0 - (8 - 4)^3$
$9 \times 8 + 5^0 - (4)^3 = 73 - 64 = 9$

To Think About

101. *Astronomy* The earth rotates once every 23 hours, 56 minutes, 4 seconds. How many seconds is that?
86,164 seconds

102. *Astronomy* The planet Saturn rotates once every 10 hours, 12 minutes. How many minutes is that? How many seconds?
612 minutes; 36,720 seconds

Cumulative Review

103. **[1.1.1]** In the number 2,038,754
   (a) What digit tells the number of ten thousands? 3
   (b) What is the value of the digit 2? 2,000,000

104. **[1.1.3]** Write in standard notation. two hundred million, seven hundred sixty-five thousand, nine hundred nine
200,765,909

105. **[1.1.3]** Write in words. 261,763,002
two hundred sixty-one million, seven hundred sixty-three thousand, two

▲ **106.** **[1.2.6 and 1.4.6]** *Geometry* New Boston High School has an athletic field that needs to be enclosed by fencing. The rectangular field is 250 feet wide and 480 feet long. How many feet of fencing are needed to surround the field? Grass needs to be planted for a new playing field for next year. What is the area in square feet of the amount of grass that must be planted? 1460 feet; 120,000 square feet

Quick Quiz 1.6

1. Write in exponent form.
$12 \times 12 \times 12 \times 12 \times 12$ 12^5

3. Perform each operation in the proper order.
$42 - 2^5 + 3 \times (9 - 6)^3$ 91

2. Evaluate. 6^4 1296

4. **Concept Check** Explain in what order you would do the steps to evaluate the expression $7 \times 6 \div 3 \times 4^2 - 2$.
Answers may vary

Classroom Quiz 1.6 You may use these problems to quiz your students' mastery of Section 1.6.

1. Write in exponent form. $15 \times 15 \times 15 \times 15$
   Ans: 15^4

2. Evaluate. 7^3 **Ans:** 343

3. Perform each operation in the proper order.
   $3 + 5^3 - 2 \times (10 - 4)^2$ **Ans:** 56

Rounding Whole Numbers

Large numbers are often expressed to the nearest hundred or to the nearest thousand, because an approximate number is "good enough" for certain uses.

Distances from the earth to other galaxies are measured in light-years. Although light really travels at 5,865,696,000,000 miles a year, we usually **round** this number to the nearest trillion and say it travels at 6,000,000,000,000 miles a year. To round a number, we first determine the place we are rounding to—in this case, trillion. Then we find which value is closest to the number that we are rounding. In this case, the number we want to round is closer to 6 trillion than to 5 trillion. How do we know the number is closer to 6 trillion than to 5 trillion?

To see which is the closest value, we may picture a **number line,** where whole numbers are represented by points on a line. To show how to use a number line in rounding, we will round 368 to the nearest hundred. 368 is between 300 and 400. When we round, we pick the hundred 368 is "closest to." We draw a number line to show 300 and 400. We also show the point midway between 300 and 400 to help us to determine which hundred 368 is closest to.

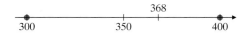

We find that the number 368 is closer to 400 than to 300, so we round 368 *up to* 400.

Let's look at another example. We will round 129 to the nearest hundred. 129 is between 100 and 200. We show this on the number line. We include the midpoint 150 as a guide.

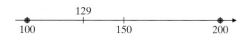

We find that the number 129 is closer to 100 than to 200, so we round 129 *down to* 100.

This leads us to the following simple rule for rounding.

Teaching Tip These rules for rounding are used uniformly in science and mathematics classes and applications. However, a particular business or bank may employ a different rule. Inform students that if rounding off is done in a business, a different rule may be employed.

> **ROUNDING A WHOLE NUMBER**
>
> 1. If the first digit to the right of the round-off place is
>   **(a)** *less than 5,* we make no change to the digit in the round-off place. (We know it is closer to the smaller number, so we round down.)
>   **(b)** *5 or more,* we increase the digit in the round-off place by 1. (We know it is closer to the larger number, so we round up.)
> 2. Then we replace the digits to the right of the round-off place by zeros.

EXAMPLE 1 Round 37,843 to the nearest thousand.

Solution

3 7, 8 4 3   According to the directions, the thousands will be the round-off place. We locate the thousands place.

3 7, ⑧4 3   We see that the first digit to the right of the round-off place is 8, which is 5 or more. We increase the thousands digit by 1, and replace all digits to the right by zero.

3 8, 0 0 0

We have rounded 37,843 to the nearest thousand: 38,000. This means that 37,843 is closer to 38,000 than to 37,000.

Practice Problem 1 Round 65,528 to the nearest thousand. 66,000

Teaching Example 1 Round 25,763 to the nearest hundred.
Ans: 25,800

EXAMPLE 2 Round 2,445,360 to the nearest hundred thousand.

Solution

2,4 4 5,3 6 0   Locate the hundred thousands round-off place.

2,4 ④5,3 6 0   The first digit to the right of this is less than 5, so round down. Do not change the hundred thousands digit.

2,4 0 0,0 0 0   Replace all digits to the right by zero.

Practice Problem 2 Round 172,963 to the nearest ten thousand. 170,000

Teaching Example 2 Round 673,598 to the nearest ten thousand.
Ans: 670,000

EXAMPLE 3 Round as indicated.

(a) 561,328 to the nearest ten   **(b)** 3,798,152 to the nearest hundred
(c) 51,362,523 to the nearest million

Solution

(a)   ↓   First locate the digit in the tens place.
   561,328   The digit to the right of the tens place is greater than 5.
   561,330   Round up.

   561,328 rounded to the nearest ten is 561,330.

(b) 3,798,152   The digit to the right of the hundreds place is 5.
   3,798,200   Round up.

   3,798,152 rounded to the nearest hundred is 3,798,200.

(c) 51,362,523   The digit to the right of the millions place is less than 5.
   51,000,000   Round down.

   51,362,523 rounded to the nearest million is 51,000,000.

Practice Problem 3 Round as indicated.

(a) 53,282 to the nearest ten 53,280
(b) 164,485 to the nearest thousand 164,000
(c) 1,365,273 to the nearest hundred thousand 1,400,000

Teaching Example 3 Round as indicated.
(a) 478,536 to the nearest thousand.
(b) 26,198,261 to the nearest million.
(c) 63,286 to the nearest ten.

Ans: (a) 479,000   **(b)** 26,000,000
(c) 63,290

EXAMPLE 4 Round 763,571.

(a) To the nearest thousand
(b) To the nearest ten thousand
(c) To the nearest million

Solution

(a) $\downarrow$
763,571 = 764,000 to the nearest thousand. The digit to the right of the thousands place is 5. We rounded up.

(b) $\downarrow$
763,571 = 760,000 to the nearest ten thousand. The digit to the right of the ten thousands place is less than 5. We rounded down.

(c) 763,571 does not have any digits for millions. If it helps, you can think of this number as 0,763,571. Since the digit to the right of the millions place is 7, we round up to obtain one million or 1,000,000.

Practice Problem 4 Round 935,682 as indicated.

(a) To the nearest thousand 936,000
(b) To the nearest hundred thousand 900,000
(c) To the nearest million 1,000,000

EXAMPLE 5 Astronomers use the parsec as a measurement of distance. One parsec is approximately 30,900,000,000,000 kilometers. Round 1 parsec to the nearest trillion kilometers.

Solution 30,900,000,000,000 km is 31,000,000,000,000 km or 31 trillion km to the nearest trillion kilometers.

Practice Problem 5 One light-year is approximately 9,460,000,000,000,000 meters. Round to the nearest hundred trillion meters. 9,500,000,000,000,000

2 Estimating the Answer to a Problem Involving Whole Numbers

Often we need to quickly check the answer of a calculation to be reasonably assured that the answer is correct. If you expected your bill to be "around $40" for the groceries you had selected and the cashier's total came to $41.89, you would probably be confident that the bill is correct and pay it. If, however, the cashier rang up a bill of $367, you would not just assume that it is correct. You would know an error had been made. If the cashier's total came to $60, you might not be certain, but you would probably suspect an error and check the calculation.

In mathematics we often **estimate,** or determine the approximate value of a calculation, if we need to do a quick check. There are many ways to estimate, but in this book we will use one simple principle of estimation. We use the symbol ≈ to mean **approximately equal to.**

PRINCIPLE OF ESTIMATION

1. Round the numbers so that there is one nonzero digit in each number.

2. Perform the calculation with the rounded numbers.

EXAMPLE 6 Estimate the sum. 163 + 237 + 846 + 922

Solution We first determine where to round each number in our problem to leave only one nonzero digit in each. In this case, we round all numbers to the nearest hundred. Then we perform the calculation with the rounded numbers.

Actual Sum	*Estimated Sum*
163	200
237	200
846	800
+ 922	+ 900
	2100

We estimate the answer to be 2100. We say the sum ≈2100. If we calculate using the exact numbers, we obtain a sum of 2168, so our estimate is quite close to the actual sum.

Practice Problem 6 Estimate the sum. 3456 + 9876 + 5421 + 1278

19,000

When we use the principle of estimation, we will not always round each number in a problem to the same place.

EXAMPLE 7 Phil and Melissa bought their first car last week. The selling price of this compact car was $8980. The dealer preparation charge was $289 and the sales tax was $449. Estimate the total cost of the car that Phil and Melissa had to pay.

Solution We round each number to have only one nonzero digit, and add the rounded numbers.

8980	9000
289	300
+ 449	+ 400
	9700

The total cost ≈$9700. (The exact answer is $9718, so we see that our answer is quite close.)

Practice Problem 7 Greg and Marcia purchased a new sofa for $697, plus $35 sales tax. The store also charged them $19 to deliver the sofa. Estimate their total cost. $760

Teaching Example 6 Estimate the sum. 2351 + 6732 + 1153 + 3689

Ans: 14,000

Teaching Example 7 Attendance at the street fair this weekend was 765 men, 837 women, and 1153 children. Estimate the total attendance.

Ans: An estimated 2600 people attended the fair.

NOTE TO STUDENT: Fully worked-out solutions to all of the Practice Problems can be found at the back of the text starting at page SP-1

Now we turn to a case where an estimate can help us discover an error.

EXAMPLE 8 Roberto added together four numbers and obtained the following result. Estimate the sum and determine if the answer seems reasonable.

$$12,456 + 17,976 + 18,452 + 32,128 \stackrel{?}{=} 61,012$$

Solution We round each number so that there is one nonzero digit. In this case, we round them all to the nearest ten thousand.

12,456	10,000
17,976	20,000
18,452	20,000
+ 32,128	+ 30,000
	80,000

Our estimate is 80,000.

This is significantly different from 61,012, so we would suspect that an error has been made. In fact, Roberto did make an error. The exact sum is actually 81,012!

Practice Problem 8 Ming did the following calculation. Estimate to see if her sum appears to be correct or incorrect.

$$11,849 + 14,376 + 16,982 + 58,151 = 81,358$$

100,000; incorrect

Next we look at a subtraction example where estimation is used.

EXAMPLE 9 The profit from Techno Industries for the first quarter of the year was $642,987,000. The profit for the second quarter was $238,890,000. Estimate how much less the profit was for the second quarter than for the first quarter.

Solution We round each number so that there is one nonzero digit. Then we subtract, using the two rounded numbers.

642,987,000	600,000,000
− 238,890,000	− 200,000,000
	400,000,000

We estimate that the profit was $400,000,000 less for the second quarter.

Practice Problem 9 The 2001 population of Florida was 16,397,426. The 2001 population of California was 34,501,728. Estimate how many more people lived in California in 2001 than in Florida. 10,000,000

We also use this principle to estimate results of multiplication and division.

EXAMPLE 10 Estimate the product. 56,789 × 529

Teaching Example 10 Estimate the product. 7746 × 312

Ans: The estimate is 2,400,000. The exact product is 2,416,752.

Solution We round each number so that there is one nonzero digit. Then we multiply the rounded numbers to obtain our estimate.

$$
\begin{array}{r}
56{,}789 \\
\times \quad 529 \\
\end{array}
\qquad
\begin{array}{r}
60{,}000 \\
\times \quad 500 \\
\hline
30{,}000{,}000
\end{array}
$$

Therefore the product is ≈30,000,000. (This is reasonably close to the exact answer of 30,041,381.)

Practice Problem 10 Estimate the product. 8945 × 7317 63,000,000

EXAMPLE 11 Estimate the answer for the following division problem.

$$23\overline{)148{,}902}$$

Teaching Example 11 Estimate the answer for the following division problem. 87,492 ÷ 46

Ans: The estimate is 1800. The exact quotient is 1902.

Solution We round each number to a number with one nonzero digit. Then we perform the division, using the two rounded numbers.

$$23\overline{)148{,}902} \qquad 20\overline{)100{,}000}^{\,5000}$$

Our estimate is 5000. (The exact answer is 6474. We see that our estimate is "in the ballpark" but is not very close to the exact answer. Remember, an estimate is just a rough approximation of the exact answer.)

Practice Problem 11 Estimate the answer for the following division problem.

$$39\overline{)75{,}342}$$ 2000

Not all division estimates come out so easily. In some cases, you may need to carry out a long-division problem of several steps just to obtain the estimate. Do not be in a hurry. Students often want to rush the steps of division. It is better to take your time and carefully do each step. This approach will be very worthwhile in the long run.

Teaching Example 12 Painting the walls and ceilings of 37 work areas in a large office building required 662 gallons of paint. Estimate the number of gallons required for each of the work areas.

Ans: We estimate that 18 gallons of paint were needed for each work area. The exact answer is slightly under 18 gallons.

EXAMPLE 12 John and Stephanie drove their car a distance of 778 miles. They used 25 gallons of gas. Estimate how many miles they can travel on 1 gallon of gas.

Solution In order to solve this problem, we need to divide 778 by 25 to obtain the number of miles John and Stephanie get with 1 gallon of gas. We round each number to a number with one nonzero digit and then perform the division, using the rounded numbers.

$$25\overline{)778}$$

$$\begin{array}{r} 26 \\ 30\overline{)800} \\ \underline{60} \\ 200 \\ \underline{180} \\ 20 \end{array} \text{ Remainder}$$

We obtain an answer of 26 with a remainder of 20. For our estimate we will use the whole number 27. Thus we estimate that the number of miles their car obtained on 1 gallon of gas was 27 miles. (This is reasonably close to the exact answer, which is just slightly more than 31 miles per gallon of gas.)

NOTE TO STUDENT: Fully worked-out solutions to all of the Practice Problems can be found at the back of the text starting at page SP-1

Practice Problem 12 The highway department purchased 58 identical trucks at a total cost of $1,864,584. Estimate the cost for one truck. $33,333

Developing Your Study Skills

How To Do Homework

Set aside time each day for your homework assignments. Make a weekly schedule and write down the times each day you will devote to doing math homework. Two hours spent studying outside class for each hour in class is usual for college courses. You may need more than that for mathematics.

Before beginning to solve your homework exercises, read your textbook very carefully. Expect to spend much more time reading a few pages of a mathematics textbook than several pages of another text. Read for complete understanding, not just for the general idea.

As you begin your homework assignments, read the directions carefully. You need to understand what is being asked. Concentrate on each exercise, taking time to solve it accurately. Rushing through your work usually results in errors. Check your answers with those given in the back of the textbook. If your answer is incorrect, check to see that you are doing the right problem. Redo the problem, watching

for errors. If it is still wrong, check with a friend. Perhaps the two of you can figure out where you are going wrong.

Also, check the examples in the textbook or in your notes for a similar exercise. Can this one be solved in the same way? Give it some thought. You may want to leave it for a while by taking a break or doing a different exercise. But come back later and try again. If you are still unable to figure it out, ask your instructor for help during office hours or in class.

Work on your assignments every day and do as many exercises as it takes for you to know what you are doing. Begin by doing all the exercises that have been assigned. If there are more available in that section of your text, then do more. When you think you have done enough exercises to fully understand the topic at hand, do a few more to be sure. This may mean that you do many more exercises than the instructor assigns, but you can never practice mathematics too much. Practice improves your skills and increases your accuracy, speed, competence, and confidence.

Verbal and Writing Skills

1. Explain the rule for rounding and provide examples.
Locate the rounding place. If the digit to the right of the rounding place is 5 or greater than 5, round up. If the digit to the right of the rounding place is less than 5, round down. *Note:* Examples provided by students will vary. Check for accuracy.

2. What happens when you round 98 to the nearest ten?
Since the digit to the right of tens is greater than 5, you round up. When you round up 9 tens, it becomes 10 tens or 100.

Round to the nearest ten.

3. 83 80   **4.** 45 50   **5.** 65 70   **6.** 57 60   **7.** 168 170   **8.** 132 130

9. 7438 7440   **10.** 2834 2830   **11.** 2961 2960   **12.** 4355 4360

Round to the nearest hundred.

13. 247 200   **14.** 661 700   **15.** 2781 2800   **16.** 1249 1200   **17.** 7692 7700   **18.** 1643 1600

Round to the nearest thousand.

19. 7621 8000   **20.** 3754 4000   **21.** 1489 1000   **22.** 515 1000   **23.** 27,863 28,000   **24.** 94,489 94,000

Applications

25. *History* The worst death rate from an earthquake was in Shaanxi, China, in 1556. That earthquake killed an estimated 832,400 people. Round this number to the nearest hundred thousand. 800,000

26. *Astronomy* One light year (the distance light travels in 1 year) measures 5,878,612,843,000 miles. Round this figure to the nearest hundred million. 5,878,600,000,000 miles

27. *Astronomy* The Hubble Space Telescope's *Guide Star Catalogue* lists 15,169,873 stars. Round this figure to the nearest million. 15,000,000 stars

28. *Geography* The point of highest elevation in the world is Mt. Everest in the country of Nepal. Mt. Everest is 29,028 feet above sea level. Round this figure to the nearest ten thousand. 30,000 feet

29. *Native American Studies* In 2005, the total number of Native Americans and Alaska Natives living in the United States was estimated to be 2,357,544. Round this figure to
(a) the nearest hundred thousand. 2,400,000
(b) the nearest thousand. 2,358,000

30. *Population Studies* The population of the United States in 2030 is projected to be 363,584,435. Round this figure to
(a) the nearest ten thousand. 363,580,000
(b) the nearest hundred million. 400,000,000

▲ **31.** *Geography* The total area of mainland China is 3,705,392 square miles, or, 9,596,960 square kilometers. For *both* square miles and square kilometers, round this figure to
(a) the nearest hundred thousand.
3,700,000 square miles, 9,600,000 square kilometers
(b) the nearest ten thousand.
3,710,000 square miles, 9,600,000 square kilometers

▲ **32.** *Geography* The area of the Pacific Ocean is 165,384,000 square kilometers. Round this figure to
(a) the nearest hundred thousand. 165,400,000
(b) the nearest ten thousand. 165,380,000

Use the principle of estimation to find an estimate for each calculation.

33. 772 + 324 + 225

$$\begin{array}{r} 800 \\ 300 \\ + 200 \\ \hline 1300 \end{array}$$

34. 186 + 509 + 872

$$\begin{array}{r} 200 \\ 500 \\ + 900 \\ \hline 1600 \end{array}$$

35. 42 + 69 + 95 + 18

$$\begin{array}{r} 40 \\ 70 \\ 100 \\ + 20 \\ \hline 230 \end{array}$$

36. $62 + 27 + 54 + 98$

$$
\begin{array}{r}
60 \\
30 \\
50 \\
+\ 100 \\
\hline
240
\end{array}
$$

37. $158{,}270 + 53{,}441 + 8701$

$$
\begin{array}{r}
200{,}000 \\
50{,}000 \\
+\ \ \ 9000 \\
\hline
259{,}000
\end{array}
$$

38. $238{,}271 + 77{,}304 + 9551$

$$
\begin{array}{r}
200{,}000 \\
80{,}000 \\
+\ \ 10{,}000 \\
\hline
290{,}000
\end{array}
$$

39. $324{,}230 - 70{,}290$

$$
\begin{array}{r}
300{,}000 \\
-\ 70{,}000 \\
\hline
230{,}000
\end{array}
$$

40. $975{,}935 - 593{,}228$

$$
\begin{array}{r}
1{,}000{,}000 \\
-\ \ 600{,}000 \\
\hline
400{,}000
\end{array}
$$

41. $842{,}512 - 78{,}234$

$$
\begin{array}{r}
800{,}000 \\
-\ \ 80{,}000 \\
\hline
720{,}000
\end{array}
$$

42. $382{,}140 - 56{,}117$

$$
\begin{array}{r}
400{,}000 \\
-\ \ 60{,}000 \\
\hline
340{,}000
\end{array}
$$

43. $33{,}261{,}378 - 18{,}199{,}276$

$$
\begin{array}{r}
30{,}000{,}000 \\
-\ 20{,}000{,}000 \\
\hline
10{,}000{,}000
\end{array}
$$

44. $89{,}263{,}000 - 54{,}198{,}635$

$$
\begin{array}{r}
90{,}000{,}000 \\
-\ 50{,}000{,}000 \\
\hline
40{,}000{,}000
\end{array}
$$

45. 47×62

$$
\begin{array}{r}
50 \\
\times\ \ 60 \\
\hline
3000
\end{array}
$$

46. 43×95

$$
\begin{array}{r}
40 \\
\times\ 100 \\
\hline
4000
\end{array}
$$

47. 1324×8

$$
\begin{array}{r}
1000 \\
\times\ \ \ 8 \\
\hline
8000
\end{array}
$$

48. 5926×3

$$
\begin{array}{r}
6000 \\
\times\ \ \ 3 \\
\hline
18{,}000
\end{array}
$$

49. $631{,}540 \times 312$

$$
\begin{array}{r}
600{,}000 \\
\times\ \ \ \ \ \ 300 \\
\hline
180{,}000{,}000
\end{array}
$$

50. $374{,}193 \times 193$

$$
\begin{array}{r}
400{,}000 \\
\times\ \ \ \ \ \ 200 \\
\hline
80{,}000{,}000
\end{array}
$$

51. $6368 \div 38$

$$
\begin{array}{r}
150 \\
40\overline{)6000}
\end{array}
$$

52. $7813 \div 22$

$$
\begin{array}{r}
400 \\
20\overline{)8000}
\end{array}
$$

53. $362{,}881 \div 39$

$$
\begin{array}{r}
10{,}000 \\
40\overline{)400{,}000}
\end{array}
$$

54. $596{,}450 \div 64$

$$
\begin{array}{r}
10{,}000 \\
60\overline{)600{,}000}
\end{array}
$$

55. $3{,}885{,}720 \div 831$

$$
\begin{array}{r}
5000 \\
800\overline{)4{,}000{,}000}
\end{array}
$$

56. $12{,}447{,}312 \div 497$

$$
\begin{array}{r}
20{,}000 \\
500\overline{)10{,}000{,}000}
\end{array}
$$

Estimate the result of each calculation. Some results are correct and some are incorrect. Which results appear to be correct? Which results appear to be incorrect?

57.

$$
\begin{array}{rr}
361 & 400 \\
522 & 500 \\
873 & 900 \\
+\ 164 & +\ 200 \\
\hline
1320 & 2000 \\
& \text{Incorrect}
\end{array}
$$

58.

$$
\begin{array}{rr}
476 & 500 \\
124 & 100 \\
516 & 500 \\
+\ 389 & +\ 400 \\
\hline
1505 & 1500 \\
& \text{Correct}
\end{array}
$$

59.

$$
\begin{array}{rr}
97{,}635 & 100{,}000 \\
52{,}123 & 50{,}000 \\
+\ 41{,}986 & +\ 40{,}000 \\
\hline
291{,}744 & 190{,}000 \\
& \text{Incorrect}
\end{array}
$$

60.

$$
\begin{array}{rr}
26{,}181 & 30{,}000 \\
47{,}998 & 50{,}000 \\
+\ 63{,}271 & +\ 60{,}000 \\
\hline
137{,}450 & 140{,}000 \\
& \text{Correct}
\end{array}
$$

61.

$$
\begin{array}{rr}
302{,}360 & 300{,}000 \\
-\ 89{,}518 & -\ 90{,}000 \\
\hline
212{,}842 & 210{,}000 \\
& \text{Correct}
\end{array}
$$

62.

$$
\begin{array}{rr}
735{,}128 & 700{,}000 \\
-\ 116{,}733 & -\ 100{,}000 \\
\hline
518{,}395 & 600{,}000 \\
& \text{Incorrect}
\end{array}
$$

63. 78,126,345
  − 48,972,103
  19,154,242

  80,000,000
  − 50,000,000
  30,000,000
  Incorrect

64. 42,765,317
  − 29,318,274
  23,447,043

  40,000,000
  − 30,000,000
  10,000,000
  Incorrect

65. 378
  × 32
  21,096

  400
  × 30
  12,000
  Incorrect

66. 512
  × 46
  20,552

  500
  × 50
  25,000
  Incorrect

67. 5896
  × 72
  424,512

  6000
  × 70
  420,000
  Correct

68. 8076
  × 89
  718,764

  8000
  × 90
  720,000
  Correct

69. 36)82,116 2281

  40)80,000 2000
  Correct

70. 52)28,912 556

  50)30,000 600
  Correct

71. 423)161,163 381

  400)200,000 500
  Correct

72. 781)477,972 612

  800)500,000 625
    4800
    2000
    1600
    4000
    4000
  Correct

Applications

▲ 73. *Geometry* Victor and Shannon just purchased a new home with a two-car garage measuring 17 feet wide and 22 feet long. Estimate the number of square feet in the garage. 400 square feet

▲ 74. *Geometry* A huge restaurant in New York City is 43 yards wide and 112 yards long. Estimate the number of square yards in the restaurant. 4000 square yards

75. *International Relations* In 2005, the populations of the three largest cities in Canada were Toronto with 5,304,600 people, Montreal with 3,635,842 people, and Vancouver with 2,208,312 people. Estimate the total population of the three cities. 11,000,000 people

76. *Financial Management* The highway departments in four towns in northwestern New York had the following budgets for snow removal for the year: $329,560, $672,940, $199,734, and $567,087. Estimate the total amount that the four towns spend for snow removal in one year. $1,800,000

77. *Business Management* The local pizzeria makes 267 pizzas on an average day. Estimate how many pizzas were made in the last 134 days. 30,000 pizzas

78. *Personal Finance* Darcy makes $68 for each shift she works. She is scheduled for 33 shifts during the next two months. Estimate how much she will earn in the next two months. $2100

79. *Transportation* In 2006, Atlanta's airport was the busiest with 976,313 flights (departures and arrivals). In the same year, the fifth busiest airport was in Las Vegas with 619,474 flights. Round each figure to the nearest ten thousand. Then estimate the difference. 360,000 flights

80. *Sports* In 1960 the average attendance at a Patriots game (officially called the Boston Patriots at that time) was 25,783. In 2007 the average attendance at a New England Patriots game was 66,789. Estimate the increase in attendance over this time period. 40,000

▲**81.** *Geography* The largest state of the United States is Alaska, with a land area of 586,412 square miles. The second largest state is Texas, with an area of 267,339 square miles. Round each figure to the nearest ten thousand. Then estimate how many square miles larger Alaska is than Texas. 320,000 square miles

▲**82.** *International Relations* The largest country in Africa is Sudan measuring 966,757 square miles. South America's largest country is Brazil measuring 3,286,470 square miles. Round each figure to the nearest hundred thousand, then estimate how many square miles larger Brazil is than Sudan. 2,300,000 square miles

To Think About

83. *Space Travel* A space probe travels at 23,560 miles per hour for a distance of 7,824,560,000 miles.

(a) How many *hours* will it take the space probe to travel that distance? (Estimate.) 400,000 hours

(b) How many *days* will it take the space probe to travel that distance? (Estimate.) 20,000 days

84. *Space Travel* A space probe travels at 28,367 miles per hour for a distance of 9,348,487,000 miles.

(a) Estimate the number of *hours* it will take the space probe to travel that distance. 300,000 hours

(b) Estimate the number of *days* it will take the space probe to travel that distance. 15,000 days

Cumulative Review *Evaluate.*

85. [1.6.2] $26 \times 3 + 20 \div 4$ 83

86. [1.6.2] $5^2 + 3^2 - (17 - 10)$ 27

87. [1.6.2] $3 \times (16 \div 4) + 8 \times 2$ 28

88. [1.6.2] $126 + 4 - (20 \div 5)^3$ 66

89. [1.4.4]
$$\begin{array}{r} 5489 \\ \times\ 67 \\ \hline 367{,}763 \end{array}$$

90. [1.5.3] $52\overline{)4524}$ 87

Quick Quiz 1.7

1. Round to the nearest hundred. 92,354 92,400

2. Round to the nearest ten thousand. 2,342,786 2,340,000

3. Use the principle of estimation to find an estimation for this calculation. $7862 \times 329{,}182$ 2,400,000,000

4. **Concept Check** Explain how to round 682,496,934 to the nearest million. Answers may vary

Classroom Quiz 1.7 You may use these problems to quiz your students' mastery of Section 1.7.

1. Round to the nearest hundred. 57,621
Ans: 57,600

2. Round to the nearest ten thousand. 2,342,786
Ans: 2,340,000

3. Use the principle of estimation to find an estimate for this calculation. $36{,}709 \times 894{,}267$
Ans: 36,000,000,000

Solving Problems Involving One Operation

When a builder constructs a new home or office building, he or she often has a *blueprint*. This accurate drawing shows the basic structure of the building. It also shows the dimensions of the structure to be built. This blueprint serves as a useful reference throughout the construction process.

Student Learning Objectives

After studying this section, you will be able to:

 Use the Mathematics Blueprint to solve problems involving one operation.

 Use the Mathematics Blueprint to solve problems involving more than one operation.

Similarly, when solving applied problems, it is helpful to have a "mathematics blueprint." This is a simple way to organize the information provided in the word problem. You record the facts you need to use and specify what you are solving for. You also record any other information that you feel will be helpful. We will use a Mathematics Blueprint for Problem Solving in this section.

Sometimes people feel totally lost when trying to solve a word problem. They sometimes say, "Where do I begin?" or "How in the world do you do this?" When you have this type of feeling, it sometimes helps to have a formal strategy or plan. Here is a plan you may find helpful:

1. *Understand the problem.*
   (a) Read the problem carefully.
   (b) Draw a picture if this helps you see the relationships more clearly.
   (c) Fill in the Mathematics Blueprint so that you have the facts and a method of proceeding in this situation.

2. *Solve and state the answer.*
   (a) Perform the calculations.
   (b) State the answer, including the unit of measure.

3. *Check.*
   (a) Estimate the answer.
   (b) Compare the exact answer with the estimate to see if your answer is reasonable.

Now exactly what does the Mathematics Blueprint for Problem Solving look like? It is a simple sheet of paper with four columns. Each column tells you something to do.

Gather the Facts—Find the numbers that you will need to use in your calculations.

What Am I Asked to Do?—Are you finding an area, a volume, a cost, the total number of people? What is it that you need to find?

How Do I Proceed?—Do you need to add items together? Do you need to multiply or divide? What types of calculations are required?

Key Points to Remember—Write down things you might forget. The length is in feet. The area is in square feet. We need the total number of something, not the intermediate totals. Whatever you need to help you, write it down in this column.

Mathematics Blueprint for Problem Solving

Gather the Facts	What Am I Asked to Do?	How Do I Proceed?	Key Points to Remember

Teaching Example 1 At the beginning of the week, Salah's checking account balance was $1167. During the week he made a deposit of $468 and wrote checks for $105, $87, $145, and $52. What was the total amount of checks written?

Ans: $389 Check: the estimate is $400.

EXAMPLE 1 Gerald made deposits of $317, $512, $84, and $161 into his checking account. He also made out checks for $100 and $125. What was the total of his deposits?

Solution

1. *Understand the problem.* First we read over the problem carefully and fill in the Mathematics Blueprint.

Mathematics Blueprint for Problem Solving

Gather the Facts	What Am I Asked to Do?	How Do I Proceed?	Key Points to Remember
We need only deposits—not checks. The **deposits** are $317, $512, $84, and $161.	Find the total of Gerald's four deposits.	I must add the four deposits to obtain the total.	Watch out! Don't use the **checks** of $100 and $125 in the calculation. We only want the total of the **deposits**.

2. *Solve and state the answer.* We need to *add* to find the sum of the deposits.

$$
\begin{array}{r}
317 \\
512 \\
84 \\
+\,161 \\
\hline
1074
\end{array}
$$

The total of the four deposits is $1074.

3. *Check.* Reread the problem. Be sure you have answered the question that was asked. Did it ask for the total of the deposits?
Yes. ✓

Is the calculation correct? You can use estimation to check. Here we round each of the deposits so that we have one nonzero digit.

$$
\begin{array}{rr}
317 & 300 \\
512 & 500 \\
84 & 80 \\
+\,161 & +\,200 \\
\hline
& 1080
\end{array}
$$

Our estimate is $1080. $1074 is close to our estimated answer of $1080. Our answer is reasonable. ✓

Thus we conclude that the total of the four deposits is $1074.

Practice Problem 1 Use the Mathematics Blueprint to solve the following problem. Diane's paycheck shows deductions of $135 for federal taxes, $28 for state taxes, $13 for FICA, and $34 for health insurance. Her gross pay (amount before deductions) is $1352. What is the total amount that is taken out of Diane's paycheck? $210

NOTE TO STUDENT: Fully worked-out solutions to all of the Practice Problems can be found at the back of the text starting at page SP-1

Mathematics Blueprint for Problem Solving

Gather the Facts	What Am I Asked to Do?	How Do I Proceed?	Key Points to Remember

Portland

Kansas
City

Teaching Example 2 A table of transportation statistics shows that the number of people who carpooled to work was 11,644,000 in 2001 and 10,057,000 in 2003. Did the number of carpoolers increase or decrease? What was the change in the number of carpoolers?

Ans: The number of carpoolers decreased by 1,587,000. Check: the estimate is 1,000,000 carpoolers.

EXAMPLE 2 Theofilos looked at his odometer before he began his trip from Portland, Oregon, to Kansas City, Kansas. He checked his odometer again when he arrived in Kansas City. The two readings are shown in the figure. How many miles did Theofilos travel?

Solution

1. **Understand the problem.** Determine what information is given.
   The mileage reading before the trip began and when the trip was over.
   What do you need to find?
   The number of miles traveled.

Mathematics Blueprint for Problem Solving

Gather the Facts	What Am I Asked to Do?	How Do I Proceed?	Key Points to Remember
At the start of the trip, the odometer read 28,353 miles. At the end of the trip, the odometer read 30,162 miles.	Find out how many miles Theofilos traveled.	I must subtract the two mileage readings.	Subtract the mileage at the start of the trip from the mileage at the end of the trip.

2. **Solve and state the answer.** We need to subtract the two mileage readings to find the difference in the number of miles. This will give us the number of miles the car traveled on this trip alone.

$$30{,}162 - 28{,}353 = 1809$$ The trip totaled 1809 miles.

3. **Check.** We estimate and compare the estimate with the preceding answer.

Kansas City	30,162 $\longrightarrow$	30,000	We subtract
Portland	28,353 $\longrightarrow$	28,000	our rounded values.
		2000	

Our estimate is 2000 miles. We compare this estimate with our answer. Our answer is reasonable. ✓

Practice Problem 2 The table on the left shows the results of the 2000 presidential race in the United States. By how many popular votes did the Democratic candidate beat the Republican candidate in that year? Why did the Democratic candidate not win the election in that year? 543,895 votes; Bush had more electoral college votes.

2000 Presidential Race, Popular Votes

Candidate	Number of Votes
Bush (R)	50,456,002
Gore (D)	50,999,897
Nader	2,882,955

Source: Federal Election Commission

Mathematics Blueprint for Problem Solving

Gather the Facts	What Am I Asked to Do?	How Do I Proceed?	Key Points to Remember

NOTE TO STUDENT: Fully worked-out solutions to all of the Practice Problems can be found at the back of the text starting at page SP-1

EXAMPLE 3   One horsepower is the power needed to lift 550 pounds a distance of 1 foot in 1 second. How many pounds can be lifted 1 foot in 1 second by 7 horsepower?

Teaching Example 3 One quart of liquid is equal to 1140 milliliters. How many milliliters are there in 3 quarts?

Ans: There are 3420 milliliters in 3 quarts. Check: the estimate is 3000 milliliters.

Solution

1. *Understand the problem.* Simplify the problem. If 1 horsepower can lift 550 pounds, how many pounds can be lifted by 7 horsepower? We draw and label a diagram.

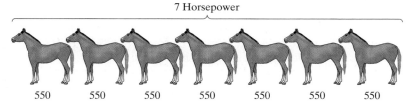

7 Horsepower

550   550   550   550   550   550   550

We use the Mathematics Blueprint to organize the information.

Mathematics Blueprint for Problem Solving

Gather the Facts	What Am I Asked to Do?	How Do I Proceed?	Key Points to Remember
One horsepower will lift 550 pounds.	Find how many pounds can be lifted by 7 horsepower.	I need to multiply 550 by 7.	I do not use the information about moving one foot in one second.

2. *Solve and state the answer.* To solve the problem we multiply the 7 horsepower by 550 pounds for each horsepower.

$$\begin{array}{r} 550 \\ \times\ \ 7 \\ \hline 3850 \end{array}$$

We find that 7 horsepower moves 3850 pounds 1 foot in 1 second. We include 1 foot in 1 second in our answer because it is part of the unit of measure.

3. *Check.* We estimate our answer. We round 550 to 600 pounds.

$$600 \times 7 = 4200 \text{ pounds}$$

Our estimate is 4200 pounds. Our calculations in step 2 gave us 3850. Is this reasonable? This answer is close to our estimate. Our answer is reasonable. ✓

Practice Problem 3   In a measure of liquid capacity, 1 gallon is 1024 fluid drams. How many fluid drams would be in 9 gallons? 9,216 fluid drams

Mathematics Blueprint for Problem Solving

Gather the Facts	What Am I Asked to Do?	How Do I Proceed?	Key Points to Remember

Teaching Example 4 Anita exercises by walking. She walks at a rate of 350 feet per minute. If Anita walked 5950 feet today, how many minutes did she walk?

Ans: Anita walked for 17 minutes. Check: the estimate is 15 minutes.

EXAMPLE 4   Laura can type 35 words per minute. She has to type an English theme that has 5180 words. How many minutes will it take her to type the theme? How many hours and how many minutes will it take her to type the theme?

Solution

1. ***Understand the problem.*** We draw a picture. Each "package" of 1 minute is 35 words. We want to know how many packages make up 5180 words.

35 words in 1 minute

35 words in 1 minute

35 words in 1 minute

35 words in 1 minute

5180 words

We use the Mathematics Blueprint to organize the information.

Mathematics Blueprint for Problem Solving			
Gather the Facts	**What Am I Asked to Do?**	**How Do I Proceed?**	**Key Points to Remember**
Laura can type 35 words per minute. She must type a paper with 5180 words.	Find out how many 35-word units are in 5180 words.	I need to divide 5180 by 35.	In converting minutes to hours, I will use the fact that 1 hour = 60 minutes.

2. ***Solve and state the answer.***

$$
\begin{array}{r}
148 \\
35\overline{)5180} \\
35 \\
\hline
168 \\
140 \\
\hline
280 \\
280 \\
\hline
0
\end{array}
$$

It will take 148 minutes.

We will change this answer to hours and minutes. Since 60 minutes = 1 hour, we divide 148 by 60. The quotient will tell us how many hours. The remainder will tell us how many minutes.

$$
\begin{array}{r}
2 \text{ R } 28 \\
60\overline{)148} \\
-120 \\
\hline
28
\end{array}
$$

Laura can type the theme in 148 minutes or 2 hours, 28 minutes.

3. *Check.* The theme has 5180 words; she can type 35 words per minute. 5180 words is approximately 5000 words.

5180 words → $\dfrac{\text{5000 words rounded to}}{\text{nearest thousand.}}$ $\begin{array}{r} 125 \\ 40\overline{)5000} \end{array}$

35 words per minute → $\dfrac{\text{40 words per minute}}{\text{rounded to nearest ten.}}$ We divide our estimated values.

Our estimate is 125 minutes. This is close to our calculated answer. Our answer is reasonable.   ✓

Practice Problem 4 Donna bought 45 shares of stock for $1620. How much did the stock cost her per share? $36 per share

NOTE TO STUDENT: Fully worked-out solutions to all of the Practice Problems can be found at the back of the text starting at page SP-1

Mathematics Blueprint for Problem Solving

Gather the Facts	What Am I Asked to Do?	How Do I Proceed?	Key Points to Remember

2 Solving Problems Involving More Than One Operation

Sometimes a chart, table, or bill of sale can be used to help us organize the data in an applied problem. In such cases, a blueprint may not be needed.

EXAMPLE 5 Cleanway Rent-A-Car bought four used luxury sedans at $21,000 each, three compact sedans at $14,000 each, and seven subcompact sedans at $8000 each. What was the total cost of the purchase?

Solution

1. *Understand the problem.* We will make an imaginary bill of sale to help us to visualize the problem.

2. *Solve and state the answer.* We do the calculation and enter the results in the bill of sale.

Teaching Example 5 The Hernandez Insurance Agency restocked its office supplies recently. The agency bought 15 reams of paper at $3 each, 2 ink cartridges at $32 each, and 4 boxes of folders at $7 each. How much did the agency pay for the office supplies?

Ans: The office supplies cost $137.

Car Fleet Sales, Inc. Hamilton, Massachusetts

Customer: *Cleanway Rent-A-Car*			
Quantity	Type of Car	Cost per Car	Amount for This Type of Car
4	Luxury sedans	$21,000	$84,000 (4 × $21,000 = $84,000)
3	Compact sedans	$14,000	$42,000 (3 × $14,000 = $42,000)
7	Subcompact sedans	$8000	$56,000 (7 × $8000 = $56,000)
		TOTAL	$182,000 (sum of the three amounts)

The total cost of all 14 cars is $182,000.

NOTE TO STUDENT: Fully worked-out solutions to all of the Practice Problems can be found at the back of the text starting at page SP-1

Teaching Example 6 Juan has investments in the stock market. Last month his stocks were worth a total of $2347. When he checked his investments this month, 2 stocks had increased in value, by $146 and $135. Three stocks had decreased in value, by $48, $86, and $93. What is the total value of his stocks this month?

Ans: Juan's stocks are now worth a total of $2401.

3. *Check.* You may use estimation to check. The check is left to the student.

Practice Problem 5 Anderson Dining Commons purchased 50 tables at $200 each, 180 chairs at $40 each, and six moving carts at $65 each. What was the total cost of the purchase? $17,590

EXAMPLE 6 Dawn had a balance of $410 in her checking account last month. She made deposits of $46, $18, $150, $379, and $22. She made out checks for $316, $400, and $89. What is her balance?

Solution

1. *Understand the problem.* We want to *add* to get a total of all deposits and *add* to get a total of all checks.

Old balance	+	total of deposits	−	total of checks	=	new balance

Mathematics Blueprint for Problem Solving

Gather the Facts	What Am I Asked to Do?	How Do I Proceed?	Key Points to Remember
Old balance: $410. New deposits: $46, $18, $150, $379, and $22. New checks: $316, $400, and $89.	Find the amount of money in the checking account after deposits are made and checks are withdrawn.	**(a)** I need to calculate the total of the deposits and the total of the checks. **(b)** I add the total deposits to the old balance. **(c)** Then I subtract the total of the checks from that result.	Deposits are added to a checking account. Checks are subtracted from a checking account.

2. *Solve and state the answer.*

First we find the total of deposits:

$$\begin{array}{r} 46 \\ 18 \\ 150 \\ 379 \\ +\ 22 \\ \hline \$615 \end{array}$$

Then the total of checks:

$$\begin{array}{r} 316 \\ 400 \\ +\ 89 \\ \hline \$805 \end{array}$$

Add the deposits to the old balance and subtract the amount of the checks.

$$\begin{array}{r|r} \text{Old balance} & 410 \\ +\text{ total deposits} & +\ 615 \\ \hline & 1025 \\ -\text{ total checks} & -\ 805 \\ \hline \text{New balance} & 220 \end{array}$$

The new balance of the checking account is $220.

3. Check. Work backward. You can add the total checks to the new balance and then subtract the total deposits. The result should be the old balance. Try it.

$$
\begin{array}{rl}
410 & \text{Old balance} \quad \checkmark \\
- \ 615 & \\
\hline
1025 & \\
+ \ 805 & \\
\hline
220 & \text{Work backward.}
\end{array}
$$

Practice Problem 6 Last month Bridget had $498 in a savings account. She made two deposits: one for $607 and one for $163. The bank credited her with $36 interest. Since last month, she has made four withdrawals: $19, $158, $582, and $74. What is her balance this month? $471

Mathematics Blueprint for Problem Solving

Gather the Facts	What Am I Asked to Do?	How Do I Proceed?	Key Points to Remember

EXAMPLE 7 When Lorenzo began his car trip, his gas tank was full and the odometer read 76,358 miles. He ended his trip at 76,668 miles and filled the gas tank with 10 gallons of gas. How many miles per gallon did he get with his car?

Solution

1. Understand the problem.

Mathematics Blueprint for Problem Solving

Gather the Facts	What Am I Asked to Do?	How Do I Proceed?	Key Points to Remember
Odometer reading at end of trip: 76,668 miles. Odometer reading at start of trip: 76,358 miles. Used on trip: 10 gallons of gas	Find the number of miles per gallon that the car obtained on the trip.	**(a)** I need to subtract the two odometer readings to obtain the number of miles traveled. **(b)** I divide the number of miles driven by the number of gallons of gas used to get the number of miles obtained per gallon of gas.	The gas tank was full at the beginning of the trip. 10 gallons fills the tank at the end of the trip.

Teaching Example 7 Alec drives a taxi. He began his day with a full tank of gas and his odometer read 103,276. At the end of the day the odometer read 103,591. Alec filled his tank with 12 gallons of gas at noon and filled it again at the end of the day with 9 gallons. How many miles per gallon did the taxi get that day?

Ans: The taxi got 15 miles per gallon that day. Check: the estimate is 15 miles per gallon.

2. *Solve and state the answer.* First we subtract the odometer readings to obtain the miles traveled.

$$
\begin{array}{r}
76{,}668 \\
-\,76{,}358 \\
\hline
310
\end{array}
$$

The trip was 310 miles.
Next we divide the miles driven by the number of gallons.

$$
\begin{array}{r}
31 \\
10\overline{)310} \\
\underline{30} \\
10 \\
\underline{10} \\
0
\end{array}
$$

Thus Lorenzo obtained 31 miles per gallon on the trip.

3. *Check.* We do not want to round to one nonzero digit here, because, if we do, the result will be zero when we subtract. Thus we will round to the nearest hundred for the values of mileage.

$$76{,}668 \longrightarrow 76{,}700$$
$$76{,}358 \longrightarrow 76{,}400$$

Now we subtract the estimated values.

$$
\begin{array}{r}
76{,}700 \\
-\,76{,}400 \\
\hline
300
\end{array}
$$

Thus we estimate the trip to be 300 miles.
Then we divide.

$$
\begin{array}{r}
30 \\
10\overline{)300}
\end{array}
$$

We obtain 30 miles per gallon for our estimate. This is very close to our calculated value of 31 miles per gallon. ✓

NOTE TO STUDENT: *Fully worked-out solutions to all of the Practice Problems can be found at the back of the text starting at page SP-1*

Practice Problem 7 Deidre took a car trip with a full tank of gas. Her trip began with the odometer at 50,698 and ended at 51,118 miles. She then filled the tank with 12 gallons of gas. How many miles per gallon did her car get on the trip? 35 miles per gallon

Mathematics Blueprint for Problem Solving

Gather the Facts	What Am I Asked to Do?	How Do I Proceed?	Key Points to Remember

In general, most students find that they are more successful at solving applied problems if they take extra time to understand the problem. This requires careful reading and thinking about what the problem is asking you to do. Use a colored pen or pencil and underline the most important facts. Draw a picture or sketch if it will help you visualize the situation. Remember, if you understand what you are solving for, your work will go much more quickly.

If you attend a traditional mathematics class that meets one or more times each week:

Developing Your Study Skills

Class Attendance

You will want to get started in the right direction by choosing to attend class every day, beginning with the first day of class. Statistics show that class attendance and good grades go together. Classroom activities are designed to enhance learning, and therefore you must be in class to benefit from them. Each day vital information and explanations are given that can help you understand concepts. Do not be deceived into thinking that you can just find out from a friend what went on in class. There is no good substitute for firsthand experience. Give yourself a push in the right direction by developing the habit of going to class every day.

If you are enrolled in an online mathematics class, a self-paced mathematics class taught in a math lab, or some other type of nontraditional class:

Developing Your Study Skills

Keeping Yourself on Schedule

In a class where you determine your own pace, you will need to commit yourself to keeping on a schedule. Follow the suggested pace provided in your course materials. Keep all your class materials organized and review them often to be sure you are doing everything that you should. If you discipline yourself to follow the suggested course schedule for the first six weeks, you will likely succeed in the class. Professor Tobey and Professor Slater both teach online mathematics courses. They have found that students usually succeed in the course as long as they do every suggested activity for the first six weeks. Make sure you succeed! Keep yourself on schedule!

Applications

You may want to use the Mathematics Blueprint for Problem Solving to help you to solve the word problems in exercises 1–34.

1. *Real Estate* Donna and Miguel want to buy a cabin for $31,500. After repairs, the total cost will be $40,300. How much will the repairs cost? $8800

▲ 2. *Geography* China has a total area of 9,596,960 square kilometers. Bodies of water account for 270,550 square kilometers. How many square kilometers of land does China have? 9,326,410 square kilometers

3. *Business Management* Paula is organizing a large two-day convention. Bert's Bagels is providing the breakfast bagels. If Paula orders 120 bakers' dozen, how many bagels is that? (There are 13 in a bakers' dozen.) 1560 bagels

4. *Business Management* There are 144 pencils in a gross. Mr. Jim Weston ordered 14 gross of pencils for the office. How many pencils did he order? 2016 pencils

5. *Consumer Affairs* A 12-ounce can of Hunts tomato sauce costs 84¢. What is the unit cost of the tomato sauce? (How much does the tomato sauce cost per ounce?) 7¢ per ounce

6. *Consumer Affairs* A 15-ounce can of Del Monte pears costs 90¢. What is the unit cost of the pears? (How much do the pears cost per ounce?) 6¢ per ounce

7. *Sports* Kimberly began running 3 years ago. She has spent $832 on 13 pairs of running shoes during this time. How much on average did each pair of shoes cost? $64

8. *Wildlife Management* There are approximately 50,000 bison living in the United States. If Northwest Trek, the animal preserve located in Mt. Rainier National Park, has 103 bison, how many bison are living elsewhere? 49,897

9. *Population Studies* In October 2006, the population of the United States reached 300,000,000. In October 2005, there were 297,020,000 people in the United States. What was the increase in population from October 2005 to October 2006? 2,980,000 people

▲ 10. *Geometry* Valleyfair, an amusement park in Minnesota, covers 26 acres. If there are 44,010 square feet in 1 acre, how many square feet of land does Valleyfair cover? 1,144,260 square feet

11. *Business Management* A games arcade has recently opened in a West Chicago neighborhood. The owners were nervous about whether it would be a success. Fortunately, the gross revenues over the last four weeks were $7356, $3257, $4777, and $4992. What was the gross revenue over these four weeks for the arcade? $20,382

12. *International Relations* The two largest cities in France are Paris, with 2,113,000 people, and Marseille, with 815,100 people. What is the difference in population between these two cities? 1,297,900 people

13. *Wildlife Management* The Federal Nigeria game preserve has 24,111 animals, 327 full-time staff, and 793 volunteers. What is the total of these three groups? How many more volunteers are there than full-time staff? 25,231; 466

14. *Geography* The longest rivers in the world are the Nile River, the Amazon River, and the Mississippi River. Their lengths are 4132 miles, 3915 miles, and 3741 miles, respectively. How many total miles do these three rivers run? What is the difference in the lengths of the Nile and the Mississippi? 11,788 miles; 391 miles

15. *World History* Every 60 minutes, the world population increases by 100,000 people. How many people will be born during the next 480 minutes? 800,000 people

16. *Personal Finance* Roberto had $2158 in his savings account six months ago. In the last six months he made four deposits: $156, $238, $1119, and $866. The bank deposited $136 in interest over the six-month period. How much does he have in the savings account at present? $4673

In exercises 17–34, more than one type of operation is required.

17. *Sports* Carmen gives golf lessons every Saturday. She charges $15 for adults, $9 for children, and $5 for club rental. Last Saturday she taught six adults and eight children, and six people needed to rent clubs. How much money did Carmen make on that day? $192

18. *Business Management* Whale Watch Excursions charges $10 for adults, $6 for children, and $7 for senior citizens. On the last trip of the day, there were five adults, seven children, and three senior citizens. How much money did the company make on this trip? $113

19. *Personal Finance* Sue Li had a balance in her checking account of $132. During the last two months she has deposited four paychecks of $715 each. She wrote two rent checks for $575 each, and wrote checks totaling $482 for other bills. When all the deposits are recorded and the checks clear, what will be the balance in her checking account? $1360

20. *Space Travel* From 1957 to 2005, the number of successful space launches totaled 4361. Of these, 2746 were launched by the Soviet Union/Russia, and 1305 were launched by the United States. How many launches were completed by other countries? 310 launches

21. *Real Estate* Diana owns 85 acres of forest land in Oregon. She rents it to a timber grower for $250 per acre per year. Her property taxes are $57 per acre. How much profit does she make on the land each year? $16,405

22. *Real Estate* Todd owns 13 acres of commercially zoned land in the city of Columbus, Ohio. He rents it to a construction company for $12,350 per acre per year. His property taxes to the city are $7362 per acre per year. How much profit does he make on the land each year? $64,844

23. *Environmental Studies* Hanna wants to determine the miles-per-gallon rating of her Chevrolet Cavalier. She filled the tank when the odometer read 14,926 miles. She then drove her car on a trip. At the end of the trip, the odometer read 15,276 miles. It took 14 gallons to fill the tank. How many miles per gallon does her car deliver? 25 miles per gallon

24. *Environmental Studies* Gary wants to determine the miles-per-gallon rating of his Geo Metro. He filled the tank when the odometer read 28,862 miles. After ten days, the odometer read 29,438 miles and the tank required 18 gallons to be filled. How many miles per gallon did Gary's car achieve? 32 miles per gallon

25. *Forestry* A beautiful piece of land in the Wilmot Nature Preserve has three times as many oak trees as birches, two times as many maples as oaks, and seven times as many pine trees as maples. If there are 18 birches on the land, how many of each of the other trees are there? How many trees are there in all?
There are 54 oak trees, 108 maple trees, and 756 pine trees. In total there are 936 trees.

26. *Business Management* The Cool Coffee Lounge in Albuquerque, New Mexico, has 27 tables, and each table has either two or four chairs. If there are a total of 94 chairs accompanying the 27 tables, how many tables have four chairs? How many tables have two chairs?
20 tables have 4 chairs; 7 tables have 2 chairs.

Use the following list to answer exercises 27–30.

Education

The following is a partial list of the primary home languages of students attending Public School 139 in Queens, New York, in the school year 2006–2007.

Language	Number of Students
Russian	226
English	183
Spanish	174
Mandarin	53
Cantonese	44
Korean	40
Hindi	29
Chinese, other dialects	21
Filipino	12
Hebrew	9
Indonesian	8
Romanian	8
Urdu	8
Dari/Farsi/Persian	7
Albanian	6
Arabic	6
Bulgarian	5
Gujarati	4

Source: Office of the Superintendent of Schools, Queens, New York.

27. How many students speak Mandarin, Cantonese, or other Chinese dialects as the primary language in their homes?
118

28. How many students speak Korean, Hindi, or Filipino as the primary language in their homes?
81

29. How many more students speak Spanish or Russian rather than English as the primary language in their homes?
217

30. How many more students speak Indonesian or Romanian rather than Albanian as the primary language in their homes?   10

Use the following bar graph to answer exercises 31–34.

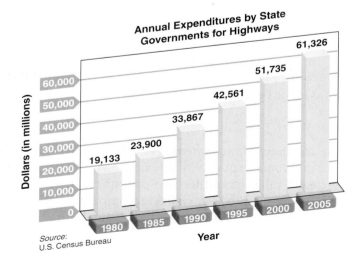

Annual Expenditures by State Governments for Highways

Dollars (in millions)

61,326
51,735
42,561
33,867
23,900
19,133

60,000
50,000
40,000
30,000
20,000
10,000
0

1980 1985 1990 1995 2000 2005
Year

Source: U.S. Census Bureau

Government Finances

31. How many more dollars were spent by state governments for highways in 1990 than in 1980?
$14,734,000,000

32. How many more dollars were spent by state governments for highways in 2000 than in 1985?
$27,835,000,000

33. If the exact same dollar increase occurs between 2000 and 2010 as occurred between 1990 and 2000, what will the expenditures by state governments for highways be in 2010?
$69,603,000,000

34. If the amount of money expended by state governments for highways remained constant for the years 2000 to 2003, how much money was spent for highways during that four-year period?
$206,940,000,000

Cumulative Review

35. [1.6.1] Evaluate. 7^3
343

36. [1.6.2] Perform in the proper order.
$3 \times 2^3 + 15 \div 3 - 4 \times 2$
$3 \times 8 + 15 \div 3 - 4 \times 2 = 24 + 5 - 8 = 21$

37. [1.4.4] Multiply. 126×38
4788

38. [1.5.3] Divide. $12\overline{)3096}$
258

39. [1.2.4] Add. $96 + 123 + 57 + 526$
802

40. [1.3.3] Subtract. $509{,}263 - 485{,}978$
23,285

41. [1.7.1] Round to the nearest thousand. 526,195,726
526,196,000

42. [1.1.3] Write this number in standard notation. Three billion, four hundred million, six hundred three thousand, twenty-five.
3,400,603,025

Quick Quiz 1.8

1. Sixteen people in a travel club chartered a bus to go to Vermont to see the fall foliage. The bill for the bus charter was $4304. How much will each club member pay if the cost is shared equally? $269

2. Maria had a balance of $471 in her checking account last month. She then deposited $198, $276, and $347. She made out checks for $49, $227, and $158. What will her new balance be? $858

3. The entire Tobey family went on a fishing charter. The cost was $11 for people 60 or older, $14 for people age 12 to 59, and $5 for children under 12. The captain counted 2 people over 60, six people age 12 to 59, and four children under 12. What was the total cost for the Tobey family members to go on the fishing charter? $126

4. Concept Check A company has purchased 38 new cars for the sales department for $836,543. Assuming that each car cost the same, explain how you would estimate the cost of each car. Answers may vary

Classroom Quiz 1.8 You may use these problems to quiz your students' mastery of Section 1.8.

1. The uniforms for the 18 members of the football team at Hamilton-Wenham Regional High School cost $11,826. The school budget has no money to pay for the uniforms. If each member of the team shares the cost equally, how much will the uniform cost each team member? **Ans:** $657

2. Mike had a balance of $64 in his checking account last month. Since then he has deposited $906 and $885. He made out checks for $29, $109, $412, and $683. What will his new balance be? **Ans:** $622

3. Susan drives a tour bus to Springfield to see a local production of "West Side Story." The fares were $6 for senior citizens, $8 for people age 12 to 64, and $3 for children under 12. Yesterday she took the trip with seven senior citizens, eight people age 12 to 64, and nine children under 12. What was the total amount she needed to collect in fares? **Ans:** $133

Putting Your Skills to Work: Use Math to Save Money

MANAGING DEBTS AND PAYMENTS

Can you imagine the joy of taking a great vacation without going into debt for it? Can you think about how great it would be to have all your debts paid off? It is a wonderful feeling! Paying off your debts and then being able to take a vacation is an excellent goal. But how is that done? Consider the story of Tracy and Max.

Facing Up to the Debt

Tracy and Max were overwhelmed with debt. Besides their ordinary living expenses, they had so much debt that they could barely make the money they earned last until the end of the month. They had little money left for extras and for having fun, and no money for a vacation. Each of their three credit cards was maxed out at $8000; they had a hospital debt of $12,000; they still owed $2000 on their car; and they had also borrowed money from friends in sums of $100 and $300.

Making a Plan

They decided to write down all of their debts, putting them in order from smallest to largest.

Then they made minimum payments on all the debts, but aimed at paying off the three smallest debts first.

1. Put the couple's debts in order from smallest to largest. Remember, there are three credit cards. $100, $300, $2000, $8000, $8000, $8000, $12,000.

The minimum payment on each credit card averaged $25 per month. They had arranged with the hospital to pay the debt off at $50 per month. Their car payment was $200 per month, and they agreed to pay each of their friends $20 per month.

2. What is the total amount of their minimum monthly payment? $365

What Tracy and Max Accomplished

Tracy and Max decided they would eliminate any extras and not spend money on fun activities so they could use this money to pay off their three smallest debts. As a result they were able to pay off the first two smallest debts in just two months while making minimum payments on all other debts. Then each month they took the $40 they would have used to pay these two small debts and applied it towards the third smallest debt. Again, they made sure they made minimum payments on all other debts.

3. How many more months will it take Tracy and Max to pay off the third smallest debt if they follow the plan stated above?
$2000 − $400 = $1600; $1600 ÷ $240 = 6.6 . . . or 7 months

After eliminating the smaller debts, they took the money they would have spent on those debts and used it on the principal of the remaining debts. In other words, they paid more than the minimum payment on the remaining debts. Because they hated the way they felt when they were in debt, they stopped using credit cards for new purchases. Max also took a temporary part-time job so they could pay off their debts faster. Finally they paid off all the debts!

Applying It to Your Life

Many debt counselors have a simple, practical suggestion for people in debt. Arrange debts in order, pay off the smallest first, and then let the consequences of that action help pay off the rest of the debts more quickly. Many people have been able to get out of debt in about two years by using this approach, other strategies for budgeting, and wise choices for living.

Chapter 1 Organizer

Topic	Procedure	Examples
Place value of numbers, p. 2.	Each digit has a value depending on location. millions / hundred thousands / ten thousands / thousands / hundreds / tens / ones	In the number 2,896,341, what place value does 9 have? ten thousands
Writing expanded notation, p. 2.	Take the number of each digit and multiply it by one, ten, hundred, thousand, ... according to its place.	Write in expanded notation. 46,235 $40,000 + 6000 + 200 + 30 + 5$
Writing whole numbers in words, p. 5.	Take the number in each period and indicate if they are (millions) (thousands) (ones) xxx, xxx, xxx	Write in words. 134,718,216. one hundred thirty-four million, seven hundred eighteen thousand, two hundred sixteen
Adding whole numbers, p. 12.	Starting with the right column, add each column separately. If a two-digit sum occurs, "carry" the left digit over to the next column to the left.	Add. $\begin{array}{r} \overset{2}{2}\,\overset{1}{5}\,8 \\ 3\,6\,7 \\ 2\,9\,1 \\ +\,4\,5\,3 \\ \hline 1\,3\,6\,9 \end{array}$
Subtracting whole numbers, p. 23.	Starting with the right column, subtract each column separately. If necessary, borrow a unit from the column to the left and bring it to the right as a "10."	Subtract. $\begin{array}{r} 1\,6,\,\overset{6}{7}\,\overset{\overset{13}{8}}{4}\,\overset{12}{2} \\ -1\,2,\,3\,9\,5 \\ \hline 4,\,3\,4\,7 \end{array}$
Multiplying several factors, p. 41.	Keep multiplying from left to right. Take each product and multiply by the next factor to the right. Continue until all factors are used once. (Since multiplication is commutative and associative, the factors can be multiplied in any order.)	Multiply. $\begin{aligned} 2 \times 9 \times 7 \times 6 \times 3 &= 18 \times 7 \times 6 \times 3 \\ &= 126 \times 6 \times 3 \\ &= 756 \times 3 \\ &= 2268 \end{aligned}$
Multiplying several-digit numbers, p. 39.	Multiply the top factor by the ones digit, then by the tens digit, then by the hundreds digit. Add the partial products together.	Multiply. $\begin{array}{r} 5\,6\,7 \\ \times\,2\,3\,8 \\ \hline 4\,5\,3\,6 \\ 1\,7\,0\,1 \\ 1\,1\,3\,4 \\ \hline 1\,3\,4,\,9\,4\,6 \end{array}$
Dividing by a two- or three-digit number, p. 52.	Figure how many times the first digit of the divisor goes into the first two digits of the dividend. To try this answer, multiply it back to see if it is too large or small. Continue each step of long division until finished.	Divide. $\begin{array}{r} 589 \\ 238\overline{)140{,}182} \\ \underline{1190} \\ 2118 \\ \underline{1904} \\ 2142 \\ \underline{2142} \\ 0 \end{array}$

Topic	Procedure	Examples
Exponent form, p. 61.	To show in short form the repeated multiplication of the same number, write the number being multiplied. (This is the base.) Write in smaller print above the line the number of times it appears as a factor. (This is the exponent.) To evaluate the exponent form, write the factor the number of times shown in the exponent. Then multiply.	Write in exponent form. $$10 \times 10 \times 10 \times 10 \times 10 \times 10 \times 10 \times 10$$ $$10^8$$ Evaluate. 6^3 $$6 \times 6 \times 6 = 216$$
Order of operations, p. 63.	1. Perform operations inside parentheses. 2. Simplify exponents. 3. Then do multiplication and division in order from left to right. 4. Then do addition and subtraction in order from left to right.	Evaluate. $$2^3 + 16 \div 4^2 \times 5 - 3$$ Raise to a power first. $$8 + 16 \div 16 \times 5 - 3$$ Then do multiplication or division from left to right. $$8 + 1 \times 5 - 3$$ $$8 + 5 - 3$$ Then do addition and subtraction. $$13 - 3 = 10$$
Rounding, p. 68.	1. If the first digit to the right of the round-off place is less than 5, the digit in the round-off place is unchanged. 2. If the first digit to the right of the round-off place is 5 or more, the digit in the round-off place is increased by 1. 3. Digits to the right of the round-off place are replaced by zeros.	Round to the nearest hundred. 56,743 $$\downarrow$$ 5 6,7④3 The digit 4 is less than 5. 56,700 Round to the nearest thousand. 128,517 $$\downarrow$$ 1 2 8,⑤1 7 The digit 5 is obviously 5 or greater. We increase the thousands digit by 1. 129,000
Estimating the answer to a calculation, p. 71.	1. Round each number so that there is one nonzero digit. 2. Perform the calculation with the rounded numbers.	Estimate the answer. $$45{,}780 \times 9453$$ First we round. $$50{,}000 \times 9000$$ Then we multiply. $$\begin{array}{r} 50{,}000 \\ \times\ \ 9{,}000 \\ \hline 450{,}000{,}000 \end{array}$$ We estimate the answer to be 450,000,000.

Procedure for Solving Applied Problems

Using the Mathematics Blueprint for Problem Solving, p. 79

In solving an applied problem, students may find it helpful to complete the following steps. You will not use all the steps all the time. Choose the steps that best fit the conditions of the problem.

1. Understand the problem.
    (a) Read the problem carefully.
    (b) Draw a picture if this helps you to visualize the situation. Think about what facts you are given and what you are asked to find.
    (c) Use the Mathematics Blueprint for Problem Solving to organize your work. Follow these four parts.
        1. Gather the facts (Write down specific values given in the problem.)
        2. What am I asked to do? (Identify what you must obtain for an answer.)
        3. Decide what calculations need to be done.
        4. Key points to remember. (Record any facts, warnings, formulas, or concepts you think will be important as you solve the problem.)

2. Solve and state the answer.
    (a) Perform the necessary calculations.
    (b) State the answer, including the unit of measure.

3. Check.
    (a) Estimate the answer to the problem. Compare this estimate to the calculated value. Is your answer reasonable?
    (b) Repeat your calculations.
    (c) Work backward from your answer. Do you arrive at the original conditions of the problem?

EXAMPLE

The Manchester highway department has just purchased two pickup trucks and three dump trucks. The cost of a pickup truck is $17,920. The cost of a dump truck is $48,670. What was the cost to purchase these five trucks?

1. *Understand the problem.*
2. *Solve and state the answer.*
    Calculate cost of pickup trucks

    $$\begin{array}{r} \$17,920 \\ \times \quad 2 \\ \hline \$35,840 \end{array}$$

    Calculate cost of dump trucks

    $$\begin{array}{r} \$48,670 \\ \times \quad 3 \\ \hline \$146,010 \end{array}$$

    Find total cost. $35,840 + $146,010 = $181,850
    The total cost of the five trucks is $181,850.

3. *Check.*
    Estimate cost of pickup trucks
    $$20,000 \times 2 = 40,000$$
    Estimate cost of dump trucks
    $$50,000 \times 3 = 150,000$$
    Total estimate
    $$40,000 + 150,000 = 190,000$$

    This is close to our calculated answer of $181,850. We determine that our answer is reasonable. ✓

Mathematics Blueprint for Problem Solving

Gather the Facts	What Am I Asked to Do?	How Do I Proceed?	Key Points to Remember
Buy 2 pickup trucks 3 dump trucks Cost Pickup: $17,920 Dump: $48,670	Find the total cost of the 5 trucks.	Find the cost of 2 pickup trucks. Find the cost of 3 dump trucks. Add to get final cost of all 5 trucks.	Multiply 2 times pickup truck cost. Multiply 3 times dump truck cost.

Chapter 1 Review Problems

If you have trouble with a particular type of exercise, review the examples in the section indicated for that group of exercises. Answers to all exercises are located in the answer key.

Section 1.1

Write in words.

1. 892   eight hundred ninety-two

2. 15,802   fifteen thousand, eight hundred two

3. 109,276   one hundred nine thousand, two hundred seventy-six

4. 423,576,055   four hundred twenty-three million, five hundred seventy-six thousand, fifty-five

Write in expanded notation.

5. 4364   4000 + 300 + 60 + 4

6. 35,414   30,000 + 5000 + 400 + 10 + 4

7. 42,166,037
40,000,000 + 2,000,000 + 100,000 + 60,000 + 6000 + 30 + 7

8. 1,305,128
1,000,000 + 300,000 + 5000 + 100 + 20 + 8

Write in standard notation.

9. nine hundred twenty-four   924

10. five thousand three hundred two   5302

11. one million, three hundred twenty-eight thousand, eight hundred twenty-eight   1,328,828

12. twenty-four million, seven hundred five thousand, one hundred twelve   24,705,112

Section 1.2

Add.

13. 76 + 39   115

14. 148 + 152   300

15. 235 + 165   400

16. 12 + 28 + 34 + 76   150

17.
```
  123
   61
    9
   84
+ 123
-----
  400
```

18.
```
  546
  254
+ 153
-----
  953
```

19.
```
  226
  134
+ 647
-----
 1007
```

20.
```
 52,134
+  7966
-------
 60,100
```

21.
```
  1356
  2892
   561
    89
+ 9805
------
14,703
```

22.
```
    26
   503
   935
  1257
+ 7861
------
10,582
```

Section 1.3

Subtract.

23.
```
  36
- 19
----
  17
```

24.
```
  54
- 48
----
   6
```

25.
```
  126
-  99
----
   27
```

26.
```
  543
- 372
----
  171
```

27.
```
 7000
-  845
-----
 6155
```

28.
```
 9000
- 5833
-----
 3167
```

29.
```
 201,340
- 120,618
--------
  80,722
```

30.
```
 320,055
- 214,237
--------
 105,818
```

31.
```
 6,325,034
-    89,023
----------
 6,236,011
```

32.
```
 5,412,022
-    79,031
----------
 5,332,991
```

98

Section 1.4

Multiply.

33. $8 \times 1 \times 9 \times 2$ 144

34. $7 \times 6 \times 0 \times 4$ 0

35. $2 \cdot 5 \cdot 10 \cdot 8$ 800

36. $4 \cdot 25 \cdot 1 \cdot 15$ 1500

37. 621×100 62,100

38. $84,312 \times 1000$
84,312,000

39. $78 \times 10,000$ 780,000

40. $563 \times 1,000,000$
563,000,000

41. $\begin{array}{r} 58 \\ \times 32 \\ \hline 1856 \end{array}$

42. $\begin{array}{r} 73 \\ \times 24 \\ \hline 1752 \end{array}$

43. $\begin{array}{r} 150 \\ \times\ 27 \\ \hline 4050 \end{array}$

44. $\begin{array}{r} 360 \\ \times\ 38 \\ \hline 13,680 \end{array}$

45. $\begin{array}{r} 709 \\ \times\ 36 \\ \hline 25,524 \end{array}$

46. $\begin{array}{r} 502 \\ \times\ 48 \\ \hline 24,096 \end{array}$

47. $\begin{array}{r} 123 \\ \times 714 \\ \hline 87,822 \end{array}$

48. $\begin{array}{r} 431 \\ \times 623 \\ \hline 268,513 \end{array}$

49. $\begin{array}{r} 1782 \\ \times\ 305 \\ \hline 543,510 \end{array}$

50. $\begin{array}{r} 2057 \\ \times\ 124 \\ \hline 255,068 \end{array}$

51. $\begin{array}{r} 3182 \\ \times\ 35 \\ \hline 111,370 \end{array}$

52. $\begin{array}{r} 2713 \\ \times\ 42 \\ \hline 113,946 \end{array}$

53. $\begin{array}{r} 1200 \\ \times 6000 \\ \hline 7,200,000 \end{array}$

54. $\begin{array}{r} 2500 \\ \times 3000 \\ \hline 7,500,000 \end{array}$

55. $\begin{array}{r} 100,000 \\ \times\ 20,000 \\ \hline 2,000,000,000 \end{array}$

56. $\begin{array}{r} 300,000 \\ \times\ 40,000 \\ \hline 12,000,000,000 \end{array}$

Section 1.5

Divide, if possible.

57. $20 \div 10$ 2

58. $40 \div 8$ 5

59. $0 \div 8$ 0

60. $12 \div 1$ 12

61. $7 \div 1$ 7

62. $0 \div 5$ 0

63. $\dfrac{81}{9}$ 9

64. $\dfrac{42}{6}$ 7

65. $\dfrac{5}{0}$ undefined

66. $\dfrac{24}{6}$ 4

67. $\dfrac{56}{8}$ 7

68. $\dfrac{63}{7}$ 9

Divide. Be sure to indicate the remainder, if one exists.

69. $6 \overline{)750}$ 125

70. $7 \overline{)875}$ 125

71. $9 \overline{)1863}$ 207

72. $4 \overline{)1236}$ 309

73. $6 \overline{)15,024}$ 2504

74. $8 \overline{)24,512}$ 3064

75. $6 \overline{)221,748}$ 36,958

76. $5 \overline{)184,605}$ 36,921

77. $8 \overline{)120,371}$ 15,046 R 3

78. $7 \overline{)250,485}$ 35,783 R 4

79. $67 \overline{)490}$ 7 R 21

80. $72 \overline{)325}$ 4 R 37

81. $21 \overline{)666}$ 31 R 15

82. $22 \overline{)319}$ 14 R 11

83. $68 \overline{)2614}$ 38 R 30

84. $53 \overline{)3202}$ 60 R 22

85. $45 \overline{)8775}$ 195

86. $35 \overline{)9030}$ 258

87. $132 \overline{)7128}$ 54

88. $204 \overline{)3876}$ 19

Section 1.6

Write in exponent form.

89. 13×13 13^2

90. $21 \times 21 \times 21$ 21^3

91. $8 \times 8 \times 8 \times 8 \times 8$ 8^5

92. $10 \times 10 \times 10 \times 10 \times 10 \times 10$ 10^6

Evaluate.

93. 2^6 64

94. 3^4 81

95. 2^7 128

96. 5^3 125

97. 7^2 49

98. 9^2 81

99. 6^3 216

100. 4^3 64

Perform each operation in proper order.

101. $7 + 2 \times 3 - 5$ 8

102. $6 \times 2 - 4 + 3$ 11

103. $2^5 + 4 - (5 + 3^2)$ 22

104. $4^3 + 20 \div (2 + 2^3)$ 66

105. $34 - 9 \div 9 \times 12$ 22

106. $2 \times 7^2 - 20 \div 1$ 78

107. $2^3 \times 5 \div 8 + 3 \times 4$ 17

108. $2^3 + 4 \times 5 - 32 \div (1 + 3)^2$ 26

109. $6 \times 3 + 3 \times 5^2 - 63 \div (5 - 2)^2$ 86

Section 1.7

Round to the nearest ten.

110. 3364   3360

111. 5895   5900

112. 15,305   15,310

113. 42,644   42,640

In exercises 114–117, round to the nearest thousand.

114. 12,350   12,000

115. 22,986   23,000

116. 675,800   676,000

117. 202,498   202,000

118. Round to the nearest hundred thousand. 4,649,320   4,600,000

119. Round to the nearest ten thousand. 9,995,312   10,000,000

Use the principle of estimation to find an estimate for each calculation.

120. $324 + 655 + 187 + 245$

```
   300
   700
   200
 + 200
  1400
```

121. $18,702 + 8331 + 36,612$

```
   20,000
    8000
 + 40,000
   68,000
```

122. $4,326,171 - 2,916,788$

```
   4,000,000
 - 3,000,000
   1,000,000
```

123. $34,950 - 15,439$

```
   30,000
 - 20,000
   10,000
```

124. 1463×5982
$$\begin{array}{r} 1000 \\ \times\ 6000 \\ \hline 6{,}000{,}000 \end{array}$$

125. $2{,}965{,}372 \times 893$
$$\begin{array}{r} 3{,}000{,}000 \\ \times\qquad 900 \\ \hline 2{,}700{,}000{,}000 \end{array}$$

126. $83{,}421 \div 24$
$$\begin{array}{r} 4{,}000 \\ 20\overline{)80{,}000} \end{array}$$

127. $876{,}321 \div 335$
$$\begin{array}{r} 3000 \\ 300\overline{)900{,}000} \end{array}$$

Section 1.8

Solve.

128. *Consumer Decisions* Professor O'Shea bought 20 dozen donut holes for the faculty meeting. How many donut holes did he buy? (There are 12 in a dozen.) 240 donut holes

129. *Computer Applications* Ward can type 25 words per minute on his computer. He typed for seven minutes at that speed. How many words did he type? 175 words

130. *Travel* In June, 2462 people visited the Renaissance Festival. There were 1997 visitors in July, and 2561 in August. How many people visited the festival during these three months? 7020 people

131. *Farming* Applepickers, Inc. bought a truck for $26,300, a car for $14,520, and a minivan for $18,650. What was the total purchase price? $59,470

132. *Aviation* A plane was flying at 14,630 feet. It flew over a mountain 4329 feet high. How many feet was it from the plane to the top of the mountain? 10,301 feet

133. *Personal Finance* Gerardo was billed $4330 for tuition, and he needs to spend $268 on books. He received a $1250 scholarship. How much will he have to pay for tuition and books after the scholarship is deducted? $3348

134. *Travel* The expedition cost a total of $32,544 for 24 paying passengers, who shared the cost equally. What was the cost per passenger? $1356

135. *Business Management* Middlebury College ordered 112 dormitory beds for $8288. What was the cost per bed? $74

136. *Personal Finance* Melissa's savings account balance last month was $810. The bank added $24 interest. Melissa deposited $105, $36, and $177. She made withdrawals of $18, $145, $250, and $461. What will be her balance this month? $278

137. *Environmental Studies* Ali began a trip on a full tank of gas with the car odometer at 56,320 miles. He ended the trip at 56,720 miles and added 16 gallons of gas to refill the tank. How many miles per gallon did he get on the trip? 25 miles per gallon

138. *Business Management* The maintenance group bought three lawn mowers at $279, four power drills at $61, and two riding tractors at $1980. What was the total purchase price for these items? $5041

139. *Business Management* Anita is opening a new café in town. She bought 15 tables at $65 each, 60 chairs for $12 each, and eight ceiling fans for $42 each. What was the total purchase price for these items? $2031

Environmental Protection *Use the following bar graph to answer exercises 140–142.*

140. How many more tons of solid waste were recovered and recycled in 1995 than in 1980?
40,500,000 tons

141. What was the greatest increase in tons of solid waste recovered and recycled in a five-year period? 21,400,000 tons, from 1990 to 1995

142. If the exact same increase in the number of tons recovered occurs from 2000 to 2010 as occurred from 1990 to 2000, how many tons of solid waste will be recovered and recycled in 2010?
93,400,000 tons

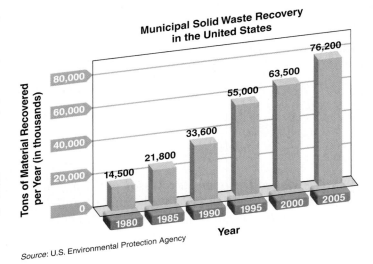

Source: U.S. Environmental Protection Agency

Mixed Practice

Perform each calculation.

143. 205 + 36 + 1983 + 60   2284

144.   56,793
          − 48,926
          _____
            7867

145. 396 × 28   11,088

146. 37)‾4773‾   129

147. Evaluate. $4 \times 12 - (12 + 9) + 2^3 \div 4$ 29

148. ***Personal Finance*** Michael Evans has $3000 in his checking account. He buys 3 computers at $699 each and 2 printers at $78 each. How much does he have remaining after the purchases?
$747

▲ **149.** ***Geometry*** Milton is building a rectangular patio in his backyard. The patio measures 22 feet by 15 feet.
  (a) How many square feet is the patio?
  330 square feet
  (b) If Milton wanted to fence in the patio, how many feet of fence would he need?
  74 feet

Remember to use your Chapter Test Prep Video CD to see the worked-out solutions to the test problems you want to review.

Write the answers.

1. Write in words. 44,007,635

2. Write in expanded notation. 26,859

3. Write in standard notation. three million, five hundred eighty-one thousand, seventy-six

Add.

4.   189
       26
       12
      528
    + 76

5.   763
      220
    + 508

6.   135,484
       2,376
      81,004
    + 100,113

Subtract.

7.   8961
    − 894

8.   501,760
    − 328,902

9.   18,400,100
    − 13,174,332

Multiply.

10. $1 \times 6 \times 9 \times 7$

11.   45
     $\times$ 96

12.   326
    $\times$592

13. 18,491
    $\times$ 7

In problems 14–16, divide. If there is a remainder, be sure to state it as part of your answer.

14. $5\overline{)15,071}$

15. $6\overline{)14,148}$

16. $37\overline{)13,024}$

17. Write in exponent form. $14 \times 14 \times 14$

18. Evaluate. 2^6

Note to Instructor: The Chapter 1 Test file in the TestGen program provides algorithms specifically matched to these problems so you can easily replicate this test for additional practice or assessment purposes.

1. forty-four million, seven thousand, six hundred thirty-five

2. $20,000 + 6000 + 800 + 50 + 9$

3. 3,581,076

4. 831

5. 1491

6. 318,977

7. 8067

8. 172,858

9. 5,225,768

10. 378

11. 4320

12. 192,992

13. 129,437

14. 3014 R 1

15. 2358

16. 352

17. 14^3

18. 64

19. ___23___

20. ___50___

21. ___79___

22. ___94,800___

23. ___6,460,000___

24. ___5,300,000___

25. ___150,000,000,000___

26. ___16,000___

27. ___$2148___

28. ___467 feet___

29. ___$127___

30. ___$292___

31. ___748,000 square feet___

32. ___46 feet___

In problems 19–21, perform each operation in proper order.

19. $5 + 6^2 - 2 \times (9 - 6)^2$ **20.** $2^4 + 3^3 + 28 \div 4$

21. $4 \times 6 + 3^3 \times 2 + 23 \div 23$

22. Round to the nearest hundred.   94,768

23. Round to the nearest ten thousand.   6,462,431

24. Round to the nearest hundred thousand.   5,278,963

Estimate the answer.

25. $4,867,010 \times 27,058$ **26.** $1423 + 3298 + 4103 + 7614$

Solve.

27. A cruise for 15 people costs $32,220. If each person paid the same amount, how much will it cost each individual?

28. The river is 602 feet wide at Big Bend Corner. A boy is in the shallow water, 135 feet from the shore. How far is the boy from the other side of the river?

29. At the bookstore, Hector bought three notebooks at $2 each, one textbook for $45, two lamps at $21 each, and two sweatshirts at $17 each. What was his total bill?

30. Patricia is looking at her checkbook. She had a balance last month of $31. She deposited $902 and $399. She made out checks for $885, $103, $26, $17, and $9. What will be her new balance?

▲ **31.** The runway at Beverly Airport needs to be resurfaced. The rectangular runway is 6800 feet long and 110 feet wide. What is the area of the runway that needs to be resurfaced?

▲ **32.** Nancy Tobey planted a vegetable garden in the backyard. However, the deer and raccoons have been stealing all the vegetables. She asked John to fence in the garden. The rectangular garden measures 8 feet by 15 feet. How many feet of fence should John purchase if he wants to enclose the garden?

CHAPTER 2

All of us have seen pictures of the Pyramids of Egypt. These amazing structures were built very carefully. Measurements had to be made that were very precise. The ancient Egyptians used an elaborate system of fractions that allowed them to make highly accurate measurements. As you master the topics of this chapter, you will master the basic skills used by the designers of the Pyramids of Egypt.

Fractions

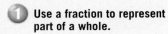

1 Using a Fraction to Represent Part of a Whole

In Chapter 1 we studied whole numbers. In this chapter we will study a fractional part of a whole number. One way to represent parts of a whole is with **fractions.** The word *fraction* (like the word *fracture*) suggests that something is being broken. In mathematics, fractions represent the part that is "broken off" from a whole. The whole can be a single object (like a whole pie) or a group (the employees of a company). Here are some examples.

Single object

$$\frac{1}{3}$$

The whole is the pie on the left. The fraction $\frac{1}{3}$ represents the shaded part of the pie, 1 of 3 pieces. $\frac{1}{3}$ is read "one-third."

A group: ACE company employs 150 men, 200 women.

$$\frac{150}{350}$$

Recipe: Applesauce
4 apples
1/2 cup sugar
1 teaspoon cinnamon

The whole is the company of 350 people (150 men plus 200 women). The fraction $\frac{150}{350}$ represents that part of the company consisting of men.

The whole is 1 whole cup of sugar. This recipe calls for $\frac{1}{2}$ cup of sugar. Notice that in many real-life situations $\frac{1}{2}$ is written as 1/2.

When we say "$\frac{3}{8}$ of a pizza has been eaten," we mean 3 of 8 equal parts of a pizza have been eaten. (See the figure.) When we write the fraction $\frac{3}{8}$, the number on the top, 3, is the **numerator,** and the number on the bottom, 8, is the **denominator.**

The numerator specifies how many parts $\rightarrow 3$
The denominator specifies the total number of parts $\rightarrow \overline{8}$

When we say, "$\frac{2}{3}$ of the marbles are red," we mean 2 marbles out of a total of 3 are red marbles.

Part we are interested in $\rightarrow 2$ numerator
Total number in the group $\rightarrow \overline{3}$ denominator

EXAMPLE 1 Use a fraction to represent the shaded or completed part of the whole shown.

(a)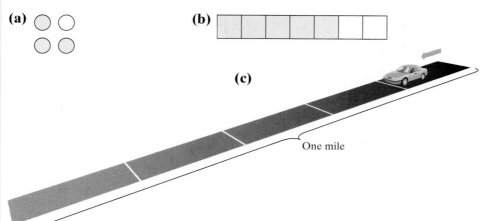

(b)

(c)

One mile

Solution

(a) Three out of four circles are shaded. The fraction is $\dfrac{3}{4}$.

(b) Five out of seven equal parts are shaded. The fraction is $\dfrac{5}{7}$.

(c) The mile is divided into five equal parts. The car has traveled 1 part out of 5 of the one-mile distance. The fraction is $\dfrac{1}{5}$.

Practice Problem 1 Use a fraction to represent the shaded part of the whole.

(a)  $\dfrac{4}{12}$ **(b)** $\dfrac{3}{6}$ **(c)** 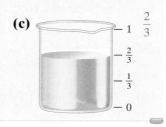 $\dfrac{2}{3}$

We can also think of a fraction as a division problem.

$$\frac{1}{3} = 1 \div 3 \qquad \text{and} \qquad 1 \div 3 = \frac{1}{3}$$

The division way of looking at fractions asks the question:

What is the result of dividing one whole into three equal parts?

Thus we can say the fraction $\frac{a}{b}$ means the same as $a \div b$. However, special care must be taken with the number 0.

Suppose that we had four equal parts and we wanted to take none of them. We would want $\frac{0}{4}$ of the parts. Since $\frac{0}{4} = 0 \div 4 = 0$, we see that $\frac{0}{4} = 0$. Any fraction with a 0 numerator equals zero.

$$\frac{0}{8} = 0 \qquad \frac{0}{5} = 0 \qquad \frac{0}{13} = 0$$

What happens when zero is in the denominator? $\frac{4}{0}$ means 4 out of 0 parts. Taking 4 out of 0 does not make sense. We say $\frac{4}{0}$ is **undefined.**

$$\frac{3}{0}, \quad \frac{7}{0}, \quad \frac{4}{0} \qquad \text{are **undefined.**}$$

We cannot have a fraction with 0 in the denominator. Since $\frac{4}{0} = 4 \div 0$, we say division by zero is *undefined.* We cannot divide by 0.

 Drawing a Sketch to Illustrate a Fraction

Drawing a sketch of a mathematical situation is a powerful problem-solving technique. The picture often reveals information not always apparent in the words.

EXAMPLE 2   Draw a sketch to illustrate.

(a) $\dfrac{7}{11}$ of an object   **(b)** $\dfrac{2}{9}$ of a group

Solution

(a) The easiest figure to draw is a rectangular bar.

We divide the bar into 11 equal parts. We then shade in 7 parts to show $\dfrac{7}{11}$.

(b) We draw 9 circles of equal size to represent a group of 9.

○○○○○○○○○

We shade in 2 of the 9 circles to show $\dfrac{2}{9}$.

Practice Problem 2   Draw a sketch to illustrate.

(a) $\dfrac{4}{5}$ of an object    **(b)** $\dfrac{3}{7}$ of a group ○○○○○○○

Recall these facts about division problems involving the number 1 and the number 0.

DIVISION INVOLVING THE NUMBER 1 AND THE NUMBER 0

1. Any nonzero number divided by itself is 1.

$$\frac{7}{7} = 1$$

2. Any number divided by 1 remains unchanged. $\dfrac{29}{1} = 29$

3. Zero may be divided by any nonzero number; the result is always zero.

$$\frac{0}{4} = 0$$

4. Division by zero is undefined. $\dfrac{3}{0}$ is undefined

③ Using Fractions to Represent Real-Life Situations

Many real-life situations can be described using fractions.

> **EXAMPLE 3**   Use a fraction to describe each situation.
>
> **(a)** A baseball player gets a hit 5 out of 12 times at bat.
> **(b)** There are 156 men and 185 women taking psychology this semester. Describe the part of the class that consists of women.
> **(c)** Robert Tobey found in the Alaska moose count that five-eighths of the moose observed were female.

Solution

(a) The baseball player got a hit $\frac{5}{12}$ of his times at bat.

(b) The total class is $156 + 185 = 341$. The fractional part that is women is 185 out of 341. Thus $\frac{185}{341}$ of the class is women.

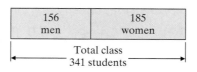

(c) Five-eighths of the moose observed were female. The fraction is $\frac{5}{8}$.

> **Practice Problem 3**   Use a fraction to describe each situation.
>
> **(a)** 9 out of the 17 players on the basketball team are on the dean's list. $\dfrac{9}{17}$
> **(b)** The senior class has 382 men and 351 women. Describe the part of the class consisting of men. $\dfrac{382}{733}$
> **(c)** John needed seven-eighths of a yard of material. $\dfrac{7}{8}$

> **EXAMPLE 4**   Wanda made 13 calls, out of which she made five sales. Albert made 17 calls, out of which he made six sales. Write a fraction that describes for both people together the number of calls in which a sale was made compared with the total number of calls.
>
> **Solution**   There are $5 + 6 = 11$ calls in which a sale was made.
>
> There were $13 + 17 = 30$ total calls.
>
> Thus $\dfrac{11}{30}$ of the calls resulted in a sale.

> **Practice Problem 4**   An inspector found that one out of seven belts was defective. She also found that two out of nine shirts were defective. Write a fraction that describes what part of all the objects examined were defective.
>
> $\dfrac{3}{16}$

Teaching Example 3 Use a fraction to describe each situation.

(a) A basketball player made seven out of twelve free-throws in the game.

(b) At the car dealership, 32 of the cars for sale are new, 25 are used. Describe the part of the cars for sale that are used.

(c) The pattern called for one-eighth of a yard of fabric.

Ans: **(a)** $\dfrac{7}{12}$ **(b)** $\dfrac{25}{57}$ **(c)** $\dfrac{1}{8}$

Teaching Example 4 13 out of 30 students in a biology class are full-time students. 15 out of 23 students in a math class are full time. Write a fraction that describes for both classes together the number of full-time students compared with the total number of students.

Ans: $\dfrac{28}{53}$

Developing Your Study Skills

Previewing New Material

Part of your study time each day should consist of looking ahead to those sections in your text that are to be covered the following day. You do not necessarily have to study and learn the material on your own, but if you survey the concepts, terminology, diagrams, and examples, the new ideas will seem more familiar to you when the instructor presents them. You can take note of concepts that appear confusing or difficult and be ready to listen carefully for your instructor's explanations. You can be prepared to ask the questions that will increase your understanding. Previewing new material enables you to see what is coming and prepares you to be ready to absorb it.

Verbal and Writing Skills

1. A __fraction__ can be used to represent part of a whole or part of a group.

2. In a fraction, the __numerator__ tells the number of parts we are interested in.

3. In a fraction, the __denominator__ tells the total number of parts in the whole or in the group.

4. Describe a real-life situation that involves fractions.

  Answers will vary. An example is: I was late 3 out of 5 times last week. I was late $\frac{3}{5}$ of the time.

Name the numerator and the denominator in each fraction.

5. $\frac{3}{5}$ N: 3 D: 5   6. $\frac{9}{11}$ N: 9 D: 11   7. $\frac{7}{8}$ N: 7 D: 8   8. $\frac{9}{10}$ N: 9 D: 10   9. $\frac{1}{17}$ N: 1 D: 17   10. $\frac{1}{15}$ N: 1 D: 15

In exercises 11–30, use a fraction to represent the shaded part of the object or the shaded portion of the set of objects.

11. $\frac{1}{3}$ 12. $\frac{1}{2}$ 13. $\frac{7}{9}$ 14. $\frac{3}{10}$

15. $\frac{3}{4}$ 16. $\frac{2}{3}$ 17. $\frac{3}{7}$ 18. $\frac{3}{8}$

19. $\frac{2}{5}$ 20. $\frac{1}{4}$ 21. $\frac{7}{10}$ 22. $\frac{4}{11}$

23. $\frac{5}{8}$ 24. $\frac{1}{8}$ 25. $\frac{4}{7}$ 26. $\frac{5}{9}$

27. $\frac{7}{8}$ 28. $\frac{7}{12}$ 29. $\frac{9}{15}$ 30. $\frac{12}{15}$

Draw a sketch to illustrate each fractional part. Object used to represent fractional parts may vary. Samples are given.

31. $\frac{1}{5}$ of an object

32. $\frac{3}{7}$ of an object

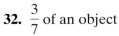

33. $\frac{3}{8}$ of an object

34. $\frac{5}{12}$ of an object

35. $\frac{7}{10}$ of an object

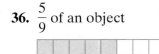

36. $\frac{5}{9}$ of an object

Applications

37. *Anthropology Class* Professor Sousa has 83 students in her anthropology lecture class. Forty-two of the students are sophomores and the others are juniors. What fraction of the class is sophomores? $\frac{42}{83}$

38. *Personal Finance* Miguel bought a notebook with a total purchase price of 98¢. Of this amount, 7¢ was sales tax. What fractional part of the total purchase price was sales tax? $\frac{7}{98}$

39. *Personal Finance* Lance bought a 100-CD jukebox for $750. Part of it was paid for with the $209 he earned parking cars for the valet service at a local wedding reception hall. What fractional part of the jukebox was paid for by his weekend earnings? $\frac{209}{750}$

40. *Personal Finance* Jillian earned $165 over the weekend at her waitressing job. She used $48 of it to repay a loan to her sister. What fractional part of her earnings did Jillian use to repay her sister? $\frac{48}{165}$

41. *Political Campaigns* The Democratic National Committee fundraising event served 122 chicken dinners and 89 roast beef dinners to its contributors. What fractional part of the guests ate roast beef? $\frac{89}{211}$

42. *Education* Bridgeton Community College has 78 full-time instructors and 31 part-time instructors. What fractional part of the faculty are part time? $\frac{31}{109}$

43. *Selling Trees* Boy Scout Troop #33 had a Christmas tree sale to raise money for a summer camping trip. In one afternoon, they sold 9 balsam firs, 12 Norwegian pines, and 5 Douglas firs. What fractional part of the trees sold were balsam firs? $\frac{9}{26}$

44. *Animal Shelters* At the local animal shelter there are 12 puppies, 25 adult dogs, 14 kittens, and 31 adult cats. What fractional part of the animals are either puppies or adult dogs? $\frac{37}{82}$

45. *Book Collection* Marie has 9 novels, 4 biographies, 12 mysteries, and 15 magazines on her bookshelf. What fractional part of the reading material is either novels or magazines? $\frac{24}{40}$

46. *Music Collection* A box of compact discs contains 5 classical CDs, 6 jazz CDs, 4 soundtracks, and 24 blues CDs. What fractional part of the total CDs is either jazz or blues? $\frac{30}{39}$

47. *Manufacturing* The West Peabody Engine Company manufactured two items last week: 101 engines and 94 lawn mowers. It was discovered that 19 engines and 3 lawn mowers were defective. Of the engines that were not defective, 40 were properly constructed but 42 were not of the highest quality. Of the lawn mowers that were not defective, 50 were properly constructed but 41 were not of the highest quality.

(a) What fractional part of all items manufactured was of the highest quality? $\frac{90}{195}$

(b) What fractional part of all items manufactured was defective? $\frac{22}{195}$

48. *Tour Bus* A Chicago tour bus held 25 women and 33 men. 12 women wore jeans. 19 men wore jeans. In the group of 25 women, a subgroup of 8 women wore sandals. In the group of 19 men, a subgroup of 10 wore sandals.

(a) What fractional part of the people on the bus wore jeans? $\frac{31}{58}$

(b) What fractional part of the women on the bus wore sandals? $\frac{8}{25}$

To Think About

49. Illustrate a real-life example of the fraction $\frac{0}{6}$.
The amount of money each of six business owners gets if the business has a profit of $0.

50. What happens when we try to illustrate a real-life example of the fraction $\frac{6}{0}$? Why?
We cannot do it. Division by zero is undefined.

Cumulative Review

51. [1.2.4] Add.
$$\begin{array}{r} 18 \\ 27 \\ 34 \\ 16 \\ 125 \\ + \ 21 \\ \hline 241 \end{array}$$

52. [1.3.3] Subtract.
$$\begin{array}{r} 56{,}203 \\ -\ 42{,}987 \\ \hline 13{,}216 \end{array}$$

53. [1.4.4] Multiply.
$$\begin{array}{r} 3178 \\ \times \ 46 \\ \hline 146{,}188 \end{array}$$

54. [1.5.3] Divide. $24\overline{)30{,}196}$ 1258 R 4

Quick Quiz 2.1

1. Use a fraction to represent the shaded part of the object. $\dfrac{4}{7}$

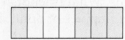

2. Silverstone Community College has 371 students taking classes on Monday night. Of those students, 204 drive a car to campus. Write a fraction that describes the part of the Monday night students who drive a car to class. $\dfrac{204}{371}$

3. At the YMCA at 10:00 P.M. last Friday 8 men were lifting weights and 5 women were lifting weights. At the same time 7 men were riding stationary bikes and 13 women were riding stationary bikes. No other people were in the gym at that time. What fractional part of the people in the gym were lifting weights? $\dfrac{13}{33}$

4. **Concept Check** One hundred twenty new businesses have opened in Springfield in the last five years. Sixty-five of them were restaurants; the remaining ones were not. Thirty new restaurants went out of business; the other new restaurants did not. Of all the new businesses that were not restaurants, 25 of them went out of business; the others did not. Explain how you can find a fraction that represents the fractional part of the new businesses that did not go out of business.

Answers may vary

Classroom Quiz 2.1 You may use these problems to quiz your students' mastery of Section 2.1.

1. Use a fraction to represent the shaded part of the object.

Ans: $\dfrac{5}{8}$

2. Westwind Bank issued 388 home mortgages last month. A total of 213 of them were for fixed-rate mortgages. What fractional part of the mortgages were for fixed-rate mortgages? **Ans:** $\dfrac{213}{388}$

3. Dr. Davidson found that all of the members of his math class drove to class each day. He learned that 3 of his students drove motorcycles, 5 of them drove trucks, 10 of them drove SUVs, and 17 of them drove cars. What fractional part of the students in the class did not drive motorcycles?

Ans: $\dfrac{32}{35}$

Writing a Number as a Product of Prime Factors

A **prime number** is a whole number greater than 1 that cannot be evenly divided except by itself and 1. If you examine all the whole numbers from 1 to 50, you will find 15 prime numbers.

THE FIRST 15 PRIME NUMBERS

2, 3, 5, 7, 11, 13, 17, 19, 23, 29, 31, 37, 41, 43, 47

A **composite number** is a whole number greater than 1 that can be divided by whole numbers other than 1 and itself. The number 12 is a composite number.

$$12 = 2 \times 6 \qquad \text{and} \qquad 12 = 3 \times 4$$

The number 1 is neither a prime nor a composite number. The number 0 is neither a prime nor a composite number.

Recall that factors are numbers that are multiplied together. Prime factors are prime numbers. To check to see if a number is prime or composite, simply divide the smaller primes (such as 2, 3, 5, 7, 11, . . .) into the given number. If the number can be divided exactly without a remainder by one of the smaller primes, it is a composite and not a prime.

Some students find the following rules helpful when deciding if a number can be divided by 2, 3, or 5.

DIVISIBILITY TESTS

1. A number is divisible by 2 if the last digit is 0, 2, 4, 6, or 8.

2. A number is divisible by 3 if the sum of the digits is divisible by 3.

3. A number is divisible by 5 if the last digit is 0 or 5.

To illustrate:

1. 478 is divisible by 2 since it ends in 8.

2. 531 is divisible by 3 since when we add the digits of 531 $(5 + 3 + 1)$ we get 9, which is divisible by 3.

3. 985 is divisible by 5 since it ends in 5.

EXAMPLE 1 Write each whole number as the product of prime factors.

(a) 12 **(b)** 60 **(c)** 168

Solution

(a) To start, write 12 as the product of any two factors. We will write 12 as 4×3.

$$12 = \quad 4 \quad \times 3 \qquad \text{Now check whether the factors are prime. If not, factor these.}$$
$$\quad\quad\quad 2 \times 2 \times 3$$
$$12 = 2 \times 2 \times 3 \qquad \text{Now all factors are prime, so 12 is completely factored.}$$

Instead of writing $2 \times 2 \times 3$, we can write $2^2 \times 3$.

Note: To start, we could write 12 as 2×6. Begin this way and follow the preceding steps. Is the product of prime factors the same? Will this always be true?

(b) We follow the same steps as in (a).

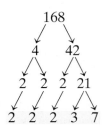

$$60 = 6 \times 10$$

$$3 \times 2 \times 2 \times 5 \quad \text{Check that all factors are prime.}$$

$$60 = 2 \times 2 \times 3 \times 5$$

Instead of writing $2 \times 2 \times 3 \times 5$, we can write $2^2 \times 3 \times 5$.

Note that in the final answer the prime factors are listed in order from least to greatest.

(c) Some students like to use a **factor tree** to help write a number as a product of prime factors as illustrated below.

$$168$$

$$4 \qquad 42$$

$$2 \quad 2 \quad 2 \quad 21$$

$$2 \quad 2 \quad 2 \quad 3 \quad 7$$

$$168 = 2 \times 2 \times 2 \times 3 \times 7$$

$$\text{or} \quad 168 = 2^3 \times 3 \times 7$$

NOTE TO STUDENT: *Fully worked-out solutions to all of the Practice Problems can be found at the back of the text starting at page SP-1*

Practice Problem 1 Write each whole number as a product of primes.

(a) 18 2×3^2 **(b)** 72 $2^3 \times 3^2$ **(c)** 400 $2^4 \times 5^2$

Suppose we started Example 1(c) by writing $168 = 14 \times 12$. Would we get the same answer? Would our answer be correct? Let's compare.

Again we will use a factor tree.

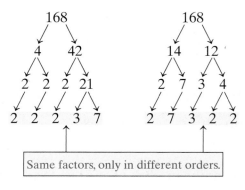

Same factors, only in different orders.

$$\text{Thus} \quad 168 = 2 \times 2 \times 2 \times 3 \times 7$$

$$\text{or} \qquad \quad = 2^3 \times 3 \times 7.$$

The order of prime factors is not important because multiplication is commutative. No matter how we start, when we factor a composite number, we always get exactly the same prime factors.

> **THE FUNDAMENTAL THEOREM OF ARITHMETIC**
> Every composite number can be written in exactly one way as a product of prime numbers.

We have seen this in our Solution to Example 1(c).

You will be able to check this theorem again in Exercises 2.2, exercises 7–26. Writing a number as a product of prime factors is also called **prime factorization.**

Reducing a Fraction to Lowest Terms

You know that $5 + 2$ and $3 + 4$ are two ways to write the same number. We say they are *equivalent* because they are *equal* to the same *value.* They are both ways of writing the value 7.

Like whole numbers, fractions can be written in more than one way. For example, $\frac{2}{4}$ and $\frac{1}{2}$ are two ways to write the same number. The value of the fractions is the same. When we use fractions, we often need to write them in another form. If we make the numerator and denominator smaller, we *simplify* the fractions.

Compare the two fractions in the drawings on the right. In each picture the shaded part is the same size. The fractions $\frac{3}{4}$ and $\frac{6}{8}$ are called **equivalent fractions.** The fraction $\frac{3}{4}$ is in **simplest form.** To see how we can change $\frac{6}{8}$ to $\frac{3}{4}$, we look at a property of the number 1.

> Any nonzero number divided by itself is 1.

$$\frac{5}{5} = \frac{17}{17} = \frac{c}{c} = 1$$

Thus, if we multiply a fraction by $\frac{5}{5}$ or $\frac{17}{17}$ or $\frac{c}{c}$ (remember, c cannot be zero), the value of the fraction is unchanged because we are multiplying by a form of 1. We can use this rule to show that $\frac{3}{4}$ and $\frac{6}{8}$ are equivalent.

$$\frac{3}{4} \times \frac{2}{2} = \frac{6}{8}$$

In general, if b and c are not zero,

$$\frac{a}{b} = \frac{a \times c}{b \times c}$$

To reduce a fraction, we find a **common factor** in the numerator and in the denominator and divide it out. In the fraction $\frac{6}{8}$, the common factor is 2.

$$\frac{6}{8} = \frac{3 \times \overset{1}{\cancel{2}}}{4 \times \underset{1}{\cancel{2}}} = \frac{3}{4}$$

$$\frac{6}{8} = \frac{3}{4}$$

For all fractions (where a, b, and c are not zero), if c is a common factor,

$$\frac{a}{b} = \frac{a \div c}{b \div c}$$

A fraction is called **simplified, reduced,** or **in lowest terms** if the numerator and the denominator have only 1 *as a common factor.*

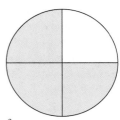

$\frac{3}{4}$ of the circle is shaded.

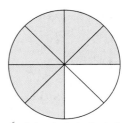

$\frac{6}{8}$ of the circle is shaded.

Teaching Example 2 Simplify (write in lowest terms).

(a) $\dfrac{18}{27}$ **(b)** $\dfrac{48}{56}$

Ans: (a) $\dfrac{2}{3}$ **(b)** $\dfrac{6}{7}$

EXAMPLE 2 Simplify (write in lowest terms).

(a) $\dfrac{15}{25}$ **(b)** $\dfrac{42}{56}$

Solution

(a) $\dfrac{15}{25} = \dfrac{15 \div 5}{25 \div 5} = \dfrac{3}{5}$ The greatest common factor is 5. Divide the numerator and the denominator by 5.

(b) $\dfrac{42}{56} = \dfrac{42 \div 14}{56 \div 14} = \dfrac{3}{4}$ The greatest common factor is 14. Divide the numerator and the denominator by 14.

Perhaps 14 was not the first common factor you thought of. Perhaps you did see the common factor 2. Divide out 2. Then look for another common factor, 7. Now divide out 7.

$$\dfrac{42}{56} = \dfrac{42 \div 2}{56 \div 2} = \dfrac{21}{28} = \dfrac{21 \div 7}{21 \div 7} = \dfrac{3}{4}$$

If we do not see large factors at first, sometimes we can simplify a fraction by dividing both numerator and denominator by a smaller common factor several times, until no common factors are left.

NOTE TO STUDENT: Fully worked-out solutions to all of the Practice Problems can be found at the back of the text starting at page SP-1

Practice Problem 2 Simplify by dividing out common factors.

(a) $\dfrac{30}{42}$ $\quad\dfrac{5}{7}$ **(b)** $\dfrac{60}{132}$ $\quad\dfrac{5}{11}$

A second method to reduce or simplify fractions is called the *method of prime factors*. We factor the numerator and the denominator into prime numbers. We then divide the numerator and the denominator by any common prime factors.

Teaching Example 3 Simplify the fractions by the method of prime factors.

(a) $\dfrac{40}{48}$ **(b)** $\dfrac{60}{102}$

Ans: (a) $\dfrac{5}{6}$ **(b)** $\dfrac{10}{17}$

EXAMPLE 3 Simplify the fractions by the method of prime factors.

(a) $\dfrac{35}{42}$ **(b)** $\dfrac{22}{110}$

Solution

(a) $\dfrac{35}{42} = \dfrac{5 \times 7}{2 \times 3 \times 7}$ We factor 35 and 42 into prime factors. The common prime factor is 7.

$\qquad = \dfrac{5 \times \overset{1}{\cancel{7}}}{2 \times 3 \times \underset{1}{\cancel{7}}}$ Now we divide out 7.

$\qquad = \dfrac{5 \times 1}{2 \times 3 \times 1} = \dfrac{5}{6}$ We multiply the factors in the numerator and denominator to write the reduced or simplified form.

Thus $\dfrac{35}{42} = \dfrac{5}{6}$, and $\dfrac{5}{6}$ is the simplified form.

(b) $\dfrac{22}{110} = \dfrac{2 \times 11}{2 \times 5 \times 11} = \dfrac{\overset{1}{\cancel{2}} \times \overset{1}{\cancel{11}}}{\underset{1}{\cancel{2}} \times 5 \times \underset{1}{\cancel{11}}} = \dfrac{1}{5}$

Practice Problem 3 Simplify the fractions by the method of prime factors.

(a) $\dfrac{120}{135}$ $\dfrac{8}{9}$

(b) $\dfrac{715}{880}$ $\dfrac{13}{16}$

NOTE TO STUDENT: *Fully worked-out solutions to all of the Practice Problems can be found at the back of the text starting at page SP-1*

③ Determining Whether Two Fractions Are Equal

After we simplify, how can we check that a reduced fraction is *equivalent* to the original fraction? If two fractions are equal, their diagonal products or **cross products** are equal. This is called the **equality test for fractions.** If $\frac{3}{4} = \frac{6}{8}$, then

$$\underset{4 \quad 8}{\overset{3 \quad 6}{\diagdown\!\!\!\!\diagup}} \longrightarrow \begin{array}{l} 4 \times 6 = 24 \\ 3 \times 8 = 24 \end{array} \longleftarrow \boxed{\text{Products are equal.}}$$

If two fractions are unequal (we use the symbol $\neq$), their *cross* products are unequal. If $\dfrac{5}{6} \neq \dfrac{6}{7}$, then

$$\underset{6 \quad 7}{\overset{5 \quad 6}{\diagdown\!\!\!\!\diagup}} \longrightarrow \begin{array}{l} 6 \times 6 = 36 \\ 5 \times 7 = 35 \end{array} \longleftarrow \boxed{\text{Products are not equal.}}$$

Since $36 \neq 35$, we know that $\dfrac{5}{6} \neq \dfrac{6}{7}$. The test can be described in this way.

EQUALITY TEST FOR FRACTIONS

For any two fractions where a, b, c, and d are whole numbers and $b \neq 0, d \neq 0$, if $\dfrac{a}{b} = \dfrac{c}{d}$, then $a \times d = b \times c$.

EXAMPLE 4 Are these fractions equal? Use the equality test.

(a) $\dfrac{2}{11} \overset{?}{=} \dfrac{18}{99}$

(b) $\dfrac{3}{16} \overset{?}{=} \dfrac{12}{62}$

Solution

(a)
$$\underset{11 \quad 99}{\overset{2 \quad 18}{\diagdown\!\!\!\!\diagup}} \longrightarrow \begin{array}{l} 11 \times 18 = 198 \\ 2 \times 99 = 198 \end{array} \longleftarrow \boxed{\text{Products are equal.}}$$

Since $198 = 198$, we know that $\dfrac{2}{11} = \dfrac{18}{99}$.

(b)
$$\underset{16 \quad 62}{\overset{3 \quad 12}{\diagdown\!\!\!\!\diagup}} \longrightarrow \begin{array}{l} 16 \times 12 = 192 \\ 3 \times 62 = 186 \end{array} \longleftarrow \boxed{\text{Products are not equal.}}$$

Since $192 \neq 186$, we know that $\dfrac{3}{16} \neq \dfrac{12}{62}$.

Teaching Example 4 Are these fractions equal? Use the equality test.

(a) $\dfrac{15}{35} = \dfrac{3}{7}$

(b) $\dfrac{7}{12} = \dfrac{86}{144}$

Ans: (a) $\dfrac{15}{35} = \dfrac{3}{7}$ (b) $\dfrac{7}{12} \neq \dfrac{86}{144}$

Practice Problem 4 Test whether the following fractions are equal.

(a) $\dfrac{84}{108} \overset{?}{=} \dfrac{7}{9}$ $\dfrac{84}{108} = \dfrac{7}{9}$

(b) $\dfrac{3}{7} \overset{?}{=} \dfrac{79}{182}$ $\dfrac{3}{7} \neq \dfrac{79}{182}$

Verbal and Writing Skills

1. Which of these whole numbers are prime?
4, 12, 11, 15, 6, 19, 1, 41, 38, 24, 5, 46   11, 19, 41, 5

2. A prime number is a whole number greater than 1 that cannot be evenly __divided__ except by itself and 1.

3. A __composite__ __number__ is a whole number greater than 1 that can be divided by whole numbers other than itself and 1.

4. Every composite number can be written in exactly one way as a __product__ of __prime__ numbers.

5. Give an example of a composite number written as a product of primes. $56 = 2 \times 2 \times 2 \times 7$

6. Give an example of equivalent (equal) fractions. $\frac{23}{135} = \frac{46}{270}$

Write each number as a product of prime factors.

7. 15 3×5 　　8. 9 3^2 　　9. 35 5×7 　　10. 8 2^3 　　11. 49 7^2

12. 30 $2 \times 3 \times 5$ 　　13. 16 2^4 　　14. 81 3^4 　　15. 55 5×11 　　16. 42 $2 \times 3 \times 7$

17. 63 $3^2 \times 7$ 　　18. 48 $2^4 \times 3$ 　　19. 84 $2^2 \times 3 \times 7$ 　　20. 125 5^3 　　21. 54 2×3^3

22. 99 $3^2 \times 11$ 　　23. 120 $2^3 \times 3 \times 5$ 　　24. 135 $3^3 \times 5$ 　　25. 184 $2^3 \times 23$ 　　26. 216 $2^3 \times 3^3$

Determine which of these whole numbers are prime. If a number is composite, write it as the product of prime factors.

27. 47 prime 　　28. 31 prime 　　29. 57 3×19 　　30. 51 3×17

31. 67 prime 　　32. 71 prime 　　33. 62 2×31 　　34. 91 7×13

35. 89 prime 　　36. 97 prime 　　37. 127 prime 　　38. 119 7×17

39. 121 11×11 　　40. 95 5×19 　　41. 129 3×43 　　42. 143 11×13

Reduce each fraction by finding a common factor in the numerator and in the denominator and dividing by the common factor.

43. $\frac{18}{27}$ $\frac{18 \div 9}{27 \div 9} = \frac{2}{3}$ 　　44. $\frac{16}{24}$ $\frac{16 \div 8}{24 \div 8} = \frac{2}{3}$ 　　45. $\frac{36}{48}$ $\frac{36 \div 12}{48 \div 12} = \frac{3}{4}$ 　　46. $\frac{28}{49}$ $\frac{28 \div 7}{49 \div 7} = \frac{4}{7}$

47. $\frac{63}{90}$ $\frac{63 \div 9}{90 \div 9} = \frac{7}{10}$ 　　48. $\frac{45}{75}$ $\frac{45 \div 15}{75 \div 15} = \frac{3}{5}$ 　　49. $\frac{210}{310}$ $\frac{210 \div 10}{310 \div 10} = \frac{21}{31}$ 　　50. $\frac{110}{140}$ $\frac{110 \div 10}{140 \div 10} = \frac{11}{14}$

Reduce each fraction by the method of prime factors.

51. $\frac{3}{15}$ $\frac{3 \times 1}{3 \times 5} = \frac{1}{5}$ 　　52. $\frac{7}{21}$ $\frac{7 \times 1}{7 \times 3} = \frac{1}{3}$ 　　53. $\frac{66}{88}$ $\frac{2 \times 3 \times 11}{2 \times 2 \times 2 \times 11} = \frac{3}{4}$ 　　54. $\frac{42}{56}$ $\frac{2 \times 3 \times 7}{2 \times 2 \times 2 \times 7} = \frac{3}{4}$

55. $\frac{30}{45}$ $\frac{2 \times 3 \times 5}{3 \times 3 \times 5} = \frac{2}{3}$ 　　56. $\frac{65}{91}$ $\frac{5 \times 13}{7 \times 13} = \frac{5}{7}$ 　　57. $\frac{60}{75}$ $\frac{2 \times 2 \times 3 \times 5}{3 \times 5 \times 5} = \frac{4}{5}$ 　　58. $\frac{42}{70}$ $\frac{2 \times 3 \times 7}{2 \times 5 \times 7} = \frac{3}{5}$

Mixed Practice   *Reduce each fraction by any method.*

59. $\dfrac{33}{36}$

$\dfrac{3 \times 11}{3 \times 12} = \dfrac{11}{12}$

60. $\dfrac{40}{96}$

$\dfrac{8 \times 5}{8 \times 12} = \dfrac{5}{12}$

61. $\dfrac{63}{108}$

$\dfrac{9 \times 7}{9 \times 12} = \dfrac{7}{12}$

62. $\dfrac{72}{132}$

$\dfrac{6 \times 12}{11 \times 12} = \dfrac{6}{11}$

63. $\dfrac{88}{121}$

$\dfrac{11 \times 8}{11 \times 11} = \dfrac{8}{11}$

64. $\dfrac{125}{200}$

$\dfrac{25 \times 5}{25 \times 8} = \dfrac{5}{8}$

65. $\dfrac{120}{200}$

$\dfrac{40 \times 3}{40 \times 5} = \dfrac{3}{5}$

66. $\dfrac{200}{300}$

$\dfrac{2 \times 100}{3 \times 100} = \dfrac{2}{3}$

67. $\dfrac{220}{260}$

$\dfrac{11 \times 20}{13 \times 20} = \dfrac{11}{13}$

68. $\dfrac{210}{390}$

$\dfrac{30 \times 7}{30 \times 13} = \dfrac{7}{13}$

Are these fractions equal? Why or why not?

69. $\dfrac{4}{16} \overset{?}{=} \dfrac{7}{28}$

$4 \times 28 \overset{?}{=} 16 \times 7$
$112 = 112$
yes

70. $\dfrac{10}{65} \overset{?}{=} \dfrac{2}{13}$

$10 \times 13 \overset{?}{=} 65 \times 2$
$130 = 130$
yes

71. $\dfrac{12}{40} \overset{?}{=} \dfrac{3}{13}$

$12 \times 13 \overset{?}{=} 40 \times 3$
$156 \neq 120$
no

72. $\dfrac{24}{72} \overset{?}{=} \dfrac{15}{45}$

$24 \times 45 \overset{?}{=} 72 \times 15$
$1080 = 1080$
yes

73. $\dfrac{23}{27} \overset{?}{=} \dfrac{92}{107}$

$23 \times 107 \overset{?}{=} 27 \times 92$
$2461 \neq 2484$
no

74. $\dfrac{70}{120} \overset{?}{=} \dfrac{41}{73}$

$70 \times 73 \overset{?}{=} 120 \times 41$
$5110 \neq 4920$
no

75. $\dfrac{27}{57} \overset{?}{=} \dfrac{45}{95}$

$27 \times 95 \overset{?}{=} 57 \times 45$
$2565 = 2565$
yes

76. $\dfrac{18}{24} \overset{?}{=} \dfrac{23}{28}$

$18 \times 28 \overset{?}{=} 24 \times 23$
$504 \neq 552$
no

77. $\dfrac{60}{95} \overset{?}{=} \dfrac{12}{19}$

$60 \times 19 \overset{?}{=} 95 \times 12$
$1140 = 1140$
yes

78. $\dfrac{21}{27} \overset{?}{=} \dfrac{112}{144}$

$21 \times 144 \overset{?}{=} 27 \times 112$
$3024 = 3024$
yes

Applications   *Reduce the fractions in your answers.*

79. *Pizza Delivery* Pizza Palace made 128 deliveries on Saturday night. The manager found that 32 of the deliveries were of more than one pizza. He wanted to study the deliveries that consisted of just one pizza. What fractional part of the deliveries were of just one pizza? $\dfrac{3}{4}$

80. *Medical Students* Medical students frequently work long hours. Susan worked a 16-hour shift, spending 12 hours in the emergency room and 4 hours in surgery. What fractional part of her shift was she in the emergency room? What fractional part of her shift was she in surgery?

$\dfrac{3}{4}$ of her shift in the emergency room

$\dfrac{1}{4}$ of her shift in surgery

81. *Teaching* Professor Nguyen found that 12 out of 96 students in his Aspects of Chemistry course failed the first exam. What fractional part of the class failed the exam? What fractional part of the class passed?

$\dfrac{1}{8}$ of the class failed; $\dfrac{7}{8}$ of the class passed

82. *Wireless Communications* William works for a wireless communications company that makes beepers and mobile phones. He inspected 315 beepers and found that 20 were defective. What fractional part of the beepers were not defective? $\dfrac{59}{63}$

83. *Personal Finance* Amelia earned $8400 during her summer vacation. She saved $6000 of her earnings for a trip to New Zealand. What fractional part of her earnings did she save for her trip?
$\frac{5}{7}$

84. *Real Estate* Monique's sister and her husband have been working two jobs each to put a down payment on a plot of land where they plan to build their house. The purchase price is $42,500. They have saved $5500. What fractional part of the cost of the land have they saved?
$\frac{11}{85}$

Education *The following data was compiled on the students attending day classes at North Shore Community College.*

Number of Students	Daily Distance Traveled from Home to College (miles)	Length of Commute
1100	0–6	Very short
1700	7–12	Short
900	13–18	Medium
500	19–24	Long
300	More than 24	Very long

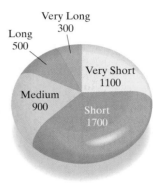

The number of students with each type of commute is displayed in the circle graph to the right.

Answer exercises 85–88 based on the preceding data. Reduce all fractions in your answers.

85. What fractional part of the student body has a short daily commute to the college? $\frac{17}{45}$

86. What fractional part of the student body has a medium daily commute to the college? $\frac{1}{5}$

87. What fractional part of the student body has a long or very long daily commute to the college? $\frac{8}{45}$

88. What fractional part of the student body has a daily commute to the college that is considered less than long? $\frac{37}{45}$

Cumulative Review

89. **[1.4.4]** Multiply. 386×425
164,050

90. **[1.5.3]** Divide. $15,552 \div 12$
1296

91. **[1.4.3]** Multiply. 3200×300
960,000

92. **[1.3.5]** ***Charities*** In 2005 the total income of the Salvation Army was $4,559,200,000. That same year the total income of the YMCA of the USA was $5,130,800,000. How much greater was the income of the YMCA than the Salvation Army? (*Source: The Christian Science Monitor* at www.csmonitor.com)
$571,600,000

Quick Quiz 2.2 Reduce each fraction.

1. $\frac{25}{35}$ $\frac{5}{7}$

2. $\frac{14}{84}$ $\frac{1}{6}$

3. $\frac{105}{40}$ $\frac{21}{8}$

4. Concept Check Explain how you would determine if the fraction $\frac{195}{231}$ can be reduced. Answers may vary

Classroom Quiz 2.2 You may use these problems to quiz your students' mastery of Section 2.2.
Reduce each fraction.

1. $\frac{77}{121}$ **Ans:** $\frac{7}{11}$

2. $\frac{42}{96}$ **Ans:** $\frac{7}{16}$

3. $\frac{135}{60}$ **Ans:** $\frac{9}{4}$

2.3 CONVERTING BETWEEN IMPROPER FRACTIONS AND MIXED NUMBERS

① Changing a Mixed Number to an Improper Fraction

We have names for different kinds of fractions. If the value of a fraction is less than 1, we say the fraction is proper.

$$\frac{3}{5}, \frac{5}{7}, \frac{1}{8} \quad \text{are called \textbf{proper fractions}.}$$

Notice that the numerator is less than the denominator. If the numerator is less than the denominator, the fraction is a proper fraction.

If the value of a fraction is greater than or equal to 1, the quantity can be written as an improper fraction or as a mixed number.

Suppose that we have 1 whole pizza and $\frac{1}{6}$ of a pizza. We could write this as $1\frac{1}{6}$. $1\frac{1}{6}$ is called a mixed number. A **mixed number** is the sum of a whole number greater than zero and a proper fraction. The notation $1\frac{1}{6}$ actually means $1 + \frac{1}{6}$. The plus sign is not usually shown.

Another way of writing $1\frac{1}{6}$ pizza is to write $\frac{7}{6}$ pizza. $\frac{7}{6}$ is called an improper fraction. Notice that the numerator is greater than the denominator. If the numerator is greater than or equal to the denominator, the fraction is an improper fraction.

$$\frac{7}{6}, \frac{6}{6}, \frac{5}{4}, \frac{8}{3}, \frac{2}{2} \quad \text{are \textbf{improper fractions}.}$$

The following chart will help you visualize these different fractions and their names.

Because in some cases improper fractions are easier to add, subtract, multiply, and divide than mixed numbers, we often change mixed numbers to improper fractions when we perform calculations with them.

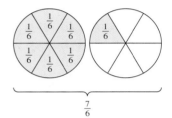

Value Less Than 1	Value Equal To 1	Value Greater Than 1	
Proper Fraction	Improper Fraction	Improper Fraction	or Mixed Number
$\frac{3}{4}$	$\frac{4}{4}$		$\frac{5}{4}$ or $1\frac{1}{4}$
$\frac{7}{8}$	$\frac{8}{8}$		$\frac{17}{8}$ or $2\frac{1}{8}$
$\frac{3}{100}$	$\frac{100}{100}$		$\frac{109}{100}$ or $1\frac{9}{100}$

Teaching Tip Some students have been "drilled" in elementary school that improper fractions are "wrong" since they are not simplified. You may need to explain that there is nothing "improper" or incorrect about improper fractions. Some mathematical problems are best left with improper fractions as answers. In mathematics we will find several places where improper fractions are needed and are appropriate.

CHANGING A MIXED NUMBER TO AN IMPROPER FRACTION

1. Multiply the whole number by the denominator of the fraction.
2. Add the numerator of the fraction to the product found in step 1.
3. Write the sum found in step 2 over the denominator of the fraction.

EXAMPLE 1   Change each mixed number to an improper fraction.

(a) $3\frac{2}{5}$ **(b)** $5\frac{4}{9}$ **(c)** $18\frac{3}{5}$

Solution

Multiply the whole number by the denominator.

Add the numerator to the product.

(a) $3\frac{2}{5} = \frac{3 \times 5 + 2}{5} = \frac{15 + 2}{5} = \frac{17}{5}$ ← Write the sum over the denominator.

(b) $5\frac{4}{9} = \frac{5 \times 9 + 4}{9} = \frac{45 + 4}{9} = \frac{49}{9}$

(c) $18\frac{3}{5} = \frac{18 \times 5 + 3}{5} = \frac{90 + 3}{5} = \frac{93}{5}$

Teaching Example 1 Change each mixed number to an improper fraction.

(a) $2\frac{7}{8}$ **(b)** $21\frac{1}{2}$ **(c)** $3\frac{5}{11}$

Ans: (a) $\frac{23}{8}$ **(b)** $\frac{43}{2}$ **(c)** $\frac{38}{11}$

NOTE TO STUDENT: Fully worked-out solutions to all of the Practice Problems can be found at the back of the text starting at page SP-1

Practice Problem 1   Change the mixed numbers to improper fractions.

(a) $4\frac{3}{7}$ $\frac{31}{7}$ **(b)** $6\frac{2}{3}$ $\frac{20}{3}$ **(c)** $19\frac{4}{7}$ $\frac{137}{7}$

② Changing an Improper Fraction to a Mixed Number

We often need to change an improper fraction to a mixed number.

CHANGING AN IMPROPER FRACTION TO A MIXED NUMBER

1. Divide the numerator by the denominator.
2. Write the quotient followed by the fraction with the remainder over the denominator.

$$\text{quotient} \frac{\text{remainder}}{\text{denominator}}$$

Teaching Example 2 Write each improper fraction as a mixed number.

(a) $\frac{11}{2}$ **(b)** $\frac{27}{5}$ **(c)** $\frac{89}{21}$ **(d)** $\frac{102}{17}$

Ans: (a) $5\frac{1}{2}$ **(b)** $5\frac{2}{5}$ **(c)** $4\frac{5}{21}$ **(d)** 6

EXAMPLE 2   Write each improper fraction as a mixed number.

(a) $\frac{13}{5}$ **(b)** $\frac{29}{7}$ **(c)** $\frac{105}{31}$ **(d)** $\frac{85}{17}$

Solution

(a) We divide the denominator 5 into 13.

$$5\overline{)13} \quad \longleftarrow \text{ quotient}$$
$$\begin{array}{r} 2 \\ 5\overline{)13} \\ \underline{10} \\ 3 \end{array} \quad \longleftarrow \text{ remainder}$$

The answer is in the form quotient $\dfrac{\text{remainder}}{\text{denominator}}$.

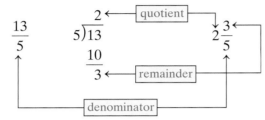

Thus $\dfrac{13}{5} = 2\dfrac{3}{5}$.

(b) $\begin{array}{r} 4 \\ 7\overline{)29} \\ \underline{28} \\ 1 \end{array}$ $\dfrac{29}{7} = 4\dfrac{1}{7}$ **(c)** $\begin{array}{r} 3 \\ 31\overline{)105} \\ \underline{93} \\ 12 \end{array}$ $\dfrac{105}{31} = 3\dfrac{12}{31}$

(d) $\begin{array}{r} 5 \\ 17\overline{)85} \\ \underline{85} \\ 0 \end{array}$ The remainder is 0, so $\dfrac{85}{17} = 5$, a whole number.

Practice Problem 2 Write as a mixed number or a whole number.

(a) $\dfrac{17}{4}$ $4\dfrac{1}{4}$ **(b)** $\dfrac{36}{5}$ $7\dfrac{1}{5}$ **(c)** $\dfrac{116}{27}$ $4\dfrac{8}{27}$ **(d)** $\dfrac{91}{13}$ 7

NOTE TO STUDENT: Fully worked-out solutions to all of the Practice Problems can be found at the back of the text starting at page SP-1

③ Reducing a Mixed Number or an Improper Fraction to Lowest Terms

Mixed numbers and improper fractions may need to be reduced if they are not in simplest form. Recall that we write the fraction in terms of prime factors. Then we look for common factors in the numerator and the denominator of the fraction. Then we divide the numerator and the denominator by the common factor.

EXAMPLE 3 Reduce the improper fraction. $\dfrac{22}{8}$

Solution

$$\frac{22}{8} = \frac{\overset{1}{\cancel{2}} \times 11}{\underset{1}{\cancel{2}} \times 2 \times 2} = \frac{11}{4}$$

Teaching Example 3 Reduce the improper fraction. $\frac{42}{8}$

Ans: $\frac{21}{4}$

Practice Problem 3 Reduce the improper fraction.

$$\frac{51}{15} \quad \frac{17}{5}$$

Teaching Example 4 Reduce the mixed number. $6\frac{18}{36}$
Ans: $6\frac{1}{2}$

EXAMPLE 4   Reduce the mixed number. $4\frac{21}{28}$

Solution   We cannot reduce the whole number 4, only the fraction $\frac{21}{28}$.

$$\frac{21}{28} = \frac{3 \times \overset{1}{\cancel{7}}}{4 \times \underset{1}{\cancel{7}}} = \frac{3}{4}$$

Therefore, $4\frac{21}{28} = 4\frac{3}{4}$.

Practice Problem 4   Reduce the mixed number.

$$3\frac{16}{80} \quad 3\frac{1}{5}$$

If an improper fraction contains a very large numerator and denominator, it is best to change the fraction to a mixed number before reducing.

Teaching Example 5 Reduce. $\frac{876}{228}$
Ans: $3\frac{16}{19}$

EXAMPLE 5   Reduce $\frac{945}{567}$ by first changing to a mixed number.

Solution

$$\begin{array}{r} 1 \\ 567\overline{)945} \\ 567 \\ \hline 378 \end{array} \quad \text{so} \quad \frac{945}{567} = 1\frac{378}{567}$$

To reduce the fraction we write

$$\frac{378}{567} = \frac{2 \times 3 \times 3 \times 3 \times 7}{3 \times 3 \times 3 \times 3 \times 7} = \frac{2 \times \overset{1}{\cancel{3}} \times \overset{1}{\cancel{3}} \times \overset{1}{\cancel{3}} \times \overset{1}{\cancel{7}}}{3 \times \underset{1}{\cancel{3}} \times \underset{1}{\cancel{3}} \times \underset{1}{\cancel{3}} \times \underset{1}{\cancel{7}}} = \frac{2}{3}$$

So $\frac{945}{567} = 1\frac{378}{567} = 1\frac{2}{3}$.

Problems like Example 5 can be done in several different ways. It is not necessary to follow these exact steps when reducing this fraction.

Practice Problem 5   Reduce $\frac{1001}{572}$ by first changing to a mixed number.

$1\frac{3}{4}$

Teaching Tip After explaining Example 5 or some similar problem, have the group do the following problem as a class activity. Have half the class reduce the fraction $\frac{663}{255}$ using this method of changing to a mixed number first. Have the remainder of the class reduce the fraction as an improper fraction. The answer is $2\frac{3}{5}$. Record which half of the class scored the greatest number of correct answers first. (As long as you divide the class into two groups of roughly equal ability, the improper fraction method will usually win.)

TO THINK ABOUT: When a Denominator Is Prime A student concluded that just by looking at the denominator he could tell that the fraction $\frac{1655}{97}$ cannot be reduced unless $1655 \div 97$ is a whole number. How did he come to that conclusion?

Note that 97 is a prime number. The only factors of 97 are 97 and 1. Therefore, *any* fraction with 97 in the denominator can be reduced only if 97 is a factor of the numerator. Since $1655 \div 97$ is not a whole number (see the following division), it is therefore impossible to reduce $\frac{1655}{97}$.

$$\begin{array}{r} 17 \\ 97\overline{)1655} \\ 97 \\ \hline 685 \\ 679 \\ \hline 6 \end{array}$$

You may explore this idea in Exercises 2.3, exercises 83 and 84.

PRACTICE   WATCH   DOWNLOAD   READ   REVIEW

Verbal and Writing Skills

1. Describe in your own words how to change a mixed number to an improper fraction.
   (a) Multiply the whole number by the denominator of the fraction.
   (b) Add the numerator of the fraction to the product formed in step (a).
   (c) Write the sum found in step (b) over the denominator of the fraction.

2. Describe in your own words how to change an improper fraction to a mixed number.
   (a) Divide the numerator by the denominator.
   (b) Write the quotient followed by the fraction with the remainder over the denominator.

Change each mixed number to an improper fraction.

3. $2\frac{1}{3}$ $\frac{7}{3}$
4. $2\frac{3}{4}$ $\frac{11}{4}$
5. $2\frac{3}{7}$ $\frac{17}{7}$
6. $3\frac{3}{8}$ $\frac{27}{8}$
7. $9\frac{2}{9}$ $\frac{83}{9}$
8. $8\frac{3}{8}$ $\frac{67}{8}$

9. $10\frac{2}{3}$ $\frac{32}{3}$
10. $15\frac{3}{4}$ $\frac{63}{4}$
11. $11\frac{3}{5}$ $\frac{58}{5}$
12. $15\frac{4}{5}$ $\frac{79}{5}$
13. $9\frac{1}{6}$ $\frac{55}{6}$
14. $41\frac{1}{2}$ $\frac{83}{2}$

15. $20\frac{1}{6}$ $\frac{121}{6}$
16. $6\frac{6}{7}$ $\frac{48}{7}$
17. $10\frac{11}{12}$ $\frac{131}{12}$
18. $13\frac{5}{7}$ $\frac{96}{7}$
19. $7\frac{9}{10}$ $\frac{79}{10}$
20. $4\frac{1}{50}$ $\frac{201}{50}$

21. $8\frac{1}{25}$ $\frac{201}{25}$
22. $12\frac{5}{6}$ $\frac{77}{6}$
23. $5\frac{5}{12}$ $\frac{65}{12}$
24. $207\frac{2}{3}$ $\frac{623}{3}$
25. $164\frac{2}{3}$ $\frac{494}{3}$
26. $33\frac{1}{3}$ $\frac{100}{3}$

27. $8\frac{11}{15}$ $\frac{131}{15}$
28. $5\frac{19}{20}$ $\frac{119}{20}$
29. $4\frac{13}{25}$ $\frac{113}{25}$
30. $5\frac{17}{20}$ $\frac{117}{20}$

Change each improper fraction to a mixed number or a whole number.

31. $\frac{4}{3}$ $1\frac{1}{3}$
32. $\frac{13}{4}$ $3\frac{1}{4}$
33. $\frac{11}{4}$ $2\frac{3}{4}$
34. $\frac{9}{5}$ $1\frac{4}{5}$
35. $\frac{15}{6}$ $2\frac{1}{2}$
36. $\frac{23}{6}$ $3\frac{5}{6}$

37. $\frac{27}{8}$ $3\frac{3}{8}$
38. $\frac{80}{5}$ 16
39. $\frac{100}{4}$ 25
40. $\frac{42}{13}$ $3\frac{3}{13}$
41. $\frac{86}{9}$ $9\frac{5}{9}$
42. $\frac{47}{2}$ $23\frac{1}{2}$

43. $\frac{70}{3}$ $23\frac{1}{3}$
44. $\frac{54}{17}$ $3\frac{3}{17}$
45. $\frac{25}{4}$ $6\frac{1}{4}$
46. $\frac{19}{3}$ $6\frac{1}{3}$
47. $\frac{57}{10}$ $5\frac{7}{10}$
48. $\frac{83}{10}$ $8\frac{3}{10}$

49. $\frac{35}{2}$ $17\frac{1}{2}$
50. $\frac{132}{11}$ 12
51. $\frac{91}{7}$ 13
52. $\frac{183}{7}$ $26\frac{1}{7}$
53. $\frac{210}{15}$ 14
54. $\frac{196}{9}$ $21\frac{7}{9}$

55. $\frac{102}{17}$ 6
56. $\frac{104}{8}$ 13
57. $\frac{175}{32}$ $5\frac{15}{32}$
58. $\frac{154}{25}$ $6\frac{4}{25}$

Reduce each mixed number.

59. $5\frac{3}{6}$ $5\frac{1}{2}$
60. $4\frac{6}{8}$ $4\frac{3}{4}$
61. $4\frac{11}{66}$ $4\frac{1}{6}$
62. $3\frac{15}{90}$ $3\frac{1}{6}$
63. $15\frac{18}{72}$ $15\frac{1}{4}$
64. $10\frac{15}{75}$ $10\frac{1}{5}$

Reduce each improper fraction.

65. $\frac{24}{6}$ 4
66. $\frac{36}{4}$ 9
67. $\frac{36}{15}$ $\frac{12}{5}$
68. $\frac{63}{45}$ $\frac{7}{5}$
69. $\frac{105}{28}$ $\frac{15}{4}$
70. $\frac{112}{21}$ $\frac{16}{3}$

Change to a mixed number and reduce.

71. $\frac{340}{126}$ $2\frac{88}{126} = 2\frac{44}{63}$
72. $\frac{390}{360}$ $1\frac{30}{360} = 1\frac{1}{12}$
73. $\frac{580}{280}$ $2\frac{20}{280} = 2\frac{1}{14}$

74. $\frac{764}{328}$ $2\frac{108}{328} = 2\frac{27}{82}$
75. $\frac{508}{296}$ $1\frac{212}{296} = 1\frac{53}{74}$
76. $\frac{2150}{1000}$ $2\frac{150}{1000} = 2\frac{3}{20}$

Applications

77. *Banner Display* The Science Museum is hanging banners all over the building to commemorate the Apollo astronauts. The art department is using $360\frac{2}{3}$ yards of starry-sky parachute fabric. Change this number to an improper fraction. $\frac{1082}{3}$ yards

78. *Sculpture* For the Northwestern University alumni homecoming, the students studying sculpture have made a giant replica of the school using $244\frac{3}{4}$ pounds of clay. Change this number to an improper fraction. $\frac{979}{4}$ pounds

79. *Environmental Studies* A Cape Cod cranberry bog was contaminated by waste from abandoned oil storage tanks at Otis Air Force Base. Damage was done to $\frac{151}{3}$ acres of land. Write this as a mixed number. $50\frac{1}{3}$ acres

80. *Theater* Waite Auditorium needs new velvet stage curtains. The manufacturer took measurements and calculated he would need $\frac{331}{4}$ square yards of fabric. Write this as a mixed number.

$82\frac{3}{4}$ square yards

81. *Cooking* The cafeteria workers at Ipswich High School cafeteria used $\frac{1131}{8}$ pounds of flour while cooking for the students last week. Write this as a mixed number.

$141\frac{3}{8}$ pounds

82. *Shelf Construction* The new Danvers Main Building at North Shore Community Colleges had several new offices for the faculty and staff. Shelving was constructed for these offices. A total of $\frac{1373}{8}$ feet of shelving was used in the construction. Write this as a mixed number. $171\frac{5}{8}$ feet

To Think About

83. Can $\frac{5687}{101}$ be reduced? Why or why not?

no; 101 is prime and is not a factor of 5687

84. Can $\frac{9810}{157}$ be reduced? Why or why not?

no; 157 is prime and not a factor of 9810

Cumulative Review

85. [1.3.3] Subtract. $1,398,210 - 1,137,963$
260,247

86. [1.7.2] Estimate the answer. $78,964 \times 229,350$
16,000,000,000

87. [1.7.2] Estimate the answer. $328,515 \div 966$
300

88. [1.8.1] ***Textbook Shipment*** Each semester college textbooks are often shipped to the bookstore in cardboard boxes that contain 24 textbooks. If 893 copies of *Basic College Mathematics* are shipped to the bookstore, how many full cartons are needed for the shipment? How many books are there in the carton that is not full?
37 full cartons are needed. There are 5 books in the carton that is not full.

Quick Quiz 2.3

1. Change to an improper fraction.

$4\frac{7}{13}$ $\frac{59}{13}$

2. Change to a mixed number.

$\frac{89}{12}$ $7\frac{5}{12}$

3. Reduce the improper fraction.

$\frac{42}{14}$ 3

4. Concept Check Explain how you change the mixed number $5\frac{6}{13}$ to an improper fraction. Answers may vary

Classroom Quiz 2.3 You may use these problems to quiz your students' mastery of Section 2.3.

1. Change to an improper fraction.

$3\frac{5}{16}$ **Ans:** $\frac{53}{16}$

2. Change to a mixed number.

$\frac{65}{11}$ **Ans:** $5\frac{10}{11}$

3. Reduce the improper fraction.

$\frac{68}{17}$ **Ans:** 4

1 Multiplying Two Fractions That Are Proper or Improper

FUDGE SQUARES

Ingredients:

2 cups sugar	1/4 teaspoon salt
4 oz chocolate	1 teaspoon vanilla
1/2 cup butter	1 cup all-purpose flour
4 eggs	1 cup nutmeats

Student Learning Objectives

After studying this section, you will be able to:

 Multiply two fractions that are proper or improper.

 Multiply a whole number by a fraction.

 Multiply mixed numbers.

Suppose you want to make an amount equal to half of what the recipe shown will produce. You would multiply the measure given for each ingredient by $\frac{1}{2}$.

$\frac{1}{2}$ of 2 cups sugar $\frac{1}{2}$ of $\frac{1}{4}$ teaspoon salt

$\frac{1}{2}$ of 4 oz chocolate $\frac{1}{2}$ of 1 teaspoon vanilla

$\frac{1}{2}$ of $\frac{1}{2}$ cup butter $\frac{1}{2}$ of 1 cup all-purpose flour

$\frac{1}{2}$ of 4 eggs $\frac{1}{2}$ of 1 cup nutmeats

We often use multiplication of fractions to describe taking a fractional part of something. To find $\frac{1}{2}$ of $\frac{3}{7}$, we multiply

$$\frac{1}{2} \times \frac{3}{7} = \frac{3}{14}.$$

We begin with a bar that is $\frac{3}{7}$ shaded. To find $\frac{1}{2}$ of $\frac{3}{7}$ we divide the bar in half and take $\frac{1}{2}$ of the shaded section. $\frac{1}{2}$ of $\frac{3}{7}$ yields 3 out of 14 squares.

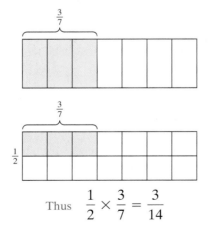

Thus $\quad \frac{1}{2} \times \frac{3}{7} = \frac{3}{14}$

When you multiply two proper fractions together, you get a smaller fraction.

To multiply two fractions, we multiply the numerators and multiply the denominators.

$$\frac{2}{3} \times \frac{5}{7} = \frac{10}{21} \begin{array}{l} \leftarrow 2 \times 5 = 10 \\ \leftarrow 3 \times 7 = 21 \end{array}$$

> **MULTIPLICATION OF FRACTIONS**
>
> In general, for all positive whole numbers a, b, c, and d,
>
> $$\frac{a}{b} \times \frac{c}{d} = \frac{a \times c}{b \times d}.$$

Teaching Example 1 Multiply.

(a) $\frac{4}{9} \times \frac{2}{7}$ (b) $\frac{6}{13} \times \frac{2}{5}$

Ans: (a) $\frac{8}{63}$ (b) $\frac{12}{65}$

EXAMPLE 1   Multiply.

(a) $\frac{3}{8} \times \frac{5}{7}$ (b) $\frac{1}{11} \times \frac{2}{13}$

Solution

(a) $\frac{3}{8} \times \frac{5}{7} = \frac{3 \times 5}{8 \times 7} = \frac{15}{56}$ (b) $\frac{1}{11} \times \frac{2}{13} = \frac{1 \times 2}{11 \times 13} = \frac{2}{143}$

Practice Problem 1   Multiply.

(a) $\frac{6}{7} \times \frac{3}{13}$ $\frac{18}{91}$ (b) $\frac{1}{5} \times \frac{11}{12}$ $\frac{11}{60}$

NOTE TO STUDENT: Fully worked-out solutions to all of the Practice Problems can be found at the back of the text starting at page SP-1

Some products may be reduced. $\frac{12}{35} \times \frac{25}{18} = \frac{300}{630} = \frac{10}{21}$

By simplifying before multiplication, the reducing can be done more easily. For a multiplication problem, a factor in the numerator can be paired with a common factor in the denominator of the same or a different fraction. We can begin by finding the prime factors in the numerators and denominators. We then divide numerator and denominator by their common prime factors.

Teaching Example 2 Simplify first, then multiply.

$$\frac{10}{27} \times \frac{6}{25}$$

Ans: $\frac{4}{45}$

EXAMPLE 2   Simplify first and then multiply. $\frac{12}{35} \times \frac{25}{18}$

Solution

$\frac{12}{35} \times \frac{25}{18} = \frac{2 \cdot 2 \cdot 3}{5 \cdot 7} \times \frac{5 \cdot 5}{2 \cdot 3 \cdot 3}$ First we find the prime factors.

$= \frac{2 \cdot 2 \cdot 3 \cdot 5 \cdot 5}{5 \cdot 7 \cdot 2 \cdot 3 \cdot 3}$ Write the product as one fraction.

$= \frac{\overset{1}{\cancel{2}} \cdot 2 \cdot \overset{1}{\cancel{3}} \cdot \overset{1}{\cancel{5}} \cdot 5}{\underset{1}{\cancel{2}} \cdot \underset{1}{\cancel{3}} \cdot 3 \cdot \underset{1}{\cancel{5}} \cdot 7}$ Arrange the factors in order and divide the numerator and denominator by the common factors.

$= \frac{10}{21}$ Multiply the remaining factors.

Practice Problem 2   Simplify first and then multiply.

$$\frac{55}{72} \times \frac{16}{33} \quad \frac{10}{27}$$

Note: Although finding the prime factors of the numerators and denominators will help you avoid errors, you can also begin these problems by dividing the numerators and denominators by larger common factors. This method will be used for the remainder of the exercises in this section of the text.

Multiplying a Whole Number by a Fraction

When multiplying a fraction by a whole number, it is more convenient to express the whole number as a fraction with a denominator of 1. We know that $5 = \frac{5}{1}$, $7 = \frac{7}{1}$, and so on.

EXAMPLE 3 Multiply.

(a) $5 \times \dfrac{3}{8}$ **(b)** $\dfrac{22}{7} \times 14$

Solution

(a) $5 \times \dfrac{3}{8} = \dfrac{5}{1} \times \dfrac{3}{8} = \dfrac{15}{8}$ or $1\dfrac{7}{8}$ **(b)** $\dfrac{22}{7} \times 14 = \dfrac{22}{\cancel{7}_{1}} \times \dfrac{\cancel{14}^{2}}{1} = \dfrac{44}{1} = 44$

Practice Problem 3 Multiply.

(a) $7 \times \dfrac{5}{13}$ $\dfrac{35}{13}$ or $2\dfrac{9}{13}$ **(b)** $\dfrac{13}{4} \times 8$ 26

EXAMPLE 4 Mr. and Mrs. Jones found that $\frac{2}{7}$ of their income went to pay federal income taxes. Last year they earned $37,100. How much did they pay in taxes?

Solution We need to find $\frac{2}{7}$ of $37,100. So we must multiply $\frac{2}{7} \times 37{,}100$.

$$\frac{2}{\cancel{7}_{1}} \times \cancel{37{,}100}^{\,5300} = \frac{2}{1} \times 5300 = 10{,}600$$

They paid $10,600 in federal income taxes.

Practice Problem 4 Fred and Linda own 98,400 square feet of land. They found that $\frac{3}{8}$ of the land is in a wetland area and cannot be used for building. How many square feet of land are in the wetland area?

36,900 square feet

 Multiplying Mixed Numbers

To multiply a fraction by a mixed number or to multiply two mixed numbers, first change each mixed number to an improper fraction.

EXAMPLE 5 Multiply.

(a) $\dfrac{5}{7} \times 3\dfrac{1}{4}$ **(b)** $20\dfrac{2}{5} \times 6\dfrac{2}{3}$ **(c)** $\dfrac{3}{4} \times 1\dfrac{1}{2} \times \dfrac{4}{7}$ **(d)** $4\dfrac{1}{3} \times 2\dfrac{1}{4}$

Solution

(a) $\dfrac{5}{7} \times 3\dfrac{1}{4} = \dfrac{5}{7} \times \dfrac{13}{4} = \dfrac{65}{28}$ or $2\dfrac{9}{28}$

(b) $20\dfrac{2}{5} \times 6\dfrac{2}{3} = \dfrac{\cancel{102}^{34}}{\cancel{5}_{1}} \times \dfrac{\cancel{20}^{4}}{\cancel{3}_{1}} = \dfrac{136}{1} = 136$

Teaching Example 3 Multiply.

(a) $8 \times \dfrac{2}{5}$ (b) $\dfrac{7}{3} \times 15$

Ans: (a) $\dfrac{16}{5}$ or $3\dfrac{1}{5}$ (b) 35

Teaching Example 4 In a math class, $\frac{5}{9}$ of the students are also enrolled in a composition course. There are 45 students in the math class. How many of these students are also taking composition?

Ans: 25 students

Teaching Example 5 Multiply.

(a) $1\dfrac{1}{9} \times \dfrac{4}{5}$ (b) $11\dfrac{2}{3} \times 3\dfrac{3}{7}$

(c) $\dfrac{5}{9} \times 1\dfrac{2}{7} \times \dfrac{3}{10}$ (d) $6\dfrac{3}{4} \times 2\dfrac{4}{9}$

Ans: (a) $\dfrac{8}{9}$ (b) 40 (c) $\dfrac{3}{14}$ (d) $\dfrac{33}{2}$ or $16\dfrac{1}{2}$

(c) $\dfrac{3}{4} \times 1\dfrac{1}{2} \times \dfrac{4}{7} = \dfrac{3}{\overset{}{\underset{1}{\cancel{4}}}} \times \dfrac{3}{2} \times \dfrac{\overset{1}{\cancel{4}}}{7} = \dfrac{9}{14}$

(d) $4\dfrac{1}{3} \times 2\dfrac{1}{4} = \dfrac{13}{\underset{1}{\cancel{3}}} \times \dfrac{\overset{3}{\cancel{9}}}{4} = \dfrac{39}{4}$ or $9\dfrac{3}{4}$

NOTE TO STUDENT: *Fully worked-out solutions to all of the Practice Problems can be found at the back of the text starting at page SP-1*

Practice Problem 5 Multiply.

(a) $2\dfrac{1}{6} \times \dfrac{4}{7}$ $\dfrac{26}{21}$ or $1\dfrac{5}{21}$ **(b)** $10\dfrac{2}{3} \times 13\dfrac{1}{2}$ 144

(c) $\dfrac{3}{5} \times 1\dfrac{1}{3} \times \dfrac{5}{8}$ $\dfrac{1}{2}$ **(d)** $3\dfrac{1}{5} \times 2\dfrac{1}{2}$ 8

Teaching Example 6 Find the area in square inches of a rectangle with width $2\dfrac{1}{2}$ inches and length $4\dfrac{2}{3}$ inches.

Ans: $11\dfrac{2}{3}$ square inches

▲ **EXAMPLE 6** Find the area in square miles of a rectangle with width $1\dfrac{1}{3}$ miles and length $12\dfrac{1}{4}$ miles.

Length = $12\dfrac{1}{4}$ miles

Width = $1\dfrac{1}{3}$ miles

Solution We find the area of a rectangle by multiplying the width times the length.

$$1\dfrac{1}{3} \times 12\dfrac{1}{4} = \dfrac{\overset{1}{\cancel{4}}}{3} \times \dfrac{49}{\underset{1}{\cancel{4}}} = \dfrac{49}{3} \quad \text{or} \quad 16\dfrac{1}{3}$$

The area is $16\dfrac{1}{3}$ square miles.

Teaching Tip Remind students that the area of both a rectangle and a square can be obtained by multiplying the length times the width or, in the case of the square, by multiplying the length of one side by itself. As a class activity, have them find the area of a square that measures $14\dfrac{1}{5}$ miles on each side. The answer is $201\dfrac{16}{25}$ square miles.

▲ **Practice Problem 6** Find the area in square meters of a rectangle with width $1\dfrac{1}{5}$ meters and length $4\dfrac{5}{6}$ meters.

$5\dfrac{4}{5}$ square meters

Teaching Example 7 Find the value of x if $\dfrac{5}{8} \cdot x = \dfrac{35}{72}$.

Ans: $x = \dfrac{7}{9}$

EXAMPLE 7 Find the value of x if

$$\dfrac{3}{7} \cdot x = \dfrac{15}{42}.$$

Solution The variable x represents a fraction. We know that 3 times one number equals 15 and 7 times another equals 42.

Since $3 \cdot 5 = 15$ and $7 \cdot 6 = 42$ we know that $\dfrac{3}{7} \cdot \dfrac{5}{6} = \dfrac{15}{42}.$

Therefore, $x = \dfrac{5}{6}.$

Practice Problem 7 Find the value of x if $\dfrac{8}{9} \cdot x = \dfrac{80}{81}.$ $x = \dfrac{10}{9}$

Multiply. Make sure all fractions are simplified in the final answer.

1. $\frac{3}{5} \times \frac{7}{11}$ $\frac{21}{55}$

2. $\frac{1}{8} \times \frac{5}{11}$ $\frac{5}{88}$

3. $\frac{3}{4} \times \frac{5}{13}$ $\frac{15}{52}$

4. $\frac{4}{7} \times \frac{3}{5}$ $\frac{12}{35}$

5. $\frac{6}{5} \times \frac{10}{12}$ 1

6. $\frac{7}{8} \times \frac{16}{21}$ $\frac{\overset{1}{\cancel{7}}}{\underset{1}{\cancel{8}}} \times \frac{\overset{2}{\cancel{16}}}{\underset{3}{\cancel{21}}} = \frac{2}{3}$

7. $\frac{5}{36} \times \frac{9}{20}$ $\frac{\overset{1}{\cancel{5}}}{\underset{4}{\cancel{36}}} \times \frac{\overset{1}{\cancel{9}}}{\underset{4}{\cancel{20}}} = \frac{1}{16}$

8. $\frac{22}{45} \times \frac{5}{11}$ $\frac{\overset{2}{\cancel{22}}}{\underset{9}{\cancel{45}}} \times \frac{\overset{1}{\cancel{5}}}{\underset{1}{\cancel{11}}} = \frac{2}{9}$

9. $\frac{12}{25} \times \frac{5}{11}$ $\frac{12}{\underset{5}{\cancel{25}}} \times \frac{\overset{1}{\cancel{5}}}{11} = \frac{12}{55}$

10. $\frac{9}{4} \times \frac{13}{27}$ $\frac{\overset{1}{\cancel{9}}}{4} \times \frac{13}{\underset{3}{\cancel{27}}} = \frac{13}{12}$ or $1\frac{1}{12}$

11. $\frac{9}{10} \times \frac{35}{12}$ $\frac{\overset{3}{\cancel{9}}}{\underset{2}{\cancel{10}}} \times \frac{\overset{7}{\cancel{35}}}{\underset{4}{\cancel{12}}} = \frac{21}{8}$ or $2\frac{5}{8}$

12. $\frac{12}{17} \times \frac{3}{24}$ $\frac{\overset{1}{\cancel{12}}}{17} \times \frac{3}{\underset{2}{\cancel{24}}} = \frac{3}{34}$

13. $8 \times \frac{3}{7}$ $\frac{8}{1} \times \frac{3}{7} = \frac{24}{7}$ or $3\frac{3}{7}$

14. $\frac{8}{9} \times 6$ $\frac{8}{\underset{3}{\cancel{9}}} \times \frac{\overset{2}{\cancel{6}}}{1} = \frac{16}{3}$ or $5\frac{1}{3}$

15. $\frac{5}{12} \times 8$ $\frac{5}{\underset{3}{\cancel{12}}} \times \frac{\overset{2}{\cancel{8}}}{1} = \frac{10}{3}$ or $3\frac{1}{3}$

16. $5 \times \frac{7}{25}$ $\frac{\overset{1}{\cancel{5}}}{1} \times \frac{7}{\underset{5}{\cancel{25}}} = \frac{7}{5}$ or $1\frac{2}{5}$

17. $\frac{4}{9} \times \frac{3}{7} \times \frac{7}{8}$ $\frac{\overset{1}{\cancel{4}}}{\underset{3}{\cancel{9}}} \times \frac{\overset{1}{\cancel{3}}}{\underset{1}{\cancel{7}}} \times \frac{\overset{1}{\cancel{7}}}{\underset{2}{\cancel{8}}} = \frac{1}{6}$

18. $\frac{8}{7} \times \frac{5}{12} \times \frac{3}{10}$ $\frac{\overset{2}{\cancel{8}}}{7} \times \frac{\overset{1}{\cancel{5}}}{\underset{3}{\cancel{12}}} \times \frac{\overset{1}{\cancel{3}}}{\underset{2}{\cancel{10}}} = \frac{1}{7}$

19. $\frac{5}{4} \times \frac{9}{10} \times \frac{8}{3}$ $\frac{\overset{1}{\cancel{5}}}{\underset{1}{\cancel{4}}} \times \frac{\overset{3}{\cancel{9}}}{\underset{2}{\cancel{10}}} \times \frac{\overset{2}{\cancel{8}}}{\underset{1}{\cancel{3}}} = 3$

20. $\frac{5}{7} \times \frac{15}{2} \times \frac{28}{15}$ $\frac{5}{\underset{1}{\cancel{7}}} \times \frac{\overset{1}{\cancel{15}}}{\underset{1}{\cancel{2}}} \times \frac{\overset{\overset{2}{\cancel{4}}}{\cancel{28}}}{\underset{1}{\cancel{15}}} = 10$

Multiply. Change any mixed number to an improper fraction before multiplying.

21. $2\frac{5}{6} \times \frac{3}{17}$ $\frac{17}{6} \times \frac{3}{17} = \frac{1}{2}$

22. $\frac{5}{6} \times 3\frac{3}{5}$ $\frac{5}{6} \times \frac{18}{5} = 3$

23. $10 \times 3\frac{1}{10}$ $\frac{10}{1} \times \frac{31}{10} = 31$

24. $12 \times 5\frac{7}{12}$ $\frac{12}{1} \times \frac{67}{12} = 67$

25. $1\frac{3}{16} \times 0$ 0

26. $0 \times 6\frac{2}{3}$ 0

27. $3\frac{7}{8} \times 1$ $3\frac{7}{8}$

28. $\frac{5}{5} \times 11\frac{5}{7}$ $11\frac{5}{7}$

29. $1\frac{1}{4} \times 3\frac{2}{3}$ $\frac{5}{4} \times \frac{11}{3} = \frac{55}{12}$ or $4\frac{7}{12}$

30. $2\frac{3}{5} \times 1\frac{4}{7}$ $\frac{13}{5} \times \frac{11}{7} = \frac{143}{35}$ or $4\frac{3}{35}$

31. $2\frac{3}{10} \times \frac{3}{5}$ $\frac{23}{10} \times \frac{3}{5} = \frac{69}{50}$ or $1\frac{19}{50}$

32. $4\frac{3}{5} \times \frac{1}{10}$ $\frac{23}{5} \times \frac{1}{10} = \frac{23}{50}$

33. $4\frac{1}{5} \times 8\frac{1}{3}$ $\frac{21}{5} \times \frac{25}{3} = 35$

34. $5\frac{1}{4} \times 4\frac{4}{7}$ $\frac{21}{4} \times \frac{32}{7} = 24$

35. $6\frac{2}{5} \times \frac{1}{4}$ $\frac{32}{5} \times \frac{1}{4} = \frac{8}{5}$ or $1\frac{3}{5}$

36. $\frac{8}{9} \times 4\frac{1}{11}$ $\frac{8}{9} \times \frac{45}{11} = \frac{40}{11}$ or $3\frac{7}{11}$

Mixed Practice *Multiply. Make sure all fractions are simplified in the final answer.*

37. $\dfrac{11}{15} \times \dfrac{35}{33}$ $\dfrac{7}{9}$

38. $\dfrac{14}{17} \times \dfrac{34}{42}$ $\dfrac{2}{3}$

39. $2\dfrac{3}{8} \times 5\dfrac{1}{3}$ $\dfrac{38}{3}$ or $12\dfrac{2}{3}$

40. $4\dfrac{3}{5} \times 3\dfrac{3}{4}$ $\dfrac{69}{4}$ or $17\dfrac{1}{4}$

Solve for x.

41. $\dfrac{4}{9} \cdot x = \dfrac{28}{81}$ $x = \dfrac{7}{9}$

42. $\dfrac{12}{17} \cdot x = \dfrac{144}{85}$ $x = \dfrac{12}{5}$

43. $\dfrac{7}{13} \cdot x = \dfrac{56}{117}$ $x = \dfrac{8}{9}$

44. $x \cdot \dfrac{11}{15} = \dfrac{77}{225}$ $x = \dfrac{7}{15}$

Applications

▲ **45. *Geometry*** A spy is running from his captors in a forest that is $8\dfrac{3}{4}$ miles long and $4\dfrac{1}{3}$ miles wide. Find the area of the forest where he is hiding. (*Hint:* The area of a rectangle is the product of the length times the width.) $37\dfrac{11}{12}$ square miles

▲ **46. *Geometry*** An area in the Midwest is a designated tornado danger zone. The land is $22\dfrac{5}{8}$ miles long and $16\dfrac{1}{2}$ miles wide. Find the area of the tornado danger zone. (*Hint:* The area of a rectangle is the product of the length times the width.)
$373\dfrac{5}{16}$ square miles

47. *Airplane Travel* A Lear jet airplane has 360 gallons of fuel. The plane averages $4\dfrac{1}{3}$ miles per gallon. How far can the plane go?   1560 miles

48. *Real Estate* Mel and Sally Hauser bought their house in 1977 for a price of $56,800. Thirty years later, in 2007, their house is worth $6\dfrac{1}{2}$ times what they paid for it. How much was Mel and Sally's house worth in 2007? $369,200

49. *Cooking* A recipe from Nanette's French cookbook for a scalloped potato tart requires $90\dfrac{1}{2}$ grams of grated cheese. How many grams of cheese would she need if she made one tart for each of her 18 cousins?   1629 grams

▲ **50. *Geometry*** The dormitory rooms in Selkirk Hall are being carpeted. Each room requires $20\dfrac{1}{2}$ square feet of carpet. If there are 30 rooms, how much carpet is needed?   615 square feet

51. *College Students* Of the 7998 students at Normandale Community College, $\dfrac{2}{3}$ of them are under 25 years of age. How many students are under 25 years of age?   5332 students

52. *Health Care* A nurse finds that of the 225 rooms at Dover Area Hospital, $\dfrac{1}{15}$ of them are occupied by surgery patients. How many rooms contain surgery patients?   15 rooms

53. *Job Search* Carlos has sent his résumé to 12,064 companies through an Internet job search service. If $\dfrac{1}{32}$ of the companies e-mail him with an invitation for an interview, how many companies will he have heard from?   377 companies

54. *Car Purchase* Russ purchased a new Buick LeSabre for $26,500. After one year the car was worth $\dfrac{4}{5}$ of the purchase price. What was the car worth after one year? $21,200

55. *Jogging* Mary jogged $4\frac{1}{4}$ miles per hour for $1\frac{1}{3}$ hours. During $\frac{1}{3}$ of her jogging time, she was jogging in the rain. How many miles did she jog in the rain? $1\frac{8}{9}$ miles

56. *College Students* There were 1340 students at the Beverly campus of North Shore Community College during the spring 2003 semester. The registrar discovered that $\frac{2}{5}$ of these students live in the city of Beverly. He further discovered that $\frac{1}{4}$ of the students living in Beverly attend classes only on Monday, Wednesday, and Friday. How many students at the Beverly campus live in the city of Beverly and attend classes only on Monday, Wednesday, and Friday? 134 students

To Think About

57. When we multiply two fractions, we look for opportunities to divide a numerator and a denominator by the same number. Why do we bother with that step? Why don't we just multiply the two numerators and the two denominators?

The step of dividing the numerator and denominator by the same number allows us to work with smaller numbers when we do the multiplication. Also, this allows us to avoid the step of having to simplify the fraction in the final answer.

58. Suppose there is an unknown fraction that has *not* been simplified (it is not reduced). You multiply this unknown fraction by $\frac{2}{5}$ and you obtain a simplified answer of $\frac{6}{35}$. How many possible values could this unknown fraction be? Give at least three possible answers.

There are an infinite number of answers. Any fraction that can be simplified to $\frac{3}{7}$ would be a correct answer. Thus three possible answers to this problem are $\frac{6}{14}, \frac{9}{21}$, or $\frac{12}{28}$.

Cumulative Review

59. **[1.5.4]** *Toll Bridge* A total of 16,399 cars used a toll bridge in January (31 days). What is the average number of cars using the bridge in one day?
529 cars

60. **[1.5.4]** *Sales* The Office of Investors Services has 15,456 calls made per month by the sales personnel. There are 42 sales personnel in the office. What is the average number of calls made per month by one salesperson? 368 calls

61. **[1.4.6]** *Commuter Driving* Gerald commutes 21 miles roundtrip between home and work. If he works 240 days a year, how many miles does he drive between work and home in one year?
5040 miles

62. **[1.4.6]** *Jet Travel* At cruising speed a new commercial jet uses 12,360 gallons of fuel per hour. How many gallons will be used in 14 hours of flying time? 173,040 gallons

Quick Quiz 2.4 Multiply.

1. $32 \times \frac{5}{16}$ 10

2. $\frac{11}{13} \times \frac{4}{5}$ $\frac{44}{65}$

3. $4\frac{1}{3} \times 2\frac{3}{4}$ $\frac{143}{12}$ or $11\frac{11}{12}$

4. **Concept Check** Explain how you would multiply the whole number 6 times the mixed number $4\frac{3}{5}$.
Answers may vary

Classroom Quiz 2.4 You may use these problems to quiz your students' mastery of Section 2.4.

Multiply.

1. $21 \times \frac{5}{7}$ **Ans:** 15

2. $\frac{13}{15} \times \frac{5}{12}$ **Ans:** $\frac{13}{36}$

3. $7\frac{2}{3} \times 1\frac{1}{5}$ **Ans:** $\frac{46}{5}$ or $9\frac{1}{5}$

Dividing Two Proper or Improper Fractions

Why would you divide fractions? Consider this problem.

• A copper pipe that is $\frac{3}{4}$ of a foot long is to be cut into $\frac{1}{4}$-foot pieces. How many pieces will there be?

To find how many $\frac{1}{4}$'s are in $\frac{3}{4}$, we divide $\frac{3}{4} \div \frac{1}{4}$. We draw a sketch.

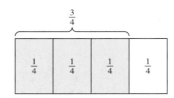

Notice that there are three $\frac{1}{4}$'s in $\frac{3}{4}$.

How do we divide two fractions? We **invert** the second fraction and multiply.

$$\frac{3}{4} \div \frac{1}{4} = \frac{3}{\cancel{4}} \times \frac{\cancel{4}^1}{1} = \frac{3}{1} = 3$$

When we invert a fraction, we interchange the numerator and the denominator. If we invert $\frac{5}{9}$, we obtain $\frac{9}{5}$. If we invert $\frac{6}{1}$, we obtain $\frac{1}{6}$. Numbers such as $\frac{5}{9}$ and $\frac{9}{5}$ are called **reciprocals** of each other.

Teaching Tip Tell students that in dividing two fractions it is always the second fraction that is inverted. Inform them that it is a very common mistake for students to erroneously invert the first fraction. If they have trouble remembering which one to flip, tell them to think of the following suggestion from a student who was a cook. "You cannot flip a one-sided pancake. It has to have two sides to flip. You cannot flip fraction number one. It has to be fraction two to flip it." This rule seems silly, but no student who has learned it ever inverts the wrong fraction!

RULE FOR DIVISION OF FRACTIONS

To divide two fractions, we invert the second fraction and multiply.

$$\frac{a}{b} \div \frac{c}{d} = \frac{a}{b} \times \frac{d}{c}$$

(when b, c, and d are not zero).

Teaching Example 1 Divide.

(a) $\frac{7}{10} \div \frac{8}{9}$ (b) $\frac{5}{8} \div \frac{25}{32}$

Ans: (a) $\frac{63}{80}$ (b) $\frac{4}{5}$

EXAMPLE 1 Divide.

(a) $\dfrac{3}{11} \div \dfrac{2}{5}$ (b) $\dfrac{5}{8} \div \dfrac{25}{16}$

Solution

(a) $\dfrac{3}{11} \div \dfrac{2}{5} = \dfrac{3}{11} \times \dfrac{5}{2} = \dfrac{15}{22}$

(b) $\dfrac{5}{8} \div \dfrac{25}{16} = \dfrac{\cancel{5}^1}{\cancel{8}_1} \times \dfrac{\cancel{16}^2}{\cancel{25}_5} = \dfrac{2}{5}$

NOTE TO STUDENT: *Fully worked-out solutions to all of the Practice Problems can be found at the back of the text starting at page SP-1*

Practice Problem 1 Divide.

(a) $\dfrac{7}{13} \div \dfrac{3}{4}$ $\dfrac{28}{39}$ (b) $\dfrac{16}{35} \div \dfrac{24}{25}$ $\dfrac{10}{21}$

Dividing a Whole Number and a Fraction

When dividing with whole numbers, it is helpful to remember that for any whole number a, $a = \dfrac{a}{1}$.

EXAMPLE 2 Divide.

(a) $\dfrac{3}{7} \div 2$ **(b)** $5 \div \dfrac{10}{13}$

Solution

(a) $\dfrac{3}{7} \div 2 = \dfrac{3}{7} \div \dfrac{2}{1} = \dfrac{3}{7} \times \dfrac{1}{2} = \dfrac{3}{14}$

(b) $5 \div \dfrac{10}{13} = \dfrac{5}{1} \div \dfrac{10}{13} = \dfrac{\overset{1}{\cancel{5}}}{1} \times \dfrac{13}{\underset{2}{\cancel{10}}} = \dfrac{13}{2}$ or $6\dfrac{1}{2}$

Practice Problem 2 Divide.

(a) $\dfrac{3}{17} \div 6$ $\dfrac{1}{34}$ **(b)** $14 \div \dfrac{7}{15}$ 30

Teaching Example 2 Divide.

(a) $\dfrac{5}{6} \div 3$ **(b)** $12 \div \dfrac{3}{5}$

Ans: (a) $\dfrac{5}{18}$ **(b)** 20

EXAMPLE 3 Divide, if possible.

(a) $\dfrac{23}{25} \div 1$ **(b)** $1 \div \dfrac{7}{5}$ **(c)** $0 \div \dfrac{4}{9}$ **(d)** $\dfrac{3}{17} \div 0$

Solution

(a) $\dfrac{23}{25} \div 1 = \dfrac{23}{25} \times \dfrac{1}{1} = \dfrac{23}{25}$

(b) $1 \div \dfrac{7}{5} = \dfrac{1}{1} \times \dfrac{5}{7} = \dfrac{5}{7}$

(c) $0 \div \dfrac{4}{9} = \dfrac{0}{1} \times \dfrac{9}{4} = \dfrac{0}{4} = 0$ Zero divided by any nonzero number is zero.

(d) $\dfrac{3}{17} \div 0$ Division by zero is undefined.

Teaching Example 3 Divide, if possible.

(a) $1 \div \dfrac{9}{8}$ **(b)** $\dfrac{31}{35} \div 1$

(c) $\dfrac{4}{9} \div 0$ **(d)** $0 \div \dfrac{2}{7}$

Ans: (a) $\dfrac{8}{9}$ **(b)** $\dfrac{31}{35}$

        (c) Division by zero is undefined.

        (d) 0

Practice Problem 3 Divide, if possible.

(a) $1 \div \dfrac{11}{13}$ $\dfrac{13}{11}$ or $1\dfrac{2}{11}$ **(b)** $\dfrac{14}{17} \div 1$ $\dfrac{14}{17}$

(c) $\dfrac{3}{11} \div 0$ Division by zero is undefined   **(d)** $0 \div \dfrac{9}{16}$ 0

SIDELIGHT: Invert and Multiply

Why do we divide by inverting the second fraction and multiplying? What is really going on when we do this? We are actually multiplying by 1. To see why, consider the following.

$$\frac{3}{7} \div \frac{2}{3} = \frac{\dfrac{3}{7}}{\dfrac{2}{3}}$$ We write the division by using another fraction bar.

$$= \frac{\dfrac{3}{7}}{\dfrac{2}{3}} \times 1$$ Any fraction can be multiplied by 1 without changing the value of the fraction. This is the fundamental rule of fractions.

$$= \frac{\dfrac{3}{7}}{\dfrac{2}{3}} \times \frac{\dfrac{3}{2}}{\dfrac{3}{2}}$$ Any nonzero number divided by itself equals 1.

$$= \frac{\dfrac{3}{7} \times \dfrac{3}{2}}{\dfrac{2}{3} \times \dfrac{3}{2}}$$ Definition of multiplication of fractions.

$$= \frac{\dfrac{3}{7} \times \dfrac{3}{2}}{1} = \frac{3}{7} \times \frac{3}{2}$$ Any number can be written as a fraction with a denominator of 1 without changing its value.

Thus

$$\frac{3}{7} \div \frac{2}{3} = \frac{3}{7} \times \frac{3}{2} = \frac{9}{14}.$$

③ Dividing Mixed Numbers

If one or more mixed numbers are involved in the division, they should be converted to improper fractions first.

Teaching Example 4 Divide.

(a) $\dfrac{3}{8} \div 1\dfrac{1}{5}$ (b) $4\dfrac{2}{3} \div 2\dfrac{5}{8}$

Ans: (a) $\dfrac{5}{16}$ (b) $\dfrac{16}{9}$ or $1\dfrac{7}{9}$

EXAMPLE 4   Divide.

(a) $3\dfrac{7}{15} \div 1\dfrac{1}{25}$ (b) $\dfrac{3}{5} \div 2\dfrac{1}{7}$

Solution

(a) $3\dfrac{7}{15} \div 1\dfrac{1}{25} = \dfrac{52}{15} \div \dfrac{26}{25} = \dfrac{\overset{2}{\cancel{52}}}{\underset{3}{\cancel{15}}} \times \dfrac{\overset{5}{\cancel{25}}}{\underset{1}{\cancel{26}}} = \dfrac{10}{3}$ or $3\dfrac{1}{3}$

(b) $\dfrac{3}{5} \div 2\dfrac{1}{7} = \dfrac{3}{5} \div \dfrac{15}{7} = \dfrac{\overset{1}{\cancel{3}}}{5} \times \dfrac{7}{\underset{5}{\cancel{15}}} = \dfrac{7}{25}$

NOTE TO STUDENT: Fully worked-out solutions to all of the Practice Problems can be found at the back of the text starting at page SP-1

Practice Problem 4   Divide.

(a) $1\dfrac{1}{5} \div \dfrac{7}{10}$ $\dfrac{12}{7}$ or $1\dfrac{5}{7}$ (b) $2\dfrac{1}{4} \div 1\dfrac{7}{8}$ $\dfrac{6}{5}$ or $1\dfrac{1}{5}$

The division of two fractions may be indicated by a wide fraction bar.

EXAMPLE 5 Divide.

(a) $\dfrac{10\frac{2}{9}}{2\frac{1}{3}}$

(b) $\dfrac{1\frac{1}{15}}{3\frac{1}{3}}$

Teaching Example 5 Divide.

(a) $\dfrac{4\frac{1}{2}}{3\frac{3}{8}}$ (b) $\dfrac{3\frac{1}{3}}{5}$

Ans: (a) $\frac{4}{3}$ or $1\frac{1}{3}$ **(b)** $\frac{2}{3}$

Solution

(a) $\dfrac{10\frac{2}{9}}{2\frac{1}{3}} = 10\frac{2}{9} \div 2\frac{1}{3} = \dfrac{92}{9} \div \dfrac{7}{3} = \dfrac{92}{\overset{}{\underset{3}{9}}} \times \dfrac{\overset{1}{3}}{7} = \dfrac{92}{21}$ or $4\frac{8}{21}$

(b) $\dfrac{1\frac{1}{15}}{3\frac{1}{3}} = 1\frac{1}{15} \div 3\frac{1}{3} = \dfrac{16}{15} \div \dfrac{10}{3} = \dfrac{\overset{8}{16}}{\underset{5}{15}} \times \dfrac{\overset{1}{3}}{\underset{5}{10}} = \dfrac{8}{25}$

Practice Problem 5 Divide.

(a) $\dfrac{5\frac{2}{3}}{7}$ $\frac{17}{21}$

(b) $\dfrac{1\frac{2}{5}}{2\frac{1}{3}}$ $\frac{3}{5}$

Some students may find Example 6 difficult at first. Read it slowly and carefully. It may be necessary to read it several times before it becomes clear.

EXAMPLE 6 Find the value of x if $x \div \frac{8}{7} = \frac{21}{40}$.

Teaching Example 6 Find the value of x if $x \div \frac{5}{8} = \frac{16}{35}$.

Ans: $x = \frac{2}{7}$

Solution First we will change the division problem to an equivalent multiplication problem.

$$x \div \frac{8}{7} = \frac{21}{40}$$

$$x \cdot \frac{7}{8} = \frac{21}{40}$$

x represents a fraction.

In the numerator, we want to know what times 7 equals 21. In the denominator, we want to know what times 8 equals 40.

$$\frac{3}{5} \cdot \frac{7}{8} = \frac{21}{40}$$

Thus $x = \frac{3}{5}$.

Practice Problem 6 Find the value of x if $x \div \frac{3}{2} = \frac{22}{36}$.

$$x = \frac{11}{12}$$

EXAMPLE 7 There are 117 milligrams of cholesterol in $4\frac{1}{3}$ cups of milk. How much cholesterol is in 1 cup of milk?

Solution We want to divide the 117 by $4\frac{1}{3}$ to find out how much is in 1 cup.

$$117 \div 4\frac{1}{3} = 117 \div \frac{13}{3} = \frac{\overset{9}{\cancel{117}}}{1} \times \frac{3}{\underset{1}{\cancel{13}}} = \frac{27}{1} = 27$$

Thus there are 27 milligrams of cholesterol in 1 cup of milk.

Practice Problem 7 A copper pipe that is $19\frac{1}{4}$ feet long will be cut into 14 equal pieces. How long will each piece be?

$1\frac{3}{8}$ feet

Take a little time to review Examples 1–7 and Practice Problems 1–7. This is important material. It is crucial to understand how to do each of these problems. Some extra time spent reviewing here will make the homework exercises go much more quickly.

Developing Your Study Skills

Why Is Review Necessary?

You master a course in mathematics by learning the concepts one step at a time. There are basic concepts like addition, subtraction, multiplication, and division of whole numbers that are considered the foundation upon which all of mathematics is built. These must be mastered first. Then the study of mathematics is built step by step upon this foundation, each step supporting the next. The process is a carefully designed procedure, so no steps can be skipped. A student of mathematics needs to realize the importance of this building process to succeed.

Because learning new concepts depends on those previously learned, students often need to take time to review. The reviewing process will strengthen the understanding and application of concepts that are weak due to lack of mastery or passage of time. Review at the right time on the right concepts can strengthen previously learned skills and make progress possible.

Timely, periodic review of previously learned mathematical concepts is absolutely necessary in order to master new concepts. You may have forgotten a concept or grown a bit rusty in applying it. Reviewing is the answer. Make use of any review sections in your textbook, whether they are assigned or not. Look back to previous chapters whenever you have forgotten how to do something. Study the examples and practice some exercises to refresh your understanding.

Be sure that you understand and can perform the computations of each new concept. This will enable you to be able to move successfully on to the next ones.

Make sure all fractions are simplified in the final answer.

Verbal and Writing Skills

1. In your own words explain how to remember that when you divide two fractions you invert the *second* fraction and multiply by the first. How can you be sure that you don't invert the *first* fraction by mistake?

Think of a simple problem like $3 \div \frac{1}{2}$. One way to think of it is, how many $\frac{1}{2}$'s can be placed in 3? For example, how many $\frac{1}{2}$-pound rocks could be put in a bag that holds 3 pounds of rocks? The answer is 6. If we inverted the first fraction by mistake, we would have $\frac{1}{3} \cdot \frac{1}{2} = \frac{1}{6}$. We know that is wrong since there are obviously several $\frac{1}{2}$-pound rocks in a bag that holds 3 pounds of rocks. The answer $\frac{1}{6}$ would make no sense.

2. Explain why $2 \div \frac{1}{3}$ is a larger number than $2 \div \frac{1}{2}$.

One way to think of it is to imagine how many $\frac{1}{3}$-pound rocks could be put in a bag that holds 2 pounds of rocks and then imagine how many $\frac{1}{2}$-pound rocks could be put in the same bag. The number of $\frac{1}{3}$-pound rocks would be larger. Therefore $2 \div \frac{1}{3}$ is a larger number.

Divide, if possible.

3. $\frac{7}{16} \div \frac{3}{4}$

$\frac{7}{16} \times \frac{4}{3} = \frac{7}{12}$

4. $\frac{3}{13} \div \frac{9}{26}$

$\frac{3}{13} \times \frac{26}{9} = \frac{2}{3}$

5. $\frac{2}{3} \div \frac{4}{27}$

$\frac{2}{3} \times \frac{27}{4} = \frac{9}{2}$ or $4\frac{1}{2}$

6. $\frac{25}{49} \div \frac{5}{7}$

$\frac{25}{49} \times \frac{7}{5} = \frac{5}{7}$

7. $\frac{7}{18} \div \frac{21}{6}$

$\frac{7}{18} \times \frac{6}{21} = \frac{1}{9}$

8. $\frac{8}{15} \div \frac{24}{35}$

$\frac{8}{15} \times \frac{35}{24} = \frac{7}{9}$

9. $\frac{5}{9} \div \frac{1}{5}$

$\frac{5}{9} \times \frac{5}{1} = \frac{25}{9}$ or $2\frac{7}{9}$

10. $\frac{3}{4} \div \frac{2}{3}$

$\frac{3}{4} \times \frac{3}{2} = \frac{9}{8}$ or $1\frac{1}{8}$

11. $\frac{4}{15} \div \frac{4}{15}$

$\frac{4}{15} \times \frac{15}{4} = 1$

12. $\frac{2}{7} \div \frac{2}{7}$

$\frac{2}{7} \times \frac{7}{2} = 1$

13. $\frac{3}{7} \div \frac{7}{3}$

$\frac{3}{7} \times \frac{3}{7} = \frac{9}{49}$

14. $\frac{11}{12} \div \frac{1}{5}$

$\frac{11}{12} \times \frac{5}{1} = \frac{55}{12}$ or $4\frac{7}{12}$

15. $\frac{4}{5} \div 1$

$\frac{4}{5} \times 1 = \frac{4}{5}$

16. $1 \div \frac{3}{7}$

$1 \times \frac{7}{3} = \frac{7}{3}$ or $2\frac{1}{3}$

17. $\frac{3}{11} \div 4$

$\frac{3}{11} \times \frac{1}{4} = \frac{3}{44}$

18. $2 \div \frac{7}{8}$

$\frac{2}{1} \times \frac{8}{7} = \frac{16}{7}$ or $2\frac{2}{7}$

19. $1 \div \frac{7}{27}$

$1 \times \frac{27}{7} = \frac{27}{7}$ or $3\frac{6}{7}$

20. $\frac{9}{16} \div 1$

$\frac{9}{16} \times 1 = \frac{9}{16}$

21. $0 \div \frac{3}{17}$

$0 \times \frac{17}{3} = 0$

22. $0 \div \frac{5}{16}$

$0 \times \frac{16}{5} = 0$

23. $\frac{18}{19} \div 0$

undefined

24. $\frac{24}{29} \div 0$

undefined

25. $8 \div \frac{4}{5}$

$\frac{8}{1} \times \frac{5}{4} = 10$

26. $16 \div \frac{8}{11}$

$\frac{16}{1} \times \frac{11}{8} = 22$

27. $\frac{7}{8} \div 4$

$\frac{7}{8} \times \frac{1}{4} = \frac{7}{32}$

28. $\frac{5}{6} \div 12$

$\frac{5}{6} \times \frac{1}{12} = \frac{5}{72}$

29. $\frac{9}{16} \div \frac{3}{4}$

$\frac{9}{16} \times \frac{4}{3} = \frac{3}{4}$

30. $\frac{3}{4} \div \frac{9}{16}$

$\frac{3}{4} \times \frac{16}{9} = \frac{4}{3}$ or $1\frac{1}{3}$

31. $3\frac{1}{4} \div 2\frac{1}{4}$

$\frac{13}{4} \times \frac{4}{9} = \frac{13}{9}$ or $1\frac{4}{9}$

32. $2\frac{2}{3} \div 4\frac{1}{3}$

$\frac{8}{3} \times \frac{3}{13} = \frac{8}{13}$

33. $6\frac{2}{5} \div 3\frac{1}{5}$

$\frac{32}{5} \times \frac{5}{16} = 2$

34. $9\frac{1}{3} \div 3\frac{1}{9}$

$\frac{28}{3} \times \frac{9}{28} = 3$

35. $6000 \div \frac{6}{5}$

$\frac{6000}{1} \times \frac{5}{6} = 5000$

36. $8000 \div \frac{4}{7}$

$\frac{8000}{1} \times \frac{7}{4} = 14{,}000$

37. $\dfrac{\frac{4}{5}}{200}$

$\frac{4}{5} \times \frac{1}{200} = \frac{1}{250}$

38. $\dfrac{\frac{5}{9}}{100}$

$\frac{5}{9} \times \frac{1}{100} = \frac{1}{180}$

39. $\dfrac{\frac{5}{8}}{\frac{25}{7}}$

$\frac{5}{8} \times \frac{7}{25} = \frac{7}{40}$

40. $\dfrac{\frac{3}{16}}{\frac{5}{8}}$

$\frac{3}{16} \times \frac{8}{5} = \frac{3}{10}$

Mixed Practice *Multiply or divide.*

41. $3\frac{1}{5} \div \frac{1}{5}$ $\frac{16}{5} \times \frac{5}{1} = 16$

42. $4\frac{3}{4} \div \frac{1}{4}$ $\frac{19}{4} \times \frac{4}{1} = 19$

43. $2\frac{1}{3} \times \frac{1}{6}$ $\frac{7}{3} \times \frac{1}{6} = \frac{7}{18}$

44. $6\frac{1}{2} \times \frac{1}{3}$ $\frac{13}{2} \times \frac{1}{3} = \frac{13}{6}$ or $2\frac{1}{6}$

45. $5\frac{1}{4} \div 2\frac{5}{8}$ $\frac{21}{4} \times \frac{8}{21} = 2$

46. $1\frac{2}{9} \div 4\frac{1}{3}$ $\frac{11}{9} \times \frac{3}{13} = \frac{11}{39}$

47. $5 \div 1\frac{1}{4}$ $\frac{5}{1} \times \frac{4}{5} = 4$

48. $7 \div 1\frac{2}{5}$ $\frac{7}{1} \times \frac{5}{7} = 5$

49. $5\frac{2}{3} \div 2\frac{1}{4}$ $\frac{17}{3} \times \frac{4}{9} = \frac{68}{27}$ or $2\frac{14}{27}$

50. $14\frac{2}{3} \div 3\frac{1}{2}$ $\frac{44}{3} \times \frac{2}{7} = \frac{88}{21}$ or $4\frac{4}{21}$

51. $\frac{7}{2} \div 3\frac{1}{2}$ $\frac{7}{2} \times \frac{2}{7} = 1$

52. $\frac{16}{3} \div 5\frac{1}{3}$ $\frac{16}{3} \times \frac{3}{16} = 1$

53. $\frac{13}{25} \times 2\frac{1}{3}$ $\frac{13}{25} \times \frac{7}{3} = \frac{91}{75}$ or $1\frac{16}{75}$

54. $\frac{11}{20} \times 4\frac{1}{2}$ $\frac{11}{20} \times \frac{9}{2} = \frac{99}{40}$ or $2\frac{19}{40}$

55. $3\frac{3}{4} \div 9$ $\frac{15}{4} \times \frac{1}{9} = \frac{5}{12}$

56. $5\frac{5}{6} \div 7$ $\frac{35}{6} \times \frac{1}{7} = \frac{5}{6}$

57. $\dfrac{5}{3\frac{1}{6}}$ $\frac{5}{1} \times \frac{6}{19} = \frac{30}{19}$ or $1\frac{11}{19}$

58. $\dfrac{8}{2\frac{1}{2}}$ $\frac{8}{1} \times \frac{2}{5} = \frac{16}{5}$ or $3\frac{1}{5}$

59. $\dfrac{0}{4\frac{3}{8}}$ $0 \times \frac{8}{35} = 0$

60. $\dfrac{5\frac{2}{5}}{0}$ undefined

61. $\dfrac{\frac{7}{12}}{3\frac{2}{3}}$ $\frac{7}{12} \times \frac{3}{11} = \frac{7}{44}$

62. $\dfrac{\frac{9}{10}}{3\frac{3}{5}}$ $\frac{9}{10} \times \frac{5}{18} = \frac{1}{4}$

63. $4\frac{2}{5} \times 2\frac{8}{11}$ $\frac{22}{5} \times \frac{30}{11} = 12$

64. $4\frac{2}{3} \times 5\frac{1}{7}$ $\frac{14}{3} \times \frac{36}{7} = 24$

Review Example 6. Then find the value of x in each of the following.

65. $x \div \frac{4}{3} = \frac{21}{20}$ $x = \frac{7}{5}$

66. $x \div \frac{2}{5} = \frac{15}{16}$ $x = \frac{3}{8}$

67. $x \div \frac{10}{7} = \frac{21}{100}$ $x = \frac{3}{10}$

68. $x \div \frac{11}{6} = \frac{54}{121}$ $x = \frac{9}{11}$

Applications *Answer each question.*

69. *Leather Factory* A leather factory in Morocco tans leather. In order to make the leather soft, it has to soak in a vat of uric acid and other ingredients. The main holding tank holds $20\frac{1}{4}$ gallons of the tanning mixture. If the mixture is distributed evenly into nine vats of equal size for the different colored leathers, how much will each vat hold? $2\frac{1}{4}$ gallons

70. *Marine Biology* A specially protected stretch of beach bordering the Great Barrier Reef in Australia is used for marine biology and ecological research. The beach, which is $7\frac{1}{2}$ miles long, has been broken up into 20 equal segments for comparison purposes. How long is each segment of the beach? $\frac{3}{8}$ mile

71. *Vehicle Travel* Bruce drove in a snowstorm to get to his favorite mountain to do some snowboarding. He traveled 125 miles in $3\frac{1}{3}$ hours. What was his average speed (in miles per hour)? $37\frac{1}{2}$ miles per hour

72. *Vehicle Travel* Roberto drove his truck to Cedarville, a distance of 200 miles, in $4\frac{1}{6}$ hours. What was his average speed (in miles per hour)?   48 miles per hour

73. *Cooking* The school cafeteria is making hamburgers for the annual Senior Day Festival. The cooks have decided that because hamburger shrinks on the grill, they will allow $\frac{2}{3}$ pound of meat for each student. If the kitchen has $38\frac{2}{3}$ pounds of meat, how many students will be fed? 58 students

74. *Making Costumes* Costumes are needed for the junior high school's "Wizard of Oz" performance. Each costume requires $4\frac{1}{3}$ yards of fabric and $151\frac{2}{3}$ yards are available. How many costumes can be made? 35 costumes

75. *Cooking* A coffee pot that holds 150 cups of coffee is being used at a company meeting. Each large Styrofoam cup holds $1\frac{1}{2}$ cups of coffee. How many large Styrofoam cups can be filled?
100 large Styrofoam cups

76. *Medicine Dosage* A small bottle of eye drops contains 16 milliliters. If the recommended use is $\frac{2}{3}$ milliliter, how many times can a person use the drops before the bottle is empty? 24 times

77. *Time Capsule* In 1907, a time capsule was placed behind a steel wall measuring $4\frac{3}{4}$ inches thick. On December 22, 2007, a special drill was used to bore through the wall and extricate the time capsule. The drill could move only $\frac{5}{6}$ inch at a time. How many drill attempts did it take to reach the other side of the steel wall? It took six drill attempts.

78. *Ink Production* Imagination Ink supplies different-colored inks for highlighter pens. Vat 1 has yellow ink, holds 150 gallons, and is $\frac{4}{5}$ full. Vat 2 has green ink, holds 50 gallons, and is $\frac{5}{8}$ full. One gallon of ink will fill 1200 pens. How many pens can be filled with the existing ink from Vats 1 and 2? 181,500 pens

To Think About *When multiplying or dividing mixed numbers it is wise to estimate your answer by rounding each mixed number to the nearest whole number.*

79. Estimate your answer to $14\frac{2}{3} \div 5\frac{1}{6}$ by rounding each mixed number to the nearest whole number. Then find the exact answer. How close was your estimate?
We estimate by dividing $15 \div 5$, which is 3. The exact value is $2\frac{26}{31}$. Our estimate is very close. It is off by only $\frac{5}{31}$.

80. Estimate your answer to $18\frac{1}{4} \times 27\frac{1}{2}$ by rounding each mixed number to the nearest whole number. Then find the exact answer. How close was your estimate? We estimate by multiplying 18×28 to obtain 504. The exact value is $501\frac{7}{8}$. Our estimate is very close. It is off by only $2\frac{1}{8}$.

Cumulative Review

81. **[1.1.3]** Write in words. 39,576,304 thirty-nine million, five hundred seventy-six thousand, three hundred four

82. **[1.1.1]** Write in expanded form. 509,270
$500,000 + 9000 + 200 + 70$

83. **[1.2.4]** Add. $126 + 34 + 9 + 891 + 12 + 27$
1099

84. **[1.1.3]** Write in standard notation. eighty-seven million, five hundred ninety-five thousand, six hundred thirty-one 87,595,631

Quick Quiz 2.5 Divide.

1. $\frac{15}{24} \div \frac{5}{6}$ $\frac{3}{4}$

2. $6\frac{1}{3} \div 2\frac{5}{12}$ $\frac{76}{29}$ or $2\frac{18}{29}$

3. $7\frac{3}{4} \div 4$ $\frac{31}{16}$ or $1\frac{15}{16}$

4. Concept Check Explain how you would divide the whole number 7 by the mixed number $3\frac{3}{5}$.
Answers may vary

Classroom Quiz 2.5 You may use these problems to quiz your students' mastery of Section 2.5.

Divide.

1. $\frac{16}{27} \div \frac{4}{13}$ **Ans:** $\frac{52}{27}$ or $1\frac{25}{27}$

2. $8\frac{1}{4} \div 3\frac{5}{6}$ **Ans:** $\frac{99}{46}$ or $2\frac{7}{46}$

3. $5\frac{1}{8} \div 3$ **Ans:** $\frac{41}{24}$ or $1\frac{17}{24}$

1. $\dfrac{3}{8}$

2. $\dfrac{8}{69}$

3. $\dfrac{5}{124}$

4. $\dfrac{1}{6}$

5. $\dfrac{1}{3}$

6. $\dfrac{1}{7}$

7. $\dfrac{7}{8}$

8. $\dfrac{4}{11}$

9. $\dfrac{11}{3}$

10. $\dfrac{46}{3}$

11. $20\dfrac{1}{4}$

12. $5\dfrac{4}{5}$

13. $2\dfrac{2}{17}$

14. $\dfrac{5}{44}$

15. $\dfrac{2}{3}$

16. $\dfrac{160}{9}$ or $17\dfrac{7}{9}$

17. 1

18. $\dfrac{1}{2}$

19. $\dfrac{69}{13}$ or $5\dfrac{4}{13}$

20. 21

How are you doing with your homework assignments in Sections 2.1 to 2.5? Do you feel you have mastered the material so far? Do you understand the concepts you have covered? Before you go further in the textbook, take some time to do each of the following problems.

2.1

1. Use a fraction to represent the shaded part of the object.

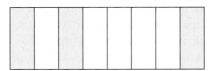

2. Frederich University had 3500 students from inside the state, 2600 students from outside the state but inside the country, and 800 students from outside the country. Write a fraction that describes the part of the student body from outside the country. Reduce the fraction.

3. An inspector checked 124 CD players. Of these, 5 were defective. Write a fraction that describes the part that was defective.

2.2

Reduce each fraction.

4. $\dfrac{3}{18}$ **5.** $\dfrac{13}{39}$ **6.** $\dfrac{16}{112}$ **7.** $\dfrac{175}{200}$ **8.** $\dfrac{44}{121}$

2.3

Change to an improper fraction.

9. $3\dfrac{2}{3}$ **10.** $15\dfrac{1}{3}$

Change to a mixed number.

11. $\dfrac{81}{4}$ **12.** $\dfrac{29}{5}$ **13.** $\dfrac{36}{17}$

2.4

Multiply.

14. $\dfrac{5}{11} \times \dfrac{1}{4}$ **15.** $\dfrac{3}{7} \times \dfrac{14}{9}$ **16.** $3\dfrac{1}{3} \times 5\dfrac{1}{3}$

2.5

Divide.

17. $\dfrac{3}{7} \div \dfrac{3}{7}$ **18.** $\dfrac{7}{16} \div \dfrac{7}{8}$ **19.** $6\dfrac{4}{7} \div 1\dfrac{5}{21}$ **20.** $12 \div \dfrac{4}{7}$

Now turn to page SA-5 for the answer to each of these problems. Each answer also includes a reference to the objective in which the problem is first taught. If you missed any of these problems, you should stop and review the Examples and Practice Problems in the referenced objective. A little review now will help you master the material in the upcoming sections of the text.

How Am I Doing? Test on Sections 2.1–2.5

Solve. Make sure all fractions are simplified in the final answer.

1. Norah answered 33 out of 40 questions correctly on her chemistry exam. Write a fraction that describes the part of the exam she answered correctly.

2. Carlos inspected the boxes that were shipped from the central warehouse. He found that 340 were the correct weight and 112 were not. Write a fraction that describes what part of the total number of the boxes were at the correct weight.

Reduce each fraction.

3. $\dfrac{19}{38}$

4. $\dfrac{40}{56}$

5. $\dfrac{24}{66}$

6. $\dfrac{125}{155}$

7. $\dfrac{50}{140}$

8. $\dfrac{84}{36}$

Change each mixed number to an improper fraction.

9. $12\dfrac{2}{3}$

10. $4\dfrac{1}{8}$

Change each improper fraction to a mixed number.

11. $\dfrac{45}{7}$

12. $\dfrac{75}{9}$

Multiply.

13. $\dfrac{3}{8} \times \dfrac{7}{11}$

14. $\dfrac{35}{16} \times \dfrac{4}{5}$

15. $18 \times \dfrac{5}{6}$

16. $\dfrac{3}{8} \times 44$

17. $2\dfrac{1}{3} \times 5\dfrac{3}{4}$

18. $24 \times 3\dfrac{1}{3}$

Divide.

19. $\dfrac{4}{7} \div \dfrac{3}{4}$

20. $\dfrac{8}{9} \div \dfrac{1}{6}$

21. $5\dfrac{1}{4} \div \dfrac{3}{4}$

22. $5\dfrac{3}{5} \div 2\dfrac{1}{3}$

1. $\dfrac{33}{40}$
2. $\dfrac{85}{113}$
3. $\dfrac{1}{2}$
4. $\dfrac{5}{7}$
5. $\dfrac{4}{11}$
6. $\dfrac{25}{31}$
7. $\dfrac{5}{14}$
8. $\dfrac{7}{3}$ or $2\dfrac{1}{3}$
9. $\dfrac{38}{3}$
10. $\dfrac{33}{8}$
11. $6\dfrac{3}{7}$
12. $8\dfrac{1}{3}$
13. $\dfrac{21}{88}$
14. $\dfrac{7}{4}$ or $1\dfrac{3}{4}$
15. 15
16. $\dfrac{33}{2}$ or $16\dfrac{1}{2}$
17. $\dfrac{161}{12}$ or $13\dfrac{5}{12}$
18. 80
19. $\dfrac{16}{21}$
20. $\dfrac{16}{3}$ or $5\dfrac{1}{3}$
21. 7
22. $\dfrac{12}{5}$ or $2\dfrac{2}{5}$

143

23. $\dfrac{63}{8}$ or $7\dfrac{7}{8}$

24. 14

25. $\dfrac{8}{3}$ or $2\dfrac{2}{3}$

26. $\dfrac{23}{8}$ or $2\dfrac{7}{8}$

27. $\dfrac{13}{16}$

28. $\dfrac{1}{14}$

29. $\dfrac{9}{32}$

30. $\dfrac{13}{15}$

31. $45\dfrac{15}{16}$ square feet

32. 4 cups

33. $46\dfrac{7}{8}$ miles

34. 16 full packages; $\dfrac{3}{8}$ lb left over

35. 51 computers

36. 24,600 gallons

37. 16 hours

38. 6 tents, 7 yards left over

39. 41 days

Mixed Practice

Perform the indicated operations. Simplify your answers.

23. $2\dfrac{1}{4} \times 3\dfrac{1}{2}$

24. $6 \times 2\dfrac{1}{3}$

25. $5 \div 1\dfrac{7}{8}$

26. $5\dfrac{3}{4} \div 2$

27. $\dfrac{13}{20} \div \dfrac{4}{5}$

28. $\dfrac{4}{7} \div 8$

29. $\dfrac{9}{22} \times \dfrac{11}{16}$

30. $\dfrac{14}{25} \times \dfrac{65}{42}$

Solve. Simplify your answer.

▲ **31.** A garden measures $5\dfrac{1}{4}$ feet by $8\dfrac{3}{4}$ feet. What is the area of the garden in square feet?

32. A recipe for two loaves of bread calls for $2\dfrac{2}{3}$ cups of flour. Lexi wants to make $1\dfrac{1}{2}$ times as much bread. How many cups of flour will she need?

33. Lisa drove $62\dfrac{1}{2}$ miles to visit a friend. Three-fourths of her trip was on the highway. How many miles did she drive on the highway?

34. The butcher prepared $12\dfrac{3}{8}$ pounds of lean ground round. He placed it in packages that held $\dfrac{3}{4}$ of a pound. How many full packages did he have? How much lean ground round was left over?

35. The college computer center has 136 computers. Samuel found that $\dfrac{3}{8}$ of them have Windows XP installed on them. How many computers have Windows XP installed on them?

36. The average household uses 82,000 gallons of water each year. About $\dfrac{3}{10}$ of this amount is used for showers and baths. How many gallons of water are used each year for showers and baths in an average household?

37. Yung Kim was paid $132 last week at his part-time job. He was paid $8\dfrac{1}{4}$ per hour. How many hours did he work last week?

38. The Outdoor Shop is making some custom tents that are very light but totally waterproof. Each tent requires $8\dfrac{1}{4}$ yards of cloth. How many tents can be made from $56\dfrac{1}{2}$ yards of cloth? How much cloth will be left over?

39. A container of vanilla-flavored syrup holds $32\dfrac{4}{5}$ ounces. Nate uses $\dfrac{4}{5}$ ounce every morning in his coffee. How many days will it take Nate to use up the container?

2.6 THE LEAST COMMON DENOMINATOR AND CREATING EQUIVALENT FRACTIONS

 Finding the Least Common Multiple (LCM) of Two Numbers

The idea of a multiple of a number is fairly straightforward.

The **multiples** of a number are the products of that number and the numbers 1, 2, 3, 4, 5, 6, 7, ...

For example, the multiples of 4 are   4, 8, 12, 16, 20, 24, 28, ...

The multiples of 5 are   5, 10, 15, 20, 25, 30, 35, ...

The **least common multiple,** or **LCM,** of two natural numbers is the smallest number that is a multiple of both.

EXAMPLE 1 Find the least common multiple of 10 and 12.

Solution

The multiples of 10 are   10,   20,   30,   40,   50,   60 ,   70, ...
The multiples of 12 are   12,   24,   36,   48,   60 ,   72,   84, ...

The first multiple that appears on both lists is the least common multiple. Thus the number 60 is the least common multiple of 10 and 12.

Practice Problem 1 Find the least common multiple of 14 and 21.   42

EXAMPLE 2 Find the least common multiple of 6 and 8.

Solution

The multiples of 6 are   6,   12,   18,   24 ,   30,   36,   42, ...
The multiples of 8 are   8,   16,   24 ,   32,   40,   48,   56, ...

The first multiple that appears on both lists is the least common multiple. Thus the number 24 is the least common multiple of 6 and 8.

Practice Problem 2 Find the least common multiple of 10 and 15.   30

Now of course we can do the problem immediately if the larger number is a multiple of the smaller number. In such cases the larger number is the least common multiple.

EXAMPLE 3 Find the least common multiple of 7 and 35.

Solution   Because $7 \times 5 = 35$, 35 is a multiple of 7.

So we can state immediately that the least common multiple of 7 and 35 is 35.

Practice Problem 3 Find the least common multiple of 6 and 54.   54

Student Learning Objectives

After studying this section, you will be able to:

 Find the least common multiple (LCM) of two numbers.

 Find the least common denominator (LCD) given two or three fractions.

 Create equivalent fractions with a least common denominator.

Teaching Example 1 Find the least common multiple of 15 and 20.

Ans: 60

NOTE TO STUDENT: Fully worked-out solutions to all of the Practice Problems can be found at the back of the text starting at page SP-1

Teaching Example 2 Find the least common multiple of 12 and 18.

Ans: 36

Teaching Example 3 Find the least common multiple of 11 and 33.

Ans: 33

Finding the Least Common Denominator (LCD) Given Two or Three Fractions

We need some way to determine which of two fractions is larger. Suppose that Marcia and Melissa each have some leftover pizza.

Marcia's Pizza
$\frac{1}{3}$ of a pizza left

Melissa's Pizza
$\frac{1}{4}$ of a pizza left

Who has more pizza left? How much more? Comparing the amounts of pizza left would be easy if each pizza had been cut into equal-sized pieces. If the original pizzas had each been cut into 12 pieces, we would be able to see that Marcia had $\frac{1}{12}$ of a pizza more than Melissa had.

Marcia's Pizza
$\left(\begin{array}{c} \text{We know that} \\ \frac{4}{12} = \frac{1}{3} \text{ by reducing.} \end{array}\right)$

Melissa's Pizza
$\left(\begin{array}{c} \text{We know that} \\ \frac{3}{12} = \frac{1}{4} \text{ by reducing.} \end{array}\right)$

The denominator 12 appears in the fractions $\frac{4}{12}$ and $\frac{3}{12}$. We call the smallest denominator that allows us to compare fractions directly the *least common denominator,* abbreviated LCD. The number 12 is the least common denominator for the fractions $\frac{1}{3}$ and $\frac{1}{4}$.

Notice that 12 is the least common multiple of 3 and 4.

> **LEAST COMMON DENOMINATOR**
>
> The **least common denominator (LCD)** of two or more fractions is the smallest number that can be divided evenly by each of the fractions' denominators.

How does this relate to least common multiples? The LCD of two fractions is the least common multiple of the two denominators.

In some problems you may be able to guess the LCD quite quickly. With practice, you can often find the LCD mentally. For example, you now know that if the denominators of two fractions are 3 and 4, the LCD is 12. For the fractions $\frac{1}{2}$ and $\frac{1}{4}$, the LCD is 4; for the fractions $\frac{1}{3}$ and $\frac{1}{6}$, the LCD is 6. We can see that if the denominator of one fraction divides without remainder into the denominator of another, the LCD of the two fractions is the larger of the denominators.

EXAMPLE 4 Determine the LCD for each pair of fractions.

(a) $\dfrac{7}{15}$ and $\dfrac{4}{5}$

(b) $\dfrac{2}{3}$ and $\dfrac{5}{27}$

Teaching Example 4 Determine the LCD for each pair of fractions.

(a) $\dfrac{7}{24}$ and $\dfrac{3}{8}$ **(b)** $\dfrac{2}{9}$ and $\dfrac{5}{36}$

Ans: (a) 24   **(b)** 36

Solution

(a) Since 5 can be divided into 15, the LCD of $\dfrac{7}{15}$ and $\dfrac{4}{5}$ is 15. (Notice that the least common multiple of 5 and 15 is 15.)

(b) Since 3 can be divided into 27, the LCD of $\dfrac{2}{3}$ and $\dfrac{5}{27}$ is 27. (Notice that the least common multiple of 3 and 27 is 27.)

Practice Problem 4 Determine the LCD for each pair of fractions.

(a) $\dfrac{3}{4}$ and $\dfrac{11}{12}$ 12

(b) $\dfrac{1}{7}$ and $\dfrac{8}{35}$ 35

NOTE TO STUDENT: Fully worked-out solutions to all of the Practice Problems can be found at the back of the text starting at page SP-1

In a few cases, the LCD is the product of the two denominators.

EXAMPLE 5 Find the LCD for $\dfrac{1}{4}$ and $\dfrac{3}{5}$.

Teaching Example 5 Find the LCD of $\frac{5}{8}$ and $\frac{2}{3}$.

Ans: 24

Solution We see that $4 \times 5 = 20$. Also, 20 is the *smallest* number that can be divided without remainder by 4 and by 5. We know this because the least common multiple of 4 and 5 is 20. So the LCD = 20.

Practice Problem 5 Find the LCD for $\dfrac{3}{7}$ and $\dfrac{5}{6}$. 42

In cases where the LCD is not obvious, the following procedure will help us find the LCD.

THREE-STEP PROCEDURE FOR FINDING THE LEAST COMMON DENOMINATOR

1. Write each denominator as the product of prime factors.
2. List all the prime factors that appear in either product.
3. Form a product of those prime factors, using each factor the greatest number of times it appears in any one denominator.

EXAMPLE 6 Find the LCD by the three-step procedure.

(a) $\dfrac{5}{6}$ and $\dfrac{4}{15}$

(b) $\dfrac{7}{18}$ and $\dfrac{7}{30}$

(c) $\dfrac{10}{27}$ and $\dfrac{5}{18}$

Teaching Tip Stress the fact that not all students will approach these problems the same way. Some students were taught in school to find the LCD, others to find the GCF, others the LCM. If a student wishes to find the LCD in a way different from the one presented in this book, that is fine as long as the student can obtain correct answers.

Solution

(a) Step 1 Write each denominator as a product of prime factors.

$$6 = 2 \times 3 \qquad 15 = 5 \times 3$$

Step 2 The LCD will contain the factors 2, 3, and 5.

$$6 = 2 \times 3 \qquad 15 = 5 \times 3$$

Step 3 LCD $= 2 \times 3 \times 5$ We form a product.

$$= 30$$

(b) Step 1 Write each denominator as a product of prime factors.

$$18 = 2 \times 9 = 2 \times 3 \times 3$$
$$30 = 3 \times 10 = 2 \times 3 \times 5$$

Step 2 The LCD will be a product containing 2, 3, and 5.

Step 3 The LCD will contain the factor 3 twice since it occurs twice in the denominator 18.

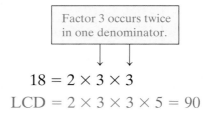

Factor 3 occurs twice in one denominator.

$$18 = 2 \times 3 \times 3$$
$$\text{LCD} = 2 \times 3 \times 3 \times 5 = 90$$

(c) Write each denominator as a product of prime factors.

$$27 = 3 \times 3 \times 3 \qquad 18 = 3 \times 3 \times 2$$

Factor 3 occurs three times.

The LCD will contain the factor 2 once but the factor 3 three times.

$$\text{LCD} = 2 \times 3 \times 3 \times 3 = 54$$

Practice Problem 6 Find the LCD for each pair of fractions.

(a) $\frac{3}{14}$ and $\frac{1}{10}$ 70   **(b)** $\frac{1}{15}$ and $\frac{7}{50}$ 150   **(c)** $\frac{3}{16}$ and $\frac{5}{12}$ 48

A similar procedure can be used for three fractions.

EXAMPLE 7 Find the LCD of $\frac{7}{12}, \frac{1}{15},$ and $\frac{11}{30}$.

Solution

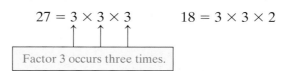

$$12 = 2 \times 2 \times 3$$
$$15 = \qquad 3 \times 5$$
$$30 = \qquad 2 \times 3 \times 5$$

$$\text{LCD} = 2 \times 2 \times 3 \times 5$$
$$= 60$$

Practice Problem 7   Find the LCD of $\dfrac{3}{49}$, $\dfrac{5}{21}$, and $\dfrac{6}{7}$.   147

Teaching Tip Ask the students to find the LCD for the fractions $\frac{3}{7}, \frac{4}{21}, \frac{7}{24}$. The correct answer is LCD $= 168$.

3 Creating Equivalent Fractions with a Least Common Denominator

In Section 2.7, we will discuss how to add fractions. We cannot add fractions that have different denominators. To change denominators, we must (1) find the LCD and (2) build up the addends—the fractions being added—into equivalent fractions that have the LCD as the denominator. We know now how to find the LCD. Let's look at how we build fractions. We know, for example, that

$$\frac{1}{2} = \frac{2}{4} = \frac{50}{100} \qquad \frac{1}{4} = \frac{25}{100} \quad \text{and} \quad \frac{3}{4} = \frac{75}{100}.$$

In these cases, we have mentally multiplied the given fraction by 1, in the form of a certain number, c, in the numerator and that same number, c, in the denominator.

$$\frac{1}{2} \times \boxed{\frac{c}{c}} = \frac{2}{4} \qquad \text{Here } c = 2, \frac{2}{2} = 1.$$

$$\frac{1}{2} \times \boxed{\frac{c}{c}} = \frac{50}{100} \qquad \text{Here } c = 50, \frac{50}{50} = 1.$$

This property is called the *building fraction property*.

> **BUILDING FRACTION PROPERTY**
>
> For whole numbers a, b, and c where $b \neq 0$, $c \neq 0$,
>
> $$\frac{a}{b} = \frac{a}{b} \times 1 = \frac{a}{b} \times \boxed{\frac{c}{c}} = \frac{a \times c}{b \times c}.$$

EXAMPLE 8   Build each fraction to an equivalent fraction with the given LCD.

(a) $\dfrac{3}{4}$, LCD $= 28$ **(b)** $\dfrac{4}{5}$, LCD $= 45$ **(c)** $\dfrac{1}{3}$ and $\dfrac{4}{5}$, LCD $= 15$

Solution

(a) $\dfrac{3}{4} \times \boxed{\dfrac{c}{c}} = \dfrac{?}{28}$ We know that $4 \times 7 = 28$, so the value c that we multiply numerator and denominator by is 7.

$$\frac{3}{4} \times \frac{7}{7} = \frac{21}{28}$$

(b) $\dfrac{4}{5} \times \boxed{\dfrac{c}{c}} = \dfrac{?}{45}$ We know that $5 \times 9 = 45$, so $c = 9$.

$$\frac{4}{5} \times \frac{9}{9} = \frac{36}{45}$$

Teaching Example 8 Build each fraction to an equivalent fraction with the given LCD.

(a) $\dfrac{7}{8}$, LCD $= 48$

(b) $\dfrac{5}{9}$, LCD $= 27$

(c) $\dfrac{4}{13}$ and $\dfrac{2}{3}$, LCD $= 39$

Ans: (a) $\dfrac{42}{48}$ **(b)** $\dfrac{15}{27}$

    (c) $\dfrac{4}{13} = \dfrac{12}{39}$ and $\dfrac{2}{3} = \dfrac{26}{39}$

(c) $\dfrac{1}{3} = \dfrac{?}{15}$ We know that $3 \times 5 = 15$, so we multiply numerator and denominator by 5.

$$\dfrac{1}{3} \times \boxed{\dfrac{5}{5}} = \dfrac{5}{15}$$

$$\dfrac{4}{5} = \dfrac{?}{15} \quad \text{We know that } 5 \times 3 = 15, \text{ so we multiply} \\ \text{numerator and denominator by 3.}$$

$$\dfrac{4}{5} \times \boxed{\dfrac{3}{3}} = \dfrac{12}{15}$$

Thus $\dfrac{1}{3} = \dfrac{5}{15}$ and $\dfrac{4}{5} = \dfrac{12}{15}$.

NOTE TO STUDENT: *Fully worked-out solutions to all of the Practice Problems can be found at the back of the text starting at page SP-1*

Practice Problem 8   Build each fraction to an equivalent fraction with the LCD.

(a) $\dfrac{3}{5}$, LCD = 40   **(b)** $\dfrac{7}{11}$, LCD = 44   **(c)** $\dfrac{2}{7}$ and $\dfrac{3}{4}$, LCD = 28

(a) $\dfrac{24}{40}$ **(b)** $\dfrac{28}{44}$ **(c)** $\dfrac{8}{28}$ and $\dfrac{21}{28}$

Teaching Example 9

(a) Find the LCD of $\frac{5}{12}$ and $\frac{7}{18}$.

(b) Build the fractions to equivalent fractions that have the LCD as their denominators.

Ans: (a) 36 **(b)** $\dfrac{15}{36}$ and $\dfrac{14}{36}$

EXAMPLE 9

(a) Find the LCD of $\dfrac{1}{32}$ and $\dfrac{7}{48}$.

(b) Build the fractions to equivalent fractions that have the LCD as their denominators.

Solution

(a) First we find the prime factors of 32 and 48.

$$32 = 2 \times 2 \times 2 \times 2 \times 2$$
$$48 = 2 \times 2 \times 2 \times 2 \times 3$$

Thus the LCD will require a factor of 2 five times and a factor of 3 one time.

$$\text{LCD} = 2 \times 2 \times 2 \times 2 \times 2 \times 3 = 96$$

(b) $\dfrac{1}{32} = \dfrac{?}{96}$ Since $32 \times 3 = 96$ we multiply by the fraction $\dfrac{3}{3}$.

$$\dfrac{1}{32} = \dfrac{1}{32} \times \boxed{\dfrac{3}{3}} = \dfrac{3}{96}$$

$$\dfrac{7}{48} = \dfrac{?}{96} \quad \text{Since } 48 \times 2 = 96, \text{ we multiply by the fraction } \dfrac{2}{2}.$$

$$\dfrac{7}{48} = \dfrac{7}{48} \times \boxed{\dfrac{2}{2}} = \dfrac{14}{96}$$

Practice Problem 9

(a) Find the LCD of $\dfrac{3}{20}$ and $\dfrac{11}{15}$.   60

(b) Build the fractions to equivalent fractions that have the LCD as their denominators. $\dfrac{9}{60}$ and $\dfrac{44}{60}$

Teaching Tip Call students' attention to the fact that some people have learned how to find the LCD of a fraction by using least common multiples. This is a good time to remind students that a good mathematician knows many ways to solve the same problem. It helps to be able to think of different approaches to solving problems in real life as well as in math.

EXAMPLE 10

(a) Find the LCD of $\dfrac{2}{125}$ and $\dfrac{8}{75}$.

(b) Build the fractions to equivalent fractions that have the LCD as their denominators.

Solution

(a) First we find the prime factors of 125 and 75.

$$125 = 5 \times 5 \times 5$$
$$75 = 5 \times 5 \times 3$$

Thus the LCD will require a factor of 5 three times and a factor of 3 one time.

$$\text{LCD} = 5 \times 5 \times 5 \times 3 = 375$$

(b) $\dfrac{2}{125} = \dfrac{?}{375}$ Since $125 \times 3 = 375$, we multiply by the fraction $\dfrac{3}{3}$.

$$\dfrac{2}{125} = \dfrac{2}{125} \times \boxed{\dfrac{3}{3}} = \dfrac{6}{375}$$

$\dfrac{8}{75} = \dfrac{?}{375}$ Since $75 \times 5 = 375$, we multiply by the fraction $\dfrac{5}{5}$.

$$\dfrac{8}{75} = \dfrac{8}{75} \times \boxed{\dfrac{5}{5}} = \dfrac{40}{375}$$

NOTE TO STUDENT: Fully worked-out solutions to all of the Practice Problems can be found at the back of the text starting at page SP-1

Practice Problem 10

(a) Find the LCD of $\dfrac{5}{64}$ and $\dfrac{3}{80}$.   320

(b) Build the fractions to equivalent fractions that have the LCD as their denominators. $\dfrac{25}{320}$ and $\dfrac{12}{320}$

Teaching Example 10

(a) Find the LCD of $\frac{3}{25}$ and $\frac{11}{15}$.

(b) Build the fractions to equivalent fractions that have the LCD as their denominators.

Ans: (a) 75   **(b)** $\dfrac{9}{75}$ and $\dfrac{55}{75}$

Find the least common multiple (LCM) for each pair of numbers.

1. 8 and 12
24

2. 6 and 9
18

3. 20 and 50
100

4. 22 and 55
110

5. 12 and 15
60

6. 18 and 30
90

7. 10 and 15
30

8. 8 and 60
120

9. 21 and 49
147

10. 25 and 35
175

Find the LCD for each pair of fractions.

11. $\dfrac{1}{5}$ and $\dfrac{3}{10}$
LCD = 10

12. $\dfrac{3}{8}$ and $\dfrac{5}{16}$
LCD = 16

13. $\dfrac{3}{7}$ and $\dfrac{1}{4}$
LCD = 28

14. $\dfrac{5}{6}$ and $\dfrac{3}{5}$
LCD = 30

15. $\dfrac{2}{5}$ and $\dfrac{3}{7}$
LCD = 35

16. $\dfrac{1}{16}$ and $\dfrac{2}{3}$
LCD = 48

17. $\dfrac{1}{6}$ and $\dfrac{5}{9}$
6 = 2 × 3
9 = 3 × 3
LCD = 18

18. $\dfrac{1}{4}$ and $\dfrac{3}{14}$
4 = 2 × 2
14 = 2 × 7
LCD = 28

19. $\dfrac{7}{12}$ and $\dfrac{14}{15}$
12 = 2 × 2 × 3
15 = 3 × 5
LCD = 60

20. $\dfrac{7}{15}$ and $\dfrac{9}{25}$
15 = 3 × 5
25 = 5 × 5
LCD = 75

21. $\dfrac{7}{32}$ and $\dfrac{3}{4}$
LCD = 32

22. $\dfrac{2}{11}$ and $\dfrac{1}{44}$
LCD = 44

23. $\dfrac{5}{10}$ and $\dfrac{11}{45}$
10 = 2 × 5
45 = 3 × 3 × 5
LCD = 90

24. $\dfrac{13}{20}$ and $\dfrac{17}{30}$
20 = 2 × 2 × 5
30 = 2 × 3 × 5
LCD = 60

25. $\dfrac{7}{16}$ and $\dfrac{17}{80}$
16 = 2 × 2 × 2 × 2
80 = 2 × 2 × 2 × 2 × 5
LCD = 80

26. $\dfrac{5}{6}$ and $\dfrac{19}{30}$
6 = 2 × 3
30 = 2 × 3 × 5
LCD = 30

27. $\dfrac{5}{21}$ and $\dfrac{8}{35}$
21 = 3 × 7
35 = 5 × 7
LCD = 105

28. $\dfrac{1}{20}$ and $\dfrac{5}{70}$
20 = 2 × 2 × 5
70 = 2 × 5 × 7
LCD = 140

29. $\dfrac{11}{24}$ and $\dfrac{7}{30}$
24 = 2 × 2 × 2 × 3
30 = 2 × 3 × 5
LCD = 120

30. $\dfrac{23}{30}$ and $\dfrac{37}{50}$
30 = 2 × 3 × 5
50 = 2 × 5 × 5
LCD = 150

Find the LCD for each set of three fractions.

31. $\dfrac{2}{3}, \dfrac{1}{2}, \dfrac{5}{6}$
LCD = 6

32. $\dfrac{1}{5}, \dfrac{1}{3}, \dfrac{7}{10}$
LCD = 30

33. $\dfrac{1}{4}, \dfrac{11}{12}, \dfrac{5}{6}$
4 = 2 × 2
12 = 2 × 2 × 3
6 = 2 × 3
LCD = 12

34. $\dfrac{21}{48}, \dfrac{1}{12}, \dfrac{3}{8}$
48 = 2 × 2 × 2 × 2 × 3
12 = 2 × 2 × 3
8 = 2 × 2 × 2
LCD = 48

35. $\dfrac{5}{11}, \dfrac{7}{12}, \dfrac{1}{6}$
11 = 11
12 = 2 × 2 × 3
6 = 2 × 3
LCD = 132

36. $\dfrac{11}{16}, \dfrac{3}{20}, \dfrac{2}{5}$
16 = 2 × 2 × 2 × 2
20 = 2 × 2 × 5
5 = 5
LCD = 80

37. $\dfrac{7}{12}, \dfrac{1}{21}, \dfrac{3}{14}$
12 = 2 × 2 × 3
21 = 3 × 7
14 = 2 × 7
LCD = 84

38. $\dfrac{1}{30}, \dfrac{3}{40}, \dfrac{7}{8}$
30 = 2 × 3 × 5
40 = 2 × 2 × 2 × 5
8 = 2 × 2 × 2
LCD = 120

39. $\dfrac{7}{15}, \dfrac{11}{12}, \dfrac{7}{8}$
15 = 3 × 5
12 = 2 × 2 × 3
8 = 2 × 2 × 2
LCD = 120

40. $\dfrac{5}{36}, \dfrac{2}{48}, \dfrac{1}{24}$
36 = 2 × 2 × 3 × 3
48 = 2 × 2 × 2 × 2 × 3
24 = 2 × 2 × 2 × 3
LCD = 144

Build each fraction to an equivalent fraction with the specified denominator. State the numerator.

41. $\dfrac{1}{3} = \dfrac{?}{9}$

3

42. $\dfrac{1}{5} = \dfrac{?}{35}$

7

43. $\dfrac{5}{7} = \dfrac{?}{49}$

35

44. $\dfrac{7}{9} = \dfrac{?}{81}$

63

45. $\dfrac{4}{11} = \dfrac{?}{55}$

20

46. $\dfrac{2}{13} = \dfrac{?}{39}$

6

47. $\dfrac{5}{12} = \dfrac{?}{96}$

40

48. $\dfrac{3}{50} = \dfrac{?}{100}$

6

49. $\dfrac{8}{9} = \dfrac{?}{108}$

96

50. $\dfrac{6}{7} = \dfrac{?}{147}$

126

51. $\dfrac{7}{20} = \dfrac{?}{180}$

63

52. $\dfrac{3}{25} = \dfrac{?}{175}$

21

The LCD of each pair of fractions is listed. Build each fraction to an equivalent fraction that has the LCD as the denominator.

53. LCD $= 36$, $\dfrac{7}{12}$ and $\dfrac{5}{9}$

$\dfrac{21}{36}$ and $\dfrac{20}{36}$

54. LCD $= 20$, $\dfrac{9}{10}$ and $\dfrac{3}{4}$

$\dfrac{18}{20}$ and $\dfrac{15}{20}$

55. LCD $= 80$, $\dfrac{5}{16}$ and $\dfrac{17}{20}$

$\dfrac{25}{80}$ and $\dfrac{68}{80}$

56. LCD $= 72$, $\dfrac{5}{24}$ and $\dfrac{7}{36}$

$\dfrac{15}{72}$ and $\dfrac{14}{72}$

57. LCD $= 20$, $\dfrac{9}{10}$ and $\dfrac{19}{20}$

$\dfrac{18}{20}$ and $\dfrac{19}{20}$

58. LCD $= 240$, $\dfrac{13}{30}$ and $\dfrac{41}{80}$

$\dfrac{104}{240}$ and $\dfrac{123}{240}$

Find the LCD. Build the fractions to equivalent fractions having the LCD as the denominator.

59. $\dfrac{2}{5}$ and $\dfrac{9}{35}$

LCD $= 35$

$\dfrac{14}{35}$ and $\dfrac{9}{35}$

60. $\dfrac{7}{9}$ and $\dfrac{35}{54}$

LCD $= 54$

$\dfrac{42}{54}$ and $\dfrac{35}{54}$

61. $\dfrac{5}{24}$ and $\dfrac{3}{8}$

LCD $= 24$

$\dfrac{5}{24}$ and $\dfrac{9}{24}$

62. $\dfrac{19}{42}$ and $\dfrac{6}{7}$

LCD $= 42$

$\dfrac{19}{42}$ and $\dfrac{36}{42}$

63. $\dfrac{8}{15}$ and $\dfrac{1}{6}$

LCD $= 30$

$\dfrac{16}{30}$ and $\dfrac{5}{30}$

64. $\dfrac{19}{20}$ and $\dfrac{7}{8}$

LCD $= 40$

$\dfrac{38}{40}$ and $\dfrac{35}{40}$

65. $\dfrac{4}{15}$ and $\dfrac{5}{12}$

LCD $= 60$

$\dfrac{16}{60}$ and $\dfrac{25}{60}$

66. $\dfrac{9}{10}$ and $\dfrac{3}{25}$

LCD $= 50$

$\dfrac{45}{50}$ and $\dfrac{6}{50}$

67. $\dfrac{5}{18}, \dfrac{11}{36}, \dfrac{7}{12}$

LCD $= 36$

$\dfrac{10}{36}, \dfrac{11}{36}, \dfrac{21}{36}$

68. $\dfrac{1}{30}, \dfrac{7}{15}, \dfrac{1}{45}$

LCD $= 90$

$\dfrac{3}{90}, \dfrac{42}{90}, \dfrac{2}{90}$

69. $\dfrac{3}{56}, \dfrac{7}{8}, \dfrac{5}{7}$

LCD $= 56$

$\dfrac{3}{56}, \dfrac{49}{56}, \dfrac{40}{56}$

70. $\dfrac{5}{9}, \dfrac{1}{6}, \dfrac{3}{54}$

LCD $= 54$

$\dfrac{30}{54}, \dfrac{9}{54}, \dfrac{3}{54}$

71. $\dfrac{5}{63}, \dfrac{4}{21}, \dfrac{8}{9}$

LCD $= 63$

$\dfrac{5}{63}, \dfrac{12}{63}, \dfrac{56}{63}$

72. $\dfrac{3}{8}, \dfrac{5}{14}, \dfrac{13}{16}$

LCD $= 112$

$\dfrac{42}{112}, \dfrac{40}{112}, \dfrac{91}{112}$

Applications

73. *Door Repair* Suppose that you wish to compare the lengths of the three portions of the given stainless steel bolt that came out of a door.

(a) What is the LCD for the three fractions?
LCD = 16

(b) Build each fraction to an equivalent fraction that has the LCD as a denominator.
$\frac{3}{16}, \frac{12}{16}, \frac{6}{16}$

74. *Plant Growth* Suppose that you want to prepare a report on the growth of a plant. The total height of the plant in the pot is recorded for each week of a three-week experiment.

(a) What is the LCD for the three fractions?
LCD = 96

(b) Build each fraction to an equivalent fraction that has the LCD for a denominator.
$\frac{15}{96}, \frac{80}{96}, \frac{84}{96}$

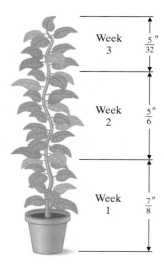

Cumulative Review

75. [1.5.3] Divide. $35\overline{)7293}$
208 R13

76. [1.4.4] Multiply. 2566×30
76,980

77. [1.6.2] Evaluate. $(5-3)^2 + 4 \times 6 - 3$
25

Quick Quiz 2.6

1. Find the least common denominator of
$\frac{5}{6}$ and $\frac{5}{21}$ 42

2. Find the least common denominator of
$\frac{27}{28}, \frac{3}{4}, \frac{19}{20}$ 140

3. Build the fraction to an equivalent fraction with the specified denominator.
$\frac{7}{26} = \frac{?}{78}$ $\frac{21}{78}$

4. Concept Check Explain how you would find the least common denominator of the fractions $\frac{5}{6}, \frac{11}{14},$ and $\frac{2}{15}$.
Answers may vary

Classroom Quiz 2.6 You may use these problems to quiz your students' mastery of Section 2.6.

1. Find the least common denominator of
$\frac{11}{14}$ and $\frac{8}{35}$
Ans: 70

2. Find the least common denominator of
$\frac{3}{5}, \frac{7}{8}, \frac{9}{10}$
Ans: 40

3. Build the fraction to an equivalent fraction with the specified denominator.
$\frac{11}{18} = \frac{?}{72}$
Ans: $\frac{44}{72}$

Adding and Subtracting Fractions with a Common Denominator

You must have common denominators (denominators that are alike) to add or subtract fractions.

If your problem has fractions without a common denominator or if it has mixed numbers, you must use what you already know about changing the form of each fraction (how the fraction looks). Only after all the fractions have a common denominator can you add or subtract.

An important distinction: You must have common denominators to add or subtract fractions, but you need not have common denominators to multiply or divide fractions.

To add two fractions that have the same denominator, add the numerators and write the sum over the common denominator.

To illustrate we use $\frac{1}{5} + \frac{2}{5} = \frac{3}{5}$. The figure shows that $\frac{1}{5} + \frac{2}{5} = \frac{3}{5}$.

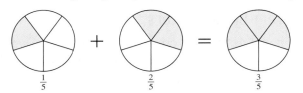

$$\frac{1}{5} \qquad \qquad \frac{2}{5} \qquad \qquad \frac{3}{5}$$

EXAMPLE 1 Add. $\frac{5}{13} + \frac{7}{13}$

Solution

$$\frac{5}{13} + \frac{7}{13} = \frac{12}{13}$$

Practice Problem 1 Add.

$$\frac{3}{17} + \frac{12}{17} \quad \frac{15}{17}$$

The answer may need to be reduced. Sometimes the answer may be written as a mixed number.

EXAMPLE 2 Add.

(a) $\frac{4}{9} + \frac{2}{9}$ **(b)** $\frac{5}{7} + \frac{6}{7}$

Solution

(a) $\frac{4}{9} + \frac{2}{9} = \frac{6}{9} = \frac{2}{3}$ **(b)** $\frac{5}{7} + \frac{6}{7} = \frac{11}{7}$ or $1\frac{4}{7}$

Practice Problem 2 Add.

(a) $\frac{1}{12} + \frac{5}{12} \quad \frac{1}{2}$ **(b)** $\frac{13}{15} + \frac{7}{15} \quad \frac{4}{3}$ or $1\frac{1}{3}$

Student Learning Objectives

After studying this section, you will be able to:

 Add and subtract fractions with a common denominator.

 Add and subtract fractions with different denominators.

Teaching Example 1 Add. $\frac{7}{11} + \frac{3}{11}$

Ans: $\frac{10}{11}$

NOTE TO STUDENT: Fully worked-out solutions to all of the Practice Problems can be found at the back of the text starting at page SP-1

Teaching Example 2 Add.

(a) $\frac{2}{15} + \frac{7}{15}$ **(b)** $\frac{7}{17} + \frac{11}{17}$

Ans: **(a)** $\frac{3}{5}$ **(b)** $\frac{18}{17}$ or $1\frac{1}{17}$

A similar rule is followed for subtraction, except that the numerators are subtracted and the result placed over the common denominator. Be sure to reduce all answers when possible.

Teaching Example 3 Subtract.

(a) $\dfrac{10}{13} - \dfrac{4}{13}$ (b) $\dfrac{18}{25} - \dfrac{3}{25}$

Ans: (a) $\dfrac{6}{13}$ (b) $\dfrac{3}{5}$

EXAMPLE 3   Subtract.

(a) $\dfrac{5}{13} - \dfrac{4}{13}$

(b) $\dfrac{17}{20} - \dfrac{3}{20}$

Solution

(a) $\dfrac{5}{13} - \dfrac{4}{13} = \dfrac{1}{13}$

(b) $\dfrac{17}{20} - \dfrac{3}{20} = \dfrac{14}{20} = \dfrac{7}{10}$

Practice Problem 3   Subtract.

(a) $\dfrac{5}{19} - \dfrac{2}{19}$ $\dfrac{3}{19}$

(b) $\dfrac{21}{25} - \dfrac{6}{25}$ $\dfrac{3}{5}$

② Adding and Subtracting Fractions with Different Denominators

If the two fractions do not have a common denominator, we follow the procedure in Section 2.6: Find the LCD and then build each fraction so that its denominator is the LCD.

Teaching Example 4 Add. $\dfrac{3}{14} + \dfrac{1}{2}$

Ans: $\dfrac{5}{7}$

EXAMPLE 4   Add. $\dfrac{7}{12} + \dfrac{1}{4}$

Solution   The LCD is 12. The fraction $\frac{7}{12}$ already has the least common denominator.

$$
\begin{array}{rcl}
\dfrac{7}{12} & = & \boxed{\dfrac{7}{12}} \\[2mm]
+\ \dfrac{1}{4} \times \dfrac{3}{3} & = & +\ \dfrac{3}{12} \\[2mm]
\hline
& & \dfrac{10}{12}
\end{array}
$$

We will need to reduce this fraction. Then we will have

$$\dfrac{7}{12} + \dfrac{1}{4} = \dfrac{7}{12} + \dfrac{3}{12} = \dfrac{10}{12} = \dfrac{5}{6}.$$

It is very important to remember to reduce our final answer.

Practice Problem 4   Add.

$$\dfrac{2}{15} + \dfrac{1}{5}\quad \dfrac{1}{3}$$

EXAMPLE 5 Add. $\dfrac{7}{20} + \dfrac{4}{15}$

Solution LCD = 60.

$$\dfrac{7}{20} \times \dfrac{3}{3} = \dfrac{21}{60} \qquad \dfrac{4}{15} \times \dfrac{4}{4} = \dfrac{16}{60}$$

Thus

$$\dfrac{7}{20} + \dfrac{4}{15} = \dfrac{21}{60} + \dfrac{16}{60} = \dfrac{37}{60}$$

Teaching Example 5 Add. $\dfrac{11}{30} + \dfrac{7}{45}$

Ans: $\dfrac{47}{90}$

Practice Problem 5 Add.

$$\dfrac{5}{12} + \dfrac{5}{16} \quad \dfrac{35}{48}$$

A similar procedure holds for the addition of three or more fractions.

EXAMPLE 6 Add. $\dfrac{3}{8} + \dfrac{5}{6} + \dfrac{1}{4}$

Solution LCD = 24.

$$\dfrac{3}{8} \times \dfrac{3}{3} = \dfrac{9}{24} \quad \dfrac{5}{6} \times \dfrac{4}{4} = \dfrac{20}{24} \quad \dfrac{1}{4} \times \dfrac{6}{6} = \dfrac{6}{24}$$

$$\dfrac{3}{8} + \dfrac{5}{6} + \dfrac{1}{4} = \dfrac{9}{24} + \dfrac{20}{24} + \dfrac{6}{24} = \dfrac{35}{24} \quad \text{or} \quad 1\dfrac{11}{24}$$

Teaching Example 6 Add. $\dfrac{5}{12} + \dfrac{4}{9} + \dfrac{1}{3}$

Ans: $1\dfrac{7}{36}$

Practice Problem 6 Add.

$$\dfrac{3}{16} + \dfrac{1}{8} + \dfrac{1}{12} \quad \dfrac{19}{48}$$

EXAMPLE 7 Subtract. $\dfrac{17}{25} - \dfrac{3}{35}$

Solution LCD = 175.

$$\dfrac{17}{25} \times \dfrac{7}{7} = \dfrac{119}{175} \quad \dfrac{3}{35} \times \dfrac{5}{5} = \dfrac{15}{175}$$

Thus

$$\dfrac{17}{25} - \dfrac{3}{35} = \dfrac{119}{175} - \dfrac{15}{175} = \dfrac{104}{175}.$$

Teaching Example 7 Subtract. $\dfrac{8}{14} - \dfrac{2}{21}$

Ans: $\dfrac{10}{21}$

Practice Problem 7 Subtract.

$$\dfrac{9}{48} - \dfrac{5}{32} \quad \dfrac{1}{32}$$

▲ **EXAMPLE 8**   John and Stephanie have a house on $\frac{7}{8}$ acre of land. They have $\frac{1}{3}$ acre of land planted with grass. How much of the land is not planted with grass?

Solution

1. **Understand the problem.** Draw a picture.

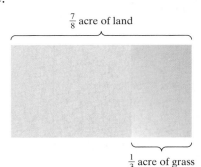

$\frac{7}{8}$ acre of land

$\frac{1}{3}$ acre of grass

We need to subtract. $\dfrac{7}{8} - \dfrac{1}{3}$

2. **Solve and state the answer.** The LCD is 24.

$$\frac{7}{8} \times \frac{3}{3} = \frac{21}{24} \qquad \frac{1}{3} \times \frac{8}{8} = \frac{8}{24}$$

$$\frac{7}{8} - \frac{1}{3} = \frac{21}{24} - \frac{8}{24} = \frac{13}{24}$$

We conclude that $\dfrac{13}{24}$ acre of the land is not planted with grass.

3. **Check.** The check is left to the student.

Practice Problem 8   Leon had $\frac{9}{10}$ gallon of cleaning fluid in the garage. He used $\frac{1}{4}$ gallon to clean the garage floor. How much cleaning fluid is left?

$\dfrac{13}{20}$ gallon

Some students may find Example 9 difficult. Read it slowly and carefully.

EXAMPLE 9   Find the value of x in the equation $x + \frac{5}{6} = \frac{9}{10}$. Reduce your answer.

Solution   The LCD for the two fractions $\dfrac{5}{6}$ and $\dfrac{9}{10}$ is 30.

$$\frac{5}{6} \times \frac{5}{5} = \frac{25}{30} \qquad \frac{9}{10} \times \frac{3}{3} = \frac{27}{30}$$

Thus we can write the equation in the equivalent form.

$$x + \frac{25}{30} = \frac{27}{30}$$

The denominators are the same. Look at the numerators. We must add 2 to 25 to get 27.

$$\frac{2}{30} + \frac{25}{30} = \frac{27}{30}$$

So $x = \frac{2}{30}$ and we reduce the fraction to obtain $x = \frac{1}{15}$.

Practice Problem 9 Find the value of x in the equation $x + \frac{3}{10} = \frac{23}{25}$.

$$x = \frac{31}{50}$$

ALTERNATE METHOD: Multiple the Denominators as a Common Denominator In all the problems in this section so far, we have combined two fractions by first finding the least common denominator. However, there is an alternate approach. You are only required to find a common denominator, not necessarily the least common denominator. One way to quickly find a common denominator of two fractions is to multiply the two denominators. However, if you use this method, the numbers will usually be larger and you will usually need to simplify the fraction in your final answer.

EXAMPLE 10 Add $\frac{11}{12} + \frac{13}{30}$ by using the product of the two denominators as a common denominator.

Teaching Example 10 Add $\frac{13}{20} + \frac{23}{30}$ by using the product of the two denominators as a common denominator.

Ans: $\frac{17}{12}$ or $1\frac{5}{12}$

Solution Using this method we just multiply the numerator and denominator of each fraction by the denominator of the other fraction. Thus no steps are needed to determine what to multiply by.

$$\frac{11}{12} \times \frac{30}{30} = \frac{330}{360} \qquad \frac{13}{30} \times \frac{12}{12} = \frac{156}{360}$$

$$\text{Thus} \quad \frac{11}{12} + \frac{13}{30} = \frac{330}{360} + \frac{156}{360} = \frac{486}{360}$$

We must reduce the fraction: $\quad \dfrac{486}{360} = \dfrac{27}{20} \quad \text{or} \quad 1\dfrac{7}{20}$

Practice Problem 10 Add $\frac{15}{16} + \frac{3}{40}$ by using the product of the two denominators as a common denominator. $\frac{81}{80}$ or $1\frac{1}{80}$

NOTE TO STUDENT: Fully worked-out solutions to all of the Practice Problems can be found at the back of the text starting at page SP-1

Some students find this alternate method helpful because you do not have to find the LCD or the number each fraction must be multiplied by. Other students find this alternate method more difficult because of errors encountered when working with large numbers or in reducing the final answer. You are encouraged to try a couple of the homework exercises by this method and make up your own mind.

Add or subtract. Simplify all answers.

1. $\dfrac{5}{9} + \dfrac{2}{9}$ $\dfrac{7}{9}$

2. $\dfrac{5}{8} + \dfrac{2}{8}$ $\dfrac{7}{8}$

3. $\dfrac{7}{18} + \dfrac{15}{18}$ $\dfrac{22}{18} = \dfrac{11}{9}$ or $1\dfrac{2}{9}$

4. $\dfrac{11}{25} + \dfrac{17}{25}$ $\dfrac{28}{25}$ or $1\dfrac{3}{25}$

5. $\dfrac{19}{20} - \dfrac{11}{20}$ $\dfrac{8}{20} = \dfrac{2}{5}$

6. $\dfrac{17}{30} - \dfrac{7}{30}$ $\dfrac{10}{30} = \dfrac{1}{3}$

7. $\dfrac{53}{88} - \dfrac{19}{88}$ $\dfrac{34}{88} = \dfrac{17}{44}$

8. $\dfrac{103}{110} - \dfrac{3}{110}$ $\dfrac{100}{110} = \dfrac{10}{11}$

Add or subtract. Simplify all answers.

9. $\dfrac{1}{3} + \dfrac{1}{2}$ $\dfrac{2}{6} + \dfrac{3}{6} = \dfrac{5}{6}$

10. $\dfrac{1}{4} + \dfrac{1}{3}$ $\dfrac{3}{12} + \dfrac{4}{12} = \dfrac{7}{12}$

11. $\dfrac{3}{10} + \dfrac{3}{20}$ $\dfrac{6}{20} + \dfrac{3}{20} = \dfrac{9}{20}$

12. $\dfrac{4}{9} + \dfrac{1}{6}$ $\dfrac{8}{18} + \dfrac{3}{18} = \dfrac{11}{18}$

13. $\dfrac{1}{8} + \dfrac{3}{4}$ $\dfrac{1}{8} + \dfrac{6}{8} = \dfrac{7}{8}$

14. $\dfrac{5}{16} + \dfrac{1}{2}$ $\dfrac{5}{16} + \dfrac{8}{16} = \dfrac{13}{16}$

15. $\dfrac{4}{5} + \dfrac{7}{20}$ $\dfrac{16}{20} + \dfrac{7}{20} = \dfrac{23}{20}$ or $1\dfrac{3}{20}$

16. $\dfrac{2}{3} + \dfrac{4}{7}$ $\dfrac{14}{21} + \dfrac{12}{21} = \dfrac{26}{21}$ or $1\dfrac{5}{21}$

17. $\dfrac{3}{10} + \dfrac{7}{100}$ $\dfrac{30}{100} + \dfrac{7}{100} = \dfrac{37}{100}$

18. $\dfrac{13}{100} + \dfrac{7}{10}$ $\dfrac{13}{100} + \dfrac{70}{100} = \dfrac{83}{100}$

19. $\dfrac{3}{10} + \dfrac{1}{6}$ $\dfrac{9}{30} + \dfrac{5}{30} = \dfrac{14}{30} = \dfrac{7}{15}$

20. $\dfrac{8}{15} + \dfrac{3}{10}$ $\dfrac{16}{30} + \dfrac{9}{30} = \dfrac{25}{30} = \dfrac{5}{6}$

21. $\dfrac{7}{8} + \dfrac{5}{12}$ $\dfrac{21}{24} + \dfrac{10}{24} = \dfrac{31}{24}$ or $1\dfrac{7}{24}$

22. $\dfrac{5}{6} + \dfrac{7}{8}$ $\dfrac{20}{24} + \dfrac{21}{24} = \dfrac{41}{24}$ or $1\dfrac{17}{24}$

23. $\dfrac{3}{8} + \dfrac{3}{10}$ $\dfrac{15}{40} + \dfrac{12}{40} = \dfrac{27}{40}$

24. $\dfrac{12}{35} + \dfrac{1}{10}$ $\dfrac{24}{70} + \dfrac{7}{70} = \dfrac{31}{70}$

25. $\dfrac{29}{18} - \dfrac{5}{9}$ $\dfrac{29}{18} - \dfrac{10}{18} = \dfrac{19}{18}$ or $1\dfrac{1}{18}$

26. $\dfrac{37}{20} - \dfrac{2}{5}$ $\dfrac{37}{20} - \dfrac{8}{20} = \dfrac{29}{20}$ or $1\dfrac{9}{20}$

27. $\dfrac{3}{7} - \dfrac{9}{21}$ $\dfrac{9}{21} - \dfrac{9}{21} = 0$

28. $\dfrac{7}{8} - \dfrac{5}{6}$ $\dfrac{21}{24} - \dfrac{20}{24} = \dfrac{1}{24}$

29. $\dfrac{5}{9} - \dfrac{5}{36}$ $\dfrac{20}{36} - \dfrac{5}{36} = \dfrac{15}{36} = \dfrac{5}{12}$

30. $\dfrac{9}{10} - \dfrac{1}{15}$ $\dfrac{27}{30} - \dfrac{2}{30} = \dfrac{25}{30} = \dfrac{5}{6}$

31. $\dfrac{5}{12} - \dfrac{7}{30}$ $\dfrac{25}{60} - \dfrac{14}{60} = \dfrac{11}{60}$

32. $\dfrac{9}{24} - \dfrac{3}{8}$ $\dfrac{9}{24} - \dfrac{9}{24} = 0$

33. $\dfrac{11}{12} - \dfrac{2}{3}$ $\dfrac{11}{12} - \dfrac{8}{12} = \dfrac{3}{12} = \dfrac{1}{4}$

34. $\dfrac{7}{10} - \dfrac{2}{5}$ $\dfrac{7}{10} - \dfrac{4}{10} = \dfrac{3}{10}$

35. $\dfrac{17}{21} - \dfrac{1}{7}$ $\dfrac{17}{21} - \dfrac{3}{21} = \dfrac{14}{21} = \dfrac{2}{3}$

36. $\dfrac{20}{25} - \dfrac{4}{5}$ $\dfrac{20}{25} - \dfrac{20}{25} = 0$

37. $\dfrac{5}{12} - \dfrac{7}{18}$ $\dfrac{15}{36} - \dfrac{14}{36} = \dfrac{1}{36}$

38. $\dfrac{7}{8} - \dfrac{1}{12}$ $\dfrac{21}{24} - \dfrac{2}{24} = \dfrac{19}{24}$

39. $\dfrac{10}{16} - \dfrac{5}{8}$ $\dfrac{10}{16} - \dfrac{10}{16} = 0$

40. $\dfrac{5}{6} - \dfrac{10}{12}$ $\dfrac{10}{12} - \dfrac{10}{12} = 0$

41. $\dfrac{23}{36} - \dfrac{2}{9}$ $\dfrac{23}{36} - \dfrac{8}{36} = \dfrac{15}{36} = \dfrac{5}{12}$

42. $\dfrac{2}{3} - \dfrac{1}{16}$ $\dfrac{32}{48} - \dfrac{3}{48} = \dfrac{29}{48}$

43. $\dfrac{1}{2} + \dfrac{2}{7} + \dfrac{3}{14}$ $\quad \dfrac{7}{14} + \dfrac{4}{14} + \dfrac{3}{14} = \dfrac{14}{14} = 1$

44. $\dfrac{7}{8} + \dfrac{5}{6} + \dfrac{7}{24}$ $\quad \dfrac{21}{24} + \dfrac{20}{24} + \dfrac{7}{24} = \dfrac{48}{24} = 2$

45. $\dfrac{5}{30} + \dfrac{3}{40} + \dfrac{1}{8}$ $\quad \dfrac{20}{120} + \dfrac{9}{120} + \dfrac{15}{120} = \dfrac{44}{120} = \dfrac{11}{30}$

46. $\dfrac{1}{12} + \dfrac{3}{14} + \dfrac{4}{21}$ $\quad \dfrac{7}{84} + \dfrac{18}{84} + \dfrac{16}{84} = \dfrac{41}{84}$

47. $\dfrac{7}{30} + \dfrac{2}{5} + \dfrac{5}{6}$ $\quad \dfrac{7}{30} + \dfrac{12}{30} + \dfrac{25}{30} = \dfrac{44}{30} = \dfrac{22}{15}$ or $1\dfrac{7}{15}$

48. $\dfrac{1}{12} + \dfrac{5}{36} + \dfrac{32}{36}$ $\quad \dfrac{3}{36} + \dfrac{5}{36} + \dfrac{32}{36} = \dfrac{40}{36} = \dfrac{10}{9}$ or $1\dfrac{1}{9}$

Study Example 9 carefully. Then find the value of x in each equation.

49. $x + \dfrac{1}{7} = \dfrac{5}{14}$ $\quad x = \dfrac{3}{14}$

50. $x + \dfrac{1}{8} = \dfrac{7}{16}$ $\quad x = \dfrac{5}{16}$

51. $x + \dfrac{2}{3} = \dfrac{9}{11}$ $\quad x = \dfrac{5}{33}$

52. $x + \dfrac{3}{4} = \dfrac{17}{18}$ $\quad x = \dfrac{7}{36}$

53. $x - \dfrac{3}{10} = \dfrac{4}{15}$ $\quad x = \dfrac{17}{30}$

54. $x - \dfrac{3}{14} = \dfrac{17}{28}$ $\quad x = \dfrac{23}{28}$

Applications

55. *Cooking* Rita is baking a cake for a dinner party. The recipe calls for $\frac{2}{3}$ cup sugar for the frosting and $\frac{3}{4}$ cup sugar for the cake. How many total cups of sugar does she need?

$1\dfrac{5}{12}$ cups

56. *Fitness Training* Kia is training for a short triathlon. On Monday she swam $\frac{1}{4}$ mile and ran $\frac{5}{6}$ mile. On Tuesday she swam $\frac{1}{2}$ mile and ran $\frac{3}{4}$ mile. How many miles has she swum so far this week? How many miles has she run so far?

$\dfrac{3}{4}$ mile; $1\dfrac{7}{12}$ miles

57. *Food Purchase* Yasmin wants to make a trail mix of nuts and dried fruit. She has $\frac{2}{3}$ pound peanuts and $\frac{1}{2}$ pound dried cranberries. She purchases $\frac{3}{4}$ pound almonds and $\frac{3}{8}$ pound raisins to mix with the peanuts and cranberries. After mixing the four ingredients how many pounds of nuts and how many pounds of dried fruit will there be in the trail mix?

$\dfrac{17}{12}$ or $1\dfrac{5}{12}$ pounds of nuts; $\dfrac{7}{8}$ pound of dried fruit

58. *Automobile Maintenance* Mandy purchased two new steel-belted all-weather radial tires for her car. The tread depth on the new tires measures $\frac{11}{32}$ of an inch. The dealer told her that when the tires have worn down and their tread depth measures $\frac{1}{8}$ of an inch, she should replace the worn tires with new ones. How much will the tread depth decrease over the useful life of the tire?

$\dfrac{7}{32}$ of an inch

59. *Power Outage* Travis typed $\frac{11}{12}$ of his book report on his computer. Then he printed out $\frac{3}{5}$ of his book report on his computer printer. Suddenly, there was a power outage, and he discovered that he hadn't saved his book report before the power went off. What fractional part of the book report was lost when the power failed?

$\dfrac{19}{60}$ of the book report

60. *Childcare* An infant's father knows that straight apple juice is too strong for his daughter. Her bottle is $\frac{1}{2}$ full, and he adds $\frac{1}{3}$ of a bottle of water to dilute the apple juice.

(a) How much is there to drink in the bottle after this addition? $\dfrac{5}{6}$ of a bottle

(b) If she drinks $\frac{2}{5}$ of the bottle, how much is left? $\dfrac{13}{30}$ of a bottle is left

61. *Food Purchase* While he was at the grocery store, Raymond purchased a box of candy for himself. On the way back to the dorm he ate $\frac{1}{4}$ of the candy. As he was putting away the groceries he ate $\frac{1}{2}$ of what was left. There are now six chocolates left in the box. How many chocolates were in the box to begin with?

16 chocolates

62. *Baking* Peter has $\frac{3}{4}$ cup of cocoa. He needs $\frac{1}{8}$ cup to make brownies, and another $\frac{1}{4}$ cup to make fudge squares. After making the brownies and the fudge, how much cocoa will Peter have left?

$\frac{3}{8}$ cup

63. *Business Management* The manager at Fit Factory Health Club was going through his files for 2007 and discovered that only $\frac{7}{10}$ of the members actually used the club. When he checked the numbers from the previous year of 2006, he found that $\frac{7}{8}$ of the members had used the club. What fractional part of the membership represents the decrease in club usage?

$\frac{7}{40}$ of the membership

Cumulative Review

64. **[2.2.2]** Reduce to lowest terms. $\frac{15}{85}$ $\frac{3}{17}$

65. **[2.2.2]** Reduce to lowest terms. $\frac{27}{207}$ $\frac{3}{23}$

66. **[2.3.2]** Change to a mixed number. $\frac{125}{14}$ $8\frac{13}{14}$

67. **[2.3.1]** Change to an improper fraction. $14\frac{3}{7}$ $\frac{101}{7}$

68. **[2.5.3]** Divide. $4\frac{1}{3} \div 1\frac{1}{2}$ $2\frac{8}{9}$

69. **[2.4.3]** Multiply. $5\frac{1}{2} \times 1\frac{3}{11}$ 7

Quick Quiz 2.7 Simplify all answers.

1. Add. $\frac{7}{16} + \frac{3}{4}$ $\frac{19}{16}$ or $1\frac{3}{16}$

2. Add. $\frac{1}{3} + \frac{5}{7} + \frac{10}{21}$ $\frac{32}{21}$ or $1\frac{11}{21}$

3. Subtract. $\frac{8}{9} - \frac{7}{15}$ $\frac{19}{45}$

4. Concept Check Explain how you would subtract the fractions $\frac{8}{9} - \frac{3}{7}$. Answers may vary

Classroom Quiz 2.7 You may use these problems to quiz your students' mastery of Section 2.7.

Simplify all answers.

1. Add. $\frac{7}{8} + \frac{7}{10}$ **Ans:** $\frac{63}{40}$ or $1\frac{23}{40}$

2. Add. $\frac{5}{24} + \frac{5}{6} + \frac{3}{8}$ **Ans:** $\frac{17}{12}$ or $1\frac{5}{12}$

3. Subtract. $\frac{2}{3} - \frac{5}{16}$ **Ans:** $\frac{17}{48}$

2.8 ADDING AND SUBTRACTING MIXED NUMBERS AND THE ORDER OF OPERATIONS

Adding Mixed Numbers

When adding mixed numbers, it is best to add the fractions together and then add the whole numbers together.

Student Learning Objectives

After studying this section, you will be able to:

1. Add mixed numbers.

2. Subtract mixed numbers.

3. Evaluate fractional expressions using the order of operations.

EXAMPLE 1 Add. $3\frac{1}{8} + 2\frac{5}{8}$

Solution

$$3 \quad \frac{1}{8}$$
$$+2 \quad \frac{5}{8}$$

Add the whole numbers. $3 + 2 = 5$ → $5 \quad \frac{6}{8}$ ← Add the fractions. $\frac{1}{8} + \frac{5}{8} = \frac{6}{8}$

$$= 5 \quad \frac{3}{4} \longleftarrow \text{Reduce } \frac{6}{8} = \frac{3}{4}$$

Practice Problem 1 Add. $5\frac{1}{12} + 9\frac{5}{12}$ $14\frac{1}{2}$

If the fraction portions of the mixed numbers do not have a common denominator, we must build the fraction parts to obtain a common denominator before adding.

EXAMPLE 2 Add. $1\frac{2}{7} + 5\frac{1}{3}$

Solution The LCD of $\frac{2}{7}$ and $\frac{1}{3}$ is 21.

$$\frac{2}{7} \times \frac{3}{3} = \frac{6}{21} \qquad \frac{1}{3} \times \frac{7}{7} = \frac{7}{21}$$

Thus $1\frac{2}{7} + 5\frac{1}{3} = 1\frac{6}{21} + 5\frac{7}{21}$.

$$1\frac{2}{7} = \quad 1 \quad \frac{6}{21}$$
$$+ 5\frac{1}{3} = + 5 \quad \frac{7}{21}$$

Add the whole numbers. $1 + 5$ → $6 \quad \frac{13}{21}$ ← Add the fractions. $\frac{6}{21} + \frac{7}{21}$

Practice Problem 2 Add. $6\frac{1}{4} + 2\frac{2}{5}$ $8\frac{13}{20}$

If the sum of the fractions is an improper fraction, we convert it to a mixed number and add the whole numbers together.

Teaching Example 1 Add. $2\frac{3}{10} + 6\frac{1}{10}$

Ans: $8\frac{2}{5}$

NOTE TO STUDENT: Fully worked-out solutions to all of the Practice Problems can be found at the back of the text starting at page SP-1

Teaching Example 2 Add. $4\frac{1}{6} + 1\frac{2}{5}$

Ans: $5\frac{17}{30}$

Teaching Example 3 Add. $3\frac{7}{12} + 4\frac{5}{8}$

Ans: $8\frac{5}{24}$

EXAMPLE 3 Add. $6\frac{5}{6} + 4\frac{3}{8}$

Solution The LCD of $\frac{5}{6}$ and $\frac{3}{8}$ is 24.

$$6 \;\boxed{\frac{5}{6} \times \frac{4}{4}}\; = \; 6 \;\boxed{\frac{20}{24}}$$

$$+ 4 \;\boxed{\frac{3}{8} \times \frac{3}{3}}\; = \; + 4 \;\boxed{\frac{9}{24}}$$

Add the whole numbers. $\longrightarrow\; 10 \;\boxed{\frac{29}{24}}\; \leftarrow$ Add the fractions.

$$= \; 10 + \boxed{1\frac{5}{24}} \quad \text{Since } \frac{29}{24} = 1\frac{5}{24}$$

$$= \; 11\frac{5}{24} \qquad \text{We add the whole numbers } 10 + 1 = 11.$$

Practice Problem 3 Add. $7\frac{1}{4} + 3\frac{5}{6}$ $\;11\frac{1}{12}$

2 Subtracting Mixed Numbers

Subtracting mixed numbers is like adding.

Teaching Example 4 Subtract. $7\frac{7}{8} - 2\frac{1}{4}$

Ans: $5\frac{5}{8}$

EXAMPLE 4 Subtract. $8\frac{5}{7} - 5\frac{5}{14}$

Solution The LCD of $\frac{5}{7}$ and $\frac{5}{14}$ is 14.

$$8 \;\boxed{\frac{5}{7} \times \frac{2}{2}}\; = \; 8\frac{10}{14}$$

$$- 5\frac{5}{14} \qquad\quad = \; -5\frac{5}{14}$$

$\boxed{\text{Subtract the whole numbers.}} \longrightarrow 3\dfrac{5}{14} \longleftarrow \boxed{\text{Subtract the fractions.}}$

Practice Problem 4 Subtract. $12\frac{5}{6} - 7\frac{5}{12}$ $\;5\frac{5}{12}$

Sometimes we must borrow before we can subtract.

Teaching Example 5 Subtract.

(a) $8\frac{1}{15} - 3\frac{5}{6}$ **(b)** $22 - 13\frac{5}{11}$

Ans: **(a)** $4\frac{7}{30}$ **(b)** $8\frac{6}{11}$

EXAMPLE 5 Subtract.

(a) $9\frac{1}{4} - 6\frac{5}{14}$ **(b)** $15 - 9\frac{3}{16}$

Solution This example is fairly challenging. Read through each step carefully. Be sure to have paper and pencil handy and see if you can verify each step.

(a) The LCD of $\frac{1}{4}$ and $\frac{5}{14}$ is 28.

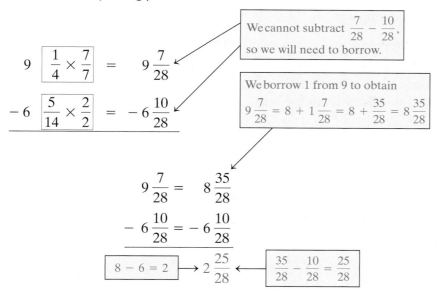

$$9 \boxed{\frac{1}{4} \times \frac{7}{7}} = 9\frac{7}{28}$$

$$-6 \boxed{\frac{5}{14} \times \frac{2}{2}} = -6\frac{10}{28}$$

We cannot subtract $\frac{7}{28} - \frac{10}{28}$, so we will need to borrow.

We borrow 1 from 9 to obtain
$$9\frac{7}{28} = 8 + 1\frac{7}{28} = 8 + \frac{35}{28} = 8\frac{35}{28}$$

$$9\frac{7}{28} = 8\frac{35}{28}$$
$$-6\frac{10}{28} = -6\frac{10}{28}$$

$$\boxed{8 - 6 = 2} \rightarrow 2\frac{25}{28} \leftarrow \boxed{\frac{35}{28} - \frac{10}{28} = \frac{25}{28}}$$

(b) The LCD = 16.

$$15 = 14\frac{16}{16} \leftarrow$$

We borrow 1 from 15 to obtain
$$15 = 14 + 1 = 14 + \frac{16}{16} = 14\frac{16}{16}$$

$$-9\frac{3}{16} = -9\frac{3}{16}$$

$$\boxed{14 - 9 = 5} \rightarrow 5\frac{13}{16} \leftarrow \boxed{\frac{16}{16} - \frac{3}{16} = \frac{13}{16}}$$

Practice Problem 5   Subtract.

(a) $9\frac{1}{8} - 3\frac{2}{3}$ $5\frac{11}{24}$ **(b)** $18 - 6\frac{7}{18}$ $11\frac{11}{18}$

EXAMPLE 6   A plumber had a pipe $5\frac{3}{16}$ inches long for a fitting under the sink. He needed a pipe that was $3\frac{7}{8}$ inches long, so he cut the pipe down. How much of the pipe did he cut off?

Solution   We will need to subtract $5\frac{3}{16} - 3\frac{7}{8}$ to find the length that was cut off.

$$5\frac{3}{16} = 5\frac{3}{16}$$

$$-3\frac{7}{8} \times \frac{2}{2} = -3\frac{14}{16}$$

$$4\frac{19}{16} \leftarrow$$

We borrow 1 from 5 to obtain
$$5\frac{3}{16} = 4 + 1\frac{3}{16} = 4 + \frac{19}{16}$$

$$-3\frac{14}{16}$$

$$\boxed{4 - 3 = 1} \rightarrow 1\frac{5}{16} \leftarrow \boxed{\frac{19}{16} - \frac{14}{16} = \frac{5}{16}}$$

The plumber had to cut off $1\frac{5}{16}$ inches of pipe.

Teaching Example 6 Molly had $13\frac{1}{3}$ yards of material. She used $4\frac{3}{4}$ yards to make a coat. How many yards of material does she have left?

Ans: $8\frac{7}{12}$ yards

Practice Problem 6 Hillary and Sam purchased $6\frac{1}{4}$ gallons of paint to paint the inside of their house. They used $4\frac{2}{3}$ gallons of paint. How much paint was left over?

$1\frac{7}{12}$ gallons

ALTERNATIVE METHOD: Add or Subtract Mixed Numbers as Improper Fractions Can mixed numbers be added and subtracted as improper fractions? Yes. Recall Example 5(a).

$$9\frac{1}{4} - 6\frac{5}{14} = 2\frac{25}{28}$$

If we write $9\frac{1}{4} - 6\frac{5}{14}$ using improper fractions, we have $\frac{37}{4} - \frac{89}{14}$. Now we build each of these improper fractions so that they both have the LCD for their denominators.

$$
\begin{aligned}
\frac{37}{4} \boxed{\times \frac{7}{7}} &= \frac{259}{28} \\
-\frac{89}{14} \boxed{\times \frac{2}{2}} &= -\frac{178}{28} \\
\hline
\frac{81}{28} &= 2\frac{25}{28}
\end{aligned}
$$

The same result is obtained as in Example 5(a). This method does not require borrowing. However, you do work with larger numbers. For more practice, see exercises 53–54.

③ Evaluating Fractional Expressions Using the Order of Operations

Recall that in Section 1.6 we discussed the order of operations when we were combining whole numbers. We will now encounter some similar problems involving fractions and mixed numbers. We will repeat here the four-step order of operations that we studied previously:

ORDER OF OPERATIONS

With grouping symbols:

Do first   **1.** Perform operations inside parentheses.

          2. Simplify any expressions with exponents.

          3. Multiply or divide from left to right.

Do last   **4.** Add or subtract from left to right.

EXAMPLE 7 Evaluate. $\dfrac{3}{4} - \dfrac{2}{3} \times \dfrac{1}{8}$

Teaching Example 7 Evaluate. $\dfrac{5}{12} \times \dfrac{4}{9} + \dfrac{1}{3}$

Ans: $\dfrac{14}{27}$

Solution

$$\dfrac{3}{4} - \dfrac{2}{3} \times \dfrac{1}{8} = \dfrac{3}{4} - \dfrac{1}{12}$$ First we must multiply $\dfrac{2}{3} \times \dfrac{1}{8}$.

$$= \dfrac{9}{12} - \dfrac{1}{12}$$ Now we subtract, but first we need to build $\dfrac{3}{4}$ to an equivalent fraction with a common denominator of 12.

$$= \dfrac{8}{12}$$ Now we can subtract $\dfrac{9}{12} - \dfrac{1}{12}$.

$$= \dfrac{2}{3}$$ Finally we reduce the fraction.

Practice Problem 7 Evaluate.

$$\dfrac{3}{5} - \dfrac{1}{15} \times \dfrac{10}{13} \quad \dfrac{107}{195}$$

EXAMPLE 8 Evaluate. $\dfrac{2}{3} \times \dfrac{1}{4} + \dfrac{2}{5} \div \dfrac{14}{15}$

Teaching Example 8 Evaluate.

$$\dfrac{3}{8} \div \dfrac{1}{2} - \dfrac{2}{3} \times \dfrac{5}{6}$$

Ans: $\dfrac{7}{36}$

Solution

$$\dfrac{2}{3} \times \dfrac{1}{4} + \dfrac{2}{5} \div \dfrac{14}{15} = \dfrac{1}{6} + \dfrac{2}{5} \div \dfrac{14}{15}$$ First we multiply $\dfrac{2}{3} \times \dfrac{1}{4}$.

$$= \dfrac{1}{6} + \dfrac{2}{5} \times \dfrac{15}{14}$$ We express the division as a multiplication problem. We invert $\dfrac{14}{15}$ and multiply.

$$= \dfrac{1}{6} + \dfrac{3}{7}$$ Now we perform the multiplication.

$$= \dfrac{7}{42} + \dfrac{18}{42}$$ We obtain equivalent fractions with an LCD of 42.

$$= \dfrac{25}{42}$$ We add the two fractions.

Practice Problem 8 Evaluate.

$$\dfrac{1}{7} \times \dfrac{5}{6} + \dfrac{5}{3} \div \dfrac{7}{6} \quad \dfrac{65}{42} \text{ or } 1\dfrac{23}{42}$$

NOTE TO STUDENT: Fully worked-out solutions to all of the Practice Problems can be found at the back of the text starting at page SP-1

Developing Your Study Skills

Problems with Accuracy

Strive for accuracy. Mistakes are often made because of human error rather than lack of understanding. Such mistakes are frustrating. A simple arithmetic or copying error can lead to an incorrect answer.

These five steps will help you cut down on errors.

1. Work carefully, and take your time. Do not rush through a problem just to get it done.

2. Concentrate on the problem. Sometimes problems become mechanical, and your mind begins to wander. You become careless and make a mistake.

3. Check your problem. Be sure that you copied it correctly from the book.

4. Check your computations from step to step. Check the solution to the problem. Does it work? Does it make sense?

5. Keep practicing new skills. Remember the old saying, "Practice makes perfect." An increase in practice results in an increase in accuracy. Many errors are due simply to lack of practice.

There is no magic formula for eliminating all errors, but these five steps will be a tremendous help in reducing them.

Add or subtract. Express the answer as a mixed number. Simplify all answers.

1. $7\frac{1}{8} + 2\frac{5}{8}$ $9\frac{3}{4}$

2. $6\frac{3}{10} + 4\frac{1}{10}$ $10\frac{2}{5}$

3. $15\frac{3}{14} - 11\frac{1}{14}$ $4\frac{1}{7}$

4. $8\frac{3}{4} - 3\frac{1}{4}$ $5\frac{1}{2}$

5. $12\frac{1}{3} + 5\frac{1}{6}$ $17\frac{1}{2}$

6. $20\frac{1}{4} + 3\frac{1}{8}$ $23\frac{3}{8}$

7. $4\frac{3}{5} + 8\frac{2}{5}$ 13

8. $8\frac{2}{9} + 7\frac{7}{9}$ 16

9. $1 - \frac{3}{7}$ $\frac{4}{7}$

10. $1 - \frac{9}{11}$ $\frac{2}{11}$

11. $1\frac{3}{4} + \frac{5}{16}$ $2\frac{1}{16}$

12. $1\frac{2}{3} + \frac{13}{18}$ $2\frac{7}{18}$

13. $5\frac{1}{6} + 4\frac{5}{18}$ $9\frac{4}{9}$

14. $6\frac{2}{5} + 7\frac{3}{20}$ $13\frac{11}{20}$

15. $8\frac{1}{4} - 8\frac{4}{16}$ 0

16. $8\frac{11}{15} - 3\frac{3}{10}$ $5\frac{13}{30}$

17. $12\frac{1}{3} - 7\frac{2}{5}$ $4\frac{14}{15}$

18. $10\frac{10}{15} - 10\frac{2}{3}$ 0

19. $30 - 15\frac{3}{7}$ $14\frac{4}{7}$

20. $25 - 14\frac{2}{11}$ $10\frac{9}{11}$

21. $3 + 4\frac{2}{5}$ $7\frac{2}{5}$

22. $8 + 2\frac{3}{4}$ $10\frac{3}{4}$

23. $14 - 3\frac{7}{10}$ $10\frac{3}{10}$

24. $19 - 5\frac{8}{9}$ $13\frac{1}{9}$

Add or subtract. Express the answer as a mixed number. Simplify all answers.

25.
$$15\frac{4}{15}$$
$$+26\frac{8}{15}$$
$$41\frac{4}{5}$$

26.
$$22\frac{1}{8}$$
$$+14\frac{3}{8}$$
$$36\frac{1}{2}$$

27.
$$6\frac{1}{6} \quad 6\frac{2}{12}$$
$$+2\frac{1}{4} \quad +2\frac{3}{12}$$
$$\overline{\quad 8\frac{5}{12}}$$

28.
$$3\frac{2}{3} \quad 3\frac{10}{15}$$
$$+4\frac{1}{5} \quad +4\frac{3}{15}$$
$$\overline{\quad 7\frac{13}{15}}$$

29.
$$3\frac{3}{4} \quad 3\frac{9}{12}$$
$$+4\frac{5}{12} \quad +4\frac{5}{12}$$
$$\overline{\quad 7\frac{14}{12} = 8\frac{1}{6}}$$

30.
$$11\frac{5}{8} \quad 11\frac{5}{8}$$
$$+13\frac{1}{2} \quad +13\frac{4}{8}$$
$$\overline{\quad 24\frac{9}{8} = 25\frac{1}{8}}$$

31.
$$47\frac{3}{10} \quad 47\frac{12}{40}$$
$$+26\frac{5}{8} \quad +26\frac{25}{40}$$
$$\overline{\quad 73\frac{37}{40}}$$

32.
$$34\frac{1}{20} \quad 34\frac{3}{60}$$
$$+45\frac{8}{15} \quad +45\frac{32}{60}$$
$$\overline{\quad 79\frac{35}{60} = 79\frac{7}{12}}$$

33.
$$19\frac{5}{6} \quad 19\frac{5}{6}$$
$$-14\frac{1}{3} \quad -14\frac{2}{6}$$
$$\overline{\quad 5\frac{3}{6} = 5\frac{1}{2}}$$

34.
$$22\frac{7}{9} \quad 22\frac{28}{36}$$
$$-16\frac{1}{4} \quad -16\frac{9}{36}$$
$$\overline{\quad 6\frac{19}{36}}$$

35.
$$6\frac{1}{12} \quad 5\frac{26}{24}$$
$$-5\frac{10}{24} \quad -5\frac{10}{24}$$
$$\overline{\quad \frac{16}{24} = \frac{2}{3}}$$

36.
$$4\frac{1}{12} \quad 3\frac{39}{36}$$
$$-3\frac{7}{18} \quad -3\frac{14}{36}$$
$$\overline{\quad \frac{25}{36}}$$

37.
$$12\frac{3}{20} \quad 11\frac{69}{60}$$
$$-7\frac{7}{15} \quad -7\frac{28}{60}$$
$$\overline{\quad 4\frac{41}{60}}$$

38.
$$8\frac{5}{12} \quad 7\frac{85}{60}$$
$$-5\frac{9}{10} \quad -5\frac{54}{60}$$
$$\overline{\quad 2\frac{31}{60}}$$

39.
$$12 \quad 11\frac{15}{15}$$
$$-3\frac{7}{15} \quad -3\frac{7}{15}$$
$$\overline{\quad 8\frac{8}{15}}$$

40.
$$40 \quad 39\frac{7}{7}$$
$$-6\frac{3}{7} \quad -6\frac{3}{7}$$
$$\overline{\quad 33\frac{4}{7}}$$

41.
$$120 \quad 119\frac{8}{8}$$
$$-17\frac{3}{8} \quad -17\frac{3}{8}$$
$$\overline{\quad 102\frac{5}{8}}$$

42.
$$98 \quad 97\frac{17}{17}$$
$$-89\frac{15}{17} \quad -89\frac{15}{17}$$
$$\overline{\quad 8\frac{2}{17}}$$

43.
$$3\frac{5}{8} \quad 3\frac{15}{24}$$
$$2\frac{2}{3} \quad 2\frac{16}{24}$$
$$+7\frac{3}{4} \quad +7\frac{18}{24}$$
$$\overline{\quad 12\frac{49}{24} = 14\frac{1}{24}}$$

44.
$$4\frac{2}{3} \quad 4\frac{40}{60}$$
$$3\frac{4}{5} \quad 3\frac{48}{60}$$
$$+6\frac{3}{4} \quad +6\frac{45}{60}$$
$$\overline{\quad 13\frac{133}{60} = 15\frac{13}{60}}$$

Applications

45. ***Mountain Biking*** Lee Hong rode his mountain bike through part of the Sangre de Cristo Mountains in New Mexico. On Wednesday he rode $20\frac{3}{4}$ miles. On Thursday he rode $22\frac{3}{8}$ miles. What was his total biking distance during those two days? $43\frac{1}{8}$ miles

46. ***Hiking*** Ryan and Omar are planning an afternoon hike. Their map shows three loops measuring $2\frac{1}{8}$ miles, $1\frac{5}{6}$ miles, and $1\frac{2}{3}$ miles. If they hike all three loops, what will their total hiking distance be? $5\frac{5}{8}$ miles

47. ***Bicycling*** Lake Harriet and Lake Calhoun have paved paths around them for runners, walkers, and bicyclists. The distance around Lake Harriet is $2\frac{4}{5}$ miles, and the distance around Lake Calhoun is $3\frac{1}{10}$ miles. The road connecting the two lakes is $\frac{1}{2}$ mile. If Lola rides her bike around both lakes, and uses the connecting road twice, how long is her bike ride? $6\frac{9}{10}$ miles

48. ***Stock Market*** Shanna purchased stock in 1985 at $\$21\frac{3}{8}$ per share. When her son was ready for college, she sold the stock in 1999 at $\$93\frac{5}{8}$ per share. How much did she make per share for her son's tuition?

$\$72\frac{1}{4}$ per share

49. ***Basketball*** Nina and Julie are the two tallest basketball players on their high school team. Nina is $69\frac{3}{4}$ inches tall and Julie is $72\frac{1}{2}$ inches tall. How many inches taller is Julie than Nina? $2\frac{3}{4}$ inches

50. ***Food Purchase*** Julio bought $3\frac{3}{4}$ pounds of roast turkey and $1\frac{2}{3}$ pounds of salami at the deli. How many more pounds of turkey than salami did he buy? $2\frac{1}{12}$ pounds

51. ***Food Purchase*** Lara needs 8 pounds of haddock for her dinner party. At the grocery store, haddock portions weighing $1\frac{3}{4}$ pounds and $2\frac{1}{6}$ pounds are placed on the scale.

(a) How many pounds of haddock are on the scale? $3\frac{11}{12}$ pounds

(b) How many more pounds of haddock does Lara need? $4\frac{1}{12}$ pounds

52. ***Medical Care*** A young man has been under a doctor's care to lose weight. His doctor wanted him to lose 46 pounds in the first three months. He lost $17\frac{5}{8}$ pounds the first month and $13\frac{1}{2}$ pounds the second month.

(a) How much did he lose during the first two months? $31\frac{1}{8}$ pounds

(b) How much would he need to lose in the third month to reach the goal? $14\frac{7}{8}$ pounds

To Think About

Use improper fractions and the Alternative Method as discussed in the text to perform each calculation.

53. $\dfrac{379}{8} + \dfrac{89}{5}$ $\quad \dfrac{1895}{40} + \dfrac{712}{40} = \dfrac{2607}{40}$ or $65\frac{7}{40}$

54. $\dfrac{151}{6} - \dfrac{130}{7}$ $\quad \dfrac{1057}{42} - \dfrac{780}{42} = \dfrac{277}{42}$ or $6\frac{25}{42}$

When adding or subtracting mixed numbers, it is wise to estimate your answer by rounding each mixed number to the nearest whole number.

55. Estimate your answer to $35\frac{1}{6} + 24\frac{5}{12}$ by rounding each mixed number to the nearest whole number. Then find the exact answer. How close was your estimate?
We estimate by adding $35 + 24$ to obtain 59. The exact answer is $59\frac{7}{12}$. Our estimate is very close. We are off by only $\frac{7}{12}$.

56. Estimate your answer to $102\frac{5}{7} - 86\frac{2}{3}$ by rounding each mixed number to the nearest whole number. Then find the exact answer. How close was your estimate?
We estimate by subtracting $103 - 87$ to obtain 16. The exact answer is $16\frac{1}{21}$. Our estimate is very close. We are off by only $\frac{1}{21}$.

Evaluate using the correct order of operations.

57. $\dfrac{6}{7} - \dfrac{4}{7} \times \dfrac{1}{3}$ $\dfrac{2}{3}$

58. $\dfrac{3}{5} - \dfrac{1}{3} \times \dfrac{6}{5}$ $\dfrac{1}{5}$

59. $\dfrac{1}{2} + \dfrac{3}{8} \div \dfrac{3}{4}$ 1

60. $\dfrac{3}{4} + \dfrac{1}{4} \div \dfrac{5}{3}$ $\dfrac{9}{10}$

61. $\dfrac{9}{10} \div \dfrac{3}{8} \times \dfrac{5}{8}$ $\dfrac{3}{2}$ or $1\dfrac{1}{2}$

62. $\dfrac{5}{12} \div \dfrac{3}{10} \times \dfrac{9}{5}$ $\dfrac{5}{2}$ or $2\dfrac{1}{2}$

63. $\dfrac{3}{5} \times \dfrac{1}{2} + \dfrac{1}{5} \div \dfrac{2}{3}$ $\dfrac{3}{5}$

64. $\dfrac{5}{6} \times \dfrac{1}{2} + \dfrac{2}{3} \div \dfrac{4}{3}$ $\dfrac{11}{12}$

65. $\left(\dfrac{3}{5} - \dfrac{3}{20}\right) \times \dfrac{4}{5}$ $\dfrac{9}{25}$

66. $\left(\dfrac{1}{3} + \dfrac{1}{6}\right) \times \dfrac{5}{11}$ $\dfrac{5}{22}$

67. $\left(\dfrac{1}{3}\right)^2 \div \dfrac{4}{9}$ $\dfrac{1}{4}$

68. $\left(\dfrac{1}{4}\right)^2 \div \dfrac{3}{4}$ $\dfrac{1}{12}$

69. $\dfrac{1}{4} \times \left(\dfrac{2}{3}\right)^2$ $\dfrac{1}{9}$

70. $\dfrac{5}{8} \times \left(\dfrac{2}{5}\right)^2$ $\dfrac{1}{10}$

71. $\dfrac{5}{6} \div \left(\dfrac{2}{3} + \dfrac{1}{6}\right)^2$ $\dfrac{6}{5}$ or $1\dfrac{1}{5}$

72. $\dfrac{4}{3} \div \left(\dfrac{3}{5} - \dfrac{3}{10}\right)^2$ $\dfrac{400}{27}$ or $14\dfrac{22}{27}$

Cumulative Review *Multiply.*

73. [1.4.3] $\begin{array}{r} 1200 \\ \times\ 400 \\ \hline 480{,}000 \end{array}$

74. [1.4.4] $\begin{array}{r} 4050 \\ \times\ 2106 \\ \hline 8{,}529{,}300 \end{array}$

Quick Quiz 2.8

1. Add. Express the answer as a mixed number.

$3\dfrac{4}{5} + 5\dfrac{3}{8}$ $9\dfrac{7}{40}$

2. Subtract. Express the answer as a mixed number.

$6\dfrac{5}{12} - 4\dfrac{7}{10}$ $1\dfrac{43}{60}$

3. Evaluate using the correct order of operations.

$\dfrac{1}{5} + \dfrac{3}{10} \div \dfrac{11}{20}$ $\dfrac{41}{55}$

4. **Concept Check** Explain how you would evaluate the following expression using the correct order of operations. $\dfrac{4}{5} - \dfrac{1}{4} \times \dfrac{2}{3}$ Answers may vary

Classroom Quiz 2.8 You may use these problems to quiz your students' mastery of Section 2.8.

1. Add. Express the answer as a mixed number.

$7\dfrac{5}{12} + 4\dfrac{11}{18}$ **Ans:** $12\dfrac{1}{36}$

2. Subtract. Express the answer as a mixed number.

$13\dfrac{2}{9} - 7\dfrac{3}{4}$ **Ans:** $5\dfrac{17}{36}$

3. Evaluate using the correct order of operations.

$\dfrac{3}{7} + \dfrac{5}{8} \div \dfrac{21}{16}$ **Ans:** $\dfrac{19}{21}$

Solving Real-Life Problems with Fractions

All problem solving requires the same kind of thinking. In this section we will combine problem-solving skills with our new computational skills with fractions. Sometimes the difficulty is in figuring out what must be done. Sometimes it is in doing the computation. Remember that *estimating* is important in problem solving. We may use the following steps.

1. **Understand the problem.**
   (a) Read the problem carefully.
   (b) Draw a picture if this helps you.
   (c) Fill in the Mathematics Blueprint.

2. **Solve.**
   (a) Perform the calculations.
   (b) State the answer, including the units of measure.

3. **Check.**
   (a) Estimate the answer. Round fractions to the nearest whole number.
   (b) Compare the exact answer with the estimate to see if your answer is reasonable.

Student Learning Objective

After studying this section, you will be able to:

1. Solve real-life problems with fractions.

▲ **EXAMPLE 1** In designing a modern offshore speedboat, the designing engineer has determined that one of the oak frames near the engine housing needs to be $26\frac{1}{8}$ inches long. At the end of the oak frame there will be $2\frac{5}{8}$ inches of insulation. Finally there will be a steel mounting that is $3\frac{3}{4}$ inches long. When all three items are assembled, how long will the oak frame and insulation and steel mounting extend?

Teaching Example 1 On Monday, the recycling center collected $5\frac{3}{10}$ tons of recyclable trash. On Tuesday, the center collected $4\frac{2}{5}$ tons of trash and on Wednesday, they collected $6\frac{1}{2}$ tons. What is the total weight of trash collected on those three days?

Ans: $16\frac{1}{5}$ tons

Solution

1. **Understand the problem.**

   We draw a picture to help us.

   Then we fill in the Mathematics Blueprint.

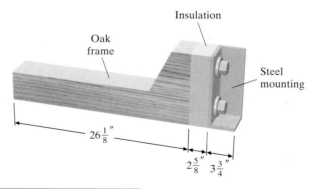

Mathematics Blueprint for Problem Solving

Gather the Facts	What Am I Asked to Do?	How Do I Proceed?	Key Points to Remember
Oak frame: $26\frac{1}{8}''$ Insulation: $2\frac{5}{8}''$ Steel mounting: $3\frac{3}{4}''$	Find the total length.	Add the lengths of the three items.	When adding mixed numbers, add the whole numbers first and then add the fractions.

2. Solve and state the answer.

Add the three amounts. $26\frac{1}{8} + 2\frac{5}{8} + 3\frac{3}{4}$

$$
\text{LCD} = 8 \quad
\begin{array}{rcl}
26\frac{1}{8} & = & 26\frac{1}{8} \\[2ex]
2\frac{5}{8} & = & 2\frac{5}{8} \\[2ex]
+ 3 \boxed{\dfrac{3}{4} \times \dfrac{2}{2}} & = & + 3\frac{6}{8} \\[1ex]
\hline
 & & 31\frac{12}{8} = 32\frac{4}{8} = 32\frac{1}{2}
\end{array}
$$

The entire assembly will be $32\frac{1}{2}$ inches.

3. Check. Estimate the sum by rounding each fraction to the nearest whole number.

Thus $\qquad 26\frac{1}{8} + 2\frac{5}{8} + 3\frac{3}{4}$

becomes $\qquad 26 + 3 + 4 = 33$

This is close to our answer, $32\frac{1}{2}$. Our answer seems reasonable.

One of the most important uses of estimation in mathematics is in the calculation of problems involving fractions. People find it easier to detect significant errors when working with whole numbers. However, the extra steps involved in the calculations with fractions and mixed numbers often distract our attention from an error that we should have detected.

Thus it is particularly critical to take the time to check your answer by estimating the results of the calculation with whole numbers. Be sure to ask yourself, is this answer reasonable? Does this answer seem realistic? Only by estimating our results with whole numbers will we be able to answer that question. It is this estimating skill that you will find more useful in your own life as a consumer and as a citizen.

NOTE TO STUDENT: Fully worked-out solutions to all of the Practice Problems can be found at the back of the text starting at page SP-1

Practice Problem 1 Nicole required the following amounts of gas for her farm tractor in the last three fill-ups: $18\frac{7}{10}$ gallons, $15\frac{2}{5}$ gallons, and $14\frac{1}{2}$ gallons. How many gallons did she need altogether?

$48\frac{3}{5}$ gallons

The word *diameter* has two common meanings. First, it means a line segment that passes through the center of and intersects a circle twice. It has its endpoints on the circle. Second, it means the *length* of this segment.

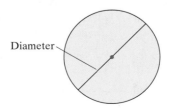

Diameter

▲ **EXAMPLE 2**   What is the inside diameter (distance across) of a cement storm drain pipe that has an outside diameter of $4\frac{1}{8}$ feet and is $\frac{3}{8}$ foot thick?

Solution

1. ***Understand the problem.*** Read the problem carefully. Draw a picture. The picture is in the margin on the right. Now fill in the Mathematics Blueprint.

Teaching Example 2 The college yearbook staff has a goal of selling 24 pages of advertising in this year's edition. In January, they sold $8\frac{2}{3}$ pages and in February, they sold $9\frac{5}{6}$ pages. How many pages of advertising do they have left to sell?

Ans: $5\frac{1}{2}$ pages

Mathematics Blueprint for Problem Solving

Gather the Facts	What Am I Asked to Do?	How Do I Proceed?	Key Points to Remember
Outside diameter is $4\frac{1}{8}$ feet. Thickness is $\frac{3}{8}$ foot on both ends of the diameter.	Find the *inside* diameter of the pipe.	Add the two measures of thickness. Then subtract this total from the outside diameter.	Since the LCD = 8, all fractions must have this denominator.

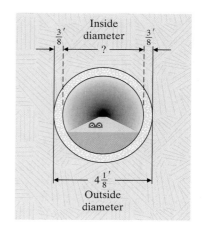

2. ***Solve and state the answer.*** Add the two thickness measurements together. Adding $\frac{3}{8} + \frac{3}{8} = \frac{6}{8}$ gives the total thickness of the pipe, $\frac{6}{8}$ foot. We will not reduce $\frac{6}{8}$ since the LCD is 8.

We subtract the total of the two thickness measurements from the outside diameter.

$$
\begin{array}{rcl}
4\frac{1}{8} & = & 3\frac{9}{8} \\
-\ \frac{6}{8} & = & -\ \frac{6}{8} \\
\hline
 & & 3\frac{3}{8}
\end{array}
$$

We borrow 1 from 4 to get $3 + 1\frac{1}{8}$ or $3\frac{9}{8}$.

The inside diameter is $3\frac{3}{8}$ feet.

3. ***Check.*** We will work backward to check. We will use the exact values. If we have done our work correctly, $\frac{3}{8}$ foot $+\ 3\frac{3}{8}$ feet $+\ \frac{3}{8}$ foot should add up to the outside diameter, $4\frac{1}{8}$ feet.

$$\frac{3}{8} + 3\frac{3}{8} + \frac{3}{8} \overset{?}{=} 4\frac{1}{8}$$

$$3\frac{9}{8} \overset{?}{=} 4\frac{1}{8}$$

$$4\frac{1}{8} = 4\frac{1}{8} \quad \checkmark$$

Our answer of $3\frac{3}{8}$ feet is correct.

▲ **Practice Problem 2**   A poster is $12\frac{1}{4}$ inches long. We want a $1\frac{3}{8}$-inch border on the top and a 2-inch border on the bottom. What is the length of the inside portion of the poster?

NOTE TO STUDENT: Fully worked-out solutions to all of the Practice Problems can be found at the back of the text starting at page SP-1

$8\frac{7}{8}$ inches

EXAMPLE 3 On Tuesday Michael earned $\$8\frac{1}{4}$ per hour working for eight hours. He also earned overtime pay, which is $1\frac{1}{2}$ times his regular rate of $\$8\frac{1}{4}$, for four hours on Tuesday. How much pay did he earn altogether on Tuesday?

Solution

1. *Understand the problem.* We draw a picture of the parts of Michael's pay on Tuesday.

Michael's earnings on Tuesday are the sum of two parts:

| Pay at regular pay rate | + | Pay at overtime pay rate | = | Total pay for the day |

Now fill in the Mathematics Blueprint.

Mathematics Blueprint for Problem Solving

Gather the Facts	What Am I Asked to Do?	How Do I Proceed?	Key Points to Remember
He works eight hours at $\$8\frac{1}{4}$ per hour. He works four hours at the overtime rate, $1\frac{1}{2}$ times the regular rate.	Find his total pay for Tuesday.	Find out how much he is paid for regular time. Find out how much he is paid for overtime. Then add the two.	The overtime rate is $1\frac{1}{2}$ multiplied by the regular rate.

2. *Solve and state the answer.* Find his overtime pay rate.

$$1\frac{1}{2} \times 8\frac{1}{4} = \frac{3}{2} \times \frac{33}{4} = \frac{\$99}{8} \text{ per hour}$$

We leave our answer as an improper fraction because we will need to multiply it by another fraction.

How much was he paid for regular time? For overtime?

For eight regular hours, he earned $8 \times 8\frac{1}{4} = \overset{2}{\cancel{8}} \times \frac{33}{\underset{1}{\cancel{4}}} = \$66.$

For four overtime hours, he earned $\overset{1}{\cancel{4}} \times \frac{99}{\underset{2}{\cancel{8}}} = \frac{99}{2} = \$49\frac{1}{2}.$

Now we add to find the total pay.

$$\begin{array}{r} \$66 \\ +\$49\frac{1}{2} \\ \hline \end{array}$$

Pay at regular pay rate

Pay at overtime rate

Michael earned $\$115\frac{1}{2}$ working on Tuesday. (This is the same as $\$115.50$, which we will use in Chapter 3.)

3. *Check.* We estimate his regular pay rate at $8 per hour.

We estimate his overtime pay rate at $1\frac{1}{2} \times 8 = \frac{3}{2} \times 8 = 12$ or $12 per hour.

$$8 \text{ hours} \times \$8 \text{ per hour} = \$64 \text{ regular pay}$$
$$4 \text{ hours} \times \$12 \text{ per hour} = \$48 \text{ overtime pay}$$

Estimated sum. $64 + $48 \approx $60 + $50 = $110

$110 is close to our calculated value, $115\frac{1}{2}$, so our answer is reasonable. ✓

Practice Problem 3 A tent manufacturer uses $8\frac{1}{4}$ yards of waterproof duck cloth to make a regular tent. She uses $1\frac{1}{2}$ times that amount to make a large tent. How many yards of cloth will she need to make 6 regular tents and 16 large tents?

$247\frac{1}{2}$ yards

▲ **EXAMPLE 4** Alicia is buying some 8-foot boards for shelving. She wishes to make two bookcases, each with three shelves. Each shelf will be $3\frac{1}{4}$ feet long.

(a) How many boards does she need to buy?

(b) How many linear feet of shelving are actually needed to build the bookcases?

(c) How many linear feet of shelving will be left over?

Solution

1. *Understand the problem.* Draw a sketch of a bookcase. Each bookcase will have three shelves. Alicia is making two such bookcases. (Alicia's boards are for the shelves, not the sides.)
Now fill in the Mathematics Blueprint.

Teaching Example 4 An office area is being repainted. There are 3 conference rooms, all the same size. Painting each conference room will require $3\frac{3}{8}$ gallons of yellow paint for the walls and $1\frac{3}{4}$ gallons of white ceiling paint.

(a) How many gallons of each type of paint must be purchased?

(b) How much of each type of paint will be left over?

Ans: (a) 11 gallons of yellow paint and 6 gallons of ceiling paint

(b) There will be $\frac{7}{8}$ gallon of yellow paint and $\frac{3}{4}$ gallon of ceiling paint left over.

Mathematics Blueprint for Problem Solving

Gather the Facts	What Am I Asked to Do?	How Do I Proceed?	Key Points to Remember
She needs three shelves for each bookcase. Each shelf is $3\frac{1}{4}$ feet long. She will make two bookcases. Shelves are cut from 8-foot boards.	Find out how many boards to buy. Find out how many feet of board are needed for shelves and how many feet will be left over.	First find out how many $3\frac{1}{4}$-foot shelves she can get from one board. Then see how many boards she needs to make all six shelves.	Each time she cuts up an 8-foot board, she will get some shelves and some leftover wood.

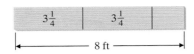

2. ***Solve and state the answer.*** We want to know how many $3\frac{1}{4}$-foot boards are in an 8-foot board. By drawing a rough sketch, we would probably guess the answer is 2. To find exactly how many $3\frac{1}{4}$-foot-long pieces are in 8 feet, we will use division.

$$8 \div 3\frac{1}{4} = \frac{8}{1} \div \frac{13}{4} = \frac{8}{1} \times \frac{4}{13} = \frac{32}{13} = 2\frac{6}{13}$$

She will get two shelves from each board, and some wood will be left over.

(a) How many boards does Alicia need to build two bookcases? For two bookcases, she needs six shelves. She will get two shelves out of each board. $6 \div 2 = 3$. She will need three 8-foot boards.

(b) How many linear feet of shelving are actually needed to build the bookcases?
She needs 6 shelves at $3\frac{1}{4}$ feet.

$$6 \times 3\frac{1}{4} = \overset{3}{\cancel{6}} \times \frac{13}{\underset{2}{\cancel{4}}} = \frac{39}{2} = 19\frac{1}{2}$$

A total of $19\frac{1}{2}$ linear feet of shelving is needed.

(c) How many linear feet of shelving will be left over?
Each time she uses one board she will have

$$8 - 3\frac{1}{4} - 3\frac{1}{4} = 8 - \left(3\frac{1}{4} + 3\frac{1}{4}\right) = 8 - 6\frac{1}{2} = 1\frac{1}{2}$$

feet left over. Each of the three boards will have $1\frac{1}{2}$ feet left over.

$$3 \times 1\frac{1}{2} = 3 \times \frac{3}{2} = \frac{9}{2} = 4\frac{1}{2}$$

A total of $4\frac{1}{2}$ linear feet of shelving will be left over.

3. ***Check.*** Work backward. See if you can check that with three 8-foot boards you
(a) can make the six shelves for the two bookcases.
(b) will use exactly $19\frac{1}{2}$ linear feet to make the shelves.
(c) will have exactly $4\frac{1}{2}$ linear feet left over.

The check is left to you.

▲ **Practice Problem 4** Michael is purchasing 12-foot boards for shelving. He wishes to make two bookcases, each with four shelves. Each shelf will be $2\frac{3}{4}$ feet long.

(a) How many boards does he need to buy? 2 boards
(b) How many linear feet of shelving are actually needed to build the bookcases? 22 feet
(c) How many linear feet of shelving will be left over? 2 feet

Another useful method for solving applied problems is called "Do a similar, simpler problem." When a problem seems difficult to understand because of the fractions, change the problem to an easier but similar

Teaching Tip Many building and construction problems are similar to Example 4. After you have covered Example 4, ask the students if they can think of similar types of problems that they have encountered in their daily lives. Often you will get some interesting problems. If they have nothing to share, give them the following problem: "A man had three rooms with wall-to-wall carpeting. One room had $12\frac{1}{2}$ square yards, a second room had $13\frac{1}{4}$ square yards, and the third room had $15\frac{7}{8}$ square yards. The carpet-cleaning company charged $3\frac{1}{2}$ per square yard or any 3 rooms cleaned for $159. The man could not determine which was a better buy. Can you?" The answer is that it is cheaper to get the $3\frac{1}{2}$-per-square-yard rate because then the cleaning job will only come to $145\frac{11}{16}$.

problem. Then decide how to solve the simpler problem and use the same steps to solve the original problem. For example:

> How many gallons of water can a tank hold if its volume is $58\frac{2}{3}$ cubic feet? (1 cubic foot holds about $7\frac{1}{2}$ gallons.)

A similar, easier problem would be: "If 1 cubic foot holds about 8 gallons and a tank holds 60 cubic feet, how many gallons of water does the tank hold?"

The easier problem can be read more quickly and seems to make more sense. Probably we will see how to solve the easier problem right away: "I can find the number of gallons by multiplying 8×60." Therefore we can solve the first problem by multiplying $7\frac{1}{2} \times 58\frac{2}{3}$ to obtain the number of gallons of water. See the next example.

EXAMPLE 5 A fishing boat traveled $69\frac{3}{8}$ nautical miles in $3\frac{3}{4}$ hours. How many knots (nautical miles per hour) did the fishing boat average?

Solution

1. **Understand the problem.** Let us think of a simpler problem. If a boat traveled 70 nautical miles in 4 hours, how many knots did it average? We would divide distance by time.

$$70 \div 4 = \text{average speed}$$

Likewise in our original problem we need to divide distance by time.

$$69\frac{3}{8} \div 3\frac{3}{4} = \text{average speed}$$

Now fill in the Mathematics Blueprint.

Mathematics Blueprint for Problem Solving

Gather the Facts	What Am I Asked to Do?	How Do I Proceed?	Key Points to Remember
Distance is $69\frac{3}{8}$ nautical miles. Time is $3\frac{3}{4}$ hours.	Find the average speed of the boat.	Divide the distance in nautical miles by the time in hours.	You must change the mixed numbers to improper fractions before dividing.

2. **Solve and state the answer.** Divide distance by time to get speed in knots.

$$69\frac{3}{8} \div 3\frac{3}{4} = \frac{555}{8} \div \frac{15}{4} = \frac{\overset{37}{\cancel{555}}}{\underset{2}{\cancel{8}}} \cdot \frac{\overset{1}{\cancel{4}}}{\underset{1}{\cancel{15}}}$$

$$= \frac{37}{2} \cdot \frac{1}{1} = \frac{37}{2} = 18\frac{1}{2}$$

The speed of the boat was $18\frac{1}{2}$ knots.

3. Check.

We estimate $69\frac{3}{8} \div 3\frac{3}{4}$.

Use $70 \div 4 = 17\frac{1}{2}$ knots

Our estimate is close to the calculated value.

Our answer is reasonable. ✓

NOTE TO STUDENT: Fully worked-out solutions to all of the Practice Problems can be found at the back of the text starting at page SP-1

Practice Problem 5 Alfonso traveled $199\frac{3}{4}$ miles in his car and used $8\frac{1}{2}$ gallons of gas. How many miles per gallon did he get? $23\frac{1}{2}$ miles per gallon

Be sure to allow extra time to read over Examples 1–5 and Practice Problems 1–5. Many students find it is helpful to study them on two different days. This allows you additional time to really understand the steps of reasoning involved.

Developing Your Study Skills

Why Study Mathematics?

Students often question the value of mathematics. They see little real use for it in their everyday lives. However, mathematics is often the key that opens the door to a better-paying job.

In our present-day technological world, many people use mathematics daily. Many vocational and professional areas—such as the fields of business, statistics, economics, psychology, finance, computer science, chemistry, physics, engineering, electronics, nuclear energy, banking, quality control, and teaching—require a certain level of expertise in mathematics. Those who want to work in these fields must be able to function at a given mathematical level. Those who cannot will not be able to enter these job areas.

So, whatever your field, be sure to realize the importance of mastering the basics of this course. It is very likely to help you advance to the career of your choice.

You may benefit from using the Mathematics Blueprint for Problem Solving when solving the following exercises.

Applications

▲ **1.** *Geometry* A triangle has three sides that measure $8\frac{1}{3}$ in., $5\frac{4}{5}$ in., and $9\frac{3}{10}$ in. What is the perimeter (total distance around) of the triangle?

$23\frac{13}{30}$ in.

2. *Automobile Travel* On Tuesday, Sally drove $10\frac{1}{2}$ miles while running errands. On Friday and Saturday, she had more errands to run and drove $6\frac{1}{3}$ miles and $12\frac{1}{4}$ miles, respectively. How many total miles did Sally drive this week while running errands? $29\frac{1}{12}$ miles

3. *Wildlife* In 2006, only 700 mountain gorillas remained in the world. Of these, about $\frac{5}{9}$ of them were living in a mountain range along the borders of Congo, Rwanda, and Uganda. How many gorillas were living in this mountain range? Round your answer to the nearest whole number. 389 gorillas

4. *Consumer Decisions* Between 2005 and 2006, prices on many electronic devices went down. The average price of a flat-panel television in 2005 was $1190. In 2006, the average price was about $\frac{4}{5}$ as much. What was the average price of a flat-panel television in 2006? $952

5. *Carpentry* A bolt extends through $\frac{3}{4}$-inch-thick plywood, two washers that are each $\frac{1}{16}$ inch thick, and a nut that is $\frac{3}{16}$ inch thick. The main body of the bolt must be $\frac{1}{2}$ inch longer than the sum of the thicknesses of plywood, washers, and nut. What is the minimum length of the bolt? $1\frac{9}{16}$ inches

6. *Carpentry* A carpenter is using an 8-foot length of wood for a frame. The carpenter needs to cut a notch in the wood that is $4\frac{7}{8}$ feet from one end and $1\frac{2}{3}$ feet from the other end. How long does the notch need to be? $1\frac{11}{24}$ feet

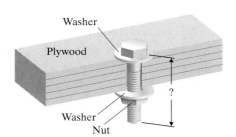

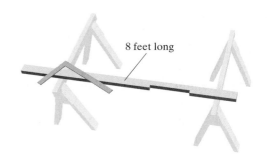

8 feet long

7. *Running a Marathon* Hank is running the Boston Marathon, which is $26\frac{1}{5}$ miles long. At $6\frac{3}{4}$ miles from the start, he meets his wife, who is cheering him on. $9\frac{1}{2}$ miles further down the marathon course, he sees some friends from his running club volunteering at a water stop. Once he passes his friends, how many more miles does Hank have left to run? $9\frac{19}{20}$ miles

8. *Carpentry* Norman Olerud makes birdhouses as a hobby. He has a long piece of lumber that measures $14\frac{1}{4}$ feet. He needs to cut it into pieces that are $\frac{3}{4}$ foot long for the birdhouse floors. How many floors will he be able to cut from the long piece? 19 floors

9. *Personal Finance* Javier earned $10\frac{1}{2}$ per hour for 8 hours of work on Saturday. His manager asked him to stay for an additional 4 hours, for which he was paid $1\frac{1}{2}$ times the regular rate. How much did Javier earn on Saturday? $147

10. *Food Purchase* For a party of the British Literature Club using all "English foods," Nancy bought a $10\frac{2}{3}$-pound wheel of Stilton cheese, to go with the pears and the apples, at $8\frac{3}{4}$ per pound. How much did the wheel of Stilton cheese cost? $93\frac{1}{3}$

▲**11.** *Geometry* How many gallons can a tank hold that has a volume of $36\frac{3}{4}$ cubic feet? (Assume that 1 cubic foot holds about $7\frac{1}{2}$ gallons.)

$275\frac{5}{8}$ gallons

▲**12.** *Geometry* A tank can hold a volume of $7\frac{1}{4}$ cubic feet. If it is filled with water, how much does the water weigh? (Assume that 1 cubic foot of water weighs $62\frac{1}{2}$ pounds.) $453\frac{1}{8}$ pounds

13. *Titanic Disaster* The night of the *Titanic* cruise ship disaster, the captain decided to run his ship at $22\frac{1}{2}$ knots (nautical miles per hour). The *Titanic* traveled at that speed for $4\frac{3}{4}$ hours before it met its tragic demise. How far did the *Titanic* travel at this excessive speed before the disaster?

$106\frac{7}{8}$ nautical miles

14. *Personal Finance* William built a porch for his neighbor and got paid \$1200. He gave $\frac{1}{10}$ of this to his brother to pay back a debt. He used $\frac{1}{3}$ of it to pay bills and used $\frac{1}{6}$ to pay his helper. How much of the \$1200 did William have left? \$480

15. *Personal Finance* Noriko earns \$660 per week. She has $\frac{1}{5}$ of her income withheld for federal taxes, $\frac{1}{15}$ of her income withheld for state taxes, and $\frac{1}{20}$ of her income withheld for medical coverage. How much per week is left for Noriko after these three deductions? \$451 per week

16. *Real Estate* Dan and Estella are saving for a down payment on a house. Their total take-home pay is \$960 per week. They have allotted $\frac{1}{4}$ of their weekly income for rent, $\frac{1}{10}$ for car insurance, and $\frac{1}{3}$ for all other expenses including groceries, clothing, entertainment, and monthly bills. How much is left per week to be saved for their down payment? \$304 per week

17. *Making Jewelry* Emily makes bracelets and sells them for $9\frac{1}{2}$. She has a long piece of wire that measures 20 feet. Each bracelet requires $\frac{3}{5}$ foot to make.
   (a) How many bracelets can Emily make from the large piece of wire? 33 bracelets
   (b) How much wire is left over? $\frac{1}{5}$ foot
   (c) If Emily sells all the bracelets, how much money will she make? \$313$\frac{1}{2}$

▲**18.** *Home Improvement* The Costellos are having new carpet and moulding installed in their sunroom. The room measures $7\frac{1}{2}$ feet by $11\frac{2}{3}$ feet.
   (a) If new carpet costs \$3 per square foot to install, how much will the new carpet cost?
   \$262$\frac{1}{2}$
   (b) The new moulding will be placed around the room where the wall and ceiling meet. How many feet of moulding will they need?
   $38\frac{1}{3}$ feet

19. *Food Purchase* Cecilia bought a loaf of sourdough bread that was made by a local gourmet bakery. The label said that the bread, plus its fancy box, weighed $18\frac{1}{2}$ ounces in total. Of this, $1\frac{1}{4}$ ounces turned out to be the weight of the ribbon. The box weighed $3\frac{1}{8}$ ounces.
   (a) How many ounces of bread did she actually buy? $14\frac{1}{8}$ ounces of bread
   (b) The box stated its net weight as $14\frac{3}{4}$ ounces. (This means that she should have found $14\frac{3}{4}$ ounces of gourmet sourdough bread in the box.) How much in error was this measurement? $\frac{5}{8}$ of an ounce

20. *Cooking* Marnie has $12\frac{1}{2}$ cups of flour. She wants to make two pies, each requiring $1\frac{1}{4}$ cups of flour, and three cakes, each requiring $2\frac{1}{8}$ cups. How much flour will be left after Marnie makes the pies and cakes? $3\frac{5}{8}$ cups

21. *Coast Guard Boat Operation* The largest Coast Guard boat stationed at San Diego can travel $160\frac{1}{8}$ nautical miles in $5\frac{1}{4}$ hours.

 (a) At how many knots is the boat traveling?

 $30\frac{1}{2}$ knots

 (b) At this speed, how long would it take the Coast Guard boat to travel $213\frac{1}{2}$ nautical miles? 7 hours

▲ **23.** *Farming* A Kansas wheat farmer has a storage bin with a capacity of $6856\frac{1}{4}$ cubic feet.

 (a) If a bushel of wheat is $1\frac{1}{4}$ cubic feet, how many bushels can the storage bin hold?

 5485 bushels

 (b) If a farmer wants to make a new storage bin $1\frac{3}{4}$ times larger, how many cubic feet will it hold? $11,998\frac{7}{16}$ cubic feet

 (c) How many bushels will the new bin hold?

 $9598\frac{3}{4}$ bushels

22. *Water Ski Boat* Russ and Norma's Mariah water ski boat can travel $72\frac{7}{8}$ nautical miles in $2\frac{3}{4}$ hours.

 (a) At how many knots is the boat traveling?

 $26\frac{1}{2}$ knots

 (b) At this speed, how long would it take their water ski boat to travel $92\frac{3}{4}$ nautical miles?

 $3\frac{1}{2}$ hours

▲ **24.** *Farming* A wheat farmer from Texas has a storage bin with a capacity of $8693\frac{1}{3}$ cubic feet.

 (a) If a bushel of wheat is $1\frac{1}{3}$ cubic feet, how many bushels can the storage bin hold?

 6520 bushels

 (b) If a farmer wants to make a new storage bin $1\frac{1}{3}$ times larger, how many cubic feet will it hold? $11,951\frac{1}{9}$ cubic feet

 (c) How many bushels will the new bin hold?

 $8693\frac{1}{3}$ bushels

Cumulative Review

25. [1.2.4] Add.
$$\begin{array}{r} 16{,}846 \\ 19{,}321 \\ +\ 8{,}078 \\ \hline 44{,}245 \end{array}$$

27. [1.4.4] Multiply.
$$\begin{array}{r} 1683 \\ \times\quad 27 \\ \hline 45{,}441 \end{array}$$

26. [1.3.3] Subtract.
$$\begin{array}{r} 209{,}364 \\ -\ 186{,}927 \\ \hline 22{,}437 \end{array}$$

28. [1.5.3] Divide. $37\overline{)13{,}172}$ 356

Quick Quiz 2.9

1. Marcia wants to put wall-to-wall carpet in her bedroom. The room measures $15\frac{3}{4}$ feet by $10\frac{2}{3}$ feet. How many square feet of carpeting does she need?

168 square feet

3. Lexi bicycled $1\frac{1}{8}$ miles from Beverly to Beverly Cove. She then traveled $1\frac{1}{2}$ miles from Beverly Cove to Chapman's Corner. Finally she traveled $2\frac{3}{4}$ miles from Chapman's Corner to Beverly Farms. How far did she travel on her bicycle? Express your answer as a mixed number. $5\frac{3}{8}$ miles

2. Ken Thompson shipped out $41\frac{3}{5}$ pounds of electrical supplies. The supplies are placed in individual packets that weigh $2\frac{3}{5}$ pounds each. How many packets did he ship out? 16 packets

4. **Concept Check** A trail to a peak on Mount Washington is $3\frac{3}{5}$ miles long. Caleb started hiking on the trail and stopped after walking $1\frac{7}{8}$ miles to take a break. Explain how you would find how far he still has to go to get to the peak. Answers may vary

Classroom Quiz 2.9 You may use these problems to quiz your students' mastery of Section 2.9.

Express all answers as mixed numbers.

1. Stephanie is in training to run in a marathon. Yesterday she ran for $2\frac{1}{3}$ hours at a speed of $4\frac{3}{4}$ miles per hour. How far did she run? **Ans:** $11\frac{1}{12}$ miles

2. At top speed, the fishing boat *Happy Days* in Key West, Florida, can travel $75\frac{3}{8}$ miles in $2\frac{1}{4}$ hours. How many miles per hour can it travel at top speed? **Ans:** $33\frac{1}{2}$ miles per hour

3. Melissa and Phil visited a ranch near Denver, CO. The back field of the ranch is in the shape of a triangle. One side is $3\frac{1}{5}$ miles long. The other two sides are $2\frac{1}{2}$ miles and $1\frac{3}{4}$ miles long. How many miles of fence are required to enclose this field? **Ans:** $7\frac{9}{20}$ miles

Putting Your Skills to Work: Use Math to Save Money

FINDING EXTRA MONEY EACH MONTH

Do you find yourself running short of money each month? Do you wish you could find a little extra cash for yourself? Is there some daily habit that costs money that maybe you could give up? Let's start with smoking cigarettes. (If you don't smoke, think of some other example, perhaps your daily cup of coffee, where you spend money.) Now consider the story of a young couple, Tricia and Jack.

Tricia and Jack both used to smoke cigarettes. Then Tricia experienced some health problems and had to go to the hospital for several days. Tricia and Jack both decided they were done with smoking. It was pretty hard to quit, and at first they just focused on how hard it was for them. But then Tricia and Jack noticed they were having more money left over at the end of the month.

They got to thinking. Where they live cigarettes cost $6 a pack, and they were both pack-a-day smokers. How much had they spent together per month on cigarettes? (Figure 30 days for an average month.)

1. **(a)** Find out how much Tricia and Jack spent in a month (30 days) on cigarettes. $360

   (b) Use your answer from (a) to find out how much they spend in 12 months on cigarettes. $4320

Tricia and Jack enjoyed smoking but knew it was bad for their health. So they decided to put the money they saved by not smoking into a savings account for something they would really enjoy.

They thought they could purchase a really nice plasma television for $2000. If they put the money they saved each month into the savings account, would there be enough money by Tricia's birthday (which is 7 months from now)?

2. **(a)** Find out if they would save enough money in 7 months for a television?   Yes

   (b) Would there be extra money for a birthday celebration dinner for Tricia? If so, how much?   Yes, there would be $520 left over for the celebration dinner.

3. If Tricia and Jack found a plasma television on sale that only costs $\frac{3}{4}$ of what the television costs in problem 2 (above), how much money would be available for the birthday dinner? If the cost of the television is $\frac{3}{4}$ of $2000, then the total would only be $1500. Thus $1020 would be left over for the birthday dinner.

Some cities and states across the U.S. are imposing taxes on the sale of cigarettes as a way to offset the cost of healthcare for people who suffer from smoking-related medical issues. These taxes make smoking cigarettes even more costly. For example in July of 2008 in New York City a pack of cigarettes cost approximately $10.00.

If Tricia and Jack lived in New York City and each smoked a pack of cigarettes per day, how much money would they spend on cigarettes for a month?

4. **(a)** Find out how much Tricia and Jack spent in 30 days on cigarettes in New York City. $600

   (b) Use your answer from (a) to find out how much Tricia and Jack would spend on cigarettes in 12 months in New York City. $7200

   (c) How much more is this amount than the amount you found in problem 1(b) above? $2880

5. Can you think of one extra expense you could eliminate so you could save money to purchase a big ticket item? Calculate the savings.   Answers may vary

Topic	Procedure	Examples
Concept of a fractional part, p. 106.	The numerator is the number of parts selected. The denominator is the number of total parts.	What part of this sketch is shaded? $\frac{7}{10}$
Prime factorization, p. 113.	Prime factorization is the writing of a number as the product of prime numbers.	Write the prime factorization of 36. $36 = 4 \times 9$ $\quad\quad 2 \times 2 \quad 3 \times 3$ $\quad\; = 2 \times 2 \times 3 \times 3$
Reducing fractions, p. 115.	1. Factor numerator and denominator into prime factors. 2. Divide out factors common to numerator and denominator.	Reduce. $\frac{54}{90}$ $\frac{54}{90} = \frac{\overset{1}{\cancel{2}} \times \overset{1}{\cancel{3}} \times \overset{1}{\cancel{3}} \times 3}{\underset{1}{\cancel{2}} \times \underset{1}{\cancel{3}} \times \underset{1}{\cancel{3}} \times 5} = \frac{3}{5}$
Changing a mixed number to an improper fraction, p. 122.	1. Multiply whole number by denominator. 2. Add product to numerator. 3. Place sum over denominator.	Write as an improper fraction. $7\frac{3}{4} = \frac{7 \times 4 + 3}{4} = \frac{28 + 3}{4} = \frac{31}{4}$
Changing an improper fraction to a mixed number, p. 122.	1. Divide denominator into numerator. 2. The quotient is the whole number. 3. The fraction is the remainder over the divisor.	Change to a mixed number. $\frac{32}{5}$ $5\overline{)32} \;= 6\frac{2}{5}$ with quotient 6, $\frac{30}{\;2}$
Multiplying fractions, p. 128.	1. Divide out common factors from the numerators and denominators whenever possible. 2. Multiply numerators. 3. Multiply denominators.	Multiply. $\frac{3}{7} \times \frac{5}{13} = \frac{15}{91}$ Multiply. $\frac{\overset{1}{\cancel{5}}}{\cancel{8}} \times \frac{\overset{2}{\cancel{16}}}{\cancel{15}} = \frac{2}{3}$
Multiplying mixed and/or whole numbers, p. 129.	1. Change any whole numbers to fractions with a denominator of 1. 2. Change any mixed numbers to improper fractions. 3. Use multiplication rule for fractions.	Multiply. $7 \times 3\frac{1}{4}$ $\frac{7}{1} \times \frac{13}{4} = \frac{91}{4}$ or $22\frac{3}{4}$
Dividing fractions, p. 134.	To divide two fractions, we invert the second fraction and multiply.	Divide. $\frac{3}{7} \div \frac{2}{9} = \frac{3}{7} \times \frac{9}{2} = \frac{27}{14}$ or $1\frac{13}{14}$
Dividing mixed numbers and/or whole numbers, p. 136.	1. Change any whole numbers to fractions with a denominator of 1. 2. Change any mixed numbers to improper fractions. 3. Use rule for division of fractions.	Divide. $8\frac{1}{3} \div 5\frac{5}{9} = \frac{25}{3} \div \frac{50}{9}$ $= \frac{\overset{1}{\cancel{25}}}{\underset{1}{\cancel{3}}} \times \frac{\overset{3}{\cancel{9}}}{\underset{2}{\cancel{50}}} = \frac{3}{2}$ or $1\frac{1}{2}$
Finding the least common denominator, p. 146.	1. Write each denominator as the product of prime factors. 2. List all the prime factors that appear in both products. 3. Form a product of those factors, using each factor the greatest number of times it appears in any denominator.	Find LCD of $\frac{1}{10}, \frac{3}{8}$, and $\frac{7}{25}$. $10 = 2 \times 5$ $8 = 2 \times 2 \times 2$ $25 = 5 \times 5$ $\text{LCD} = 2 \times 2 \times 2 \times 5 \times 5 = 200$

(Continued on next page)

Topic	Procedure	Examples
Building fractions, *p. 149.*	1. Find how many times the original denominator can be divided into the new denominator. 2. Multiply that value by numerator and denominator of original fraction.	Build $\frac{5}{7}$ to an equivalent fraction with a denominator of 42. First we find $7\overline{)42}\,^{6}$. Then we multiply the numerator and denominator by 6. $$\frac{5}{7} \times \frac{6}{6} = \frac{30}{42}$$
Adding or subtracting fractions with a common denominator, *p. 155.*	1. Add or subtract the numerators. 2. Keep the common denominator.	Add. $\dfrac{3}{13} + \dfrac{5}{13} = \dfrac{8}{13}$ Subtract. $\dfrac{15}{17} - \dfrac{12}{17} = \dfrac{3}{17}$
Adding or subtracting fractions without a common denominator, *p. 156.*	1. Find the LCD of the fractions. 2. Build each fraction, if needed, to obtain the LCD in the denominator. 3. Follow the steps for adding and subtracting fractions with the same denominator.	Add. $\dfrac{1}{4} + \dfrac{3}{7} + \dfrac{5}{8}$ LCD = 56 $$\frac{1}{4} \times \frac{14}{14} + \frac{3}{7} \times \frac{8}{8} + \frac{5}{8} \times \frac{7}{7}$$ $$= \frac{14}{56} + \frac{24}{56} + \frac{35}{56} = \frac{73}{56} \text{ or } 1\frac{17}{56}$$
Adding mixed numbers, *p. 163.*	1. Change fractional parts to equivalent fractions with LCD as a denominator, if needed. 2. Add whole numbers and fractions separately. 3. If improper fractions occur, change to mixed numbers and simplify.	Add. $6\frac{3}{4} + 2\frac{5}{8}$ $\begin{aligned} 6\;\boxed{\tfrac{3}{4} \times \tfrac{2}{2}} &= 6\frac{6}{8} \\ +\,2\frac{5}{8} &= +2\frac{5}{8} \\ \hline & 8\frac{11}{8} = 9\frac{3}{8} \end{aligned}$
Subtracting mixed numbers, *p. 164.*	1. Change fractional parts to equivalent fractions with LCD as a denominator, if needed. 2. If necessary, borrow from whole number to subtract fractions. 3. Subtract whole numbers and fractions separately.	Subtract. $8\frac{1}{5} - 4\frac{2}{3}$ $\begin{aligned} 8\;\boxed{\tfrac{1}{5} \times \tfrac{3}{3}} &= 8\frac{3}{15} = 7\frac{18}{15} \\ -\,4\;\boxed{\tfrac{2}{3} \times \tfrac{5}{5}} &= -4\frac{10}{15} = -4\frac{10}{15} \\ \hline & \qquad\qquad\quad 3\frac{8}{15} \end{aligned}$

Order of Operations *p. 166.*	With grouping symbols:
	
	 Do first 1. Perform operations inside parentheses.
	            2. Simplify any expressions with exponents.
	            3. Multiply or divide from left to right.
	 Do last 4. Add or subtract from left to right.
		$\dfrac{5}{6} \div \left(\dfrac{4}{5} - \dfrac{7}{15}\right)$ First combine numbers inside the parentheses.
		
		 $\dfrac{5}{6} \div \left(\dfrac{12}{15} - \dfrac{7}{15}\right)$ Transform $\dfrac{4}{5}$ to equivalent fraction $\dfrac{12}{15}$.
		
		 $\dfrac{5}{6} \div \dfrac{1}{3}$ Subtract the two fractions inside the parentheses and reduce.
		
		 $\dfrac{5}{6} \times \dfrac{3}{1}$ Invert the second fraction and multiply.
		
		 $\dfrac{5}{2}$ or $2\dfrac{1}{2}$ Simplify.

Procedure for Solving Applied Problems

Using the Mathematics Blueprint for Problem Solving, p. 171

In solving an applied problem with fractions, students may find it helpful to complete the following steps. You will not use all the steps all of the time. Choose the steps that best fit the conditions of the problem.

1. **Understand the problem.**

   (a) Read the problem carefully.

   (b) Draw a picture if this helps you to visualize the situation. Think about what facts you are given and what you are asked to find.

   (c) It may help to write a similar, simpler problem to get started and to determine what operation to use.

   (d) Use the Mathematics Blueprint for Problem Solving to organize your work. Follow these four parts.

      1. Gather the facts (Write down specific values given in the problem.)
      2. What am I asked to do? (Identify what you must obtain for an answer.)
      3. How do I proceed? (Decide what calculations need to be done.)
      4. Key points to remember (Record any facts, warnings, formulas, or concepts you think will be important as you solve the problem.)

2. **Solve and state the answer.**

   (a) Perform the necessary calculations.

   (b) State the answer, including the unit of measure.

3. **Check.**

   (a) Estimate the answer to the problem. Compare this estimate to the calculated value. Is your answer reasonable?

   (b) Repeat your calculations.

   (c) Work backward from your answer. Do you arrive at the original conditions of the problem?

EXAMPLE

A wire is $95\frac{1}{3}$ feet long. It is cut up into smaller, equal-sized pieces, each $4\frac{1}{3}$ feet long. How many pieces will there be?

1. **Understand the problem.**

   Draw a picture of the situation.

   How will we find the number of pieces?

   Now we will use a simpler problem to clarify the idea. A wire 100 feet long is cut up into smaller pieces each 4 feet long. How many pieces will there be? We readily see that we would divide 100 by 4. Thus in our original problem we should divide $95\frac{1}{3}$ feet by $4\frac{1}{3}$ feet. This will tell us the number of pieces. Now we fill in the Mathematics Blueprint (see below).

2. **Solve and state the answer.**

   We need to divide $95\frac{1}{3} \div 4\frac{1}{3}$.

   $$\frac{286}{3} \div \frac{13}{3} = \frac{\overset{22}{\cancel{286}}}{\underset{1}{\cancel{3}}} \times \frac{\overset{1}{\cancel{3}}}{\underset{1}{\cancel{13}}} = \frac{22}{1} = 22$$

   There will be 22 pieces of wire.

3. **Check.**

   Estimate. Rounded to the nearest ten, $95\frac{1}{3} \approx 100$.

   Rounded to the nearest integer, $4\frac{1}{3} \approx 4$.

   $$100 \div 4 = 25$$

   This is close to our estimate. Our answer is reasonable. ✔

Mathematics Blueprint for Problem Solving

Gather the Facts	What Am I Asked to Do?	How Do I Proceed?	Key Points to Remember
Wire is $95\frac{1}{3}$ feet. It is cut into equal pieces $4\frac{1}{3}$ feet long.	Determine how many pieces of wire there will be.	Divide $95\frac{1}{3}$ by $4\frac{1}{3}$.	Change mixed numbers to improper fractions before carrying out the division.

Chapter 2 Review Problems

Be sure to simplify all answers.

Section 2.1

Use a fraction to represent the shaded part of each object.

1. $\frac{3}{8}$

2. $\frac{5}{12}$

In exercises 3 and 4, draw a sketch to illustrate each fraction.

3. $\frac{4}{7}$ of an object   Answers will vary.

4. $\frac{7}{10}$ of a group   Answers will vary.

5. **Quality Control** An inspector looked at 80 semi-conductors and found 9 of them defective. What fractional part of these items was defective?
    $\frac{9}{80}$

6. **Education** The dean asked the 100 freshmen if they would be staying in the dorm over the holidays. A total of 87 said they would not. What fractional part of the freshmen said they would not?
    $\frac{87}{100}$

Section 2.2

Express each number as a product of prime factors.

7. 54 2×3^3

8. 120 $2^3 \times 3 \times 5$

9. 168 $2^3 \times 3 \times 7$

Determine which of the following numbers are prime. If a number is composite, express it as the product of prime factors.

10. 59   prime

11. 78 $2 \times 3 \times 13$

12. 167   prime

Reduce each fraction.

13. $\frac{12}{42}$ $\frac{2}{7}$

14. $\frac{13}{52}$ $\frac{1}{4}$

15. $\frac{27}{72}$ $\frac{3}{8}$

16. $\frac{26}{34}$ $\frac{13}{17}$

17. $\frac{168}{192}$ $\frac{7}{8}$

18. $\frac{51}{105}$ $\frac{17}{35}$

Section 2.3

Change each mixed number to an improper fraction.

19. $4\frac{3}{8}$ $\frac{35}{8}$

20. $15\frac{3}{4}$ $\frac{63}{4}$

21. $5\frac{2}{7}$ $\frac{37}{7}$

22. $6\frac{3}{5}$ $\frac{33}{5}$

Change each improper fraction to a mixed number.

23. $\frac{45}{8}$ $5\frac{5}{8}$

24. $\frac{100}{21}$ $4\frac{16}{21}$

25. $\frac{53}{7}$ $7\frac{4}{7}$

26. $\frac{74}{9}$ $8\frac{2}{9}$

27. Reduce and leave your answer as a mixed number.

$3\dfrac{15}{55}$ $3\dfrac{3}{11}$

28. Reduce and leave your answer as an improper fraction.

$\dfrac{234}{16}$ $\dfrac{117}{8}$

29. Change to a mixed number and then reduce.

$\dfrac{132}{32}$ $4\dfrac{1}{8}$

Section 2.4

Multiply.

30. $\dfrac{4}{7} \times \dfrac{5}{11}$ $\dfrac{20}{77}$

31. $\dfrac{7}{9} \times \dfrac{21}{35}$ $\dfrac{7}{15}$

32. $12 \times \dfrac{3}{7} \times 0$ 0

33. $\dfrac{3}{5} \times \dfrac{2}{7} \times \dfrac{10}{27}$ $\dfrac{4}{63}$

34. $12 \times 8\dfrac{1}{5}$ $\dfrac{492}{5}$ or $98\dfrac{2}{5}$

35. $5\dfrac{1}{4} \times 4\dfrac{6}{7}$ $\dfrac{51}{2}$ or $25\dfrac{1}{2}$

36. $5\dfrac{1}{8} \times 3\dfrac{1}{5}$ $\dfrac{82}{5}$ or $16\dfrac{2}{5}$

37. $36 \times \dfrac{4}{9}$ 16

38. ***Stock Market*** In 1999, one share of stock cost $\$37\dfrac{5}{8}$. How much money did 18 shares cost?

$\$677\dfrac{1}{4}$

▲ **39.** ***Geometry*** The O'Gara's new family room addition measures $13\dfrac{1}{2}$ feet long by $9\dfrac{2}{3}$ feet wide. Find the area of the addition.

$\dfrac{261}{2}$ or $130\dfrac{1}{2}$ square feet

Section 2.5

Divide, if possible.

40. $\dfrac{3}{7} \div \dfrac{2}{5}$ $\dfrac{15}{14}$ or $1\dfrac{1}{14}$

41. $\dfrac{3}{5} \div \dfrac{1}{10}$ 6

42. $1200 \div \dfrac{5}{8}$ 1920

43. $900 \div \dfrac{3}{5}$ 1500

44. $5\dfrac{3}{4} \div 11\dfrac{1}{2}$ $\dfrac{1}{2}$

45. $\dfrac{20}{2\dfrac{1}{2}}$ 8

46. $0 \div 3\dfrac{7}{5}$ 0

47. $4\dfrac{2}{11} \div 3$ $\dfrac{46}{33}$ or $1\dfrac{13}{33}$

▲ **48.** ***Floor Carpeting*** Each roll of carpet covers $28\dfrac{1}{2}$ square yards. The community center has 342 square yards of flooring to carpet. How many rolls are needed?   12 rolls

49. There are 420 calories in $2\dfrac{1}{4}$ cans of grape soda. How many calories are in 1 can of soda?

$\dfrac{560}{3}$ or $186\dfrac{2}{3}$ calories

Section 2.6

Find the LCD for each pair of fractions.

50. $\dfrac{7}{14}$ and $\dfrac{3}{49}$ 98

51. $\dfrac{13}{20}$ and $\dfrac{3}{25}$ 100

52. $\dfrac{5}{18}, \dfrac{1}{6}, \dfrac{7}{45}$ 90

Build each fraction to an equivalent fraction with the specified denominator.

53. $\dfrac{3}{7} = \dfrac{?}{56}$ $\dfrac{24}{56}$

54. $\dfrac{11}{24} = \dfrac{?}{72}$ $\dfrac{33}{72}$

55. $\dfrac{8}{15} = \dfrac{?}{150}$ $\dfrac{80}{150}$

56. $\dfrac{17}{18} = \dfrac{?}{198}$ $\dfrac{187}{198}$

Section 2.7

Add or subtract.

57. $\dfrac{9}{14} - \dfrac{5}{14}$　$\dfrac{2}{7}$

58. $\dfrac{1}{2} + \dfrac{1}{3} + \dfrac{1}{4}$　$\dfrac{13}{12}$ or $1\dfrac{1}{12}$

59. $\dfrac{4}{7} + \dfrac{7}{9}$　$\dfrac{85}{63}$ or $1\dfrac{22}{63}$

60. $\dfrac{7}{8} - \dfrac{3}{5}$　$\dfrac{11}{40}$

61. $\dfrac{7}{30} + \dfrac{2}{21}$　$\dfrac{23}{70}$

62. $\dfrac{5}{18} + \dfrac{7}{10}$　$\dfrac{44}{45}$

63. $\dfrac{15}{16} - \dfrac{13}{24}$　$\dfrac{19}{48}$

64. $\dfrac{14}{15} - \dfrac{3}{25}$　$\dfrac{61}{75}$

Section 2.8

Evaluate using the correct order of operations.

65. $8 - 2\dfrac{3}{4}$　$5\dfrac{1}{4}$

66. $6 - \dfrac{5}{9}$　$\dfrac{49}{9}$ or $5\dfrac{4}{9}$

67. $3 + 5\dfrac{2}{3}$　$8\dfrac{2}{3}$

68. $9\dfrac{3}{7} + 13$　$22\dfrac{3}{7}$

69. $3\dfrac{3}{8} + 2\dfrac{3}{4}$　$\dfrac{49}{8}$ or $6\dfrac{1}{8}$

70. $5\dfrac{11}{16} - 2\dfrac{1}{5}$　$\dfrac{279}{80}$ or $3\dfrac{39}{80}$

71. $\dfrac{3}{5} \times \dfrac{1}{2} + \dfrac{2}{5} \div \dfrac{2}{3}$　$\dfrac{9}{10}$

72. $\left(\dfrac{4}{5} - \dfrac{1}{2}\right)^2 \times \dfrac{10}{3}$　$\dfrac{3}{10}$

73. *Jogging* Bob jogged $1\dfrac{7}{8}$ miles on Monday, $2\dfrac{3}{4}$ miles on Tuesday, and $4\dfrac{1}{10}$ miles on Wednesday. How many miles did he jog on these three days?

$8\dfrac{29}{40}$ miles

74. *Fuel Economy* When it was new, Ginny Sue's car got $28\dfrac{1}{6}$ miles per gallon. It now gets $1\dfrac{5}{6}$ miles per gallon less. How far can she drive now if the car has $10\dfrac{3}{4}$ gallons in the tank?

$283\dfrac{1}{12}$ miles

Section 2.9

75. *Cooking* A recipe calls for $3\dfrac{1}{3}$ cups of sugar and $4\dfrac{1}{4}$ cups of flour. How much sugar and how much flour would be needed for $\dfrac{1}{2}$ of that recipe?

$1\dfrac{2}{3}$ cups sugar, $2\dfrac{1}{8}$ cups flour

76. *Fuel Economy* Rafael travels in a car that gets $24\dfrac{1}{4}$ miles per gallon. He has $8\dfrac{1}{2}$ gallons of gas in the gas tank. Approximately how far can he drive?

$206\dfrac{1}{8}$ miles

77. *Construction* How many lengths of pipe $3\dfrac{1}{5}$ inches long can be cut from a pipe 48 inches long?　15 lengths

78. *Automobile Maintenance* A car radiator holds $15\dfrac{3}{4}$ liters. If it contains $6\dfrac{1}{8}$ liters of antifreeze and the rest is water, how much is water?

$9\dfrac{5}{8}$ liters

79. *Reading Speed* Tim found that he can read 5 pages of his biology book in $32\dfrac{1}{2}$ minutes. He has three chapters to read over the weekend. The first is 12 pages, the second is 9 pages, and the third is 14 pages. How long will it take him?

$227\dfrac{1}{2}$ minutes or 3 hours and $47\dfrac{1}{2}$ minutes

80. *Personal Finance* Tatiana earns $\$9\dfrac{1}{2}$ per hour for regular pay and $1\dfrac{1}{2}$ times that rate of pay for overtime. On Saturday she worked eight hours at regular pay and four hours at overtime. How much did she earn on Saturday?　$133

81. *Stock Market* George bought 70 shares of stock in 2001 at $15\frac{3}{4}$ a share. He sold all the shares in 2003 for $24 each. How much did George make when he sold his shares?

$577\frac{1}{2}$

82. *Carpentry* A 3-inch bolt passes through $1\frac{1}{2}$ inches of pine board, a $\frac{1}{16}$-inch washer, and a $\frac{1}{8}$-inch nut. How many inches does the bolt extend beyond the board, washer, and nut if the head of the bolt is $\frac{1}{4}$ inch long?

$1\frac{1}{16}$ inch

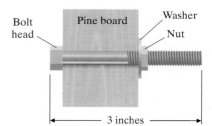

83. *Budgeting* Francine has a take-home pay of $880 per month. She gives $\frac{1}{10}$ of it to her church, spends $\frac{1}{2}$ of it for rent and food, and spends $\frac{1}{8}$ of it on electricity, heat, and telephone. How many dollars per month does she have left for other things? $242

84. *Cost of Auto Travel* Manuel's new car used $18\frac{2}{5}$ gallons of gas on a 460-mile trip.
   (a) How many miles can his car travel on 1 gallon of gas? 25 miles per gallon
   (b) How much did his trip cost him in gasoline expense if the average cost of gasoline was $3\frac{1}{5}$ per gallon? $58\frac{22}{25}$

Mixed Practice

Perform each calculation or each requested operation.

85. Reduce. $\dfrac{27}{63}$ $\frac{3}{7}$

86. $\dfrac{7}{15} + \dfrac{11}{25}$ $\frac{68}{75}$

87. $4\dfrac{1}{3} - 2\dfrac{11}{12}$ $1\frac{5}{12}$

88. $\dfrac{36}{49} \times \dfrac{14}{33}$ $\frac{24}{77}$

89. $4\dfrac{1}{4} \div \dfrac{3}{2}$ $\frac{17}{6}$ or $2\frac{5}{6}$

90. $\left(\dfrac{4}{7}\right)^3$ $\frac{64}{343}$

91. $\dfrac{3}{8} \div \dfrac{1}{10}$ $\frac{15}{4}$ or $3\frac{3}{4}$

92. $5\dfrac{1}{2} \times 18$ 99

93. $150 \div 3\dfrac{1}{8}$ 48

Note to Instructor: The Chapter 2 Test file in
the TestGen program provides algorithms
specifically matched to these problems so
you can easily replicate this test for additional practice or assessment purposes.

How Am I Doing? Chapter 2 Test

*Remember to use your Chapter Test Prep Video CD to see the worked-out
solutions to the test problems you want to review.*

Solve.

1. Use a fraction to represent the shaded part of the object.

2. A basketball star shot at the hoop 388 times. The ball went in 311 times.
Write a fraction that describes the part of the time that his shots went in.

Reduce each fraction.

3. $\dfrac{18}{42}$ **4.** $\dfrac{15}{70}$ **5.** $\dfrac{225}{50}$

6. Change to an improper fraction. $6\dfrac{4}{5}$

7. Change to a mixed number. $\dfrac{145}{14}$

Multiply.

8. $42 \times \dfrac{2}{7}$ **9.** $\dfrac{7}{9} \times \dfrac{2}{5}$ **10.** $2\dfrac{2}{3} \times 5\dfrac{1}{4}$

Divide.

11. $\dfrac{7}{8} \div \dfrac{5}{11}$ **12.** $\dfrac{12}{31} \div \dfrac{8}{13}$

13. $7\dfrac{1}{5} \div 1\dfrac{1}{25}$ **14.** $5\dfrac{1}{7} \div 3$

Find the least common denominator of each set of fractions.

15. $\dfrac{5}{12}$ and $\dfrac{7}{18}$ **16.** $\dfrac{3}{16}$ and $\dfrac{1}{24}$ **17.** $\dfrac{1}{4}, \dfrac{3}{8}, \dfrac{5}{6}$

18. Build the fraction to an equivalent fraction with the specified
denominator. $\dfrac{5}{12} = \dfrac{?}{72}$

1. $\dfrac{3}{5}$

2. $\dfrac{311}{388}$

3. $\dfrac{3}{7}$

4. $\dfrac{3}{14}$

5. $\dfrac{9}{2}$

6. $\dfrac{34}{5}$

7. $10\dfrac{5}{14}$

8. 12

9. $\dfrac{14}{45}$

10. 14

11. $\dfrac{77}{40}$ or $1\dfrac{37}{40}$

12. $\dfrac{39}{62}$

13. $\dfrac{90}{13}$ or $6\dfrac{12}{13}$

14. $\dfrac{12}{7}$ or $1\dfrac{5}{7}$

15. 36

16. 48

17. 24

18. $\dfrac{30}{72}$

Evaluate using the correct order of operations.

19. $\dfrac{7}{9} - \dfrac{5}{12}$

20. $\dfrac{2}{15} + \dfrac{5}{12}$

21. $\dfrac{1}{4} + \dfrac{3}{7} + \dfrac{3}{14}$

22. $8\dfrac{3}{5} + 5\dfrac{4}{7}$

23. $18\dfrac{6}{7} - 13\dfrac{13}{14}$

24. $\dfrac{2}{9} \div \dfrac{8}{3} \times \dfrac{1}{4}$

25. $\left(\dfrac{1}{2} + \dfrac{1}{3} \right) \times \dfrac{7}{5}$

Answer each question.

▲ **26.** Erin needs to find the area of her kitchen so she knows how much tile to purchase. The room measures $16\frac{1}{2}$ feet by $9\frac{1}{3}$ feet. How many square feet is the kitchen?

27. A butcher has $18\frac{2}{3}$ pounds of steak that he wishes to place into packages that average $2\frac{1}{3}$ pounds each. How many packages can he make?

28. From central parking it is $\frac{9}{10}$ of a mile to the science building. Bob started at central parking and walked $\frac{1}{5}$ of a mile toward the science building. He stopped for coffee. When he finished, how much farther did he have to walk to reach the science building?

29. Robin jogged $4\frac{1}{8}$ miles on Monday, $3\frac{1}{6}$ miles on Tuesday, and $6\frac{3}{4}$ miles on Wednesday. How far did she jog on those three days?

30. Mr. and Mrs. Samuel visited Florida and purchased 120 oranges. They gave $\frac{1}{4}$ of them to relatives, ate $\frac{1}{12}$ of them in the hotel, and gave $\frac{1}{3}$ of them to friends. They shipped the rest home to Illinois.
(a) How many oranges did they ship?
(b) If it costs 24¢ for each orange to be shipped to Illinois, what was the total shipping bill?

31. A candle company purchased $48\frac{1}{8}$ pounds of wax to make specialty candles. It takes $\frac{5}{8}$ pound of wax to make one candle. The owners of the business plan to sell the candles for $12 each. The specialty wax cost them $2 per pound.
(a) How many candles can they make?
(b) How much does it cost to make one candle?
(c) How much profit will they make if they sell all of the candles?

19. $\dfrac{13}{36}$

20. $\dfrac{11}{20}$

21. $\dfrac{25}{28}$

22. $14\dfrac{6}{35}$

23. $4\dfrac{13}{14}$

24. $\dfrac{1}{48}$

25. $\dfrac{7}{6}$ or $1\dfrac{1}{6}$

26. 154 square feet

27. 8 packages

28. $\dfrac{7}{10}$ mile

29. $14\dfrac{1}{24}$ miles

30. (a) 40 oranges

(b) $9\dfrac{3}{5}$

31. (a) 77 candles

(b) $1\dfrac{1}{4}$

(c) $827\dfrac{3}{4}$

One-half of this test is based on Chapter 1 material. The remainder is based on material covered in Chapter 2.

1. Write in words. 84,361,208

2. Add.　　235
　　　　　　　152
　　　　　　　　95
　　　　　　+ 78

3. Add.　　156,200
　　　　　　　364,700
　　　　　　+198,320

4. Subtract.　　5718
　　　　　　　− 3643

5. Subtract.　　1,000,361
　　　　　　　−　983,145

6. Multiply.　　126
　　　　　　　×　38

7. Multiply.　　70,000
　　　　　　　×　　12

8. Divide. $7\overline{)32,606}$

9. Divide. $15\overline{)4631}$

10. Evaluate. 7^2

11. Round to the nearest thousand. 6,037,452

12. Perform the operations in their proper order. $6 \times 2^3 + 12 \div (4 + 2)$

13. For his new job, Ellis bought four dress shirts for $25 each, two pairs of pants for $36 each, and a pair of shoes for $65. What was his total bill?

14. Leslie had a balance of $64 in her checking account. She deposited $1160. She made checks out for $516, $199, and $203. What will be her new balance?

15. A supermarket survey found that of the 112 people that went grocery shopping on Friday morning, 83 were women. Write the fractions that describe the part of the shoppers that was women and the part that was men.

16. Reduce. $\dfrac{28}{52}$

17. Write as an improper fraction. $18\dfrac{3}{4}$

Answer column:

1. eighty-four million, three hundred sixty-one thousand, two hundred eight

2. 560

3. 719,220

4. 2075

5. 17,216

6. 4788

7. 840,000

8. 4658

9. 308 R 11

10. 49

11. 6,037,000

12. 50

13. $237

14. $306

15. $\dfrac{83}{112}$ were women; $\dfrac{29}{112}$ were men

16. $\dfrac{7}{13}$

17. $\dfrac{75}{4}$

18. Write as a mixed number. $\dfrac{100}{7}$

19. Multiply. $3\dfrac{1}{2} \times 4\dfrac{2}{3}$

20. Divide. $\dfrac{44}{49} \div 2\dfrac{13}{21}$

21. Find the least common denominator of $\dfrac{5}{8}$ and $\dfrac{7}{10}$.

Evaluate using the correct order of operations.

22. $\dfrac{7}{18} + \dfrac{20}{27}$

23. $2\dfrac{1}{8} + 6\dfrac{3}{4}$

24. $12\dfrac{1}{5} - 4\dfrac{2}{3}$

25. $\dfrac{1}{2} \times \dfrac{2}{3} + \dfrac{1}{4} \div \dfrac{3}{2}$

Answer each question.

26. Marcos is on a special diet and exercise plan to lose weight. By the end of May, his goal is to have lost 15 pounds. In March he lost $5\dfrac{1}{2}$ pounds, and in April he lost $6\dfrac{3}{4}$ pounds. How many pounds must he lose in May to reach his goal?

27. Melinda traveled $221\dfrac{2}{5}$ miles on 9 gallons of gas. How many miles per gallon did her car get?

28. A biscuit recipe requires $1\dfrac{3}{4}$ cups of flour. Marcia wants to make $2\dfrac{1}{2}$ times the recipe. How much flour will she need? If she uses this amount from a new bag of flour containing 12 cups, how much will be left?

29. A space probe travels at 28,356 miles per hour for 2142 hours. Estimate how many miles it travels.

30. To raise money for cancer research, the local YMCA sponsored a road race. Contributions totaled $960. One-sixth of this amount was used for refreshments and T-shirts for the participants. How much did the refreshments and T-shirts cost?

18.	$14\dfrac{2}{7}$
19.	$\dfrac{49}{3}$ or $16\dfrac{1}{3}$
20.	$\dfrac{12}{35}$
21.	40
22.	$\dfrac{61}{54}$ or $1\dfrac{7}{54}$
23.	$\dfrac{71}{8}$ or $8\dfrac{7}{8}$
24.	$\dfrac{113}{15}$ or $7\dfrac{8}{15}$
25.	$\dfrac{1}{2}$
26.	$2\dfrac{3}{4}$ pounds
27.	$24\dfrac{3}{5}$ miles per gallon
28.	$\dfrac{35}{8}$ or $4\dfrac{3}{8}$ cups; $7\dfrac{5}{8}$ cups
29.	60,000,000 miles
30.	$160

Decimals

CHAPTER

3

The rising cost of gasoline is of great concern to every driver in the country. But what is the best bargain when you purchase gasoline? How can you use math to save you money at the pump? Should you always go to the station with the lowest price? Turn to page 251 and you may be surprised at some of the answers.

3.1 USING DECIMAL NOTATION

Writing a Word Name for a Decimal Fraction

In Chapter 2 we discussed *fractions*—the set of numbers such as $\frac{1}{2}, \frac{2}{3}, \frac{1}{10}, \frac{6}{7}$, $\frac{18}{100}$, and so on. Now we will take a closer look at **decimal fractions**—that is, fractions with 10, 100, 1000, and so on, in the denominator, such as $\frac{1}{10}, \frac{18}{100}$, and $\frac{43}{1000}$.

Why, of all fractions, do we take special notice of these? Our hands have 10 digits. Our U.S. money system is based on the dollar, which has 100 equal parts, or cents. And the international system of measurement called the *metric system* is based on 10 and powers of 10.

As with other numbers, these decimal fractions can be written in different ways (forms). For example, the shaded part of the whole in the following drawing can be written:

in words (one-tenth)
in fractional form $\left(\frac{1}{10}\right)$
in decimal form (0.1)

All mean the same quantity, namely 1 out of 10 equal parts of the whole. We'll see that when we use decimal notation, computations can be easily done based on the old rules for whole numbers and a few new rules about where to place the decimal point. In a world where calculators and computers are commonplace, many of the fractions we encounter are decimal fractions. A decimal fraction is a fraction whose denominator is a power of 10.

$\frac{7}{10}$ is a decimal fraction. $\frac{89}{10^2} = \frac{89}{100}$ is a decimal fraction.

Decimal fractions can be written with numerals in two ways: fractional form or decimal form. Some decimal fractions are shown in decimal form below.

Fractional Form		Decimal Form
$\frac{3}{10}$	=	0.3
$\frac{59}{100}$	=	0.59
$\frac{171}{1000}$	=	0.171

The zero in front of the decimal point is not actually required. We place it there simply to make sure that we don't miss seeing the decimal point. A number written in decimal notation has three parts.

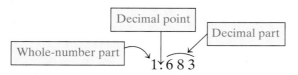

When a number is written in decimal form, the first digit to the right of the decimal point represents tenths, the next digit hundredths, the next digit thousandths, and so on. 0.9 means nine tenths and is equivalent to $\frac{9}{10}$. 0.51 means fifty-one hundredths and is equivalent to $\frac{51}{100}$. Some decimals

Student Learning Objectives

After studying this section, you will be able to:

 Write a word name for a decimal fraction.

 Change from fractional notation to decimal notation.

 Change from decimal notation to fractional notation.

Teaching Tip Students often wonder if they should write .45 rather than 0.45 because they think the 0 is unnecessary. Remind them that the extra zero in front of a decimal point reduces the chances for error. Also, many calculators display this extra 0.

are larger than 1. For example, 1.683 means one and six hundred eighty-three thousandths. It is equivalent to $1\frac{683}{1000}$. Note that the word *and* is used to indicate the decimal point. A place-value chart is helpful.

Decimal Place Values

Hundreds	Tens	Ones	Decimal point	Tenths	Hundredths	Thousandths	Ten-thousandths
100	10	1	"and"	$\frac{1}{10}$	$\frac{1}{100}$	$\frac{1}{1000}$	$\frac{1}{10,000}$
1	5	6	.	2	8	7	4

So, we can write 156.2874 in words as one hundred fifty-six and two thousand eight hundred seventy-four ten-thousandths. We say ten-thousandths because it is the name of the last decimal place on the right.

Teaching Example 1 Write a word name for each decimal.

(a) 0.8   (b) 0.0134

(c) 0.04   (d) 7.678

Ans:

(a) eight tenths

(b) one hundred thirty-four ten-thousandths

(c) four hundredths

(d) seven and six hundred seventy-eight thousandths

NOTE TO STUDENT: Fully worked-out solutions to all of the Practice Problems can be found at the back of the text starting at page SP-1

EXAMPLE 1 Write a word name for each decimal.

(a) 0.79   **(b)** 0.5308   **(c)** 1.6   **(d)** 23.765

Solution

(a) 0.79 = seventy-nine hundredths

(b) 0.5308 = five thousand three hundred eight ten-thousandths

(c) 1.6 = one and six tenths

(d) 23.765 = twenty-three and seven hundred sixty-five thousandths

Practice Problem 1 Write a word name for each decimal.

(a) 0.073   **(b)** 4.68   **(c)** 0.0017   **(d)** 561.78

(a) seventy-three thousandths   **(b)** four and sixty-eight hundredths   **(c)** seventeen ten-thousandths   **(d)** five hundred sixty-one and seventy-eight hundredths

Sometimes, decimals are used where we would not expect them. For example, we commonly say that there are 365 days in a year, with 366 days in every fourth year (or leap year). However, this is not quite correct. In fact, from time to time further adjustments need to be made to the calendar to adjust for these inconsistencies. Astronomers know that a more accurate measure of a year is called a **tropical year** (measured from one equinox to the next). Rounded to the nearest hundred-thousandth, 1 tropical year = 365.24122 days. This is read "three hundred sixty-five and twenty-four thousand, one hundred twenty-two hundred-thousandths." This approximate value is a more accurate measurement of the amount of time it takes the earth to complete one orbit around the sun.

Note the relationship between fractions and their equivalent numbers' decimal forms.

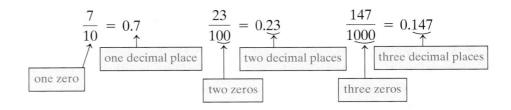

Decimal notation is commonly used with money. When writing a check, we often write the amount that is less than 1 dollar, such as 23¢, as $\frac{23}{100}$ dollar.

Teaching Tip Additional coverage on balancing a checkbook can be found in the Consumer Finance appendix.

EXAMPLE 2 Write a word name for the amount on a check made out for $672.89.

Teaching Example 2 Write a word name for the amount on a check made out for $75.03.

Ans: seventy-five and $\frac{3}{100}$ dollars

Solution Six hundred seventy-two and $\frac{89}{100}$ dollars

Practice Problem 2 Write a word name for the amount of a check made out for $7863.04. seven thousand eight hundred sixty-three and $\frac{4}{100}$ dollars

② Changing from Fractional Notation to Decimal Notation

It is helpful to be able to write decimals in both decimal notation and fractional notation. First we illustrate changing a fraction with a denominator of 10, 100, or 1000 into decimal form.

EXAMPLE 3 Write as a decimal.

(a) $\frac{8}{10}$ **(b)** $\frac{74}{100}$ **(c)** $1\frac{3}{10}$ **(d)** $2\frac{56}{1000}$

Teaching Example 3 Write as a decimal.

(a) $\frac{23}{100}$ **(b)** $4\frac{378}{1000}$

(c) $6\frac{2}{10}$ **(d)** $\frac{59}{1000}$

Ans: **(a)** 0.23   **(b)** 4.378
      (c) 6.2   **(d)** 0.059

Solution

(a) $\frac{8}{10} = 0.8$ **(b)** $\frac{74}{100} = 0.74$ **(c)** $1\frac{3}{10} = 1.3$ **(d)** $2\frac{56}{1000} = 2.056$

Note: In part (d), we need to add a zero before the digits 56. Since there are three zeros in the denominator, we need three decimal places in the decimal number.

Practice Problem 3 Write as a decimal.

(a) $\frac{9}{10}$ 0.9   **(b)** $\frac{136}{1000}$ 0.136   **(c)** $2\frac{56}{100}$ 2.56   **(d)** $34\frac{86}{1000}$ 34.086

③ Changing from Decimal Notation to Fractional Notation

EXAMPLE 4 Write in fractional notation.

(a) 0.51   **(b)** 18.1   **(c)** 0.7611   **(d)** 1.363

Teaching Example 4 Write in fractional notation.

(a) 0.7   **(b)** 10.67

(c) 1.491   **(d)** 0.0023

Ans:

(a) $\frac{7}{10}$ **(b)** $10\frac{67}{100}$

(c) $1\frac{491}{1000}$ **(d)** $\frac{23}{10,000}$

Solution

(a) $0.51 = \frac{51}{100}$ **(b)** $18.1 = 18\frac{1}{10}$ **(c)** $0.7611 = \frac{7611}{10,000}$ **(d)** $1.363 = 1\frac{363}{1000}$

Practice Problem 4 Write in fractional notation.

(a) 0.37 $\frac{37}{100}$ **(b)** 182.3 $182\frac{3}{10}$ **(c)** 0.7131 $\frac{7131}{10,000}$ **(d)** 42.019 $42\frac{19}{1000}$

When we convert from decimal form to fractional form, we reduce whenever possible.

Teaching Example 5 Write in fractional notation. Reduce whenever possible

(a) 3.08   (b) 0.204

(c) 5.48   (d) 0.0627

Ans:

(a) $3\frac{2}{25}$ (b) $\frac{51}{250}$

(c) $5\frac{12}{25}$ (d) $\frac{627}{10,000}$

EXAMPLE 5 Write in fractional notation. Reduce whenever possible.

(a) 2.6   (b) 0.38   (c) 0.525   (d) 361.007

Solution

(a) $2.6 = 2\dfrac{6}{10} = 2\dfrac{3}{5}$ (b) $0.38 = \dfrac{38}{100} = \dfrac{19}{50}$

(c) $0.525 = \dfrac{525}{1000} = \dfrac{105}{200} = \dfrac{21}{40}$

(d) $361.007 = 361\dfrac{7}{1000}$ (cannot be reduced)

Practice Problem 5 Write in fractional notation. Reduce whenever possible.

(a) 8.5 $8\frac{1}{2}$ (b) 0.58 $\frac{29}{50}$ (c) 36.25 $36\frac{1}{4}$ (d) 106.013 $106\frac{13}{1000}$

Teaching Example 6 A chemist found that the concentration of chlorine in a water sample was 12 parts per hundred thousand. What fraction would represent the concentration of chlorine?

Ans: $\dfrac{3}{25,000}$

EXAMPLE 6 A chemist found that the concentration of lead in a water sample was 5 parts per million. What fraction would represent the concentration of lead?

Solution Five parts per million means 5 parts out of 1,000,000. As a fraction, this is $\frac{5}{1,000,000}$. We can reduce this by dividing numerator and denominator by 5. Thus

$$\frac{5}{1,000,000} = \frac{1}{200,000}.$$

The concentration of lead in the water sample is $\frac{1}{200,000}$.

Teaching Tip You may want to show the class a few simple problems like 5 parts per thousand, or 15 parts per ten thousand, before doing the type of numbers contained in Example 6.

Practice Problem 6 A chemist found that the concentration of PCBs in a water sample was 2 parts per billion. What fraction would represent the concentration of PCBs?

$\dfrac{1}{500,000,000}$

Developing Your Study Skills

Steps Toward Success in Mathematics

Mathematics is a building process, mastered one step at a time. The foundation of this process is formed by a few basic requirements. Those who are successful in mathematics realize the absolute necessity for building a study of mathematics on the firm foundation of these six minimum requirements.

1. Attend class every day.
2. Read the textbook.
3. Take notes in class.
4. Do assigned homework every day.
5. Get help immediately when needed.
6. Review regularly.

If you are in an online class or self-paced class, do some of your math assignment on five days during each week.

Verbal and Writing Skills

1. Describe a decimal fraction and provide examples.
A decimal fraction is a fraction whose denominator is a power of 10. $\frac{23}{100}$ and $\frac{563}{1000}$ are decimal fractions.

2. What word is used to describe the decimal point when writing the word name for a decimal that is greater than one?
The word that describes the decimal point is *and*.

3. What is the name of the last decimal place on the right for the decimal 132.45678?
hundred-thousandths

4. When writing $82.75 on a check, we write 75¢ as $\frac{75}{100}$

Write a word name for each decimal.

5. 0.57 fifty-seven hundredths

6. 0.78 seventy-eight hundredths

7. 3.8 three and eight tenths

8. 12.4 twelve and four tenths

9. 7.013 seven and thirteen thousandths

10. 2.056 two and fifty-six thousandths

11. 28.0037 twenty-eight and thirty-seven ten-thousandths

12. 54.0013 fifty-four and thirteen ten-thousandths

Write a word name as you would on a check.

13. $124.20 one hundred twenty-four and $\frac{20}{100}$ dollars

14. $510.31 five hundred ten and $\frac{31}{100}$ dollars

15. $1236.08 one thousand, two hundred thirty-six and $\frac{8}{100}$ dollars

16. $5304.05 five thousand three hundred four and $\frac{5}{100}$ dollars

17. $12,015.45 twelve thousand fifteen and $\frac{45}{100}$ dollars

18. $20,000.67 twenty thousand and $\frac{67}{100}$ dollars

Write in decimal notation.

19. seven tenths 0.7

20. six tenths 0.6

21. ninety-six hundredths 0.96

22. eighteen hundredths 0.18

23. four hundred eighty-one thousandths 0.481

24. twenty-two thousandths 0.022

25. six thousand one hundred fourteen millionths 0.006114

26. one thousand three hundred eighteen millionths 0.001318

Write each fraction as a decimal.

27. $\frac{7}{10}$ 0.7

28. $\frac{3}{10}$ 0.3

29. $\frac{76}{100}$ 0.76

30. $\frac{84}{100}$ 0.84

31. $\frac{1}{100}$ 0.01

32. $\frac{6}{100}$ 0.06

33. $\frac{53}{1000}$ 0.053

34. $\frac{328}{1000}$ 0.328

35. $\frac{2403}{10,000}$ 0.2403

36. $\frac{7794}{10,000}$ 0.7794

37. $10\frac{9}{10}$ 10.9

38. $5\frac{3}{10}$ 5.3

39. $84\frac{13}{100}$ 84.13

40. $52\frac{77}{100}$ 52.77

41. $3\frac{529}{1000}$ 3.529

42. $2\frac{23}{1000}$ 2.023

43. $235\frac{104}{10,000}$ 235.0104

44. $116\frac{312}{10,000}$ 116.0312

Write in fractional notation. Reduce whenever possible.

45. 0.02 $\frac{2}{100} = \frac{1}{50}$

46. 0.05 $\frac{5}{100} = \frac{1}{20}$

47. 3.6 $3\frac{6}{10} = 3\frac{3}{5}$

48. 8.9 $8\frac{9}{10}$

49. 7.41 $7\frac{41}{100}$

50. 15.75 $15\frac{75}{100} = 15\frac{3}{4}$

51. 12.625 $12\frac{625}{1000} = 12\frac{5}{8}$

52. 29.875 $29\frac{875}{1000} = 29\frac{7}{8}$

53. 7.0615 $7\frac{615}{10,000} = 7\frac{123}{2000}$

54. 4.0016 $4\frac{16}{10,000} = 4\frac{1}{625}$

55. 8.0108 $8\frac{108}{10,000} = 8\frac{27}{2500}$

56. 7.0605 $7\frac{605}{10,000} = 7\frac{121}{2000}$

57. 235.1254

$$235\frac{1254}{10,000} = 235\frac{627}{5000}$$

58. 581.2406

$$581\frac{2406}{10,000} = 581\frac{1203}{5000}$$

59. 0.0125

$$\frac{125}{10,000} = \frac{1}{80}$$

60. 0.3375

$$\frac{3375}{10,000} = \frac{27}{80}$$

Applications

61. *Cigarette Use* The highest use of cigarettes in the United States takes place in Kentucky. In 2005, 30,600 out of every 100,000 men age 18 or older who lived in Kentucky were smokers. That same year, 26,900 out of every 100,000 women age 18 or older who lived in Kentucky were smokers.

(a) What fractional part of the male population in Kentucky were smokers?

(b) What fractional part of the female population in Kentucky were smokers? Be sure to express these fractions in reduced form. (*Source:* www.cdc.gov)

(a) $\dfrac{153}{500}$ (b) $\dfrac{269}{1000}$

62. *Cigarette Use* The lowest use of cigarettes in the United States takes place in Utah. In 2005, 13,700 out of every 100,000 men age 18 or older who lived in Utah were smokers. That same year, 9300 out of every 100,000 women age 18 or older who lived in Utah were smokers.

(a) What fractional part of the male population in Utah were smokers?

(b) What fractional part of the female population in Utah were smokers? Be sure to express these fractions in reduced form. (*Source:* www.cdc.gov)

(a) $\dfrac{137}{1000}$ (b) $\dfrac{93}{1000}$

63. *Bald Eagle Eggs* American bald eagles have been fighting extinction due to environmental hazards such as DDT, PCBs, and dioxin. The problem is with the food chain. Fish or rodents consume contaminated food and/or water. Then the eagles ingest the poison, which in turn affects the durability of the eagles' eggs. It takes only 4 parts per million of certain chemicals to ruin an eagle egg; write this number as a fraction in lowest terms. (In 1994 the bald eagle was removed from the endangered species list.)

$$\frac{4}{1,000,000} = \frac{1}{250,000}$$

64. *Turtle Eggs* Every year turtles lay eggs on the islands of South Carolina. Unfortunately, due to illegal polluting, a lot of the eggs are contaminated. If the turtle eggs contain more than 2 parts per one hundred million of chemical pollutants, they will not hatch and the population will continue to head toward extinction. Write the preceding amount of chemical pollutants as a fraction in the lowest terms.

$$\frac{2}{100,000,000} = \frac{1}{50,000,000}$$

Cumulative Review

65. **[1.2.4]** Add.

$$\begin{array}{r} 207 \\ 54 \\ 123 \\ 86 \\ + \ 55 \\ \hline 525 \end{array}$$

66. **[1.3.3]** Subtract.

$$\begin{array}{r} 12,843 \\ - \ 11,905 \\ \hline 938 \end{array}$$

67. **[1.7.1]** Round to the nearest *hundred*.

56,758   56,800

68. **[1.7.1]** Round to the nearest *thousand*.

8,069,482   8,069,000

Quick Quiz 3.1

1. Write a word name for the decimal. 5.367

five and three hundred sixty-seven thousandths

2. Write as a decimal. $\dfrac{523}{10,000}$ 0.0523

3. Write in fractional notation. Reduce your answer.

12.58 $12\dfrac{29}{50}$

4. **Concept Check** Explain how you know how many zeros to put in your answer if you need to write $\dfrac{953}{100,000}$ as a decimal. Answers may vary

Classroom Quiz 3.1   You may use these problems to quiz your students' mastery of Section 3.1.

1. Write a word name for the decimal.

9.158   **Ans:** nine and one hundred fifty-eight thousandths

2. Write as a decimal. $\dfrac{692}{10,000}$

Ans: 0.0692

3. Write in fractional notation. Reduce your answer. 26.85

Ans: $26\dfrac{17}{20}$

1 Comparing Decimals

All of the numbers we have studied have a specific order. To illustrate this order, we can place the numbers on a **number line.** Look at the number line in the margin. Each number has a specific place on it. The arrow points in the direction of increasing value. Thus, if one number is to the right of a second number, it is larger, or greater, than that number. Since 5 is to the right of 2 on the number line, we say that 5 is greater than 2. We write $5 > 2$.

Since 4 is to the left of 6 on the number line, we say that 4 is less than 6. We write $4 < 6$. The symbols ">" and "<" are called **inequality symbols.**

$a < b$ is read "a is less than b."

$a > b$ is read "a is greater than b."

We can assign exactly one point on the number line to each decimal number. When two decimal numbers are placed on a number line, the one farther to the right is the larger. Thus we can say that $3.4 > 2.7$ and $4.3 > 4.0$. We can also say that $0.5 < 1.0$ and $1.8 < 2.2$. Why?

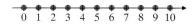

To compare or order decimals, we compare each digit.

> **COMPARING TWO NUMBERS IN DECIMAL NOTATION**
>
> 1. Start at the left and compare corresponding digits. If the digits are the same, move one place to the right.
>
> 2. When two digits are different, the larger number is the one with the larger digit.

EXAMPLE 1 Write an inequality statement with 0.167 and 0.166.

Solution The numbers in the tenths place are the same. They are both 1.

$$0.\overset{\downarrow}{1}6\,7 \qquad 0.\overset{\downarrow}{1}6\,6$$

The numbers in the hundredths place are the same. They are both 6.

$$0.1\,\overset{\downarrow}{6}\,7 \qquad 0.1\,\overset{\downarrow}{6}\,6$$

The numbers in the thousandths place differ.

$$0.1\,6\,\overset{\downarrow}{7} \qquad 0.1\,6\,\overset{\downarrow}{6}$$

Since $7 > 6$, we know that $0.167 > 0.166$.

Practice Problem 1 Write an inequality statement with 5.74 and 5.75.

$5.74 < 5.75$

Student Learning Objectives

After studying this section, you will be able to:

1 Compare decimals.

2 Place decimals in order from smallest to largest.

3 Round decimals to a specified decimal place.

Teaching Tip This is a good time to stress the use of the number line. Remind the students that they can make any scale they want for a number line. It is just as easy to make a number line with units labeled 1.34, 1.35, 1.36, 1.37, 1.38, etc., as it is to make a number line with units labeled 1, 2, 3, 4, 5, 6, etc. This will help them whenever they compare decimals. The number line can be expanded to show decimal numbers between whole numbers.

Teaching Example 1 Write an inequality statement with 2.032 and 2.031.

Ans: $2.032 > 2.031$

NOTE TO STUDENT: Fully worked-out solutions to all of the Practice Problems can be found at the back of the text starting at page SP-1

Whenever necessary, extra zeros can be written to the right of the last digit—that is, to the right of the decimal point—without changing the value of the decimal. Thus

$$0.56 = 0.56000 \quad \text{and} \quad 0.7768 = 0.77680.$$

The zero to the left of the decimal point is optional. Thus $0.56 = .56$. Both notations are used. You are encouraged to place a zero to the left of the decimal point so that you don't miss the decimal point when you work with decimals.

EXAMPLE 2 Fill in the blank with one of the symbols <, =, or >.

$$0.77 \_\_\_\_ 0.777$$

Solution We begin by adding a zero to the first decimal.

$$0.77\underline{0} \qquad 0.77\underline{7}$$

We see that the tenths and hundredths digits are equal. But the thousandths digits differ. Since $0 < 7$, we have $0.770 < 0.777$.

Practice Problem 2 Fill in the blank with one of the symbols <, =, or >.

$$0.894 \; \underline{>} \; 0.89$$

② Placing Decimals in Order from Smallest to Largest

Which is the heaviest—a puppy that weighs 6.2 ounces, a puppy that weighs 6.28 ounces, or a puppy that weighs 6.028 ounces? Did you choose the puppy that weighs 6.28 ounces? You are correct.

You can place two or more decimals in order. If you are asked to order the decimals from smallest to largest, look for the smallest decimal and place it first.

EXAMPLE 3 Place the following five decimal numbers in order from smallest to largest.

$$1.834, \quad 1.83, \quad 1.381, \quad 1.38, \quad 1.8$$

Solution First we add zeros to make the comparison easier.

$$1.834, \quad 1.830, \quad 1.381, \quad 1.380, \quad 1.800$$

Now we rearrange with smallest first.

$$1.380, \quad 1.381, \quad 1.800, \quad 1.830, \quad 1.834$$

Practice Problem 3 Place the following five decimal numbers in order from smallest to largest.

$$2.45, \quad 2.543, \quad 2.46, \quad 2.54, \quad 2.5$$

2.45, 2.46, 2.5, 2.54, 2.543

Rounding Decimals to a Specified Decimal Place

Sometimes in calculations involving money, we see numbers like $386.432 and $29.5986. To make these useful, we usually round them to the nearest cent. $386.432 is rounded to $386.43. $29.5986 is rounded to $29.60. A general rule for rounding decimals follows.

Teaching Tip You may want to ask the students if they are familiar with other rules for rounding numbers. Particularly if you have students from other countries or cultures, you may see some unusual rounding rules. Remind them that this rule for rounding decimals is used uniformly in mathematics courses in the United States and that they will need to be able to use it.

> **ROUNDING DECIMALS**
>
> 1. Find the decimal place (units, tenths, hundredths, and so on) to which rounding is required.
> 2. If the first digit to the right of the given place value is less than 5, drop it and all digits to the right of it.
> 3. If the first digit to the right of the given place value is 5 or greater, increase the number in the given place value by one. Drop all digits to the right of this place.

EXAMPLE 4 Round 156.37 to the nearest tenth.

Solution   156.3 7

└──── We find the tenths place.

Note that 7, the next place to the right, is greater than 5. We round up to 156.4 and drop the digits to the right. The answer is 156.4.

Teaching Example 4 Round 15.259 to the nearest hundredth.

Ans: 15.26

Practice Problem 4 Round 723.88 to the nearest tenth.   723.9

NOTE TO STUDENT: Fully worked-out solutions to all of the Practice Problems can be found at the back of the text starting at page SP-1

EXAMPLE 5 Round to the nearest thousandth.

(a) 0.06358   **(b)** 128.37448

Solution

(a) 0.06 3 58

└──── We locate the thousandths place.

Note that the digit to the right of the thousandths place is 5. We round up to 0.064 and drop all the digits to the right.

(b) 128.37 4 48

└──── We locate the thousandths place.

Note that the digit to the right of the thousandths place is less than 5. We round to 128.374 and drop all the digits to the right.

Teaching Example 5 Round to the nearest thousandth.
(a) 3.0258   **(b)** 0.70328

Ans: (a) 3.026   **(b)** 0.703

Practice Problem 5 Round to the nearest thousandth.

(a) 12.92647 12.926   **(b)** 0.007892 0.008

Remember that rounding up to the next digit in a position may result in several digits being changed.

EXAMPLE 6 Round to the nearest hundredth. Fred and Linda used 203.9964 kilowatt-hours of electricity in their house in May.

203.9 9 64

⎿——— We locate the hundredths place.

Solution Since the digit to the right of the hundredths place is greater than 5, we round up. This affects the next two positions. Do you see why? The result is 204.00 kilowatt-hours. Notice that we have the two zeros to the right of the decimal place to show we have rounded to the nearest hundredth.

Practice Problem 6 Round to the nearest tenth. Last month the college gymnasium used 15,699.953 kilowatt-hours of electricity. 15,700.0

Sometimes we round a decimal to the nearest whole number. For example, when writing figures on income tax forms, a taxpayer may round all figures to the nearest dollar.

EXAMPLE 7 To complete her income tax return, Marge needs to round these figures to the nearest whole dollar.

Medical bills $779.86   Taxes $563.49
Retirement contributions $674.38   Contributions to charity $534.77

Solution Round the amounts.

	Original Figure	*Rounded to Nearest Dollar*
Medical bills	$779.86	$780
Taxes	$563.49	$563
Retirement	$674.38	$674
Charity	$534.77	$535

Practice Problem 7 Round the following figures to the nearest whole dollar.

Medical bills $375.50   Taxes $971.39
Retirement contributions $980.49   Contributions to charity $817.65

$376, $980, $971, $818

CAUTION: Why is it so important to consider only *one* digit to the right of the desired round-off position? What is wrong with rounding in steps? Suppose that Mark rounds 1.349 to the nearest tenth in steps. First he rounds 1.349 to 1.35 (nearest hundredth). Then he rounds 1.35 to 1.4 (nearest tenth). What is wrong with this reasoning?

```
                          1.349
   +——+——+——+——+——+—H—+——+——+——+——+
 1.30 1.31 1.32 1.33 1.34 1.35 1.36 1.37 1.38 1.39 1.40
```

To round 1.349 to the nearest tenth, we ask if 1.349 is closer to 1.3 or to 1.4. It is closer to 1.3. Mark got 1.4, so he is not correct. He "rounded in steps" by first moving to 1.35, thus increasing the error and moving in the wrong direction. To control rounding errors, we consider *only* the first digit to the right of the decimal place to which we are rounding.

Fill in the blank with one of the symbols <, =, or >.

1. 1.3 $>$ 1.29   **2.** 2.6 $>$ 2.58   **3.** 0.34 $=$ 0.340   **4.** 72.54 $<$ 72.56

5. 18.92 $<$ 18.93   **6.** 0.460 $=$ 0.46   **7.** 0.00043 $>$ 0.0004   **8.** 0.0037 $<$ 0.036

9. 1.002 $<$ 1.0021   **10.** 2.0056 $<$ 2.006   **11.** 126.34 $>$ 125.35   **12.** 406.78 $<$ 407.75

13. 0.888 $<$ 0.8888   **14.** 0.666 $<$ 0.6666   **15.** 0.777 $>$ 0.7077   **16.** 0.555 $>$ 0.5505

17. $\dfrac{72}{1000}$ $=$ 0.072   **18.** $\dfrac{54}{1000}$ $=$ 0.054   **19.** $\dfrac{8}{10}$ $>$ 0.08   **20.** $\dfrac{5}{100}$ $>$ 0.005

Arrange each set of decimals from smallest to largest.

21. 12.6, 12.8, 12.65   **22.** 18.32, 18.038, 18.04   **23.** 0.0071, 0.05, 0.007
12.6, 12.65, 12.8   18.038, 18.04, 18.32   0.007, 0.0071, 0.05

24. 0.0025, 0.0052, 0.002   **25.** 8.4, 8.39, 8.41, 8.31   **26.** 5.1, 5.01, 5.23, 5.02
0.002, 0.0025, 0.0052   8.31, 8.39, 8.4, 8.41   5.01, 5.02, 5.1, 5.23

27. 26.034, 26.003, 26.04, 26.033   **28.** 33.082, 33.02, 33.088, 33.079
26.003, 26.033, 26.034, 26.04   33.02, 33.079, 33.082, 33.088

29. 18.006, 18.060, 18.066, 18.606, 18.065   **30.** 15.020, 15.002, 15.001, 15.018, 15.0019
18.006, 18.060, 18.065, 18.066, 18.606   15.001, 15.0019, 15.002, 15.018, 15.020

Round to the nearest tenth.

31. 6.92 6.9   **32.** 8.35 8.4   **33.** 28.98 29.0   **34.** 47.94 47.9

35. 578.064 578.1   **36.** 454.99 455.0   **37.** 2176.83 2176.8   **38.** 4082.74 4082.7

Round to the nearest hundredth.

39. 26.032 26.03   **40.** 47.071 47.07   **41.** 36.997 37.00   **42.** 24.999 25.00

43. 156.1749 156.17 **44.** 283.8441 283.84 **45.** 2786.706 2786.71 **46.** 4609.285 4609.29

Round to the nearest indicated place.

47. 7.8155; thousandths 7.816   **48.** 8.10263; thousandths 8.103

49. 0.05951; ten-thousandths 0.0595   **50.** 0.063148; ten-thousandths 0.0631

51. 12.0157823; hundred-thousandths 12.01578 **52.** 15.4159266; hundred-thousandths 15.41593

53. 135.564; nearest whole number 136   **54.** 389.645; nearest whole number 390

Round to the nearest dollar.

55. $788.42 $788   **56.** $912.75 $913   **57.** $15,020.50 $15,021   **58.** $20,159.48 $20,159

Round to the nearest cent.

59. $96.3357 $96.34   **60.** $42.9261 $42.93   **61.** $5783.716 $5783.72   **62.** $3928.649 $3928.65

Applications

63. *Baseball* During the 2006 baseball season, the winning percentages of the New York Yankees and the Seattle Mariners were 0.59876 and 0.48148, respectively. Round these values to the nearest thousandth. 0.599; 0.481

64. *Sales Tax* Bryan purchased a CD for himself and a toy for his daughter. The sales tax calculated on the CD was $1.2593 and the sales tax on the toy was $1.7143. Round these values to the nearest cent. $1.26; $1.71

65. *Astronomy* The number of days in a year is 365.24122. Round this value to the nearest hundredth. 365.24

66. *Mathematics History* The numbers π and e are approximately equal to 3.14159 and 2.71828, respectively. We will be using π later in this textbook. You will encounter e in higher level mathematics courses. Round these values to the nearest hundredth. 3.14; 2.72

To Think About

67. Arrange in order from smallest to largest.

$$0.61, 0.062, \frac{6}{10}, 0.006, 0.0059,$$

$$\frac{6}{100}, 0.0601, 0.0519, 0.0612$$

0.0059, 0.006, 0.0519, $\frac{6}{100}$, 0.0601, 0.0612, 0.062, $\frac{6}{10}$, 0.61

68. Arrange in order from smallest to largest.

$$1.05, 1.512, \frac{15}{10}, 1.0513, 0.049,$$

$$\frac{151}{100}, 0.0515, 0.052, 1.051$$

0.049, 0.0515, 0.052, 1.05, 1.051, 1.0513, $\frac{15}{10}$, $\frac{151}{100}$, 1.512

69. A person wants to round 86.23498 to the nearest hundredth. He first rounds 86.23498 to 86.2350. He then rounds to 86.235. Finally, he rounds to 86.24. What is wrong with his reasoning? You should consider only one digit to the right of the decimal place that you wish to round to. 86.23498 is closer to 86.23 than to 86.24.

70. *Personal Finance* Fred is checking the calculations on his monthly bank statement. An interest charge of $16.3724 was rounded to $16.38. An interest charge of $43.7214 was rounded to $43.73. What rule does the bank use for rounding off to the nearest cent? The bank rounds up for any fractional part of a cent.

Cumulative Review

71. **[2.8.1]** Add. $3\frac{1}{4} + 2\frac{1}{2} + 6\frac{3}{8}$ $12\frac{1}{8}$

72. **[2.8.2]** Subtract. $27\frac{1}{5} - 16\frac{3}{4}$ $10\frac{9}{20}$

73. **[1.3.5]** *Car Travel* Mary drove her Dodge Caravan on a trip. At the start of the trip, the odometer (which measures distance) read 46,381. At the end of the trip, it read 47,073. How many miles long was the trip? 692 miles

74. **[1.7.2]** *Boat Sales* Don's New and Used Watercraft sold four boats one weekend for $18,650, $2490, $835, and $9845. Estimate the total amount of the sales. $31,800

Quick Quiz 3.2

1. Arrange from smallest to largest:

4.56,   4.6,   4.056,   4.559   4.056,   4.559,   4.56,   4.6

2. Round to the nearest hundredth. 27.1782   27.18

3. Round to the nearest thousandth. 155.52525 155.525

4. **Concept Check** Explain how you would round 34.958365 to the nearest ten-thousandth. Answers may vary

Classroom Quiz 3.2 You may use these problems to quiz your students' mastery of Section 3.2.

1. Arrange from smallest to largest: 7.7,   7.67,   7.76,   7.067   **Ans:** 7.067, 7.67, 7.7, 7.76

2. Round to the nearest hundredth.   58.2637   **Ans:** 58.26

3. Round to the nearest thousandth.   122.78658   **Ans:** 122.787

Adding Decimals

We often add decimals when we check the addition of our bill at a restaurant or at a store. We can relate addition of decimals to addition of fractions. For example,

$$\frac{3}{10} + \frac{6}{10} = \frac{9}{10} \quad \text{and} \quad 1\frac{1}{10} + 2\frac{8}{10} = 3\frac{9}{10}.$$

These same problems can be written more efficiently as decimals.

$$
\begin{array}{r}
0.3 \\
+\ 0.6 \\
\hline
0.9
\end{array}
\qquad
\begin{array}{r}
1.1 \\
+\ 2.8 \\
\hline
3.9
\end{array}
$$

The steps to follow when adding decimals are listed in the following box.

ADDING DECIMALS

1. Write the numbers to be added vertically and line up the decimal points. Extra zeros may be placed to the right of the decimal points if needed.
2. Add all the digits with the same place value, starting with the right column and moving to the left.
3. Place the decimal point of the sum in line with the decimal points of the numbers added.

EXAMPLE 1 Add.

(a) $2.8 + 5.6 + 3.2$

(b) $158.26 + 200.07 + 315.98$

(c) $5.3 + 26.182 + 0.0007 + 624$

Solution

(a)
$$
\begin{array}{r}
\overset{1}{2.8} \\
5.6 \\
+\ 3.2 \\
\hline
11.6
\end{array}
$$

(b)
$$
\begin{array}{r}
\overset{11\ 2}{158.26} \\
200.07 \\
+\ 315.98 \\
\hline
674.31
\end{array}
$$

(c)
$$
\begin{array}{r}
\overset{1}{5.3000} \\
26.1820 \\
0.0007 \\
+\ 624.0000 \\
\hline
655.4827
\end{array}
$$
Extra zeros have been added to make the problem easier.
Note: The decimal point is understood to be to the right of the digit 4.

Practice Problem 1 Add.

(a)
$$
\begin{array}{r}
9.8 \\
3.6 \\
+\ 5.4
\end{array}
$$

(b)
$$
\begin{array}{r}
300.72 \\
163.75 \\
+\ 291.08
\end{array}
$$

(c) $8.9 + 37.056 + 0.0023 + 945$

(a) 18.8   (b) 755.55   (c) 990.9583

SIDELIGHT: Adding in Extra Zeros

When we add decimals like 3.1 + 2.16 + 4.007, we may write in zeros, as shown:

$$
\begin{array}{r}
3.100 \\
2.160 \\
+ \ 4.007 \\
\hline
9.267
\end{array}
$$

What are we really doing here? What is the advantage of adding these extra zeros?

"Decimals" means "decimal fractions." If we look at the number as fractions, we see that we are actually using the property of multiplying a fraction by 1 in order to obtain common denominators. Look at the problem this way:

$$
\left.
\begin{array}{l}
3.1 \quad = 3\dfrac{1}{10} \\[2mm]
2.16 \quad = 2\dfrac{16}{100} \\[2mm]
4.007 = 4\dfrac{7}{1000}
\end{array}
\right\}
$$

The least common denominator is 1000. To obtain the common denominator for the first two fractions, we multiply.

$$
\left.
\begin{array}{r}
3 \ \dfrac{1}{10} \times \dfrac{100}{100} = 3\dfrac{100}{1000} \\[2mm]
2 \ \dfrac{16}{100} \times \dfrac{10}{10} = 2\dfrac{160}{1000} \\[2mm]
+4 \ \dfrac{7}{1000} \qquad = 4\dfrac{7}{1000}
\end{array}
\right\}
$$

Once we obtain a common denominator, we can add the three fractions.

$$
9\dfrac{267}{1000} = 9.267
$$

This is the answer we arrived at earlier using the decimal form for each number. Thus writing in zeros in a decimal fraction is really an easy way to transform fractions to equivalent fractions with a common denominator. Working with decimal fractions is easier than working with other fractions.

The final digit of most odometers measures tenths of a mile. The odometer reading shown in the odometer on the left is 38,516.2 miles.

Calculator

Adding Decimals

The calculator can be used to verify your work. You can use your calculator to add decimals. To find 23.08 + 8.53 + 9.31 enter:

23.08 [+] 8.53 [+]

9.31 [=]

Display:

[40.92]

Teaching Example 2 On Monday morning, Karina's odometer read 53,278.4. She traveled 216.7 miles that week. What was the odometer reading at the end of the week?

Ans: 53,495.1

EXAMPLE 2 Barbara checked her odometer before the summer began. It read 49,645.8 miles. She traveled 3852.6 miles that summer in her car. What was the odometer reading at the end of the summer?

Solution

$$
\begin{array}{r}
\overset{11 \quad 1}{49{,}645.8} \\
+ \ 3852.6 \\
\hline
53{,}498.4
\end{array}
$$

The odometer read 53,498.4 miles.

Practice Problem 2 A car odometer read 93,521.8 miles before a trip of 1634.8 miles. What was the final odometer reading? 95,156.6 miles

EXAMPLE 3 During his first semester at Tarrant County Community College, Kelvey deposited checks into his checking account in the amounts of $98.64, $157.32, $204.81, $36.07, and $229.89. What was the sum of his five checks?

Solution

$$
\begin{array}{r}
\overset{2\,3\,2\ 2}{\$\ 98.64} \\
157.32 \\
204.81 \\
36.07 \\
+\ \ 229.89 \\
\hline
\$726.73
\end{array}
$$

Teaching Example 3 Narda's checking account balance on Monday was $1075.28. On Tuesday, she deposited a check for $102.68 and on Wednesday, she deposited a check for $56.07. What was her checking account balance after her second deposit?

Ans: $1234.03

Practice Problem 3 During the spring semester, Will deposited the following checks into his account: $80.95, $133.91, $256.47, $53.08, and $381.32. What was the sum of his five checks? $905.73

 Subtracting Decimals

It is important to see the relationship between the decimal form of a mixed number and the fractional form of a mixed number. This relationship helps us understand why calculations with decimals are done the way they are. Recall that when we subtract mixed numbers with common denominators, sometimes we must borrow from the whole number.

$$
\begin{array}{rcr}
5\dfrac{1}{10} & = & 4\dfrac{11}{10} \\[2mm]
-\ 2\dfrac{7}{10} & = & -\ 2\dfrac{7}{10} \\[2mm]
\hline
& & 2\dfrac{4}{10}
\end{array}
$$

We could write the same problem in decimal form:

$$
\begin{array}{r}
\overset{4\ 11}{\cancel{5}.\cancel{1}} \\
-\ 2.7 \\
\hline
2.4
\end{array}
$$

Subtraction of decimals is thus similar to subtraction of fractions (we get the same result), but it's usually easier to subtract with decimals than to subtract with fractions.

SUBTRACTING DECIMALS

1. Write the decimals to be subtracted vertically and line up the decimal points. Additional zeros may be placed to the right of the decimal point if not all numbers have the same number of decimal places.

2. Subtract all digits with the same place value, starting with the right column and moving to the left. Borrow when necessary.

3. Place the decimal point of the difference in line with the decimal point of the two numbers being subtracted.

EXAMPLE 4 Subtract.

(a)
$$\begin{array}{r} 84.8 \\ -\ 27.3 \\ \hline \end{array}$$

(b)
$$\begin{array}{r} 1076.320 \\ -\ 983.518 \\ \hline \end{array}$$

Solution

(a)
$$\begin{array}{r} {}^{7}\ {}^{14} \\ \cancel{8}\ \cancel{4}.8 \\ -\ 2\ 7.3 \\ \hline 5\ 7.5 \end{array}$$

(b)
$$\begin{array}{r} {}^{9} \\ \cancel{10}\ {}^{17}\ {}^{5}\ \ {}^{13}\ {}^{1}\ {}^{10} \\ \cancel{1}\ \cancel{0}\ \cancel{7}\ \cancel{6}.\cancel{3}\ \cancel{2}\ \cancel{0} \\ -\ \ \ 9\ 8\ 3.5\ 1\ 8 \\ \hline 9\ 2.8\ 0\ 2 \end{array}$$

Practice Problem 4 Subtract.

(a)
$$\begin{array}{r} 38.8 \\ -\ 26.9 \end{array}\quad 11.9$$

(b)
$$\begin{array}{r} 2034.908 \\ -\ 1986.325 \end{array}\quad 48.583$$

When the two numbers being subtracted do not have the same number of decimal places, write in zeros as needed.

EXAMPLE 5 Subtract.

(a) $12 - 8.362$

(b) $156.381 - 99.82$

Solution

(a)
$$\begin{array}{r} {}^{11}\ \ {}^{9}\ {}^{9} \\ \cancel{1}\ \ \cancel{0}\ \cancel{0}\ {}^{10} \\ \cancel{1}\ 2.\cancel{0}\ \cancel{0}\ \cancel{0} \\ -\ \ \ \ 8.3\ 6\ 2 \\ \hline 3.6\ 3\ 8 \end{array}$$

(b)
$$\begin{array}{r} {}^{14}\ {}^{15} \\ \cancel{4}\ \cancel{5}\ \ {}^{13} \\ \cancel{1}\ \cancel{5}\ \cancel{6}.\cancel{3}\ 8\ 1 \\ -\ \ \ 9\ 9.8\ 2\ 0 \\ \hline 5\ 6.5\ 6\ 1 \end{array}$$

Practice Problem 5 Subtract.

(a) $19 - 12.579$ 6.421

(b) $283.076 - 96.38$ 186.696

EXAMPLE 6 On Tuesday, Don Ling filled the gas tank in his car. The odometer read 56,098.5. He drove for four days. The next time he filled the tank, the odometer read 56,420.2. How many miles had he driven?

Solution

$$\begin{array}{r} {}^{11}\ {}^{9} \\ {}^{3}\ \cancel{1}\ \cancel{10}\ \ {}^{12} \\ 5\ 6,\cancel{4}\ \cancel{2}\ \cancel{0}.\cancel{2} \\ -\ 5\ 6,0\ 9\ 8.5 \\ \hline 3\ 2\ 1.7 \end{array}$$

He had driven 321.7 miles.

Practice Problem 6 Abdul had his car oil changed when his car odometer read 82,370.9 miles. When he changed the oil again, the odometer read 87,160.1 miles. How many miles did he drive between oil changes? 4,789.2 miles

EXAMPLE 7 Find the value of x if $x + 3.9 = 14.6$.

Solution Recall that the letter x is a variable. It represents a number that is added to 3.9 to obtain 14.6. We can find the number x if we calculate $14.6 - 3.9$.

$$\begin{array}{r} \overset{3\ 16}{1\cancel{4}.\cancel{6}} \\ -\ \ 3.9 \\ \hline 10.7 \end{array}$$

Thus $x = 10.7$.

Check. Is this true? If we replace x by 10.7, do we get a true statement?

$$x + 3.9 = 14.6$$
$$10.7 + 3.9 \overset{?}{=} 14.6$$
$$14.6 = 14.6 \ \checkmark$$

Practice Problem 7 Find the value of x if $x + 10.8 = 15.3$. $x = 4.5$

Teaching Example 7 Find the value of x if $x + 15.4 = 21.3$.

Ans: $x = 5.9$

NOTE TO STUDENT: Fully worked-out solutions to all of the Practice Problems can be found at the back of the text starting at page SP-1

Adding and subtracting decimals is an important part of life. When you are recording deposits at the bank, reconciling your checkbook, or completing your income tax forms, you are adding and subtracting decimals. Be sure to learn to do it accurately. In the homework exercises, always check your answers with the Answers section in the back of the book. Making sure you have the correct answers is very important.

Developing Your Study Skills

Making a Friend in the Class

Attempt to make a friend in your class. You may find that you enjoy sitting together and drawing support and encouragement from each other. Exchange phone numbers so you can call each other whenever you get stuck in your work. Set up convenient times to study together on a regular basis, to do homework, and to review for exams.

You must not depend on a friend or fellow student to tutor you, do your work for you, or in any way be responsible for your learning. However, you will learn from each other as you seek to master the course. Studying with a friend and comparing notes, methods, and solutions can be very helpful. And it can make learning mathematics a lot more fun!

Add.

1. 57.1 + 19.7
76.8

2. 78.3 + 29.4
107.7

3. 384.25 + 209.65
593.9

4. 193.42 + 768.78
962.2

5.   13.4
7.6
+ 275.2
296.2

6.   176.5
8.4
+ 22.5
207.4

7.   4.71
+ 8.05
12.76

8.   9.284
+ 5.77
15.054

9.   4.9637
28.12
+ 3.645
36.7287

10.   7.0276
3.451
+ 16.98
27.4586

11.   12.
3.62
+ 51.8
67.42

12.   13.
4.52
+ 63.7
81.22

13. 108.36 + 14.3 + 85.12 + 28
235.78

14. 215.45 + 48 + 30.77 + 15.8
310.02

15. 753.61 + 28.75 + 162.3 + 100.5 + 67
1112.16

16. 432.51 + 16.08 + 892.1 + 301.2 + 84
1725.89

Applications *In exercises 17 and 18, calculate the perimeter of each triangle.*

17.
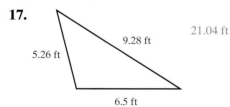
9.28 ft
5.26 ft
6.5 ft
21.04 ft

18.
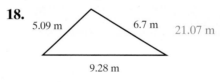
5.09 m   6.7 m
9.28 m
21.07 m

19. *Weight Loss* Lamar is losing weight by walking each evening after dinner. During the first week in February he lost 1.75 pounds. During the second, third, and fourth weeks, he lost 2.5 pounds, 1.55 pounds, and 2.8 pounds, respectively. How many total pounds did Lamar lose in February?
8.6 pounds

20. *Health* Olivia knows she needs to drink more water while at work. One day during her morning break she drank 7.15 ounces. At lunch she drank 12.45 ounces and throughout the afternoon she drank 10.75 ounces. How many total ounces of water did she drink? 30.35 ounces

21. *Beach Vacation* Mick and Keith have arrived in Miami and are going to the beach. They buy sunblock for $4.99, beverages for $12.50, sandwiches for $11.85, towels for $28.50, bottled water for $3.29, and two novels for $16.99. After they got what they needed, what was Mick and Keith's bill for their day at the beach? $78.12

22. *Consumer Mathematics* Anika bought school supplies at the campus bookstore. She purchased a calculator for $37.25, pens for $5.89, a T-shirt for $13.95, and notebooks for $10.49. The amount of sales tax was $4.05. What was the total of Anika's bill including tax? $71.63

23. *Truck Travel* A truck odometer read 46,276.0 miles before a trip of 778.9 miles. What was the final odometer reading?   47,054.9

24. *Car Travel* Jane traveled 1723.1 miles. The car odometer at the beginning of the trip read 23,195.0 miles. What was the final odometer reading?   24,918.1

Personal Banking In exercises 25 and 26, a portion of a bank checking account deposit slip is shown. Add the numbers to determine the total deposit. The line drawn between the dollars and the cents column serves as the decimal point.

25.

26.

Subtract.

27. 12.8 − 9.3
3.5

28. 15.8 − 6.7
9.1

29. 35.75 − 9.82
25.93

30. 84.33 − 8.09
76.24

31. 126 − 76.22
49.78

32. 209 − 81.54
127.46

33.   586.513
  − 78.2
508.313

34.   243.967
  − 84.2
159.767

35.   220.9
  − 85.47
135.43

36.   181.9
  − 62.23
119.67

37.   24.0079
  − 19.3614
4.6465

38.   52.0708
  − 41.9312
10.1396

39.   8
  − 1.263
6.737

40.   12
  − 7.981
4.019

41.   7362.14
  − 6173.07
1189.07

42.   4986.71
  − 3615.93
1370.78

43.   1.5
  − 0.0365
1.4635

44.   2.8
  − 0.07763
2.72237

Mixed Practice *Add or subtract.*

45. 123.621 + 52.96
176.581

46. 241.983 + 75.48
317.463

47. 98.3 − 56.71
41.59

48. 79.2 − 45.93
33.27

49. 0.0763 + 2 + 3.16
5.2363

50. 18 − 2.75
15.25

51. 197.600 − 124.375
73.225

52. 382.700 − 291.927
90.773

Applications

53. World Records The heaviest apple on record was grown in Japan in 2005 and weighed 4.0678 pounds. The heaviest lemon was grown in Israel in 2003 and weighed 11.583 pounds. How much heavier was the lemon than the apple? (*Source:* www.guinessworldrecords.com)   7.5152 pounds

54. Health At her 4-month checkup, baby Grace weighed 7.675 kilograms. When she was born, she weighed 3.7 kilograms. How much weight has Grace gained since she was born?
3.975 kilograms

55. Telescope A child's beginner telescope is priced at $79.49. The price of a certain professional telescope is $37,026.65. How much more does the professional telescope cost?
$36,947.16

56. Automobile Travel During their spring break vacation, Jeff and Manuel drove from their college in Boise, Idaho, to San Diego, California, and back. When they began the trip, the odometer of their rental car read 12,265.4 miles. When they returned the car, the odometer read 14,537.9 miles. How many miles did they drive?
2272.5 miles

57. Taxi Trip Malcolm took a taxi from John F. Kennedy Airport in New York to his hotel in the city. His fare was $47.70 and he tipped the driver $7.00. How much change did Malcolm get back if he gave the driver a $100 bill? $45.30

58. Personal Banking Nathan took $200 out of the ATM. He bought snow boots for $65.49, pet supplies for $27.75, and a bouquet of flowers for $18.95. How much money does he have left?
$87.81

59. Electric Wire Construction An insulated wire measures 12.62 centimeters. The last 0.98 centimeter of the wire is exposed. How long is the part of the wire that is not exposed?

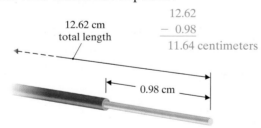

```
  12.62
- 0.98
  11.64 centimeters
```

12.62 cm
total length

0.98 cm

60. Plumbing The outside radius of a pipe is 9.39 centimeters. The inside radius is 7.93 centimeters. What is the thickness of the pipe?

```
  9.39
- 7.93
  1.46 centimeters
```

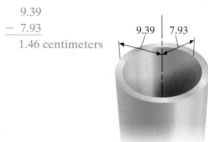

9.39   7.93

61. Medical Research A cancer researcher is involved in an important experiment. She is trying to determine how much of an anticancer drug is necessary for a Stage I (nonhuman or animal) test. She pours 2.45 liters of the experimental anticancer formula in one container and 1.35 liters of a reactive liquid in another. She then pours the contents of one container into the other. If 0.85 liter is expected to evaporate during the process, how much liquid will be left?   2.95 liters

62. Rainforest Loss Everyone is becoming aware of the rapid loss of Earth's rainforests. Mexico's rainforests have one of the highest deforestation rates in the world. In 1997, there were 39.7 million hectares of rainforest in Mexico. By 2006, it had lost approximately 4.64 million hectares. How many hectares of rainforest did Mexico have in 2006? (A hectare is equal to 10,000 square meters.) (*Source:* www.geography.ndo.co.uk)
35.06 million hectares

The federal water safety standard requires that drinking water contain no more than 0.015 milligram of lead per liter of water. (Source: Environmental Protection Agency)

63. *Well Water Safety* Carlos and Maria had the well that supplies their home analyzed for safety. A sample of well water contained 0.0089 milligram of lead per liter of water. What is the difference between their sample and the federal safety standard? Is it safe for them to drink the water? 0.0061 milligram; yes.

64. *City Water Safety* Fred and Donna use water provided by the city for the drinking water in their home. A sample of their tap water contained 0.023 milligram of lead per liter of water. What is the difference between their sample and the federal safety standard? Is it safe for them to drink the water? 0.008 milligram; no.

Income of Industries *The following table shows the income of the United States by industry. Use this table for exercises 65–68. Write each answer as a decimal and as a whole number. The table values are recorded in billions of dollars.*

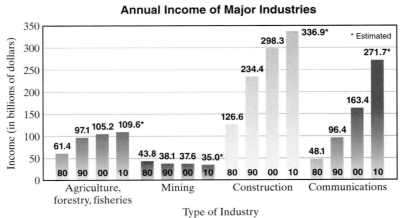

Annual Income of Major Industries

Source: Bureau of Labor Statistics

65. How many more dollars were earned in mining in 1980 than in 2000? $6.2 billion; $6,200,000,000

66. How many more dollars were earned in construction in 2000 than in 1980? $171.7 billion; $171,700,000,000

67. In 2010, how many more dollars will be earned in communications than in agriculture, forestry, and fisheries? $162.1 billion; $162,100,000,000

68. In 1990, how many more dollars were earned in communications than in mining? $58.3 billion; $58,300,000,000 dollars

To Think About *Mr. Jensen made up the following shopping list of items he needs and the cost of each item. Use the list to answer exercises 69 and 70.*

can cranberry sauce	$0.99
hot dog relish	$0.79
ranch salad dressing	$1.47
large can solid white tuna	$2.29
can tomato soup	$0.68
can sliced peaches	$1.26
large jar tomato sauce	$1.65
large box Cheerios	$3.79
medium box Raisin Bran	$2.63
medium jar peanut butter	$2.19

69. *Grocery Shopping* Mr. Jensen goes to the store to buy the following items from his list: Raisin Bran, ranch salad dressing, sliced peaches, hot dog relish, and peanut butter. He has a ten-dollar bill. Estimate the cost of buying these items by first rounding the cost of each item to the nearest ten cents. Does he have enough money to buy all of them? Find the exact cost of these items. How close was your estimate?

$8.40; yes, $8.34; very close: the estimate was off by 6¢

70. *Grocery Shopping* The next day the Jensens' daughter, Brenda, goes to the store to buy the following items from the list: Cheerios, tomato sauce, peanut butter, white tuna, tomato soup, and cranberry sauce. She has fifteen dollars. Estimate the cost of buying these items by first rounding the cost of each item to the nearest ten cents. Does she have enough money to buy all of them? Find the exact cost of these items. How close was your estimate?

$11.70; yes; $11.59; very close: the estimate was off by 11¢

Find the value of x.

71. $x + 7.1 = 15.5$
$x = 8.4$

72. $x + 4.8 = 23.1$
$x = 18.3$

73. $156.9 + x = 200.6$
$x = 43.7$

74. $210.3 + x = 301.2$
$x = 90.9$

75. $4.162 = x + 2.053$
$x = 2.109$

76. $7.076 = x + 5.602$
$x = 1.474$

Cumulative Review *Multiply.*

77. [1.4.2]
$\begin{array}{r} 2536 \\ \times\ \ \ \ 8 \\ \hline 20{,}288 \end{array}$

78. [1.4.4]
$\begin{array}{r} 827 \\ \times\ \ 59 \\ \hline 48{,}793 \end{array}$

79. [2.4.2] $\frac{1}{4} \times 100$
25

80. [2.4.2] $800 \times \frac{1}{2}$
400

Quick Quiz 3.3

1. Add. $53.261 + 1.9 + 17.82$ 72.981

2. Subtract. $5.2608 - 3.0791$ 2.1817

3. Subtract. $59.6 - 3.925$ 55.675

4. Concept Check Explain how you perform the correct borrowing and correct use of the decimal point if you subtract $567.45 - 345.9872$. Answers may vary

Classroom Quiz 3.3 You may use these problems to quiz your students' mastery of Section 3.3.

1. Add. $9.8 + 71.562 + 19.39$ **Ans:** 100.752

2. Subtract. $9.0702 - 4.9631$ **Ans:** 4.1071

3. Subtract. $68.2 - 5.793$ **Ans:** 62.407

3.4 MULTIPLYING DECIMALS

1 Multiplying a Decimal by a Decimal or a Whole Number

We learned previously that the product of two fractions is the product of the numerators over the product of the denominators. For example,

$$\frac{3}{10} \times \frac{7}{100} = \frac{21}{1000}$$

In decimal form this product would be written

$$\underset{\substack{\text{one}\\\text{decimal}\\\text{place}}}{0.3} \times \underset{\substack{\text{two}\\\text{decimal}\\\text{places}}}{0.07} = \underset{\substack{\text{three}\\\text{decimal}\\\text{places}}}{0.021}$$

MULTIPLICATION OF DECIMALS

1. Multiply the numbers just as you would multiply whole numbers.
2. Find the sum of the number of decimal places in the two factors.
3. Place the decimal point in the product so that the product has the same number of decimal places as the sum in step 2. You may need to write zeros to the left of the number found in step 1.

Now use these steps to do the preceding multiplication problem.

EXAMPLE 1 Multiply. 0.07×0.3

Solution

$$
\begin{array}{r}
0.07 \\
\times\ 0.3 \\
\hline
0.021
\end{array}
$$

2 decimal places
1 decimal place
3 decimal places in product $(2 + 1 = 3)$

Practice Problem 1 Multiply. 0.09×0.6 0.054

When performing the calculation, it is usually easier to place the factor with the smallest number of nonzero digits underneath the other factor.

EXAMPLE 2 Multiply.

(a) 0.38×0.26

(b) 12.64×0.572

Solution

(a)

$$
\begin{array}{r}
0.38 \\
\times\ 0.26 \\
\hline
228 \\
76 \\
\hline
0.0988
\end{array}
$$

2 decimal places
2 decimal places

4 decimal places
$(2 + 2 = 4)$

Note that we need to insert a zero before the 988.

(b)

$$
\begin{array}{r}
12.64 \\
\times 0.572 \\
\hline
2528 \\
8848 \\
6\ 320 \\
\hline
7.23008
\end{array}
$$

2 decimal places
3 decimal places

5 decimal places
$(2 + 3 = 5)$

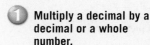

Student Learning Objectives

After studying this section, you will be able to:

1 Multiply a decimal by a decimal or a whole number.

2 Multiply a decimal by a power of 10.

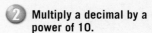

Calculator

Multiplying Decimals

You can use your calculator to multiply a decimal by a decimal. To find 0.08×1.53 enter:

$$0.08 \boxed{\times}\ 1.53 \boxed{=}$$

Display:

$$\boxed{0.1224}$$

Teaching Example 1 Multiply. 0.04×0.9

Ans: 0.036

Teaching Example 2 Multiply.

(a) 0.17×0.45 **(b)** 0.27×13.186

Ans: **(a)** 0.0765 **(b)** 3.56022

217

Practice Problem 2 Multiply.

(a) 0.47×0.28 0.1316

(b) 0.436×18.39 8.01804

When multiplying decimal fractions by a whole number, you need to remember that a whole number has no decimal places.

Teaching Example 3 Multiply. 0.273×56

Ans: 15.288

EXAMPLE 3 Multiply. 5.261×45

Solution

$$
\begin{array}{r}
5.261 \\
\times \quad 45 \\
\hline
26\ 305 \\
210\ 44 \\
\hline
236.745
\end{array}
$$

    3 decimal places
    0 decimal places

    3 decimal places $(3 + 0 = 3)$

Practice Problem 3 Multiply. 0.4264×38 16.2032

Teaching Example 4 A photograph is 7.3 centimeters wide and 12.8 centimeters long. What is the area of the photograph in square centimeters?

Ans: 93.44 square centimeters

▲ **EXAMPLE 4** Uncle Roger's rectangular front lawn measures 50.6 yards wide and 71.4 yards long. What is the area of the lawn in square yards?

Solution Since the lawn is rectangular, we will use the fact that to find the area of a rectangle we multiply the length by the width.

$$
\begin{array}{r}
71.4 \\
\times\ 50.6 \\
\hline
42\ 84 \\
3570\ 0 \\
\hline
3612.84
\end{array}
$$

    1 decimal place
    1 decimal place

    2 decimal places

The area of the lawn is 3612.84 square yards.

71.4 yards

50.6 yards

▲ **Practice Problem 4** A rectangular computer chip measures 1.26 millimeters wide and 2.3 millimeters long. What is the area of the chip in square millimeters?   2.898 square millimeters

❷ Multiplying a Decimal by a Power of 10

Observe the following pattern.

Teaching Tip Multiplication of a decimal by a power of 10 is a simple idea. However, it will be totally new to some students. Be sure to do an example such as this one: Multiply $123.8765 \times 100,000 = 12,387,650$ "in your head." Ask students what rule allowed them to do that problem. You will usually get several enthusiastic suggestions from the class. Students like the idea of this type of shortcut.

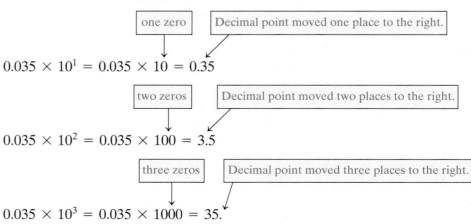

| one zero | Decimal point moved one place to the right. |

$0.035 \times 10^1 = 0.035 \times 10 = 0.35$

| two zeros | Decimal point moved two places to the right. |

$0.035 \times 10^2 = 0.035 \times 100 = 3.5$

| three zeros | Decimal point moved three places to the right. |

$0.035 \times 10^3 = 0.035 \times 1000 = 35.$

MULTIPLICATION OF A DECIMAL BY A POWER OF 10

To multiply a decimal by a power of 10, move the decimal point to the right the same number of places as the number of zeros in the power of 10.

EXAMPLE 5 Multiply.

(a) 2.671×10 **(b)** 37.85×100

Solution

(a) $2.671 \times 10 \qquad = 26.71$

one zero Decimal point moved one place to the right.

(b) $37.85 \times 100 \qquad = 3785.$

two zeros Decimal point moved two places to the right.

Practice Problem 5 Multiply.

(a) 0.0561×10 0.561 **(b)** 1462.37×100 146,237

Sometimes it is necessary to add extra zeros before placing the decimal point in the answer.

EXAMPLE 6 Multiply.

(a) 4.8×1000 **(b)** $0.076 \times 10,000$

Solution

(a) $4.8 \times 1000 \qquad = 4800.$

three zeros Decimal point moved three places to the right. Two extra zeros were needed.

(b) $0.076 \times 10,000 \qquad = 760.$

four zeros Decimal point moved four places to the right. One extra zero was needed.

Practice Problem 6 Multiply.

(a) 0.26×1000 260 **(b)** $5862.89 \times 10,000$ 58,628,900

If the number that is a power of 10 is in exponent form, move the decimal point to the right the same number of places as the number that is the exponent.

Teaching Example 5 Multiply.
(a) 5.047×10 **(b)** 62.35×100
Ans: (a) 50.47 **(b)** 6235

Teaching Example 6 Multiply.
(a) 21.6×100 **(b)** $9.15 \times 10,000$
Ans: (a) 2160 **(b)** 91,500

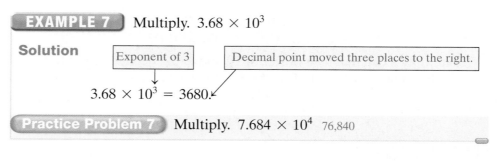

EXAMPLE 7 Multiply. 3.68×10^3

Solution

| Exponent of 3 | Decimal point moved three places to the right. |

$$3.68 \times 10^3 = 3680.$$

Practice Problem 7 Multiply. 7.684×10^4 76,840

SIDELIGHT: Moving the Decimal Point

Can you devise a quick rule to use when multiplying a decimal fraction by $\frac{1}{10}, \frac{1}{100}, \frac{1}{1000}$, and so on? How is it like the rules developed in this section? Consider a few examples:

Original Problem	Change Fraction to Decimal	Decimal Multiplication	Observation
$86 \times \frac{1}{10}$	86×0.1	$\begin{array}{r} 86 \\ \times\ 0.1 \\ \hline 8.6 \end{array}$	Decimal point moved one place to the left.
$86 \times \frac{1}{100}$	86×0.01	$\begin{array}{r} 86 \\ \times\ 0.01 \\ \hline 0.86 \end{array}$	Decimal point moved two places to the left.
$86 \times \frac{1}{1000}$	86×0.001	$\begin{array}{r} 86 \\ \times\ 0.001 \\ \hline 0.086 \end{array}$	Decimal point moved three places to the left.

Can you think of a way to describe a rule that you could use in solving this type of problem without going through all the foregoing steps?

You use multiplying by a power of 10 when you convert a larger unit of measure to a smaller unit of measure in the metric system.

EXAMPLE 8 Change 2.96 kilometers to meters.

Solution Since we are going from a larger unit of measure to a smaller one, we multiply. There are 1000 meters in 1 kilometer. Multiply 2.96 by 1000.

$$2.96 \times 1000 = 2960$$

2.96 kilometers is equal to 2960 meters.

Practice Problem 8 Change 156.2 kilometers to meters. 156,200 meters

TO THINK ABOUT: Names Used to Describe Large Numbers

Often when reading the newspaper or watching television news shows, we hear words like 3.46 trillion or 67.8 billion. These are abbreviated notations that are used to describe large numbers. When you encounter these numbers, you can change them to standard notation by multiplication of the appropriate value.

For example, if someone says that the population of China is 1.31 billion people, we can write 1.31 billion = 1.31×1 billion = $1.31 \times 1,000,000,000 = 1,310,000,000$. If someone says the population of Chicago is 2.92 million people, we can write

$$2.92 \text{ million} = 2.92 \times 1 \text{ million} = 2.92 \times 1,000,000 = 2,920,000.$$

Multiply.

Verbal and Writing Skills

1. Explain in your own words how to determine where to put the decimal point in the answer when you multiply 0.67 × 0.08.

Each factor has two decimal places. You add the number of decimal places to get four decimal places. You multiply 67 × 8 to obtain 536. Now you must place the decimal point four places to the left in your answer. The result is 0.0536.

2. Explain in your own words how to determine where to put the decimal point in the answer when you multiply 3.45 × 0.9.

The first factor has two decimal places. The second factor has one. You add the number of decimal places to get three decimal places. You multiply 345 × 9 to obtain 3105. Now you must place the decimal three places to the left in your answer. The result is 3.105.

3. Explain in your own words how to determine where to put the decimal point in the answer when you multiply 0.0078 × 100.

When you multiply a number by 100 you move the decimal point two places to the right. The answer is 0.78.

4. Explain in your own words how to determine where to put the decimal point in the answer when you multiply 5.0807 by 1000.

When you multiply a number by 1000 you move the decimal point three places to the right. The answer is 5080.7.

5.
$$\begin{array}{r} 0.6 \\ \times 0.2 \\ \hline 0.12 \end{array}$$

6.
$$\begin{array}{r} 0.9 \\ \times 0.3 \\ \hline 0.27 \end{array}$$

7.
$$\begin{array}{r} 0.12 \\ \times\ 0.5 \\ \hline 0.06 \end{array}$$

8.
$$\begin{array}{r} 0.17 \\ \times\ 0.4 \\ \hline 0.068 \end{array}$$

9.
$$\begin{array}{r} 0.0036 \\ \times\ 0.8 \\ \hline 0.00288 \end{array}$$

10.
$$\begin{array}{r} 0.067 \\ \times\ 0.07 \\ \hline 0.00469 \end{array}$$

11.
$$\begin{array}{r} 452 \\ \times 0.12 \\ \hline 54.24 \end{array}$$

12.
$$\begin{array}{r} 316 \\ \times 0.24 \\ \hline 75.84 \end{array}$$

13.
$$\begin{array}{r} 0.043 \\ \times 0.012 \\ \hline 0.000516 \end{array}$$

14.
$$\begin{array}{r} 0.037 \\ \times 0.011 \\ \hline 0.000407 \end{array}$$

15.
$$\begin{array}{r} 10.97 \\ \times\ 0.06 \\ \hline 0.6582 \end{array}$$

16.
$$\begin{array}{r} 18.07 \\ \times\ 0.05 \\ \hline 0.9035 \end{array}$$

17.
$$\begin{array}{r} 3423 \\ \times\ 0.8 \\ \hline 2738.4 \end{array}$$

18.
$$\begin{array}{r} 5119 \\ \times\ 0.7 \\ \hline 3583.3 \end{array}$$

19.
$$\begin{array}{r} 2.163 \\ \times 0.008 \\ \hline 0.017304 \end{array}$$

20.
$$\begin{array}{r} 1.892 \\ \times 0.007 \\ \hline 0.013244 \end{array}$$

21.
$$\begin{array}{r} 0.7613 \\ \times\ 1009 \\ \hline 768.1517 \end{array}$$

22.
$$\begin{array}{r} 0.6178 \\ \times\ 5004 \\ \hline 3091.4712 \end{array}$$

23.
$$\begin{array}{r} 2350 \\ \times\ 3.6 \\ \hline 8460 \end{array}$$

24.
$$\begin{array}{r} 3720 \\ \times\ 8.1 \\ \hline 30{,}132 \end{array}$$

25. 4.57 × 11.8
53.926

26. 73.2 × 2.45
179.34

27. 0.001 × 6523.7
6.5237

28. 0.01 × 826.75
8.2675

Applications

29. *Car Payments* Kenny is making payments on his Ford Escort of $155.40 per month for the next 60 months. How much will he have spent in car payments after he sends in his final payment? $9324

30. *Food Purchase* Each carton of ice cream contains 1.89 liters. Paul stocked his freezer with 25 cartons. How many total liters of ice cream did he buy? 47.25 liters

31. *Personal Income* Mei Lee works for a forest and conservation company and earns $12.35 per hour for a 40-hour week. How much does she earn in one week? (The average wage in 2005 for U.S. forest and conservation workers was $11.19 per hour.) (*Source:* www.bls.gov) $494

32. *Personal Income* Barry works as a fitness trainer and earns $14.75 per hour for a 40-hour week. How much does he earn in one week? (The average wage in 2005 for U.S. fitness trainers/aerobics instructors was $14.93 per hour.) (*Source:* www.bls.gov) $590

▲ **33.** *Geometry* Ralph and Darlene are getting new carpet in their bedroom and need to find how many square feet they need to purchase. The dimensions of their rectangular bedroom are 15.5 feet and 19.2 feet. What is the area of the room in square feet? 297.6 square feet

▲ **34.** *Geometry* Sal is having his driveway paved by a company that charges by the square yard. Sal's driveway measures 8.6 yards by 17.5 yards. How many square yards is his driveway? 150.5 square yards

35. *Student Loan* Dwight is paying off a student loan at Westmont College with payments of $36.90 per month for the next 18 months. How much will he pay off during the next 18 months? $664.20

36. *Car Payments* Marcia is making car payments to Highfield Center Chevrolet of $230.50 per month for 16 more months. How much will she pay for car payments in the next 16 months? $3688.00

37. *Fuel Efficiency* Steve's car gets approximately 26.4 miles per gallon. His gas tank holds 19.5 gallons. Approximately how many miles can he travel on a full tank of gas? 514.8 miles

38. *Fuel Efficiency* Caleb's 4 × 4 truck gets approximately 18.6 miles per gallon. His gas tank holds 19.5 gallons. Approximately how many miles can he travel on a full tank of gas? Compare this to your answer in exercise 37. 362.7 miles; Steve can travel 152.1 miles farther than Caleb on a tank of gas.

Multiply.

39. 2.86×10
28.6

40. 1.98×10
19.8

41. 52.125×100
5212.5

42. 86.375×100
8637.5

43. 22.615×1000
22,615

44. 34.105×1000
34,105

45. $5.60982 \times 10,000$
56,098.2

46. $1.27986 \times 10,000$
12,798.6

47. $17,561.44 \times 10^2$
1,756,144

48. 7163.241×10^2
716,324.1

49. 816.32×10^3
816,320

50. 763.49×10^4
7,634,900

Applications

51. *Metric Conversion* To convert from meters to centimeters, multiply by 100. How many centimeters are in 5.932 meters? 593.2 centimeters

52. *Metric Conversion* One meter is about 39.36 inches. About how many inches are in 100 meters? 3936 inches

53. *Metric Conversion* One meter is about 3.281 feet. How many feet are in 1000 meters? 3281 feet

54. *Stock Market* Jeremy bought 1000 shares of stock each worth $1.45. How much did Jeremy spend on the stock? $1450

55. *Personal Finance* In April, Ellen received a $925.75 tax refund. She decided to spend the money on some gifts. She spent $95.00 on her parents' anniversary gift, $47.50 on each of her two cousins' graduation gifts, and $39.25 on each of her three nieces' birthday gifts. How much money does she have left over? $618.00

56. *Pet Cats* Tomba is a beautiful orange tabby cat. When he was found by the side of the road, he was three weeks old and weighed 0.95 lb. At the age of three months, he weighed 2.85 lb. At the age of nine months, he weighed 6.30 lb; at one year, he weighed 11.7 lb. Today, Tomba the cat is $1\frac{1}{2}$ years old, and weighs 15.75 lb.

(a) How much weight did he gain? 14.8 lb

(b) If the veterinarian wants him to lose 0.25 lb per week until he weighs 13.5 lb, how long will it take? nine weeks

▲**57.** *Geometry* The college is purchasing new carpeting for the learning center. What is the price of a carpet that is 19.6 yards wide and 254.2 yards long if the cost is $12.50 per square yard?

$$\begin{array}{r} 254.2 \\ \times\ 19.6 \\ \hline 4982.32 \text{ square yards} \end{array} \qquad \begin{array}{r} 4982.32 \\ \times\ 12.5 \\ \hline \$62,279.00 \end{array}$$

58. *Jewelry Store Operations* A jewelry store purchased long lengths of gold chain, which will be cut and made into necklaces and bracelets. The store purchased 3220 grams of gold chain at $3.50 per gram.

(a) How much did the jewelry store spend? $11,270

(b) If they sell a 28-gram gold necklace for $17.75 per gram, how much profit will they make on the necklace? $399

To Think About

59. State in your own words a rule for mental multiplication by 0.1, 0.01, 0.001, 0.0001, and so on.
To multiply by numbers such as 0.1, 0.01, 0.001, and 0.0001, count the number of decimal places in this first number. Then, in the other number, move the decimal point to the left from its present position the same number of decimal places as were in the first number.

60. State in your own words a rule for mental multiplication by 0.2, 0.02, 0.002, 0.0002, and so on.
To multiply by numbers such as 0.2, 0.02, 0.002, and 0.0002, first double the second number. Then move the decimal point to the left using the rule stated in exercise 59.

Cumulative Review *Divide. Be sure to include any remainder as part of your answer.*

61. [1.5.3] $20\overline{)4080}$
204

62. [1.5.3] $35\overline{)7035}$
201

63. [1.5.3] $48\overline{)6099}$
127 R 3

64. [1.5.3] $124\overline{)56,024}$
451 R 100

Pets in the United States *The total number of pets owned in the United States for the year 2005 is given in the bar graph below. Use the graph to answer exercises 65–68.*

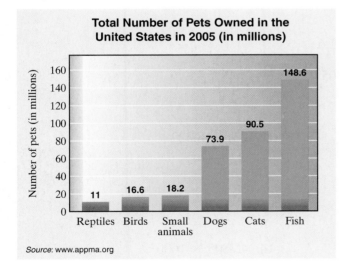

Total Number of Pets Owned in the United States in 2005 (in millions)

Reptiles 11, Birds 16.6, Small animals 18.2, Dogs 73.9, Cats 90.5, Fish 148.6

Number of pets (in millions)

Source: www.appma.org

65. [3.3.2] How many more pet cats are there in the United States than pet dogs?
16.6 million or 16,600,000

66. [3.3.2] How many more pet dogs are there in the United States than pet birds?
57.3 million or 57,300,000

67. [3.3.1] How many more pet fish are there in the United States than pet dogs, small animals, birds, and reptiles combined?
28.9 million or 28,900,000

68. [3.3.1] How many more pet cats and dogs combined are there in the United States than pet fish?
15.8 million or 15,800,000

Quick Quiz 3.4

1. Multiply. 0.76×0.04 0.0304

2. Multiply. 25.6×0.128 3.2768

3. Multiply. 5.162×10^4 51,620

4. **Concept Check** Explain how you know where to put the decimal point in the answer when you multiply 3.45×9.236. Answers may vary

Classroom Quiz 3.4 You may use these problems to quiz your students' mastery of Section 3.4.

1. Multiply 0.05×0.93 **Ans:** 0.0465

2. Multiply 15.7×0.198 **Ans:** 3.1086

3. Multiply 9.186×10^2 **Ans:** 918.6

1.	forty-seven and eight hundred thirteen thousandths
2.	0.0567
3.	$4\dfrac{9}{100}$
4.	$\dfrac{21}{40}$
5.	1.59, 1.6, 1.601, 1.61
6.	123.5
7.	8.0654
8.	17.99
9.	19.45
10.	27.191
11.	10.59
12.	7.671
13.	0.3501
14.	4780.5
15.	37.96
16.	7.85
17.	6.874
18.	0.00000312

How are you doing with your homework assignments in Sections 3.1 to 3.4? Do you feel you have mastered the material so far? Do you understand the concepts you have covered? Before you go further in the textbook, take some time to do each of the following problems.

3.1

1. Write a word name for the decimal. 47.813

2. Express as a decimal. $\dfrac{567}{10,000}$

Write as a fraction or a mixed number. Reduce whenever possible.

3. 4.09

4. 0.525

3.2

5. Place the set of numbers in the proper order from smallest to largest.
1.6, 1.59, 1.61, 1.601

6. Round to the nearest tenth. 123.49268

7. Round to the nearest ten thousandth. 8.065447

8. Round to the nearest hundredth. 17.98523

3.3

Add.

9. $5.12 + 4.7 + 8.03 + 1.6$

10. $24.613 + 0.273 + 2.305$

Subtract.

11.
$$\begin{array}{r} 42.16 \\ -\ 31.57 \\ \hline \end{array}$$

12. $26 - 18.329$

3.4

Multiply.

13.
$$\begin{array}{r} 11.67 \\ \times\ 0.03 \\ \hline \end{array}$$

14. 4.7805×1000

15. 0.0003796×10^5

16. 3.14×2.5

17. 982×0.007

18. 0.00052×0.006

Now turn to page SA-7 for the answer to each of these problems. Each answer also includes a reference to the objective in which the problem is first taught. If you missed any of these problems, you should stop and review the Examples and Practice Problems in the referenced objective. A little review now will help you master the material in the upcoming sections of the text.

Dividing a Decimal by a Whole Number

When you divide a decimal by a whole number, place the decimal point for the quotient directly above the decimal point in the dividend. Then divide as if the numbers were whole numbers.

To divide 26.8 by 4, we place the decimal point of our answer (the quotient) directly *above* the decimal point in the dividend.

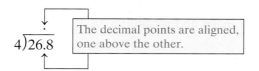

$$4\overline{)26.8}$$ The decimal points are aligned, one above the other.

Then we divide as if we were dividing whole numbers.

$$
\begin{array}{r}
6.7 \\
4\overline{)26.8} \\
\underline{24} \\
2\,8 \\
\underline{2\,8} \\
0
\end{array}
$$

The quotient is 6.7.

The quotient to a problem may have all digits to the right of the decimal point. In some cases you will have to put a zero in the quotient as a "place holder." Let's divide 0.268 by 4.

$$
\begin{array}{r}
0.067 \\
4\overline{)0.268} \\
\underline{24} \\
28 \\
\underline{28} \\
0
\end{array}
$$

Note that we must have a zero after the decimal point in 0.067.

EXAMPLE 1 Divide.

(a) $9\overline{)0.3204}$ **(b)** $14\overline{)36.12}$

Solution

(a)
$$
\begin{array}{r}
0.0356 \\
9\overline{)0.3204} \\
\underline{27} \\
50 \\
\underline{45} \\
54 \\
\underline{54} \\
0
\end{array}
$$

Note the zero *after* the decimal point.

(b)
$$
\begin{array}{r}
2.58 \\
14\overline{)36.12} \\
\underline{28} \\
81 \\
\underline{70} \\
112 \\
\underline{112} \\
0
\end{array}
$$

Practice Problem 1 Divide.

(a) $7\overline{)1.806}$ 0.258   **(b)** $16\overline{)0.0928}$ 0.0058

Student Learning Objectives

After studying this section, you will be able to:

 Divide a decimal by a whole number.

 Divide a decimal by a decimal.

Teaching Tip This is a good time to remind students that they will be very glad they learned these three words: *divisor, dividend, quotient.* They are referred to frequently in mathematics. You may want to put a diagram on the board as a reminder:

$$\text{divisor}\overline{)\text{dividend}}^{\,\text{quotient}}$$

Teaching Example 1 Divide.
(a) $0.3504 \div 6$ (b) $1.644 \div 12$

Ans: (a) 0.0584   (b) 0.137

NOTE TO STUDENT: Fully worked-out solutions to all of the Practice Problems can be found at the back of the text starting at page SP-1

Teaching Tip It is helpful for most students to see a variety of division problems with decimals where the divisor is an integer and to master the ability to perform each division before going onward. Be sure to carefully discuss Examples 1 and 2 or similar problems before moving on to the case where the divisor is a decimal fraction.

Teaching Example 2 Divide and round to the nearest tenth.

$$37.8 \div 23$$

Ans: 1.6

Some division problems do not yield a remainder of zero. In such cases, we may be asked to round the answer to a specified place. To round when dividing, we carry out the division until our answer contains a digit that is one place to the right of that to which we intend to round. Then we round our answer to the specified place. For example, to round to the nearest thousandth, we carry out the division to the ten-thousandths place. In some division problems, you will need to write in zeros at the end of the dividend so that this division can be carried out.

EXAMPLE 2 Divide and round the quotient to the nearest thousandth.

$$12.67 \div 39$$

Solution We will carry out our division to the ten-thousandths place. Then we will round our answer to the nearest thousandth.

```
         0.3248
     39)12.6700  ←    Two extra zeros are written
        11 7            here to carry out the division
        ────            to the required place.
          97
          78
        ────
         190
         156
        ────
         340
         312
        ────
          28    Note that the remainder is not zero.
```

Now we round 0.3248 to 0.325. The answer is rounded to the nearest thousandth.

Practice Problem 2 Divide and round the quotient to the nearest hundredth. $23.82 \div 46$ 0.52

Teaching Example 3 Jorge filled his car's gas tank with 13 gallons of gas. He paid $30.81. How much did he pay per gallon?

Ans: $2.37 per gallon

EXAMPLE 3 Maria paid $5.92 for 16 pounds of tomatoes. How much did she pay per pound?

Solution The cost of one pound of tomatoes equals the total cost, $5.92, divided by 16 pounds. Thus we will divide.

```
         0.37       Maria paid
     16)5.92        $0.37 per pound
        4 8         for the tomatoes.
       ────
        112
        112
       ────
          0
```

Practice Problem 3 Won Lin will pay off his auto loan for $3538.75 over 19 months. If the monthly payments are equal, how much will he pay each month? $186.25

Dividing a Decimal by a Decimal

When the divisor is not a whole number, we can convert the division problem to an equivalent problem that has a whole number as a divisor. Think about the reasons why this procedure will work. We will ask you about it after you study Examples 4 and 5.

> **DIVIDING A DECIMAL BY A DECIMAL**
>
> 1. Make the divisor a whole number by moving the decimal point to the right. Mark that position with a caret ($_\wedge$). Count the number of places the decimal point moved.
> 2. Move the decimal point in the dividend to the right the same number of places. Mark that position with a caret.
> 3. Place the decimal point of your answer directly above the caret marking the decimal point of the dividend.
> 4. Divide as with whole numbers.

EXAMPLE 4 (a) Divide. $0.08\overline{)1.632}$ (b) Divide. $1.352 \div 0.026$

Teaching Example 4 Divide.
(a) $0.441 \div 0.3$ (b) $4.05 \div 0.15$
Ans: (a) 1.47   (b) 27

Solution

(a) $0.08.\overline{)1.63.2}$ Move each decimal point two places to the right.

Place the decimal point of the answer directly above the caret.

$0.08_\wedge\overline{)1.63_\wedge2}$ Mark the new position by a caret ($_\wedge$).

$$\begin{array}{r} 20.4 \\ 0.08_\wedge\overline{)1.63_\wedge2} \\ \underline{16} \\ 3\ 2 \\ \underline{3\ 2} \\ 0 \end{array}$$ Perform the division.

The answer is 20.4.

(b) $\begin{array}{r} 52. \\ 0.026_\wedge\overline{)1.352_\wedge} \\ \underline{1\ 30} \\ 52 \\ \underline{52} \\ 0 \end{array}$ Move each decimal point three places to the right and mark the new position by a caret.

The answer is 52.

Practice Problem 4 Divide.

(a) $0.09\overline{)0.1008}$ 1.12   (b) $1.702 \div 0.037$ 46

NOTE TO STUDENT: Fully worked-out solutions to all of the Practice Problems can be found at the back of the text starting at page SP-1

Teaching Tip Sometimes students find that they need to work out some additional examples like Example 5 until they are sure of themselves. If you see a need for this, have them do 0.03154 ÷ 0.019 (the answer is 1.66) and 1.2999 ÷ 0.0007 (the answer is 1857).

TO THINK ABOUT: The Multiplicative Identity Why do we move the decimal point to the right in the divisor and the dividend? What rule allows us to do this? How do we know the answer will be valid? We are actually using the property that multiplication of a fraction by 1 leaves the fraction unchanged. This is called the *multiplicative identity*. Let us examine Example 4(b) again. We will write 1.352 ÷ 0.026 as a fraction.

$$\frac{1.352}{0.026} \times 1 \qquad \text{Multiplication of a fraction by 1 does not change the value of the fraction.}$$

$$= \frac{1.352}{0.026} \times \frac{1000}{1000} \qquad \text{We know that } \frac{1000}{1000} = 1.$$

$$= \frac{1352}{26} \qquad \text{Multiplication by 1000 can be done by moving the decimal point three places to the right.}$$

$$= 52 \qquad \text{Divide the whole numbers.}$$

Thus in Example 4(b) when we moved the decimal point three places to the right in the divisor and the dividend, we were actually creating an equivalent fraction where the numerator and the denominator of the original fraction were multiplied by 1000.

Teaching Example 5 Divide.

(a) 0.00798 ÷ 2.1 (b) 6.48 ÷ 0.054

Ans: (a) 0.0038 (b) 120

EXAMPLE 5 Divide.

(a) $1.7\overline{)0.0323}$

(b) $0.0032\overline{)7.68}$

Solution

(a)
$$\begin{array}{r} 0.019 \\ 1.7_\wedge\overline{)0.0_\wedge 323} \\ \underline{17} \\ 153 \\ \underline{153} \\ 0 \end{array}$$

Move the decimal point in the divisor and dividend one place to the right and mark that position with a caret.

(b)
$$\begin{array}{r} 2400. \\ 0.0032_\wedge\overline{)7.6800_\wedge} \\ \underline{64} \\ 1\,28 \\ \underline{1\,28} \\ 000 \end{array}$$

Note that two extra zeros are needed in the dividend as we move the decimal point four places to the right.

Practice Problem 5 Divide.

(a) $1.8\overline{)0.0414}$ 0.023

(b) $0.0036\overline{)8.316}$ 2310

Teaching Example 6

(a) Find 23.4 ÷ 5.3 rounded to the nearest hundredth.

(b) Find 2.17 ÷ 1.16 rounded to the nearest tenth.

Ans: (a) 4.42 (b) 1.9

EXAMPLE 6

(a) Find $2.9\overline{)431.2}$ rounded to the nearest tenth.

(b) Find $2.17\overline{)0.08}$ rounded to the nearest thousandth.

Solution

(a)
$$
\begin{array}{r}
14\,8.68 \\
2.9_\wedge\overline{)431.2_\wedge 00} \\
\underline{29} \\
141 \\
\underline{116} \\
25\,2 \\
\underline{23\,2} \\
2\,0\,0 \\
\underline{1\,7\,4} \\
2\,60 \\
\underline{2\,32} \\
28
\end{array}
$$

Calculate to the hundredths place and round the answer to the nearest tenth.

The answer rounded to the nearest tenth is 148.7.

(b)
$$
\begin{array}{r}
0.0368 \\
2.17_\wedge\overline{)0.08_\wedge 0000} \\
\underline{6\,51} \\
1\,490 \\
\underline{1\,302} \\
1880 \\
\underline{1736} \\
144
\end{array}
$$

Calculate to the ten-thousandths place and then round the answer. Rounding 0.0368 to the nearest thousandth, we obtain 0.037.

Calculator

Dividing Decimals

You can use your calculator to divide a decimal by a decimal. To find $21.38\overline{)54.53}$ rounded to the nearest hundredth, enter:

| 54.53 | ÷ | 21.38 | = |

Display:

| 2.5505145 |

This is an approximation. Some calculators will round to eight digits. The answer rounded to the nearest hundredth is 2.55.

Practice Problem 6

(a) Find $3.8\overline{)521.6}$ rounded to the nearest tenth.   137.3

(b) Find $8.05\overline{)0.17}$ rounded to the nearest thousandth.   0.021

EXAMPLE 7   John drove his 1997 Cavalier 420.5 miles to Chicago. He used 14.5 gallons of gas on the trip. How many miles per gallon did his car get on the trip?

Solution   To find miles per gallon we need to divide the number of miles, 420.5, by the number of gallons, 14.5.

$$
\begin{array}{r}
29. \\
14.5_\wedge\overline{)420.5_\wedge} \\
\underline{290} \\
130\,5 \\
\underline{130\,5} \\
0
\end{array}
$$

John's car achieved 29 miles per gallon on the trip to Chicago.

Practice Problem 7   Sarah rented a large truck to move to Boston. She drove 454.4 miles yesterday. She used 28.5 gallons of gas on the trip. How many miles per gallon did the rental truck get? Round to the nearest tenth.

15.9 miles per gallon

EXAMPLE 8   Find the value of n if $0.8 \times n = 2.68$.

Solution   Here 0.8 is multiplied by some number n to obtain 2.68. What is this number n? If we divide 2.68 by 0.8, we will find the value of n.

$$
\begin{array}{r}
3.35 \\
0.8_\wedge\overline{)2.6_\wedge 80} \\
\underline{2\,4} \\
2\,8 \\
\underline{2\,4} \\
40 \\
\underline{40} \\
0
\end{array}
$$

Thus the value of n is 3.35.

Teaching Example 7 On a long hike, Karen walked 10.4 miles in 3.25 hours. How many miles did she walk each hour?

Ans: 3.2 miles each hour

Teaching Example 8 Find the value of n if $0.13 \times n = 0.325$.

Ans: $n = 2.5$

Check. Is this true? Are we sure the value of $n = 3.35$?
We substitute the value of $n = 3.35$ into the equation to see if it makes the statement true.

$$0.8 \times n \quad = 2.68$$
$$0.8 \times 3.35 \overset{?}{=} 2.68$$
$$2.68 = 2.68 \quad \checkmark \quad \text{Yes, it is true.}$$

Practice Problem 8 Find the value of n if $0.12 \times n = 0.696$. $n = 5.8$

Teaching Example 9 Use the bar graph to find the average level of sulfur dioxide for the two years 1980 and 2000. By how much does this average differ from the five-year average?

Ans: 7.715 million tons; 0.441 million tons

EXAMPLE 9 The level of sulfur dioxide emissions in the air has slowly been decreasing over the last 20 years, as can be seen in the accompanying bar graph. Find the average amount of sulfur dioxide emissions in the air over these five specific years.

Solution

First we take the sum of the five years.

$$
\begin{array}{r}
9.37 \\
9.30 \\
8.68 \\
7.37 \\
+\ 6.06 \\
\hline
40.78
\end{array}
$$

Then we divide by five to obtain the average.

$$
\begin{array}{r}
8.156 \\
5\overline{)40.780} \\
\underline{40} \\
7 \\
\underline{5} \\
28 \\
\underline{25} \\
30 \\
\underline{30} \\
0
\end{array}
$$

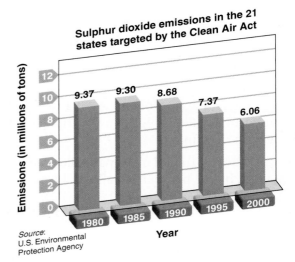

Sulphur dioxide emissions in the 21 states targeted by the Clean Air Act

Source: U.S. Environmental Protection Agency

Thus the yearly average is 8.156 million tons of sulfur dioxide emissions in these 21 states.

Practice Problem 9 Use the accompanying bar graph to find the average level of sulfur dioxide for the three years: 1985, 1990, and 1995. By how much does the three-year average differ from the five-year average?

8.45 million tons; 0.294 million tons

Developing Your Study Skills

Exam Time: How To Review

Reviewing adequately for an exam enables you to bring together the concepts you have learned over several sections. For your review, you will need to do the following:

1. Reread your textbook. Make a list of any terms, rules, or formulas you need to know for the exam. Be sure you understand them all.

2. Reread your notes. Go over returned homework and quizzes. Redo the problems you missed.

3. Practice some of each type of problem covered in the chapter(s) you are to be tested on. In fact, it is a good idea

to construct a practice test of your own and then discuss it with a friend from class.

4. Use the end-of-chapter materials provided in your textbook. Read carefully through the Chapter Organizer. Do the Chapter Review Problems. Take the Chapter Test. When you are finished, check your answers. Redo any problems you missed.

5. Get help if any concepts give you difficulty.

Divide until there is a remainder of zero.

1. $6\overline{)12.6}$ (2.1)

2. $8\overline{)17.28}$ (2.16)

3. $4\overline{)71.32}$ (17.83)

4. $6\overline{)83.16}$ (13.86)

5. $7\overline{)73.64}$ (10.52)

6. $8\overline{)168.48}$ (21.06)

7. $0.6\overline{)81.9}$ (136.5)

8. $0.5\overline{)32.15}$ (64.3)

9. $0.2706 \div 0.05$ 5.412

10. $0.6092 \div 0.08$ 7.615

11. $153.7 \div 2.9$ 53

12. $75.6 \div 3.6$ 21

13. $68.4 \div 3.8$ 18

14. $728 \div 5.6$ 130

15. $40.30 \div 0.31$ 130

Divide and round your answer to the nearest tenth.

16. $8\overline{)44}$ (5.5)

17. $9\overline{)47.31}$ (5.3)

18. $1.8\overline{)4.16}$ (2.3)

19. $1.9\overline{)2.36}$ (1.2)

20. $0.95\overline{)32.067}$ (33.8)

21. $0.85\overline{)41.901}$ (49.3)

Divide and round your answer to the nearest hundredth.

22. $4\overline{)263.82}$ (65.96)

23. $5\overline{)471.03}$ (94.21)

24. $1.7\overline{)20.8}$ (12.24)

25. $1.8\overline{)24.41}$ (13.56)

26. $24\overline{)3.126}$ (0.13)

27. $35\overline{)7.369}$ (0.21)

Divide and round your answer to the nearest thousandth.

28. $8\overline{)0.2019}$ (0.025)

29. $7\overline{)0.5681}$ (0.081)

30. $0.69\overline{)8.45}$ (12.246)

31. $0.87\overline{)79.40}$ (91.264)

Divide and round your answer to the nearest whole number.

32. $12\overline{)1396}$ (116)

33. $19\overline{)2341}$ (123)

34. $0.0024\overline{)0.2168}$ (90)

35. $0.0046\overline{)0.981}$ (213)

Applications

36. *Travel in Mexico* Rhett and Liza are traveling in Mexico, where distances on the highway are given in kilometers. There are approximately 1.6 kilometers in one mile. They see a sign that reads "Mexico City: 342 km." How many miles is it to Mexico City? 213.75 miles

37. *Computer Payments* The Miller family wants to use the latest technology to access the Internet from their home television system. The equipment needed to upgrade their existing equipment will cost $992.76. If the Millers make 12 equal monthly payments, how much will they pay per month? $82.73

38. *Lasagna Dinner* Four students sit down to their weekly lasagna dinner. At one end of the table, there is a bottle containing 67.6 ounces of a popular soft drink. At the other end of the table is a bottle that contains 33.6 ounces of water.

   (a) If the students share the soft drink and water equally, how many ounces of liquid will each student drink? 25.3 ounces

   (b) At the last minute, another student is asked to join the group. How many ounces of liquid will each of the five students share? 20.24 ounces

39. *Fuel Efficiency* Wally owns a Dodge Caliber that travels 360 miles on 13.2 gallons of gas. How many miles per gallon does it achieve? (Round your answer to the nearest tenth.)
approximately 27.3 miles per gallon

40. *Costs of a Ski Trip* The church youth group went on a ski trip. The ski resort charged the group $1200 for 32 lift tickets. How much was each ticket? $37.50

41. *Flower Sales* Andrea makes Mother's Day bouquets each year for extra income. This year her goal is to make $300. If she sells each bouquet for $12.50, how many bouquets must she sell to reach her goal? 24 bouquets

42. *Outdoor Deck Payments* Demitri had a contractor build an outdoor deck for his back porch. He now has $1131.75 to pay off, and he agreed to pay $125.75 per month. How many more payments on the outdoor deck must he make? 9 payments

43. *Wedding Reception Costs* For their wedding reception, Sharon and Richard spent $1865.50 on food and drinks. If the caterer charged them $10.25 per person, how many guests did they have? 182 guests

44. *Record Rainfall*
(a) Using the chart below, find the average amount of precipitation for the months April, May, and June. 35.58 in.
(b) On average, how much more precipitation does Mount Waialeale get per day in April than in March? (Use 30 days in a month, and round to the nearest thousandth.) 0.684 in.

45. *Quality Inspection* Yoshi is working as an inspector for a company that makes snowboards. A Mach 1 snowboard weighs 3.8 kilograms. How many of these snowboards are contained in a box in which the contents weigh 87.40 kilograms? If the box is labeled CONTENTS: 24 SNOWBOARDS, how great an error was made in packing the box?
23 snowboards; the error was in putting one less snowboard in the box than was required

Month	Average Amount of Precipitation in Mount Waialeale, Hawaii, for January–June
January	24.78 in.
February	24.63 in.
March	27.24 in.
April	47.75 in.
May	28.34 in.
June	30.65 in.

Source: www.wrcc.dri.edu

Find the value of n.

46. $0.5 \times n = 3.55$
$n = 7.1$

47. $0.3 \times n = 9.66$
$n = 32.2$

48. $1.7 \times n = 129.2$
$n = 76$

49. $1.3 \times n = 1267.5$
$n = 975$

50. $n \times 0.063 = 2.835$
$n = 45$

51. $n \times 0.098 = 4.312$
$n = 44$

To Think About *Multiply the numerator and denominator of each fraction by 10,000. Then divide the numerator by the denominator. Is the result the same if we divided the original numerator by the original denominator? Why?*

yes; multiplying the numerator and denominator by 10,000 is the same as multiplying by $\frac{10,000}{10,000}$, which is 1

52. $\frac{3.8702}{0.0523} \times \frac{10,000}{10,000} = \frac{38,702}{523} = 74$

53. $\frac{2.9356}{0.0716} \times \frac{10,000}{10,000} = \frac{29,356}{716} = 41$

Cumulative Review

54. [2.8.1] Add. $\frac{3}{8} + 2\frac{4}{5}$ $\frac{127}{40}$ or $3\frac{7}{40}$

55. [2.8.2] Subtract. $2\frac{13}{16} - 1\frac{7}{8}$ $\frac{15}{16}$

56. [2.4.3] Multiply. $3\frac{1}{2} \times 2\frac{1}{6}$ $\frac{91}{12}$ or $7\frac{7}{12}$

57. [2.5.3] $7\frac{1}{2} \div \frac{1}{2}$ 15

Most Damaging Hurricanes *The amount of property damage for the five most destructive hurricanes to hit the United States is represented in the following bar graph. Use the bar graph to answer exercises 58–61.*

Dollar amounts given in year 2000 dollars.
Source: www.cement.org

58. [3.3.2] How much more property damage occurred during Hurricane Andrew than Hurricane Hugo? $25.21 billion or $25,210,000,000

59. [3.3.2] How much more property damage occurred during Hurricane Hugo than Hurricane Betsy? $1.22 billion or $1,220,000,000

60. [3.5.2] How many times more property damage occurred during Hurricane Katrina than Hurricane Andrew?   about 3.6 times

61. [3.5.2] How many times more property damage occurred during Hurricane Katrina than Hurricane Betsy?   about 14.8 times

Quick Quiz 3.5

1. Divide. $0.07\overline{)0.04606}$ 0.658

2. Divide. $0.52\overline{)1.69416}$ 3.258

3. Divide and round to the nearest hundredth.
$8\overline{)52.643}$ 6.58

4. Concept Check Explain how you would know where to place the decimal point in the answer if you divide $0.173 \div 0.578$.   Answers may vary

Classroom Quiz 3.5   You may use these problems to quiz your students' mastery of Section 3.5.

1. Divide. $0.09\overline{)0.5625}$ **Ans:** 6.25

2. Divide. $0.48\overline{)82.56}$ **Ans:** 172

3. Divide and round your answer to the nearest hundredth.
$7\overline{)17.69}$ **Ans:** 2.53

Student Learning Objectives

After studying this section, you will be able to:

 1 Convert a fraction to a decimal.

 2 Use the order of operations with decimals.

Springfield
$2\frac{1}{2}$ miles

Springfield
2.5 miles

Teaching Tip Sometimes a student will not see the difference in the three results discussed in Converting a Fraction to an Equivalent Decimal. You may want to give an example of each one right next to the rule.

(a) $\frac{5}{8} = 0.625$. The remainder becomes zero.

(b) $\frac{1}{3} = 0.333\ldots$. The remainder repeats itself.

(c) Rounded to the nearest thousandth, $\frac{13}{19} = 0.684$. The desired number of decimal places is achieved.

1 Converting a Fraction to a Decimal

A number can be expressed in two equivalent forms: as a fraction or as a decimal.

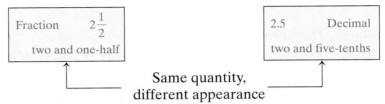

Fraction	$2\frac{1}{2}$
	two and one-half

2.5	Decimal
two and five-tenths	

Same quantity, different appearance

Every decimal in this chapter can be expressed as an equivalent fraction. For example,

Decimal form $\Rightarrow$ fraction form

$$0.75 = \frac{75}{100} \quad \text{or} \quad \frac{3}{4}$$

$$0.5 = \frac{5}{10} \quad \text{or} \quad \frac{1}{2}$$

$$2.5 = 2\frac{5}{10} = 2\frac{1}{2} \quad \text{or} \quad \frac{5}{2}.$$

And every fraction can be expressed as an equivalent decimal, as we will learn in this section. For example,

Fraction form $\Rightarrow$ decimal form

$$\frac{1}{5} = 0.20 \quad \text{or} \quad 0.2$$

$$\frac{3}{8} = 0.375$$

$$\frac{5}{11} = 0.4545\ldots \text{. (The "45" keeps repeating.)}$$

Some of these decimal equivalents are so common that people find it helpful to memorize them. You would be wise to memorize the following equivalents:

$$\frac{1}{2} = 0.5 \qquad \frac{1}{4} = 0.25 \qquad \frac{1}{5} = 0.2 \qquad \frac{1}{10} = 0.1.$$

We previously studied how to convert some fractions with a denominator of 10, 100, 1000, and so on to decimal form. For example, $\frac{3}{10} = 0.3$ and $\frac{7}{100} = 0.07$. We need to develop a procedure to write other fractions, such as $\frac{3}{8}$ and $\frac{5}{16}$, in decimal form.

CONVERTING A FRACTION TO AN EQUIVALENT DECIMAL

Divide the denominator into the numerator until

(a) the remainder becomes zero, or

(b) the remainder repeats itself, or

(c) the desired number of decimal places is achieved.

EXAMPLE 1 Write as an equivalent decimal.

(a) $\dfrac{3}{8}$

(b) $\dfrac{31}{40}$ of a second

Divide the denominator into the numerator until the remainder becomes zero.

Solution

(a)
$$
\begin{array}{r}
0.375 \\
8\overline{)3.000} \\
\underline{2\ 4} \\
60 \\
\underline{56} \\
40 \\
\underline{40} \\
0
\end{array}
$$

(b)
$$
\begin{array}{r}
0.775 \\
40\overline{)31.000} \\
\underline{28\ 0} \\
3\ 00 \\
\underline{80} \\
200 \\
\underline{200} \\
0
\end{array}
$$

Therefore, $\dfrac{3}{8} = 0.375$.

Therefore, $\dfrac{31}{40} = 0.775$ of a second.

Practice Problem 1 Write as an equivalent decimal.

(a) $\dfrac{5}{16}$ 0.3125

(b) $\dfrac{11}{80}$ 0.1375

Athletes' times in Olympic events, such as the 100-meter dash, are measured to the nearest hundredth of a second. Future Olympic athletes' times will be measured to the nearest thousandth of a second.

Decimals such as 0.375 and 0.775 are called **terminating decimals.** When converting $\frac{3}{8}$ to 0.375 or $\frac{31}{40}$ to 0.775, the division operation eventually yields a remainder of zero. Other fractions yield a repeating pattern. For example, $\frac{1}{3} = 0.3333\ldots$ and $\frac{2}{3} = 0.6666\ldots$ have a pattern of repeating digits. Decimals that have a digit or a group of digits that repeats are called **repeating decimals.** We often indicate the repeating pattern with a bar over the repeating group of digits:

$$0.\,3333\ldots = 0.\overline{3} \qquad 0.\,74\ 74\ 74\ldots = 0.\overline{74}$$
$$0.\,218\ 218\ 218\ldots = 0.\overline{218} \qquad 0.\,8942\ 8942\ldots = 0.\overline{8942}$$

If when converting fractions to decimal form the remainder repeats itself, we know that we have a repeating decimal.

EXAMPLE 2 Write as an equivalent decimal.

(a) $\dfrac{5}{11}$

(b) $\dfrac{13}{22}$

(c) $\dfrac{5}{37}$

Solution

(a)
$$11\overline{)5.0000}$$ quotient 0.4545

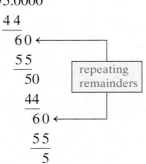

Thus $\dfrac{5}{11} = 0.4545\ldots = 0.\overline{45}$.

(b)
$$22\overline{)13.00000}$$ quotient 0.59090

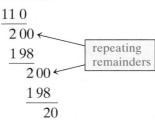

Thus $\dfrac{13}{22} = 0.5909090\ldots = 0.5\overline{90}$.

Notice that the bar is over the digits
9 and 0 but *not* over the digit 5.

(c)
$$37\overline{)5.0000}$$ quotient 0.1351

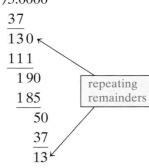

Thus $\dfrac{5}{37} = 0.135135\ldots = 0.\overline{135}$.

Practice Problem 2   Write as an equivalent decimal.

(a) $\dfrac{7}{11}$ $0.\overline{63}$

(b) $\dfrac{8}{15}$ $0.5\overline{3}$

(c) $\dfrac{13}{44}$ $0.29\overline{54}$

Calculator

 Fraction to Decimal

You can use a calculator to change
$\dfrac{5}{8}$ to a decimal.

Enter:

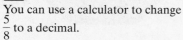

$$5 \;\boxed{\div}\; 8 \;\boxed{=}$$

The display should read

$$\boxed{0.625}$$

Try the following.

(a) $\dfrac{17}{25}$ **(b)** $\dfrac{2}{9}$

(c) $\dfrac{13}{10}$ **(d)** $\dfrac{15}{19}$

Note: 0.78947368 is an
approximation for $\dfrac{15}{19}$. Some
calculators round to only eight
places.

EXAMPLE 3   Write as an equivalent decimal.

(a) $3\dfrac{7}{15}$ **(b)** $\dfrac{20}{11}$

Solution

(a) $3\dfrac{7}{15}$ means $3 + \dfrac{7}{15}$

$$15\overline{)7.000}$$ quotient 0.466
$$\begin{array}{r} 0.466 \\ 15\overline{)7.000} \\ \underline{6\,0} \\ 100 \\ \underline{90} \\ 100 \\ \underline{90} \\ 10 \end{array}$$

Thus $\dfrac{7}{15} = 0.4\overline{6}$ and $3\dfrac{7}{15} = 3.4\overline{6}$.

(b)
$$\begin{array}{r} 1.818 \\ 11\overline{)20.000} \\ \underline{11} \\ 9\,0 \\ \underline{8\,8} \\ 20 \\ \underline{11} \\ 90 \\ \underline{88} \\ 2 \end{array}$$

Thus $\dfrac{20}{11} = 1.818181\ldots = 1.\overline{81}$.

Practice Problem 3   Write as an equivalent decimal.

(a) $2\dfrac{11}{18}$ $2.6\overline{1}$

(b) $\dfrac{28}{27}$ $1.\overline{037}$

In some cases, the pattern of repeating is quite long. For example,

$$\frac{1}{7} = 0.142857142857\ldots = 0.\overline{142857}$$

Such problems are often rounded to a certain value.

EXAMPLE 4 Express $\frac{5}{7}$ as a decimal rounded to the nearest thousandth.

Solution

$$
\begin{array}{r}
0.7142 \\
7\overline{)5.0000} \\
\underline{4\,9} \\
10 \\
\underline{7} \\
30 \\
\underline{28} \\
20 \\
\underline{14} \\
6
\end{array}
$$

Rounding to the nearest thousandth, we round 0.7142 to 0.714. (In repeating form, $\frac{5}{7} = 0.714285714285\ldots = 0.\overline{714285}$.)

Practice Problem 4 Express $\frac{19}{24}$ as a decimal rounded to the nearest thousandth. 0.792

Recall that we studied placing two decimals in order in Section 3.2. If we are required to place a fraction and a decimal in order, it is usually easiest to change the fraction to decimal form and then compare the two decimals.

EXAMPLE 5 Fill in the blank with one of the symbols <, =, or >.

Solution

$$\frac{7}{16} \underline{\hspace{1cm}} 0.43$$

Now we divide to find the decimal equivalent of $\frac{7}{16}$.

$$
\begin{array}{r}
0.4375 \\
16\overline{)7.0000} \\
\underline{64} \\
60 \\
\underline{48} \\
120 \\
\underline{112} \\
80 \\
\underline{80} \\
0
\end{array}
$$

Teaching Example 4 Express $\frac{3}{14}$ as a decimal rounded to the nearest hundredth.

Ans: 0.21

NOTE TO STUDENT: Fully worked-out solutions to all of the Practice Problems can be found at the back of the text starting at page SP-1

Teaching Example 5 Fill in the blank with one of the symbols <, =, or >.

$$\frac{9}{11} \underline{\hspace{1cm}} 0.82$$

Ans: $\frac{9}{11} < 0.82$

Now in the thousandths place $7 > 0$, so we know

$$0.43\ 7\ 5 > 0.43\ 0\ 0.$$

Therefore, $\dfrac{7}{16} > 0.43$.

> **Practice Problem 5** Fill in the blank with one of the symbols $<$, $=$, or $>$.
>
> $$\dfrac{5}{8} \underline{\ <\ } 0.63$$

 Using the Order of Operations with Decimals

The rules for order of operations that we discussed in Section 1.6 apply to operations with decimals.

> **ORDER OF OPERATIONS**
>
> Do first | **1.** Perform operations inside parentheses.
> | **2.** Simplify any expressions with exponents.
> | **3.** Multiply or divide from left to right.
> Do last | **4.** Add or subtract from left to right.

Sometimes exponents are used with decimals. In such cases, we merely evaluate using repeated multiplication.

$$(0.2)^2 = 0.2 \times 0.2 = 0.04$$
$$(0.2)^3 = 0.2 \times 0.2 \times 0.2 = 0.008$$
$$(0.2)^4 = 0.2 \times 0.2 \times 0.2 \times 0.2 = 0.0016$$

EXAMPLE 6 Evaluate. $(0.3)^3 + 0.6 \times 0.2 + 0.013$

Solution First we need to evaluate $(0.3)^3 = 0.3 \times 0.3 \times 0.3 = 0.027$. Thus

$(0.3)^3 + 0.6 \times 0.2 + 0.013$

$= 0.027 + 0.6 \times 0.2 + 0.013$

$= 0.027 + 0.12 + 0.013$ ⟵ When addends have a different number of decimal places, writing the problem in column form makes adding easier.

$$\begin{array}{r} 0.027 \\ 0.120 \\ +\ 0.013 \\ \hline 0.160 \end{array}$$

$= 0.16$

> **Practice Problem 6** Evaluate. $0.3 \times 0.5 + (0.4)^3 - 0.036$ 0.178

In the next example all four steps of the rules for order of operations will be used.

EXAMPLE 7 Evaluate. $(8 - 0.12) \div 2^3 + 5.68 \times 0.1$

Solution

$(8 - 0.12) \div 2^3 + 5.68 \times 0.1$

$= 7.88 \div 2^3 + 5.68 \times 0.1$ First do subtraction inside the parentheses.

$= 7.88 \div 8 + 5.68 \times 0.1$ Simplify the expression with an exponent.

$= 0.985 + 0.568$ From left to right do division and multiplication.

$= 1.553$ Add the final two numbers.

Practice Problem 7 Evaluate. $6.56 \div (2 - 0.36) + (8.5 - 8.3)^2$ 4.04

Teaching Example 7 Evaluate.

$0.25 \div 5 - (0.2)^2 + (4 - 0.85)$

Ans: 3.16

NOTE TO STUDENT: Fully worked-out solutions to all of the Practice Problems can be found at the back of the text starting at page SP-1

Take the time to review these seven Examples and seven Practice Problems. This is an important skill to master. Some careful review will help you to work the homework exercises much more quickly and accurately.

Developing Your Study Skills

Keep Trying

We live in a highly technical world, and you cannot afford to give up on the study of mathematics. Dropping mathematics may prevent you from entering certain career fields that you may find interesting. You may not have to take math courses as high-level as calculus, but such courses as intermediate algebra, finite math, college algebra, and trigonometry may be necessary. Learning mathematics can open new doors for you.

Learning mathematics is a process that takes time and effort. You will find that regular study and daily practice are necessary to strengthen your skills and to help you grow academically. This process will lead you toward success in mathematics. Then, as you become more successful, your confidence in your ability to do mathematics will grow.

Verbal and Writing Skills

1. 0.75 and $\frac{3}{4}$ are different ways to express the __same quantity__.

2. To convert a fraction to an equivalent decimal, divide the __denominator__ into the numerator.

3. Why is $0.\overline{8942}$ called a repeating decimal?
The digits 8942 repeat.

4. The order of operations for decimals is the same as the order of operations for whole numbers. Write the steps for the order of operations.
1. Perform operations inside parentheses.
2. Simplify any expressions with exponents.
3. Multiply or divide from left to right.
4. Add or subtract from left to right.

Write as an equivalent decimal. If a repeating decimal is obtained, use notation such as $0.\overline{7}$, $0.\overline{16}$, or $0.\overline{245}$.

5. $\frac{1}{4}$ 0.25

6. $\frac{3}{4}$ 0.75

7. $\frac{4}{5}$ 0.8

8. $\frac{2}{5}$ 0.4

9. $\frac{1}{8}$ 0.125

10. $\frac{3}{8}$ 0.375

11. $\frac{7}{20}$ 0.35

12. $\frac{3}{40}$ 0.075

13. $\frac{31}{50}$ 0.62

14. $\frac{23}{25}$ 0.92

15. $\frac{9}{4}$ 2.25

16. $\frac{14}{5}$ 2.8

17. $2\frac{7}{8}$ 2.875

18. $3\frac{13}{16}$ 3.8125

19. $5\frac{3}{16}$ 5.1875

20. $2\frac{5}{12}$ 2.41$\overline{6}$

21. $\frac{2}{3}$ 0.$\overline{6}$

22. $\frac{5}{6}$ 0.8$\overline{3}$

23. $\frac{5}{11}$ 0.$\overline{45}$

24. $\frac{7}{11}$ 0.$\overline{63}$

25. $3\frac{7}{12}$ 3.58$\overline{3}$

26. $7\frac{1}{3}$ 7.$\overline{3}$

27. $4\frac{2}{9}$ 4.$\overline{2}$

28. $8\frac{7}{9}$ 8.$\overline{7}$

Write as an equivalent decimal or a decimal approximation. Round your answer to the nearest thousandth if needed.

29. $\frac{4}{13}$ 0.308

30. $\frac{8}{17}$ 0.471

31. $\frac{19}{21}$ 0.905

32. $\frac{20}{21}$ 0.952

33. $\frac{7}{48}$ 0.146

34. $\frac{5}{48}$ 0.104

35. $\frac{57}{28}$ 2.036

36. $\frac{15}{7}$ 2.143

37. $\frac{21}{52}$ 0.404

38. $\frac{1}{38}$ 0.026

39. $\frac{17}{18}$ 0.944

40. $\frac{5}{13}$ 0.385

41. $\frac{22}{7}$ 3.143

42. $\frac{17}{14}$ 1.214

43. $3\frac{9}{19}$ 3.474

44. $4\frac{11}{17}$ 4.647

Fill in the blank with one of the symbols $<$, $=$, or $>$.

45. $\frac{7}{8}$ $<$ 0.88

46. $\frac{10}{11}$ $>$ 0.9

47. 0.07 $>$ $\frac{1}{16}$

48. 0.9 $<$ $\frac{15}{16}$

Applications

49. *New York Stock Exchange* One day in February 2007, the value of one share of Ann Taylor stock decreased by $\frac{7}{25}$ of a dollar. Write the amount of decrease as a decimal. 0.28

50. *New York Stock Exchange* One day in February 2007, the value of one share of DuPont stock increased by $\frac{19}{50}$ of a dollar. Write the amount of increase as a decimal. 0.38

51. *U.S. Women's Shoe Sizes* A size 7 women's shoe measures 9.31 inches and a size $7\frac{1}{2}$ measures $9\frac{1}{2}$ inches. What is the difference in length between a size 7 and a size $7\frac{1}{2}$ shoe? 0.19 inch

52. *U.S. Men's Shoe Sizes* A size $9\frac{1}{2}$ men's shoe measures $10\frac{1}{2}$ inches and a size 10 measures 10.69 inches. What is the difference in length between a size $9\frac{1}{2}$ and a size 10 shoe? 0.19 inch

53. *Safety Regulations* Federal safety regulations specify that the slots between the bars on a baby's crib must not be more than $2\frac{3}{8}$ inches. One crib's slots measured 2.4 inches apart. Is this too wide? If so, by how much?
yes; it is 0.025 inch too wide.

54. *Manufacturing* To manufacture a circuit board, Rick must program a computer to place a piece of thin plastic atop a circuit board. For the current to flow through the circuit, the top plastic piece must form a border of exactly $\frac{1}{16}$ inch with the circuit board. A few circuit boards were made with a border of 0.055 inch by accident. Is this border too small or too large? By how much? too small; 0.0075 inch

Evaluate.

55. $2.4 + (0.5)^2 - 0.35$
$2.4 + 0.25 - 0.35 = 2.3$

56. $9.6 + 3.6 - (0.4)^2$
$9.6 + 3.6 - 0.16 = 13.04$

57. $2.3 \times 3.2 - 5 \times 0.8$
$7.36 - 4 = 3.36$

58. $9.6 \div 3 + 0.21 \times 6$
$3.2 + 1.26 = 4.46$

59. $12 \div 0.03 - 50 \times (0.5 + 1.5)^3$
$400 - 400 = 0$

60. $61.95 \div 1.05 - 2 \times (1.7 + 1.3)^3$
$59 - 54 = 5$

61. $(1.1)^3 + 2.6 \div 0.13 + 0.083$
$1.331 + 20 + 0.083 = 21.414$

62. $(1.1)^3 + 8.6 \div 2.15 - 0.086$
$1.331 + 4 - 0.086 = 5.245$

63. $(14.73 - 14.61)^2 \div (1.18 + 0.82)$
$0.0144 \div 2 = 0.0072$

64. $(32.16 - 32.02)^2 \div (2.24 + 1.76)$
$0.0196 \div 4 = 0.0049$

65. $(0.5)^3 + (3 - 2.6) \times 0.5$
0.325

66. $(0.6)^3 + (7 - 6.3) \times 0.07$
0.265

67. $(0.76 + 4.24) \div 0.25 + 8.6$
28.6

68. $(2.4)^2 + 3.6 \div (1.2 - 0.7)$
12.96

Evaluate.

69. $(1.6)^3 + (2.4)^2 + 18.666 \div 3.05 + 4.86$
     $4.096 + 5.76 + 6.12 + 4.86 = 20.836$

70. $5.9 \times 3.6 \times 2.4 - 0.1 \times 0.2 \times 0.3 \times 0.4$
     $50.976 - 0.0024 = 50.9736$

Write as a decimal. Round your answer to six decimal places.

71. $\dfrac{5236}{8921}$
     0.586930

72. $\dfrac{17,359}{19,826}$
     0.875567

To Think About

73. Subtract. $0.\overline{16} - 0.00\overline{16}$

   (a) What do you obtain?

   (b) Now subtract $0.\overline{16} - 0.01\overline{6}$. What do you obtain?

   (c) What is different about these results?

(a) $0.1616\overline{16}$
   $\underline{-\ 0.001616}$
   0.16

(b) $0.1616\overline{16}$
   $\underline{-\ 0.016666}$
   $0.1449\overline{49}$

(c) (b) is a repeating and (a) is a nonrepeating decimal.

74. Subtract. $1.\overline{89} - 0.01\overline{89}$

   (a) What do you obtain?

   (b) Now subtract $1.\overline{89} - 0.18\overline{9}$. What do you obtain?

   (c) What is different about these results?

(a) $1.8989\overline{89}$
   $\underline{-\ 0.018989}$
   1.88

(b) 1.89898989
   $\underline{-\ 0.18999999}$
   1.70898989

(c) (b) is a repeating and (a) is a nonrepeating decimal.

Cumulative Review

75. **[2.9.1]** *Boating Dock* John and Nancy put in a new dock at the end of Tobey Lane. A pipe at the end of the dock supports the dock and is driven deep into the mud and sand at the bottom of Eel Pond. The pipe is 25 feet long. Half of the pipe is above the surface of the water at low tide. The pipe is driven $6\frac{3}{4}$ feet deep into the mud and sand. How deep is the water at the end of the dock at low tide?

$5\dfrac{3}{4}$ feet deep

76. **[2.9.1]** *Tidal Fluctuation* Fisherman's Wharf in Digby, Nova Scotia, has an average tidal range of $25\frac{4}{5}$ feet. These huge tidal ranges require considerable ingenuity in the design of docks and ramps for boats. If the water is $6\frac{1}{2}$ feet deep at low tide at the end of Fisherman's Wharf during an average low tide, how deep is the water at the same location during an average high tide? (*Source:* Nova Scotia Board of Tourism)

$32\dfrac{3}{10}$ feet

Quick Quiz 3.6

1. Write as an equivalent decimal. $3\dfrac{9}{16}$ 3.5625

2. Write as an equivalent decimal. Round your answer to the nearest hundredth. $\dfrac{5}{17}$ 0.29

3. Perform the operations in the proper order.
   $(0.7)^2 + 1.92 \div 0.3 - 0.79$ 6.1

4. **Concept Check** Explain how you would perform the operations in the calculation
   $45.78 - (3.42 - 2.09)^2 \times 0.4.$ Answers may vary

Classroom Quiz 3.6 You may use these problems to quiz your students' mastery of Section 3.6.

1. Write as an equivalent decimal. $4\dfrac{7}{16}$ **Ans:** 4.4375

2. Write as an equivalent decimal. Round your answer to the nearest thousandth. $\dfrac{13}{18}$ **Ans:** 0.722

3. Perform the operations in the proper order.
   $(0.6)^2 + 0.82 \div 0.2 - 1.93$ **Ans:** 2.53

3.7 ESTIMATING AND SOLVING APPLIED PROBLEMS INVOLVING DECIMALS

① Estimating Sums, Differences, Products, and Quotients of Decimals

When we encounter real-life applied problems, it is important to know if an answer is reasonable. A car may get 21.8 miles per gallon. However, a car will not get 218 miles per gallon. Neither will a car get 2.18 miles per gallon. To avoid making an error in solving applied problems, it is wise to make an estimate. The most useful time to make an estimate is at the end of solving the problem, in order to see if the answer is reasonable.

There are several different rules for estimating. Not all mathematicians agree what is the best method for estimating in each case. Most students find that a very quick and simple method to estimate is to round each number so that there is one nonzero digit. Then perform the calculation. We will use that approach in this section of the book. However, you should be aware that there are other valid approaches. Your instructor may wish you to use another method.

Student Learning Objectives

After studying this section, you will be able to:

 Estimate sums, differences, products, and quotients of decimals.

 Solve applied problems using operations with decimals.

EXAMPLE 1 Estimate.

(a) 184,987.09 + 676,393.95

(b) 0.00782 − 0.00358

(c) 145.87 × 78.323

(d) 138.85 ÷ 5.887

Solution In each case we will round to one nonzero digit to estimate.

(a) 184,987.09 + 676,393.95 ≈ 200,000 + 700,000 = 900,000

(b) 0.00782 − 0.00358 ≈ 0.008 − 0.004 = 0.004

(c) 145.87 × 78.323 ≈

$$
\begin{array}{r}
100 \\
\times\ \ 80 \\
\hline
8000
\end{array}
$$

Thus 145.87 × 78.323 ≈ 8000

(d) $138.85 \div 5.887 \approx 6)\overline{100}$

$$
\begin{array}{r}
16 \\
6)\overline{100} \\
\underline{6} \\
40 \\
\underline{36} \\
4
\end{array}
$$

$138.85 \div 5.887 \approx 6)\overline{100} = 16\frac{4}{6} \approx 17$

Thus 138.85 ÷ 5.887 ≈ 17

Here we round the answer to the nearest whole number.

Practice Problem 1 Round to one nonzero digit. Then estimate the result of the indicated calculation.

(a) 385.98 + 875.34   1300

(b) 0.0932 − 0.0579   0.03

(c) 5876.34 × 0.087   540

(d) 46,873 ÷ 8.456   6250

Teaching Example 1 Estimate.

(a) 183.47 + 736.1

(b) 0.00367 − 0.00218

(c) 6.4978 × 8.05534

(d) 578.01 ÷ 12.539

Ans: **(a)** 900   **(b)** 0.002   **(c)** 48   **(d)** 60

NOTE TO STUDENT: *Fully worked-out solutions to all of the Practice Problems can be found at the back of the text starting at page SP-1*

Take a few minutes to review Example 1. Be sure you can perform these estimation steps. We will use this type of estimation to check our work in the applied problems in this section.

 ## Solving Applied Problems Using Operations with Decimals

We use the basic plan of solving applied problems that we discussed in Section 1.8 and Section 2.9. Let us review how we analyze applied-problem situations.

1. *Understand the problem.*
2. *Solve and state the answer.*
3. *Check.*

In the United States for almost all jobs where you are paid an hourly wage, if you work more than 40 hours in one week, you should be paid overtime. The overtime rate is 1.5 times the normal hourly rate, for the extra hours worked in that week. The next problem deals with overtime wages.

Teaching Example 2 The Speedy Delivery Company charges $1.30 per pound for the first 5 pounds and 1.5 times that rate for every pound over 5 pounds. A medical supply company wants to ship a 6.2-pound package. How much will the shipping cost be?

Ans: $8.84

EXAMPLE 2 A laborer is paid $7.38 per hour for a 40-hour week and 1.5 times that wage for any hours worked beyond the standard 40. If he works 47 hours in a week, what will he earn?

Solution

1. *Understand the problem.*

Mathematics Blueprint for Problem Solving

Gather the Facts	What Am I Asked to Do?	How Do I Proceed?	Key Points to Remember
He works 47 hours. He gets paid $7.38 per hour for 40 hours. He gets paid 1.5 × $7.38 per hour for 7 hours.	Find the earnings of the laborer if he works 47 hours in one week.	Add the earnings of 40 hours at $7.38 per hour to the earnings of 7 hours at overtime pay.	Multiply 1.5 × $7.38 to find the pay he earns for overtime.

2. *Solve and state the answer.*

We want to compute his regular pay and his overtime pay and add the results.

$$\text{Regular pay} + \text{Overtime pay} = \text{Total pay}$$

Regular pay: Calculate his pay for 40 hours of work.

$$\begin{array}{r} 7.38 \\ \times\ \ 40 \\ \hline 295.20 \end{array}$$

He earns $295.20 at $7.38 per hour.

Overtime pay: Calculate his overtime pay rate. This is 7.38 × 1.5.

$$
\begin{array}{r}
7.38 \\
\times\ 1.5 \\
\hline
3\ 690 \\
7\ 38\ \\
\hline
11.070
\end{array}
$$

He earns $11.07 per hour in overtime.

Calculate how much he earned doing 7 hours of overtime work.

$$
\begin{array}{r}
11.07 \\
\times\qquad 7 \\
\hline
77.49
\end{array}
$$

For 7 overtime hours he earns $77.49.

Total pay: Add the two amounts.

$$
\begin{array}{r}
\$295.20 \\
+\quad 77.49 \\
\hline
\$372.69
\end{array}
$$

Regular 40-hour-week earnings
Overtime earnings
Total earnings

The total earnings of the laborer for a 47-hour workweek will be $372.69.

3. Check.

Estimate his regular pay.

$$40 \times \$7 = \$280$$

Estimate his overtime rate of pay, and then his overtime pay.

$$2 \times \$7 = \$14$$
$$7 \times \$10 = \$70$$

Then add.

$$
\begin{array}{r}
\$280 \\
+\quad 70 \\
\hline
\$350
\end{array}
$$

Our estimate of $350 is close to our answer of $372.69. Our answer is reasonable. ✓

NOTE TO STUDENT: *Fully worked-out solutions to all of the Practice Problems can be found at the back of the text starting at page SP-1*

Practice Problem 2 Melinda works for the phone company as a line repair technician. She earns $9.36 per hour. She worked 51 hours last week. If she gets time and a half for all hours worked above 40 hours per week, how much did she earn last week? $528.84

Teaching Example 3 A lumber company has a stock of boards that are each 4.2 meters long. A customer needs boards that are 0.84 meter long.

(a) How many 0.84-meter boards can be cut from each 4.2-meter board?

(b) If the boards are to be sold for $2.63 per meter, how much will each of the smaller boards cost? (Round to the nearest cent.)

Ans: (a) 5 **(b)** $2.21

EXAMPLE 3 A chemist is testing 36.85 liters of cleaning fluid. She wishes to pour it into several smaller containers that each hold 0.67 liter of fluid. (a) How many containers will she need? (b) If each liter of this fluid costs $3.50, how much does the cleaning fluid in one container cost? (Round your answer to the nearest cent.)

Mathematics Blueprint for Problem Solving

Gather the Facts	What Am I Asked to Do?	How Do I Proceed?	Key Points to Remember
The total amount of cleaning fluid is 36.85 liters. Each small container holds 0.67 liter. Each liter of fluid costs $3.50.	(a) Find out how many containers the chemist needs. (b) Find the cost of cleaning fluid in each small container.	(a) Divide the total, 36.85 liters, by the amount in each small container, 0.67 liter, to find the number of containers. (b) Multiply the cost of one liter, $3.50, by the amount of liters in one container, 0.67.	If you are not clear as to what to do at any stage of the problem, then do a similar, simpler problem.

Solution

(a) How many containers will the chemist need?

She has 36.85 liters of cleaning fluid and she wants to put it into several equal-sized containers each holding 0.67 liter. Suppose we are not sure what to do. Let's do a similar, simpler problem. If we had 40 liters of cleaning fluid and we wanted to put it into little containers each holding 2 liters, what would we do? Since the little containers would only hold 2 liters, we would need 20 containers. We know that $40 \div 2 = 20$. So we see that, in general, we divide the total number of liters by the amount in the small container. Thus $36.85 \div 0.67$ will give us the number of containers in this case.

$$
\begin{array}{r}
55. \\
0.67_\wedge \overline{)36.85_\wedge} \\
\underline{33\ 5} \\
3\ 35 \\
\underline{3\ 35}
\end{array}
$$

The chemist will need 55 containers to hold this amount of cleaning fluid.

(b) How much does the cleaning fluid in each container cost? Each container will hold only 0.67 liter. If one liter costs $3.50, then to find the cost of one container we multiply $0.67 \times \$3.50$.

$$
\begin{array}{r}
3.50 \\
\times\ 0.67 \\
\hline
2450 \\
2100 \\
\hline
2.3450
\end{array}
$$

We round our answer to the nearest cent. Thus each container would cost $2.35.

Check.

(a) Is it really true that 55 containers each holding 0.67 liter will hold a total of 36.85 liters? To check, we multiply.

$$
\begin{array}{r}
55 \\
\times\ 0.67 \\
\hline
385 \\
330 \\
\hline
36.85 \quad \checkmark
\end{array}
$$

(b) One liter of cleaning fluid costs $3.50. We would expect the cost of 0.67
    liter to be less than $3.50. $2.35 is less than $3.50. ✓
        We use estimation to check more closely.

$$
\begin{array}{rcr}
\$3.50 & \longrightarrow & \$4.00 \\
\times \quad 0.67 & \longrightarrow & \times \quad 0.7 \\
\hline
 & & \$2.800
\end{array}
$$

$2.80 is fairly close to $2.35. Our answer is reasonable. ✓

Practice Problem 3 A butcher divides 17.4 pounds of prime steak
into small equal-sized packages. Each package contains 1.45 pounds of
prime steak. (a) How many packages of steak will he have? (b) Prime
steak sells for $4.60 per pound. How much will each package of prime
steak cost? **(a)** 12   **(b)** $6.67

*NOTE TO STUDENT: Fully worked-out
solutions to all of the Practice Problems
can be found at the back of the text
starting at page SP-1*

MyMathLab   PRACTICE   WATCH   DOWNLOAD   READ   REVIEW

In exercises 1–10, first round each number to one nonzero digit. Then perform the calculation using the rounded numbers to obtain an estimate.

1. 238,598,980 + 487,903,870 700,000,000

2. 5,927,000 + 9,983,000 16,000,000

3. 56,789.345 − 33,875.125 30,000

4. 6949.45 − 1432.88 6000

5. 12,638 × 0.7892 8000

6. 47,225 × 0.463 25,000

7. 879.654 ÷ 56.82 15

8. 34.5684 ÷ 0.55 50

9. *Car Sales* Last year the sales of Honda Accords at Hopkins Honda totaled $11,760,770. If this represented a purchase of 483 Accords, estimate the average price per car. $20,000

10. *Boat Sales* Last year the sales of boats in Massachusetts totaled $865,987,273.45. If this represented a purchase of 55,872 boats, estimate the average price per boat. $15,000

Applications *Estimate an answer to each of the following by rounding each number first, then perform the actual calculation.*

11. *Currency Conversion* Kristy is taking a trip to Denmark. Before she leaves, she checks the newspaper and finds that every U.S. dollar is equal to 5.68 kroner (Danish currency). If Kristy takes $525 on her trip, how many kroner will she receive when she does the exchange? 2982 kroner

▲ 12. *Football Field Dimensions* The dimensions of a professional football field, including the end zones, are about 48.8 meters wide by 109.7 meters long. What is the area of a professional football field? 5353.36 square meters

▲ 13. *Geometry* Juan and Gloria are having their roof reshingled and need to determine its area in square feet. The dimensions of the roof are 48.3 feet by 56.9 feet. What is the area of the roof in square feet? 2748.27 square feet

14. *Baby Formula* A large can of infant formula contains 808 grams of powder. To prepare a bottle, 35.2 grams are needed. How many bottles can be prepared from the can? Round to the nearest whole number. about 23 bottles

15. *Cooking* Hans is making gourmet chocolate in Switzerland. He has 11.52 liters of liquid white chocolate that will be poured into molds that hold 0.12 liter each. How many individual molds can Hans make with his 11.52 liters of liquid white chocolate? 96 molds

16. *Food Purchase* David bought MacIntosh apples and Anjou pears at the grocery store for a fruit salad. At the checkout counter, the apples weighed 2.7 pounds and the pears weighed 1.8 pounds. If the apples cost $1.29 per pound and the pears cost $1.49 per pound, how much did David spend on fruit? (Round your answer to the nearest cent.) $6.17

17. ***Hawaii Rainfall*** One year in Mount Waialeale, Hawaii, considered the "rainiest place in the world," the yearly rainfall totaled 11.68 meters. The next year, the yearly rainfall on this mountain totaled 10.42 meters. The third year it was 12.67 meters. On average, how much rain fell on Mount Waialeale, Hawaii, per year? 11.59 meters

18. ***Auto Travel*** Emma and Jennie took a trip in their Ford Taurus from Saskatoon, Saskatchewan, to Calgary, Alberta, in Canada to check out the glacier lakes. When they left, their odometer read 54,089. When they returned home, the odometer read 55,401. They used 65.6 gallons of gas. How many miles per gallon did they get on the trip?

20 miles per gallon

19. ***Food Portions*** A jumbo bag of potato chips contains 18 ounces of chips. The recommended serving is 0.75 ounce. How many servings are in the jumbo bag? 24 servings

20. ***Telephone Costs*** Sylvia's telephone company offers a special rate of $0.23 per minute on calls made to the Philippines during certain parts of the day. If Sylvia makes a 28.5-minute call to the Philippines at this special rate, how much will it cost? $6.56

21. ***Consumer Mathematics*** The local Police Athletic League raised enough money to renovate the local youth hall and turn it into a coffeehouse/activity center so that there is a safe place to hang out. The room that holds the Ping-Pong table needs 43.9 square yards of new carpeting. The entryway needs 11.3 square yards, and the stage/seating area needs 63.4 square yards. The carpeting will cost $10.65 per square yard. What will be the total bill for carpeting these three areas of the coffeehouse? $1263.09

22. ***Painting Costs*** Kevin has a job as a house painter. One family needs its kitchen, family room, and hallway painted. The respective amounts needed are 2.7 gallons, 3.3 gallons, and 1.8 gallons. If paint costs $7.40 per gallon, how much will Kevin need to spend on paint to do the job? $57.72

23. ***Overtime Pay*** Lucy earns $8.50 per hour at the neighborhood café. She earns time and a half (1.5 times the hourly wage) for each hour she works on a holiday. Lucy worked eight hours each day for six days, then worked eight hours on New Year's Day. How much did she earn for that week? $510

24. ***Electrician's Pay*** An electrician is paid $14.30 per hour for a 40-hour week. She is paid time and a half for overtime (1.5 times the hourly wage) for every hour more than 40 hours worked in the same week. If she works 48 hours in one week, what will she earn for that week? $743.60

25. ***Rainforest Loss*** In 1997, Brazil had 2.943 million square kilometers of rainforest. Each year, approximately 0.018 million square kilometer is lost to deforestation and development. By 2007, how many square kilometers of rainforest remained in Brazil? (*Source:* www.geography.ndo. co.uk) 2.763 million or 2,763,000 square kilometers

26. ***Consumer Mathematics*** At the beginning of each month, Raul withdraws $100 for small daily purchases. This month he spent $18.50 on bus fares, $42.75 on coffee and snacks, and $21.25 on news magazines. How much did Raul have left at the end of the month? $17.50

27. *Car Payments* Charlie borrowed $11,500 to purchase a new car. His loan requires him to pay $288.65 each month over the next 60 months (five years). How much will he pay over the five years? How much more will he pay back than the amount of the loan? $17,319; $5819

28. *House Payments* Mel and Sally borrowed $140,000 to buy their new home. They make monthly payments to the bank of $764.35 to repay the loan. They will be making these payments for the next 30 years. How much money will they pay to the bank in the next 30 years? How much more will they pay back than they borrowed?
$275,166; $135,166

29. *Drinking Water Safety* The EPA standard for safe drinking water is a maximum of 1.3 milligrams of copper per liter of water. A study was conducted on a sample of 7 liters of water drawn from Jeff Slater's house. The analysis revealed 8.06 milligrams of copper in the sample. Is the water safe or not? By how much?
yes; by 0.149 milligram per liter

30. *Drinking Water Safety* The EPA standard for safe drinking water is a maximum of 0.015 milligram of lead per liter of water. A study was conducted on 6 liters of water from West Towers Dormitory. The analysis revealed 0.0795 milligram of lead in the sample. Is the water safe or not? By how much?
yes; by 0.00175 milligram per liter

31. *Jet Travel* A jet fuel tank containing 17,316.8 gallons is being emptied at the rate of 126.4 gallons per minute. How many minutes will it take to empty the tank?   137 minutes

32. *Monopoly Game* In a New Jersey mall, the average price of a Parker Brothers Monopoly game is $11.50. The Alfred Dunhill Company made a special commemorative set for $25,000,000.00. Instead of plastic houses and hotels, you can buy and trade gold houses and silver hotels! How many regular Monopoly games could you purchase for the price of one special commemorative set?   2,173,913 games

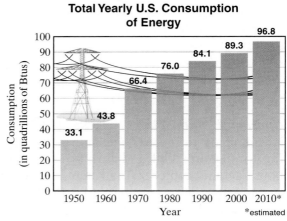

Source: U.S. Department of Energy

Energy Consumption *Use the bar graph to answer exercises 33–36.*

33. How many more Btu were consumed in the United States during 2000 than in 1970?
22.9 quadrillion Btu

34. What was the greatest increase in consumption of energy in a 10-year period? When did it occur?
22.6 quadrillion Btu; from 1960 to 1970

35. What was the average consumption of energy per year in the United States for the years 1950, 1960, and 1970? Write your answer in quadrillion Btu and then write your answer in Btu. (Remember that a quadrillion is 1000 trillion.)
approximately 47.8 quadrillion Btu;
47,800,000,000,000,000 Btu

36. What will be the average consumption of energy per year in the United States for the years 1990, 2000, and 2010? Write your answer in quadrillion Btu and then write your answer in Btu. (Remember that a quadrillion is 1000 trillion.)
approximately 90.1 quadrillion Btu;
90,100,000,000,000,000 Btu

Cumulative Review *Calculate.*

37. [2.8.3] $\dfrac{4}{7} + \dfrac{1}{2} \times \dfrac{2}{3}$

$\dfrac{19}{21}$

38. [2.8.3] $\dfrac{3}{19} + \dfrac{5}{38} - \dfrac{2}{19}$

$\dfrac{7}{38}$

39. [2.4.1] $\dfrac{7}{25} \times \dfrac{15}{42}$

$\dfrac{1}{10}$

40. [2.5.3] $2\dfrac{2}{3} \div \dfrac{1}{3}$

8

Quick Quiz 3.7

1. The rainfall for Springfield last year was 1.23 inches in March, 2.58 inches in April, and 3.67 inches in May. Normally that city gets 8.5 inches during those three months. How much less rain was received during those three months compared to the normal rainfall amount?   1.02 inches

2. Melissa and Phil started on a trip to the mountains with their Honda CRV. Their odometer read 87,569.2 miles at the start of the trip and 87,929.2 miles at the end of the trip. They used 15.5 gallons of gas on the trip. How many miles per gallon did they achieve with their car? (Round to the nearest tenth.)   23.2 miles per gallon

3. Chris Smith is making car payments of $275.50 for the next 36 months to pay off a car loan for a new Saturn. He borrowed $8000 from a bank to purchase the car.

   How much will he make in car payments over the next three years? How much more will he pay back than the original amount of the loan? $9918; $1918

4. **Concept Check** Explain how you would solve the following problem. The Classic Chocolate Company has 24.7 pounds of chocolate. They wish to place them in individual boxes that each hold 1.3 pounds of chocolate. How many boxes will they need?   Answers may vary

Classroom Quiz 3.7   You may use these problems to quiz the students' mastery of Section 3.7.

1. The snowfall in Pine City last year was 22.5 inches in December, 32.7 inches in January, and 26.9 inches in February. Normally, Pine City would receive 90.5 inches of snow during those three months. How much less snow was received than would normally be expected for that time period?   **Ans:** 8.9 inches

2. Greg and Marcia took a trip in their Dodge Caravan to Honeyrock camp. The odometer read 45,678.2 miles at the start of the trip and 46,228.2 miles at the end of the trip. They obtained 24 miles per gallon on the trip. How many gallons of gas did they use? (Round your answer to the nearest tenth.)   **Ans:** 22.9 gallons

3. Joel worked last week for 40 hours at his normal pay rate of $9.50 per hour. Then he worked overtime for 17 hours and was paid 1.5 times his normal pay rate for the overtime hours. How much did he earn last week?   **Ans:** $622.25

Putting Your Skills to Work: Use Math to Save Money

GAS PRICES

It's July 2008 in Stockton, California, and Sam needs to put gas in his car. He is on a street that has an ARCO gas station and a SHELL station. Sam will use his debit/credit card to pay for the gas. The ARCO station is charging $4.43 per gallon of gas while the SHELL station is charging $4.55 per gallon.

   If Sam's goal is to save money it would seem obvious that he should go to ARCO, right? But Sam knows from experience it's not that simple.

He knows that ARCO will charge an extra $0.45 as an "ATM Transaction Fee" in addition to the gas he buys.

1. If Sam plans on buying just **one gallon** of gas, which gas station should he choose?   SHELL

2. If Sam plans on buying **three gallons** of gas, which gas station should he choose?   SHELL

3. If Sam plans on buying **four gallons** of gas, which gas station should he choose?   ARCO

4. If Sam plans on buying **ten gallons** of gas, which gas station should he choose?   ARCO

5. How many gallons of gas would Sam need to buy for the cost to be **exactly the same** at the two gas stations? Consider the results of Question 2 and Question 3 when formulating your answer.   3.75 gallons

6. Does the station where you normally get gas charge the same for cash or credit?   Answers may vary

7. Do you know if the station charges an "ATM transaction fee"?   Answers may vary

8. Has the increase in gas prices caused you to change your driving habits? If so, please explain.   Answers may vary

Topic	Procedure	Examples
Word names for decimals, p. 196.	 Hundreds — 3 Tens — 4 Ones — 1 Decimal point — . Tenths — 6 Hundredths — 7 Thousandths — 8 Ten-thousandths — 3	The word name for 341.6783 is three hundred forty-one and six thousand seven hundred eighty-three ten-thousandths.
Writing a decimal as a fraction, p. 197.	**1.** Read the decimal in words. **2.** Write it in fraction form. **3.** Reduce if possible.	Write 0.36 as a fraction. **1.** 0.36 is read "thirty-six hundredths." **2.** Write the fractional form. $\frac{36}{100}$ **3.** Reduce. $\frac{36}{100} = \frac{9}{25}$

Determining which of two decimals is larger, p. 201.	**1.** Start at the left and compare corresponding digits. Write in extra zeros if needed.
	2. When two digits are different, the larger number is the one with the larger digit.
		Which is larger? 0.138 or 0.13
		  0.138 ? 0.130
		  8 > 0
		So 0.138 > 0.130.

Rounding decimals, p. 203.	**1.** Locate the place (units, tenths, hundredths, etc.) to which rounding is required. **2.** If the first digit to the right of the given place value is less than 5, drop it and all the digits to the right of it. **3.** If the first digit to the right of the given place value is 5 or greater, increase the number in the given place value by one. Drop all digits to the right.	Round to the nearest hundredth: 0.8652 0.87 Round to the nearest thousandth: 0.21648 0.216
Adding and subtracting decimals, p. 207.	**1.** Write the numbers vertically and line up the decimal points. Extra zeros may be written to the right of the decimal points after the nonzero digits if needed. **2.** Add or subtract all the digits with the same place value, starting with the right column, moving to the left. Use carrying or borrowing as needed. **3.** Place the decimal point of the result in line with the decimal points of all the numbers added or subtracted.	Add. $36.3 + 8.007 + 5.26$ $\begin{array}{r} \overset{1}{3}6.300 \\ 8.007 \\ + \ 5.260 \\ \hline 49.567 \end{array}$ Subtract. $82.5 - 36.843$ $\begin{array}{r} 82.500 \\ - \ 36.843 \\ \hline 45.657 \end{array}$
Multiplying decimals, p. 217.	**1.** Multiply the numbers just as you would multiply whole numbers. **2.** Find the sum of the number of decimal places in the two factors. **3.** Place the decimal point in the product so that the product has the same number of decimal places as the sum in step 2. You may need to insert zeros to the left of the number found in step 1.	Multiply. $\begin{array}{r} 0.2 \\ \times \ 0.6 \\ \hline 0.12 \end{array}$ $\begin{array}{r} 0.3174 \\ \times \quad 0.8 \\ \hline 0.25392 \end{array}$ $\begin{array}{r} 0.0064 \\ \times \quad 0.21 \\ \hline 64 \\ 128 \\ \hline 0.001344 \end{array}$ $\begin{array}{r} 1364 \\ \times \quad 0.7 \\ \hline 954.8 \end{array}$
Multiplying a decimal by a power of 10, p. 219.	Move the decimal point to the right the same number of places as there are zeros in the power of 10 or the same number of places as the exponent on the 10. (Sometimes it is necessary to write extra zeros before placing the decimal point in the answer.)	Multiply. $5.623 \times 10 = 56.23$ $0.597 \times 10^4 = 5970$ $0.0082 \times 1000 = 8.2$ $0.075 \times 10^6 = 75,000$ $28.93 \times 10^2 = 2893$
Dividing by a decimal, p. 227.	**1.** Make the divisor a whole number by moving the decimal point to the right. Mark that position with a caret ($\wedge$). **2.** Move the decimal point in the dividend to the right the same number of places. Mark that position with a caret. **3.** Place the decimal point of your answer directly above the caret in the dividend. **4.** Divide as with whole numbers.	Divide. **(a)** $0.06\overline{)0.162}$ **(b)** $0.003\overline{)85.8}$ **(a)** $\begin{array}{r} 2.7 \\ 0.06_\wedge\overline{)0.16_\wedge 2} \\ \underline{12} \\ 42 \\ \underline{42} \\ 0 \end{array}$ **(b)** $\begin{array}{r} 28\,600. \\ 0.003_\wedge\overline{)85.800_\wedge} \\ \underline{6} \\ 25 \\ \underline{24} \\ 18 \\ \underline{18} \\ 0 \end{array}$

Topic	Procedure	Examples
Converting a fraction to a decimal, p. 234.	Divide the denominator into the numerator until 1. the remainder is zero, or 2. the decimal repeats itself, or 3. the desired number of decimal places is achieved.	Find the decimal equivalent. (a) $\dfrac{13}{22}$ (b) $\dfrac{5}{7}$, rounded to the nearest ten-thousandth (c) $\dfrac{13}{22}$ $$\begin{array}{r} 0.5909 \\ 22\overline{)13.0000} \\ 110 \\ \hline 200 \\ 198 \\ \hline 200 \\ 198 \\ \hline 2 \end{array}$$ (a) $$\begin{array}{r} 0.71428 \\ 7\overline{)5.00000} \\ 49 \\ \hline 10 \\ 7 \\ \hline 30 \\ 28 \\ \hline 20 \\ 14 \\ \hline 60 \\ 56 \\ \hline 4 \end{array}$$ (b) 0.71428 rounded to the nearest ten-thousandth is 0.7143. (c) $\dfrac{13}{22} = 0.5\overline{90}$ or $0.5909090\ldots$
Order of operations with decimal numbers, p. 238.	Same as order of operations of whole numbers. 1. Perform operations inside parentheses. 2. Simplify any expressions with exponents. 3. Multiply or divide from left to right. 4. Add or subtract from left to right.	Evaluate. $(0.4)^3 + 1.26 \div 0.12 - 0.12 \times (1.3 - 1.1)$ $= (0.4)^3 + 1.26 \div 0.12 - 0.12 \times 0.2$ $= 0.064 + 1.26 \div 0.12 - 0.12 \times 0.2$ $= 0.064 + 10.5 - 0.024$ $= 10.564 - 0.024$ $= 10.54$

Procedure for Solving Applied Problems

Using the Mathematics Blueprint for Problem Solving, p. 244

In solving a real-life problem with decimals, students may find it helpful to complete the following steps. You will not use all the steps all of the time. Choose the steps that best fit the conditions of the problem.

1. Understand the problem.

   (a) Read the problem carefully.

   (b) Draw a picture if it helps you visualize the situation. Think about what facts you are given and what you are asked to find.

   (c) It may help to write a similar, simpler problem to get started and to determine what operation to use.

   (d) Use the Mathematics Blueprint for Problem Solving to organize your work. Follow these four parts.

      1. Gather the Facts (Write down specific values given in the problem.)

      2. What Am I Asked to Do? (Identify what you must obtain for an answer.)

      3. How Do I Proceed? (Determine what calculations need to be done.)

      4. Key Points to Remember (Record any facts, warnings, formulas, or concepts you think will be important as you solve the problem.)

2. Solve and state the answer.

   (a) Perform the necessary calculations.

   (b) State the answer, including the units of measure.

3. Check.

   (a) Estimate the answer to the problem. Compare this estimate to the calculated value. Is your answer reasonable?

   (b) Repeat your calculations.

   (c) Work backward from your answer. Do you arrive at the original conditions of the problem?

EXAMPLE

▲ Fred has a rectangular living room that is 3.5 yards wide and 6.8 yards long. He has a hallway that is 1.8 yards wide and 3.5 yards long. He wants to carpet each area using carpeting that costs $12.50 per square yard. What will the carpeting cost him? *Understand the problem.*

It is helpful to draw a sketch.

6.8 yd

| Living Room | 3.5 yd |

3.5 yd

| Hallway | 1.8 yd |

(*continued on next page*)

Procedure for Solving Applied Problems (*continued*)

Mathematics Blueprint for Problem Solving

Gather the Facts	What Am I Asked to Do?	How Do I Proceed?	Key Points to Remember
Living room: 6.8 yards by 3.5 yards Hallway: 3.5 yards by 1.8 yards Cost of carpet: $12.50 per square yard	Find out what the carpeting will cost Fred.	Find the area of each room. Add the two areas. Multiply the total area by $12.50.	Multiply the length by the width to get the area of the room. Remember, area is measured in square yards.

To find the area of each room, we multiply the dimensions for each room.

Living room 6.8 × 3.5 = 23.80 square yards

Hallway 3.5 × 1.8 = 6.30 square yards

Add the two areas.

```
  23.80
+  6.30
  30.10  square yards
```

Multiply the total area by the cost per square yard.

30.1 × 12.50 = $376.25

Estimate to check. You may be able to do some of this mentally.

7 × 4 = 28 square yards   4 × 2 = 8 square yards

```
  28
+  8
  36 square yards
```

36 × 10 = $360 $360 is close to $376.25. ✓

Chapter 3 Review Problems

Section 3.1

Write a word name for each decimal.

1. 13.672   thirteen and six hundred seventy-two thousandths

2. 0.00084   eighty-four hundred-thousandths

Write as a decimal.

3. $\frac{7}{10}$ 0.7

4. $\frac{81}{100}$ 0.81

5. $1\frac{523}{1000}$ 1.523

6. $\frac{79}{10,000}$ 0.0079

Write as a fraction or a mixed number.

7. 0.17 $\frac{17}{100}$

8. 0.036 $\frac{9}{250}$

9. 34.24 $34\frac{6}{25}$

10. 1.00025 $1\frac{1}{4000}$

Section 3.2

Fill in the blank with <, =, or >.

11. $2\frac{9}{100}$ = 2.09

12. 0.716 > 0.706

13. $\frac{65}{100}$ < 0.655

14. 0.824 > 0.804

In exercises 15–18, arrange each set of decimal numbers from smallest to largest.

15. 0.981, 0.918, 0.98, 0.901   0.901, 0.918, 0.98, 0.981

16. 5.62, 5.2, 5.6, 5.26, 5.59   5.2, 5.26, 5.59, 5.6, 5.62

17. 0.419, 0.49, 0.409, 0.491   0.409, 0.419, 0.49, 0.491

18. 2.36, 2.3, 2.362, 2.302   2.3, 2.302, 2.36, 2.362

19. Round to the nearest tenth. 0.613   0.6

20. Round to the nearest hundredth. 19.2076   19.21

21. Round to the nearest ten-thousandth. 9.85215   9.8522

22. Round to the nearest dollar. $156.48 $156

Section 3.3

23. Add.

$$\begin{array}{r} 9.6 \\ 11.5 \\ 21.8 \\ + 34.7 \\ \hline 77.6 \end{array}$$

24. Add.

$$\begin{array}{r} 2.5 \\ 32.7 \\ 116.94 \\ + 0.67 \\ \hline 152.81 \end{array}$$

25. Subtract.

$$\begin{array}{r} 17.03 \\ - 2.448 \\ \hline 14.582 \end{array}$$

26. Subtract.

$$\begin{array}{r} 182.422 \\ - 68.55 \\ \hline 113.872 \end{array}$$

Section 3.4

In exercises 27–32, multiply.

27.

$$\begin{array}{r} 0.098 \\ \times \ 0.032 \\ \hline 0.003136 \end{array}$$

28.

$$\begin{array}{r} 126.83 \\ \times \quad 7 \\ \hline 887.81 \end{array}$$

29.

$$\begin{array}{r} 78 \\ \times \ 5.2 \\ \hline 405.6 \end{array}$$

30.

$$\begin{array}{r} 7053 \\ \times \ 0.34 \\ \hline 2398.02 \end{array}$$

31. 0.000613×10^3 0.613

32. 1.2354×10^5 123,540

33. *Food Cost* Roast beef was on sale for $3.49 per pound. How much would 2.5 pounds cost? Round to the nearest cent. $8.73

Section 3.5

In exercises 34–36, divide until there is a remainder of zero.

34. $0.07\overline{)0.0001806}$ 0.00258

35. $5.2\overline{)191.36}$ 36.8

36. $8\overline{)1863.2}$ 232.9

37. Divide and round your answer to the nearest tenth.

$$1.3\overline{)746.75}$$ 574.4

38. Divide and round your answer to the nearest thousandth.

$$0.06\overline{)0.003539}$$ 0.059

Section 3.6

Write as an equivalent decimal.

39. $\dfrac{11}{12}$ $0.91\overline{6}$

40. $\dfrac{17}{20}$ 0.85

41. $1\dfrac{5}{6}$ $1.8\overline{3}$

42. $\dfrac{19}{16}$ 1.1875

Write as a decimal rounded to the nearest thousandth.

43. $\dfrac{11}{14}$ 0.786

44. $\dfrac{10}{29}$ 0.345

45. $2\dfrac{5}{17}$ 2.294

46. $3\dfrac{9}{23}$ 3.391

Evaluate by doing the operations in proper order.

47. $2.3 \times 1.82 + 3 \times 5.12$ 19.546

48. $2.175 \div 0.15 \times 10 + 27.32$ 172.32

49. $3.57 - (0.4)^3 \times 2.5 \div 5$ 3.538

50. $2.4 \div (2 - 1.6)^2 + 8.13$ 23.13

Mixed Practice

Calculate.

51. $2398.26 - 1959.07$ 439.19

52. $32.15 \times 0.02 \times 10^2$ 64.3

53. $1.809 - 0.62 + 3.27$ 4.459

54. $2.0792 \div 2.3$
0.904

55. $8 \div 0.4 + 0.1 \times (0.2)^2$
20.004

56. $(3.8 - 2.8)^3 \div (0.5 + 0.3)$
1.25

Applications

Section 3.7

Solve each problem.

57. *Football Tickets* At a large football stadium there are 2,600 people in line for tickets. In the first two minutes the computer is running slowly and tickets 228 people. Then the computer stops. For the next 2.5 minutes, the computer runs at medium speed and tickets 388 people per minute. For the next three minutes the computer runs at full speed and tickets 430 people per minute. Then the computer stops. How many people still have not received their tickets? 112 people

58. *Fuel Efficiency* Phil drove to the mountains. His odometer read 26,005.8 miles at the start, and 26,325.8 miles at the end of the trip. He used 12.9 gallons of gas on the trip. How many miles per gallon did his car get? (Round your answer to the nearest tenth.)
24.8 miles per gallon

59. *Car Payments* Robert is considering buying a car and making installment payments of $189.60 for 48 months. The cash price of the car is $6930.50. How much extra does he pay if he uses the installment plan instead of buying the car with one payment?
$2170.30

60. *Comparing Job Salaries* Mr. Zeno has a choice of working as an assistant manager at ABC Company at $315.00 per week or receiving an hourly salary of $8.26 per hour at the XYZ company. He learned from several previous assistant managers at both companies that they usually worked 38 hours per week. At which company will he probably earn more money? ABC Company

61. *Drinking Water Safety* The EPA standard for safe drinking water is a maximum of 0.002 milligram of mercury in one liter of water. The town wells at Winchester were tested. The test was done on 12 liters of water. The entire 12-liter sample contained 0.03 milligram of mercury. Is the water safe or not? By how much does it differ from the standard? no; by 0.0005 milligram per liter

62. *Infant Head Size* It is common for infants to have their heads measured during the first year of life. At two months, Will's head measured 40 centimeters. There are 2.54 centimeters in one inch. How many inches was this measurement? Round to the nearest hundredth.
15.75 inches

63. *Geometry* Dick Wright's new rectangular garden measures 18.3 feet by 9.6 feet. He needs to install wire fence on all four sides. **(a)** 55.8 feet
  (a) How many feet of fence does he need?
  (b) The number of bags of wood chips Dick buys depends on the area of the garden. What is the area? 175.68 square feet

64. *Geometry* Bill Tupper's rectangular driveway needs to be resurfaced. It is 75.5 feet long and 18.5 feet wide. How large is the area of the driveway? 1396.75 square feet

65. *Travel Distances* The following strip map shows the distances in miles between several local towns in Pennsylvania. How much longer is the distance from Coudersport to Gaines than the distance from Galeton to Wellsboro? 6.1 miles

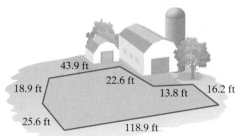

66. *Geometry* A farmer in Vermont has a field with an irregular shape. The distances are marked on the diagram. There is no fence but there is a path on the edge of the field. How long is the walking path around the field? 259.9 feet

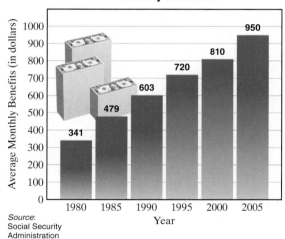

67. *Car Payments* Marcia and Greg purchased a new car. For the next five years they will be making monthly payments of $212.50. Their bank has offered to give them a loan at a smaller interest rate so that they would make monthly payments of only $199.50. The bank would charge them $285.00 to reissue their car loan. How much would it cost them to keep their original loan? How much would it cost them if they took the new loan from the bank? Should they make the change or keep the original loan?
$12,750.00; $12,255.00; they should change to the new loan

Social Security Benefits Use the following bar graph to answer exercises 68–73. Round all answers to the nearest cent.

68. How much did the average monthly social security benefit increase from 1985 to 1995? $241.00

69. How much did the average monthly social security benefit increase from 1995 to 2005? $230.00

70. What was the average daily social security benefit in 1980? (Assume 30 days in a month.) $11.37

71. What was the average daily social security benefit in 2005? (Assume 30 days in a month.) $31.67

72. If the average daily social security benefit increases by the same amount from 2005 to 2020 as it did from 1990 to 2005, what will be the average daily social security benefit in 2020? (Assume 30 days in a month.) $43.23

73. If the average daily social security benefit increases by the same amount from 2005 to 2015 as it did from 1995 to 2005, what will be the average daily social security benefit in 2015? (Assume 30 days in a month.) $39.33

Average Monthly Social Security Benefits

Average Monthly Benefits (in dollars)

1980: 341
1985: 479
1990: 603
1995: 720
2000: 810
2005: 950

Year

Source:
Social Security
Administration

CHAPTER

TEST PREP
VIDEO CD

Note to Instructor: The Chapter 3 Test file in the TestGen program provides algorithms specifically matched to these problems so you can easily replicate this test for additional practice or assessment purposes.

1.	twelve and forty-three thousandths
2.	0.3977
3.	$7\frac{3}{20}$
4.	$\frac{261}{1000}$
5.	2.19, 2.9, 2.907, 2.91
6.	78.66
7.	0.0342
8.	99.698
9.	37.53
10.	0.0979
11.	71.155
12.	0.5817
13.	2189
14.	0.1285
15.	47
16.	$1.\overline{2}$
17.	0.875
18.	1.487
19.	6.1952
20.	$26.95
21.	18.8 miles per gallon
22.	3.43 centimeters less
23.	$390.55

How Am I Doing? Chapter 3 Test

Remember to use your Chapter Test Prep Video CD to see the worked-out solutions to the test problems you want to review.

1. Write a word name for the decimal. 12.043

2. Write as a decimal. $\frac{3977}{10,000}$

In questions 3 and 4, write in fractional notation. Reduce whenever possible.

3. 7.15

4. 0.261

5. Arrange from smallest to largest. 2.19, 2.91, 2.9, 2.907

6. Round to the nearest hundredth. 78.6562

7. Round to the nearest ten-thousandth. 0.0341752

Add.

8.
$$\begin{array}{r} 96.2 \\ 1.348 \\ + \ 2.15 \\ \hline \end{array}$$

9. 17 + 2.1 + 16.8 + 0.04 + 1.59

Subtract.

10.
$$\begin{array}{r} 1.0075 \\ - \ 0.9096 \\ \hline \end{array}$$

11. 72.3 − 1.145

Multiply.

12.
$$\begin{array}{r} 8.31 \\ \times \ 0.07 \\ \hline \end{array}$$

13. 2.189×10^3

Divide.

14. $0.08\overline{)0.01028}$

15. $0.69\overline{)32.43}$

Write as a decimal.

16. $\frac{11}{9}$

17. $\frac{7}{8}$

In questions 18 and 19, perform the operations in the proper order.

18. $(0.3)^3 + 1.02 \div 0.5 - 0.58$

19. $19.36 \div (0.24 + 0.26) \times (0.4)^2$

20. Peter put 8.5 gallons of gas in his car. The price per gallon is $3.17. How much did Peter spend on gas? Round to the nearest cent.

21. Frank traveled from the city to the shore. His odometer read 42,620.5 miles at the start and 42,780.5 at the end of the trip. He used 8.5 gallons of gas. How many miles per gallon did his car achieve? Round to the nearest tenth.

22. The rainfall for March in Central City was 8.01 centimeters; for April, 5.03 centimeters; and for May, 8.53 centimeters. The normal rainfall for these three months is 25 centimeters. How much less rain fell during these three months than usual; that is, how does this year's figure compare with the figure for normal rainfall?

23. Wendy is earning $7.30 per hour in her new job as a teller trainee at the Springfield National Bank. She earns 1.5 times that amount for every hour over 40 hours she works in one week. She was asked to work 49 hours last week. How much did she earn last week?

Approximately one-half of this test is based on Chapter 3 material. The remainder is based on material covered in Chapters 1 and 2.

1. Write in words. 38,056,954

2. Add. 156,028
301,579
+ 21,980

3. Subtract. 1,091,000
− 1,036,520

4. Multiply. 589
× 67

5. Divide. $15\overline{)4740}$

6. Evaluate. $20 \div 4 + 2^5 - 7 \times 3$

7. Reduce. $\dfrac{18}{45}$

8. Add. $5\dfrac{3}{8} + 2\dfrac{11}{12}$

9. Subtract. $\dfrac{23}{35} - \dfrac{2}{5}$

10. Evaluate. $\dfrac{5}{16} \times \dfrac{4}{5} + \dfrac{9}{10} \times \dfrac{2}{3}$

11. Divide. $52 \div 3\dfrac{1}{4}$

12. Divide. $1\dfrac{3}{8} \div \dfrac{5}{12}$

13. Estimate. $58,216 \times 438,207$

14. Write as a decimal. $\dfrac{39}{1000}$

15. Arrange from smallest to largest. 2.1, 20.1, 2.01, 2.12, 2.11

16. Round to the nearest thousandth. 26.07984

17. Add. 3.126
8.4
10.33
+ 0.09

18. Subtract. 28.1
− 14.982

19. Multiply. 28.7×0.05

20. Multiply. 0.1823×1000

21. Divide. $0.06\overline{)0.06348}$

22. Write as a decimal. $\dfrac{13}{16}$

23. Perform the operations in the correct order.
$1.44 \div 0.12 + (0.3)^3 + 1.57$

▲ 24. Dr. Bob Wells has a small square garden that measures 10.5 feet on each side.
(a) What is the area of this garden?
(b) What is the perimeter of this garden?

25. Sue's savings account balance is $199.36. This month she earned interest of $1.03. She deposited $166.35 and $93.50. She withdrew money three times, in the amounts of $90.00, $37.49, and $137.18. What will her balance be at the start of next month?

26. Russ and Norma Camp borrowed some money from the bank to purchase a new car. They are paying off the car loan at the rate of $320.50 per month. At the end of the loan period they will have paid $19,230.00 to the bank. How many months will it take to pay off this car loan?

1.	thirty-eight million, fifty-six thousand, nine hundred fifty-four
2.	479,587
3.	54,480
4.	39,463
5.	316
6.	16
7.	$\dfrac{2}{5}$
8.	$8\dfrac{7}{24}$
9.	$\dfrac{9}{35}$
10.	$\dfrac{17}{20}$
11.	16
12.	$\dfrac{33}{10}$ or $3\dfrac{3}{10}$
13.	24,000,000,000
14.	0.039
15.	2.01, 2.1, 2.11, 2.12, 20.1
16.	26.080
17.	21.946
18.	13.118
19.	1.435
20.	182.3
21.	1.058
22.	0.8125
23.	13.597
24. (a)	110.25 square feet
(b)	42 feet
25.	$195.57
26.	60 months

Ratio and Proportion

We all know that too much fast food is not good for us. Many people eat fast food nearly every day, but they may be unaware of just how many calories they are consuming. Do you know how many calories are in some common fast foods? How much exercise is necessary to burn off these calories? The mathematics you learn in this chapter will enable you to answer those questions.

① Using a Ratio to Compare Two Quantities with the Same Units

Assume that you earn 13 dollars an hour and your friend earns 10 dollars per hour. The *ratio* 13 : 10 compares what you and your friend make. This ratio means that for every 13 dollars you earn, your friend earns 10. The *rate* you are paid is 13 dollars per hour, which compares 13 dollars to 1 hour. In this section we see how to use both ratios and rates to solve many everyday problems.

Suppose that we want to compare an object weighing 20 pounds to an object weighing 23 pounds. The ratio of their weights would be 20 to 23. We may also write this as $\frac{20}{23}$. A **ratio** is the comparison of two quantities that have the *same units*.

A commonly used video display for a computer has a horizontal dimension of 14 inches and a vertical dimension of 10 inches. The ratio of the horizontal dimension to the vertical dimension is 14 to 10. In reduced form we would write that as 7 to 5. We can express the ratio three ways.

> We can write "the ratio of 7 to 5."
> We can write 7 : 5 using a colon.
> We can write $\frac{7}{5}$ using a fraction.

All three notations are valid ways to compare 7 to 5. Each is read as "7 to 5."

> We always want to write a ratio in simplest form. A ratio is in **simplest form** when the two numbers do not have a common factor and both numbers are whole numbers.

Student Learning Objectives

After studying this section, you will be able to:

① Use a ratio to compare two quantities with the same units.

② Use a rate to compare two quantities with different units.

CLICK & READ
WebNews
The latest news updated by the minute!
Heatwave Continues in Texas

EXAMPLE 1 Write in simplest form. Express your answer as a fraction.

(a) the ratio of 15 hours to 20 hours
(b) the ratio of 36 hours to 30 hours
(c) 125 : 150

Solution

(a) $\frac{15}{20} = \frac{3}{4}$ **(b)** $\frac{36}{30} = \frac{6}{5}$ **(c)** $\frac{125}{150} = \frac{5}{6}$

Notice that in each case the two numbers *do* have a common factor. When we form the fraction—that is, the ratio—we take the extra step of *reducing* the fraction. However, improper fractions *are not* changed to mixed numbers.

Practice Problem 1 Write in simplest form. Express your answer as a fraction.

(a) the ratio of 36 feet to 40 feet $\frac{9}{10}$
(b) the ratio of 18 feet to 15 feet $\frac{6}{5}$
(c) 220 : 270 $\frac{22}{27}$

Teaching Example 1 Write in simplest form. Express your answer as a fraction.

(a) the ratio of 12 pounds to 18 pounds
(b) the ratio of 40 pounds to 30 pounds
(c) 350 : 400

Ans: **(a)** $\frac{2}{3}$ **(b)** $\frac{4}{3}$ **(c)** $\frac{7}{8}$

Teaching Tip You will need to remind students to reduce fractions to lowest terms when expressing ratios. They often forget this step or reduce only partially.

EXAMPLE 2   Martin earns $350 weekly. However, he takes home only $250 per week in his paycheck.

$350.00 gross pay   (what Martin earns)

$\left.\begin{array}{l} 45.00 \ \text{ withheld for federal tax} \\ 20.00 \ \text{ withheld for state tax} \\ 35.00 \ \text{ withheld for retirement} \end{array}\right\}$ $\left(\begin{array}{l}\text{what is taken out} \\ \text{of Martin's earnings}\end{array}\right)$

$250.00 take-home pay   (what Martin has left)

(a) What is the ratio of the amount withheld for federal tax to gross pay?

(b) What is the ratio of the amount withheld for state tax to the amount withheld for federal tax?

Solution

(a) The ratio of the amount withheld for federal tax to gross pay is

$$\frac{45}{350} = \frac{9}{70}.$$

(b) The ratio of the amount withheld for state tax to the amount withheld for federal tax is

$$\frac{20}{45} = \frac{4}{9}.$$

Practice Problem 2   Recently President Burton conducted a survey of students at North Shore Community College who use the Internet. He wanted to determine how many of the students use the college Internet provider versus how many use AOL, MSN, or other commercial Internet providers. The results of his survey are shown in the circle graph.

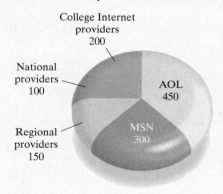

College Internet providers 200

National providers 100

AOL 450

Regional providers 150

MSN 300

(a) Write the ratio of the number of students who use the college Internet provider to the number of students who use AOL.

(b) Write the ratio of the number of students who use MSN to the total number of students who use the Internet.

(a) $\dfrac{4}{9}$ **(b)** $\dfrac{1}{4}$

TO THINK ABOUT: Mach Numbers Perhaps you have heard statements like "a certain jet plane travels at Mach 2.2." What does that mean? A Mach number is a ratio that compares the velocity (speed) of an object to the velocity of sound. Sound travels at about 330 meters per second. The Mach number is written in decimal form.

What is the Mach number of a jet traveling at 690 meters per second?

$$\text{Mach number of jet} = \frac{690 \text{ meters per second}}{330 \text{ meters per second}} = \frac{69}{33} = \frac{23}{11}$$

Dividing this out, we obtain

$$2.09090909\ldots \text{ or } 2.\overline{09}$$

Rounded to the nearest tenth, the Mach number of the jet is 2.1.
Exercises 73 and 74 in 4.1 Exercises deal with Mach numbers.

Using a Rate to Compare Two Quantities with Different Units

A **rate** is a comparison of two quantities with *different units*. Usually, to avoid misunderstanding, we express a rate as a reduced or simplified fraction with the units included.

EXAMPLE 3 Recently an automobile manufacturer spent $946,000 for a 48-second television commercial shown on a national network. What is the rate of dollars spent to seconds of commercial time?

Solution The rate is $\dfrac{946{,}000 \text{ dollars}}{48 \text{ seconds}} = \dfrac{59{,}125 \text{ dollars}}{3 \text{ seconds}}$.

Practice Problem 3 A farmer is charged a $44 storage fee for every 900 tons of grain he stores. What is the rate of the storage fee in dollars to tons of grain?

$$\frac{11 \text{ dollars}}{225 \text{ tons}}$$

Teaching Example 3 The recipe for cooking a sauce calls for 8 teaspoons of butter for each 12 ounces of milk. What is the rate of teaspoons of butter to ounces of milk?

Ans: $\dfrac{2 \text{ teaspoons}}{3 \text{ ounces}}$

Often we want to know the rate for a single unit, which is the unit rate. A **unit rate** is a rate in which the denominator is the number 1. Often we need to divide the numerator by the denominator to obtain this value.

EXAMPLE 4 A car traveled 301 miles in seven hours. Find the unit rate.

Solution $\frac{301}{7}$ can be simplified. We find $301 \div 7 = 43$.

Thus

$$\frac{301 \text{ miles}}{7 \text{ hours}} = \frac{43 \text{ miles}}{1 \text{ hour}}$$

The denominator is 1. We write our answer as 43 miles/hour. The fraction line is read as the word *per*, so our answer here is read "43 miles per hour." *Per* means "for every," so a rate of 43 miles per hour means 43 miles traveled for every hour traveled.

Teaching Example 4 Sopheac filled his car's gas tank after driving 240 miles. The gasoline cost him $31.20. Find the unit rate of cents per mile.

Ans: 13 cents/mile

Practice Problem 4 A car traveled 212 miles in four hours. Find the unit rate. 53 miles/hour

NOTE TO STUDENT: Fully worked-out solutions to all of the Practice Problems can be found at the back of the text starting at page SP-1

EXAMPLE 5 A grocer purchased 200 pounds of apples for $68. He sold the 200 pounds of apples for $86. How much profit did he make per pound of apples?

Solution

$$\begin{array}{rl} \$86 & \text{selling price} \\ -\ 68 & \text{cost} \\ \hline \$18 & \text{profit} \end{array}$$

The rate that compares profit to pounds of apples sold is $\dfrac{18 \text{ dollars}}{200 \text{ pounds}}$. We will find $18 \div 200$.

$$\begin{array}{r} 0.09 \\ 200\overline{)18.00} \\ \underline{18\ 00} \\ 0 \end{array}$$

The unit rate of profit is $0.09 per pound.

Practice Problem 5 A retailer purchased 120 nickel-cadmium batteries for flashlights for $129.60. She sold them for $170.40. What was her profit per battery? $0.34

EXAMPLE 6 Hamburger at a local butcher shop is packaged in large and extra-large packages. A large package costs $7.86 for 6 pounds and an extra-large package is $10.08 for 8 pounds.

(a) What is the unit rate in dollars per pound for each size package?

(b) How much per pound does a consumer save by buying the extra-large package?

Solution

(a) $\dfrac{7.86 \text{ dollars}}{6 \text{ pounds}} = \$1.31/\text{pound}$ for the large package

$\dfrac{10.08 \text{ dollars}}{8 \text{ pounds}} = \$1.26/\text{pound}$ for the extra-large package

(b) $\begin{array}{r} \$1.31 \\ -\ 1.26 \\ \hline \$0.05 \end{array}$ A person saves $0.05/pound by buying the extra-large package.

Practice Problem 6 A 12-ounce package of Fred's favorite cereal costs $2.04. A 20-ounce package of the same cereal costs $2.80.

(a) What is the unit rate in cost per ounce of each size of cereal package?

(b) How much per ounce would Fred save by buying the larger size?

(a) 12 ounce: $0.17/ounce, 20 ounce: $0.14/ounce   (b) $0.03/ounce

PRACTICE   WATCH   DOWNLOAD   READ   REVIEW

Verbal and Writing Skills

1. A _____ratio_____ is a comparison of two quantities that have the same units.

2. A rate is a comparison of two quantities that have _____different_____ units.

3. The ratio 5 : 8 is read _____5 to 8_____.

4. Marion compares the number of loaves of bread she bakes to the number of pounds of flour she needs to make the bread. Is this a ratio or a rate? Why? a rate; it compares different units: loaves of bread to pounds of flour

Write in simplest form. Express your answer as a fraction.

5. 6 : 18
$\frac{1}{3}$

6. 8 : 20
$\frac{2}{5}$

7. 21 : 18
$\frac{7}{6}$

8. 50 : 35
$\frac{10}{7}$

9. 150 : 225
$\frac{2}{3}$

10. 360 : 480
$\frac{3}{4}$

11. 165 to 90
$\frac{11}{6}$

12. 135 to 120
$\frac{9}{8}$

13. 60 to 72
$\frac{5}{6}$

14. 55 to 77
$\frac{5}{7}$

15. 28 to 42
$\frac{2}{3}$

16. 21 to 98
$\frac{3}{14}$

17. 32 to 20
$\frac{8}{5}$

18. 90 to 54
$\frac{5}{3}$

19. 8 ounces to 12 ounces
$\frac{2}{3}$

20. 50 years to 85 years
$\frac{10}{17}$

21. 39 kilograms to 26 kilograms
$\frac{3}{2}$

22. 255 meters to 15 meters
$\frac{17}{1}$

23. $75 to $95
$\frac{15}{19}$

24. $54 to $78
$\frac{9}{13}$

25. 312 yards to 24 yards
$\frac{13}{1}$

26. 91 tons to 133 tons
$\frac{13}{19}$

27. $2\frac{1}{2}$ pounds to $4\frac{1}{4}$ pounds
$\frac{10}{17}$

28. $4\frac{1}{3}$ feet to $5\frac{2}{3}$ feet
$\frac{13}{17}$

Personal Finance *Use the following table to answer exercises 29–32.*

ROBIN'S WEEKLY PAYCHECK

Total (Gross) Pay	Federal Withholding	State Withholding	Retirement	Insurance	Savings Contribution	Take-Home Pay
$285	$35	$20	$28	$16	$21	$165

29. What is the ratio of take-home pay to total (gross) pay? $\frac{165}{285} = \frac{11}{19}$

30. What is the ratio of retirement to insurance? $\frac{28}{16} = \frac{7}{4}$

31. What is the ratio of federal withholding to take-home pay? $\frac{35}{165} = \frac{7}{33}$

32. What is the ratio of retirement to total (gross) pay? $\frac{28}{285}$

Useful Life of an Automobile *An automobile insurance company prepared the following analysis for its clients. Use this table for exercises 33–36.*

33. What is the ratio of sedans that lasted two years or less to the total number of sedans?

$$\frac{205}{1225} = \frac{41}{245}$$

34. What is the ratio of sedans that lasted more than six years to the total number of sedans?

$$\frac{315}{1225} = \frac{9}{35}$$

35. What is the ratio of the number of sedans that lasted six years or less but more than four years to the number of sedans that lasted two years or less?

$$\frac{450}{205} = \frac{90}{41}$$

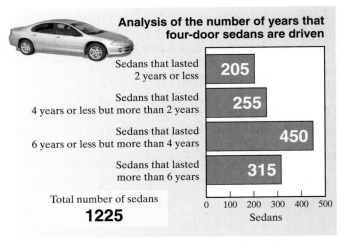

Analysis of the number of years that four-door sedans are driven

	Sedans
Sedans that lasted 2 years or less	205
Sedans that lasted 4 years or less but more than 2 years	255
Sedans that lasted 6 years or less but more than 4 years	450
Sedans that lasted more than 6 years	315

Total number of sedans **1225**

36. What is the ratio of the number of sedans that lasted more than six years to the number of sedans that lasted four years or less but more than two years?

$$\frac{315}{255} = \frac{21}{17}$$

37. ***Basketball*** A basketball team scored a total of 704 points during one season. Of these, 44 points were scored by making one-point free throws. What is the ratio of free-throw points to total points?

$$\frac{1}{16}$$

38. ***Sales Tax*** When Michael bought his home theater sound system, he paid $34 in tax. The total cost was $714. What is the ratio of tax to total cost?

$$\frac{1}{21}$$

Write as a rate in simplest form.

39. $42 for 12 pairs of socks

$$\frac{\$7}{2 \text{ pairs of socks}}$$

40. $50 for 15 deli sandwiches

$$\frac{\$10}{3 \text{ sandwiches}}$$

41. $170 for 12 bushes

$$\frac{\$85}{6 \text{ bushes}}$$

42. 98 pounds for 22 people

$$\frac{49 \text{ pounds}}{11 \text{ people}}$$

43. $114 for 12 CDs

$$\frac{\$19}{2 \text{ CDs}}$$

44. $150 for 12 house plants

$$\frac{\$25}{2 \text{ plants}}$$

45. 6150 revolutions for every 15 miles

$$\frac{410 \text{ revolutions}}{1 \text{ mile}} \text{ or } 410 \text{ rev/mile}$$

46. 9540 revolutions for every 18 miles

$$\frac{530 \text{ revolutions}}{1 \text{ mile}} \text{ or } 530 \text{ rev/mile}$$

47. $330,000 for 12 employees

$$\frac{\$27,500}{1 \text{ employee}} \text{ or } \$27,500/\text{employee}$$

48. $156,000 for 24 people

$$\frac{\$6500}{1 \text{ person}} \text{ or } \$6500/\text{person}$$

Write as a unit rate.

49. Earn $600 in 40 hours

$$\frac{\$600}{40 \text{ hours}} = \$15/\text{hr}$$

50. Earn $315 in 35 hours

$$\frac{\$315}{35 \text{ hours}} = \$9/\text{hr}$$

51. Travel 192 miles on 12 gallons of gas

$$\frac{192 \text{ miles}}{12 \text{ gallons}} = 16 \text{ mi/gal}$$

52. Travel 322 miles on 14 gallons of gas

$$\frac{322 \text{ miles}}{14 \text{ gallons}} = 23 \text{ mi/gal}$$

▲**53.** 1120 people in 16 square miles
70 people/sq mi

▲**54.** 3600 people in 24 square miles
150 people/sq mi

55. 840 books for 12 libraries
70 books/library

56. 930 points in 15 games
62 points/game

57. Travel 297 miles in 4.5 hours
66 mi/hour

58. Travel 374 miles in 5.5 hours
68 mi/hour

59. 475 patients for 25 doctors
19 patients/doctor

60. 375 trees planted on 15 acres
25 trees/acre

61. 60 eggs from 12 chickens
5 eggs/chicken

62. 78 children in 26 families
3 children/family

Applications

63. *Stock Market* $3870 was spent for 129 shares of Mattel stock. Find the cost per share.
$30/share

64. *Stock Market* $6150 was spent for 150 shares of Polaroid stock. Find the cost per share.
$41/share

65. *Toy Store Profit* A toy store owner purchased 90 puppets for $1080. She sold them for $1485. How much profit did she make per puppet?
$4.50 profit per puppet

66. *Clothing Store Profit* The manager of an outdoor clothing store ordered 40 pairs of hiking boots for $2400. The boots will sell for a total of $3560. How much profit will the store make per pair? $29 profit per pair

67. *Food Cost* A 16-ounce box of dry pasta costs $1.28. A 24-ounce box of the same pasta costs $1.68. **(b)** 1¢ per ounce or $0.01 per ounce

  (a) What is the cost per ounce of each box of pasta? $0.08/oz small box; $0.07/oz large box
  (b) How much does the educated consumer save by buying the larger box?
  (c) How much does the consumer save by buying 2 large boxes instead of 3 small boxes?
  The consumer saves $0.48.

68. *Food Cost* A 16-ounce can of beef stew costs $2.88. A 26-ounce can of the same beef stew costs $4.16.

  (a) What is the cost per ounce of each can of stew?
  (b) How much does the consumer save per ounce by buying the larger can?
  (a) smaller can   larger can
  $\dfrac{\$2.88}{16 \text{ ounces}} = \$0.18/\text{oz}$ $\dfrac{\$4.16}{26 \text{ ounces}} = \$0.16/\text{oz}$
  (b) 2¢ per ounce or $0.02/oz

69. *Moose Population Density* Dr. Robert Tobey completed a count of two herds of moose in the central regions of Alaska. He recorded 3978 moose on the North Slope and 5520 moose on the South Slope. There are 306 acres on the North Slope and 460 acres on the South Slope.

  (a) How many moose per acre were found on the North Slope? 13 moose
  (b) How many moose per acre were found on the South Slope? 12 moose
  (c) In which region are the moose more closely crowded together? North Slope

70. *Australia Population Density* In Melbourne, Australia, 27,900 people live in the suburb of St. Kilda and 38,700 live in the suburb of Caulfield. The area of St. Kilda is 6500 acres. The area of Caulfield is 9200 acres. Round your answers to the nearest tenth. **(a)** 4.3 people **(b)** 4.2 people

  (a) How many people per acre live in St. Kilda?
  (b) How many people per acre live in Caulfield?
  (c) Which suburb is more crowded? St. Kilda

71. ***Stock Market***
   (a) Ms. Handley bought 350 shares of Home Depot stock for $14,332.50. How much did she pay per share? $40.95
   (b) Mr. Johnston bought 210 shares of Office Max stock for $11,088. How much did he pay per share? $52.80
   (c) How much more per share did Mr. Johnston pay than Ms. Handley? $11.85

72. ***Baseball Statistics*** In 2006, Ryan Howard of the Philadelphia Phillies hit 58 home runs with 581 "at-bats." In the same year, David Ortiz of the Boston Red Sox hit 54 home runs with 558 at-bats.
   (a) What are the rates of at-bats per home run for Ryan and David? Round to the nearest tenth. Ryan Howard, 10.0; David Ortiz, 10.3
   (b) Which person hit home runs more often? Ryan Howard

To Think About *For exercises 73 and 74, recall that the speed of sound is about 330 meters per second. (See the To Think About discussion on pages 262 and 263.) Round your answers to the nearest tenth.*

73. ***Jet Speed*** A jet plane was originally designed to fly at 750 meters per second. It was modified to fly at 810 meters per second. By how much was its Mach number increased? increased by Mach 0.2

74. ***Rocket Speed*** A rocket was first flown at 1960 meters per second. It proved unstable and unreliable at that speed. It is now flown at a maximum of 1920 meters per second. By how much was its Mach number decreased? decreased by Mach 0.1

Cumulative Review *Calculate.*

75. **[2.8.1]** $2\frac{1}{4} + \frac{3}{8}$ $2\frac{5}{8}$

76. **[2.5.1]** $\frac{5}{7} \div \frac{3}{21}$ 5

77. **[2.8.3]** $\frac{3}{5} \times \frac{5}{8} - \frac{2}{3} \times \frac{1}{4}$ $\frac{5}{24}$

78. **[2.8.2]** $3\frac{1}{16} - 2\frac{1}{24}$ $1\frac{1}{48}$

▲ **79.** **[3.7.2]** ***Geometry*** A room 12 yards × 5.2 yards had a carpet installed. The bill was $764.40. What was the cost of the installed carpet per square yard? $12.25/square yard

80. **[1.8.2]** ***Electronic Game Store Profit*** An electronics superstore bought 1050 computer games for $23 each. How much did the store pay in all for these games? The store sold the games for $39 each. How much profit did the store make? $24,150; $16,800

Quick Quiz 4.1

1. Write as a ratio in simplest form.
   51 to 85 $\frac{3}{5}$

2. Write as a rate in simplest form. Express your answer as a fraction.
   1700 square feet for 55 pounds $\frac{340 \text{ square feet}}{11 \text{ pounds}}$

3. Write as a unit rate. Round to the nearest hundredth if necessary.
   462 trees planted on 17 acres   27.18 trees/acre

4. **Concept Check** At a company picnic, there were 663 cans of soda for 231 people. Explain how you would write that as a rate in simplest form. Answers may vary

Classroom Quiz 4.1 You may use these problems to quiz your students' mastery of Section 4.1.

1. Write as a ratio in simplest form.
   26 to 96   **Ans:** $\frac{13}{48}$

2. Write as a rate in simplest form. Express your answer as a fraction.
   128 pounds for 36 people   **Ans:** $\frac{32 \text{ pounds}}{9 \text{ people}}$

3. Write as a unit rate. Round to the nearest hundredth if necessary.
   592 patients for 27 doctors   **Ans:** 21.93 patients/doctor

① Writing a Proportion

A **proportion** states that two ratios or two rates are equal. For example, $\frac{5}{8} = \frac{15}{24}$ is a proportion and $\frac{7\text{ feet}}{8\text{ dollars}} = \frac{35\text{ feet}}{40\text{ dollars}}$ is also a proportion. A proportion can be read two ways. The proportion $\frac{5}{8} = \frac{15}{24}$ can be read "five eighths equals fifteen twenty-fourths," or it can be read "five *is to* eight *as* fifteen *is to* twenty-four."

EXAMPLE 1 Write the proportion 5 is to 7 as 15 is to 21.

Solution
$$\frac{5}{7} = \frac{15}{21}$$

Practice Problem 1 Write the proportion 6 is to 8 as 9 is to 12.

$$\frac{6}{8} = \frac{9}{12}$$

EXAMPLE 2 Write a proportion to express the following: If four rolls of wallpaper measure 300 feet, then eight rolls of wallpaper will measure 600 feet.

Solution When you write a proportion, order is important. Be sure that the similar units for the rates are in the same position in the fractions.

$$\frac{4\text{ rolls}}{300\text{ feet}} = \frac{8\text{ rolls}}{600\text{ feet}}$$

Practice Problem 2 Write a proportion to express the following: If it takes two hours to drive 72 miles, then it will take three hours to drive 108 miles.

$$\frac{2\text{ hours}}{72\text{ miles}} = \frac{3\text{ hours}}{108\text{ miles}}$$

Student Learning Objectives

After studying this section, you will be able to:

① Write a proportion.

② Determine whether a statement is a proportion.

Teaching Example 1 Write the proportion 4 is to 5 as 12 is to 15.

Ans: $\dfrac{4}{5} = \dfrac{12}{15}$

Teaching Example 2 Write the proportion to express the following: If 3 quarts of paint will cover 165 square feet, then 9 quarts of paint will cover 495 square feet.

Ans: $\dfrac{3\text{ quarts}}{165\text{ square feet}} = \dfrac{9\text{ quarts}}{495\text{ square feet}}$

NOTE TO STUDENT: Fully worked-out solutions to all of the Practice Problems can be found at the back of the text starting at page SP-1

② Determining Whether a Statement Is a Proportion

By definition, a proportion states that two ratios are equal. $\frac{2}{7} = \frac{4}{14}$ is a proportion because $\frac{2}{7}$ and $\frac{4}{14}$ are equivalent fractions. You might say that $\frac{2}{7} = \frac{4}{14}$ is a *true* statement. It is easy enough to see that $\frac{2}{7} = \frac{4}{14}$ is true. However, is $\frac{4}{14} = \frac{6}{21}$ true? Is $\frac{4}{14} = \frac{6}{21}$ a proportion? To determine whether a statement is a proportion, we use the equality test for fractions.

> **EQUALITY TEST FOR FRACTIONS**
>
> For any two fractions where $b \neq 0$ and $d \neq 0$,
>
> $$\frac{a}{b} = \frac{c}{d} \text{ if and only if } a \times d = b \times c.$$

Thus, to see if $\dfrac{4}{14} = \dfrac{6}{21}$, we can multiply.

$$\frac{4}{14} \bowtie \frac{6}{21} \qquad \begin{array}{l} 14 \times 6 = 84 \\ 4 \times 21 = 84 \end{array} \leftarrow \boxed{\begin{array}{l}\text{The cross}\\\text{products}\\\text{are equal.}\end{array}}$$

$$\frac{4}{14} = \frac{6}{21} \text{ is true. } \frac{4}{14} = \frac{6}{21} \text{ is a proportion.}$$

This method is called finding **cross products.**

EXAMPLE 3 Determine which equations are proportions.

(a) $\frac{14}{18} \overset{?}{=} \frac{35}{45}$ (b) $\frac{16}{21} \overset{?}{=} \frac{174}{231}$

Solution

(a) $\frac{14}{18} \overset{?}{=} \frac{35}{45}$

$$18 \times 35 = 630$$

$\frac{14}{18} \times \frac{35}{45}$ The cross products are equal.   Thus $\frac{14}{18} = \frac{35}{45}$. This is a proportion.

$$14 \times 45 = 630$$

(b) $\frac{16}{21} \overset{?}{=} \frac{174}{231}$

$$21 \times 174 = 3654$$

$\frac{16}{21} \times \frac{174}{231}$ The cross products are not equal.   Thus $\frac{16}{21} \neq \frac{174}{231}$. This is not a proportion.

$$16 \times 231 = 3696$$

Practice Problem 3 Determine which equations are proportions.

(a) $\frac{10}{18} \overset{?}{=} \frac{25}{45}$ This is a proportion.   (b) $\frac{42}{100} \overset{?}{=} \frac{22}{55}$ This is not a proportion.

Proportions may involve fractions or decimals.

EXAMPLE 4 Determine which equations are proportions.

(a) $\frac{5.5}{7} \overset{?}{=} \frac{33}{42}$ (b) $\frac{5}{8\frac{3}{4}} \overset{?}{=} \frac{40}{72}$

Solution

(a) $\frac{5.5}{7} \overset{?}{=} \frac{33}{42}$

$$7 \times 33 = 231$$

$\frac{5.5}{7} \times \frac{33}{42}$ The cross products are equal.   Thus $\frac{5.5}{7} = \frac{33}{42}$. This is a proportion.

$$5.5 \times 42 = 231$$

(b) $\dfrac{5}{8\frac{3}{4}} \overset{?}{=} \dfrac{40}{72}$

First we multiply $8\frac{3}{4} \times 40 = \dfrac{35}{\cancel{4}} \times \cancel{40}^{10} = 35 \times 10 = 350$

$$8\frac{3}{4} \times 40 = 350$$

$\dfrac{5}{8\frac{3}{4}} \diagdown \dfrac{40}{72}$ The cross products are not equal.   Thus $\dfrac{5}{8\frac{3}{4}} \neq \dfrac{40}{72}$. This is not a proportion.

$$5 \times 72 = 360$$

Practice Problem 4 Determine which equations are proportions.

(a) $\dfrac{2.4}{3} \overset{?}{=} \dfrac{12}{15}$ This is a proportion.

(b) $\dfrac{2\frac{1}{3}}{6} \overset{?}{=} \dfrac{14}{38}$ This is not a proportion.

**NOTE TO STUDENT:** *Fully worked-out solutions to all of the Practice Problems can be found at the back of the text starting at page SP-1*

EXAMPLE 5 **(a)** Is the rate $\dfrac{\$86}{13 \text{ tons}}$ equal to the rate $\dfrac{\$79}{12 \text{ tons}}$?

(b) Is the rate $\dfrac{3 \text{ American dollars}}{2 \text{ British pounds}}$ equal to the rate $\dfrac{27 \text{ American dollars}}{18 \text{ British pounds}}$?

Solution

(a) We want to know whether $\dfrac{86}{13} = \dfrac{79}{12}$.

$$13 \times 79 = 1027$$

$\dfrac{86}{13} \diagdown \dfrac{79}{12}$ The cross products are not equal.   Thus the two rates are not equal. This is not a proportion.

$$86 \times 12 = 1032$$

(b) We want to know whether $\dfrac{3}{2} = \dfrac{27}{18}$.

$$2 \times 27 = 54$$

$\dfrac{3}{2} \diagdown \dfrac{27}{18}$ The cross products are equal.   Thus the two rates are equal. This is a proportion.

$$3 \times 18 = 54$$

Teaching Example 5

(a) Is the rate $\dfrac{95 \text{ instructors}}{2000 \text{ students}}$ equal to the rate $\dfrac{150 \text{ instructors}}{3000 \text{ students}}$?

(b) Is the rate $\dfrac{\$2.50}{3 \text{ lb}}$ equal to the rate $\dfrac{\$10}{12 \text{ lb}}$?

Ans: (a) The two rates are not equal. This is not a proportion.

(b) The two rates are equal. This is a proportion.

Teaching Tip Some teachers like to use the statement "the proportion is false" for Example 5a, and the statement "the proportion is true" for Example 5b. That is an alternate way of describing the two possibilities.

Practice Problem 5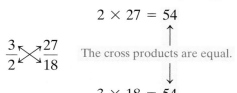

(a) Is the rate $\dfrac{1260 \text{ words}}{7 \text{ pages}}$ equal to the rate $\dfrac{3530 \text{ words}}{20 \text{ pages}}$?

(b) Is the rate $\dfrac{2 \text{ American dollars}}{11 \text{ French francs}}$ equal to the rate $\dfrac{16 \text{ American dollars}}{88 \text{ French francs}}$?

(a) The rates are not equal. This is not a proportion.
(b) The rates are equal. This is a proportion.

Verbal and Writing Skills

1. A proportion states that two ratios or rates are ___equal___ .

2. Explain in your own words how we use the equality test for fractions to determine if a statement is a proportion. Give an example. Answers may vary. Check explanations and examples for accuracy.

Write a proportion.

3. 6 is to 8 as 3 is to 4.

$$\frac{6}{8} = \frac{3}{4}$$

4. 12 is to 10 as 6 is to 5.

$$\frac{12}{10} = \frac{6}{5}$$

5. 20 is to 36 as 5 is to 9.

$$\frac{20}{36} = \frac{5}{9}$$

6. 120 is to 15 as 160 is to 20.

$$\frac{120}{15} = \frac{160}{20}$$

7. 220 is to 11 as 400 is to 20.

$$\frac{220}{11} = \frac{400}{20}$$

8. $2\frac{1}{2}$ is to 10 as $7\frac{1}{2}$ is to 30.

$$\frac{2\frac{1}{2}}{10} = \frac{7\frac{1}{2}}{30}$$

9. $4\frac{1}{3}$ is to 13 as $5\frac{2}{3}$ is to 17.

$$\frac{4\frac{1}{3}}{13} = \frac{5\frac{2}{3}}{17}$$

10. 5.5 is to 10 as 11 is to 20.

$$\frac{5.5}{10} = \frac{11}{20}$$

11. 6.5 is to 14 as 13 is to 28.

$$\frac{6.5}{14} = \frac{13}{28}$$

Applications *Write a proportion.*

12. *Cooking* When Jenny makes rice in her steamer, she mixes 2 cups of rice with 3 cups of water. To make 8 cups of rice, she needs 12 cups of water.

$$\frac{2 \text{ cups rice}}{3 \text{ cups water}} = \frac{8 \text{ cups rice}}{12 \text{ cups water}}$$

13. *Cartography* A cartographer (a person who makes maps) uses a scale of 3 inches to represent 40 miles. 27 inches would then represent 360 miles.

$$\frac{3 \text{ inches}}{40 \text{ miles}} = \frac{27 \text{ inches}}{360 \text{ miles}}$$

14. *Reading Speed* If Marcella can read 32 pages of her novel in 2 hours, she can read 80 pages in 5 hours.

$$\frac{32 \text{ pages}}{2 \text{ hours}} = \frac{80 \text{ pages}}{5 \text{ hours}}$$

15. *Tips for Valets* Stephen works as a valet parker. If he earns $40 in tips for parking 12 cars, he should earn $60 for parking 18 cars.

$$\frac{\$40}{12 \text{ cars}} = \frac{\$60}{18 \text{ cars}}$$

16. *Food Cost* If 20 pounds of pistachio nuts cost $75, then 30 pounds will cost $112.50.

$$\frac{20 \text{ pounds}}{\$75} = \frac{30 \text{ pounds}}{\$112.50}$$

17. *Education* If three credit hours at El Paso Community College cost $525, then seven credit hours should cost $1225.

$$\frac{3 \text{ hours}}{\$525} = \frac{7 \text{ hours}}{\$1225}$$

18. *Education* When Ridgewood Community College had 1200 students enrolled, 24 mathematics sections were offered. This year there are 1450 students enrolled, so 29 mathematics sections should be offered.

$$\frac{1200 \text{ students}}{24 \text{ sections}} = \frac{1450 \text{ students}}{29 \text{ sections}}$$

19. *Teaching Ratio* There are 3 teaching assistants for every 40 children in the elementary school. If we have 280 children, then we will have 21 teaching assistants.

$$\frac{3 \text{ teaching assistants}}{40 \text{ children}} = \frac{21 \text{ teaching assistants}}{280 \text{ children}}$$

20. *Lawn Care* If 16 pounds of fertilizer cover 1520 square feet of lawn, then 19 pounds of fertilizer should cover 1805 square feet of lawn.

$$\frac{16 \text{ pounds}}{1520 \text{ square feet}} = \frac{19 \text{ pounds}}{1805 \text{ square feet}}$$

21. *Restaurants* When New City had 4800 people, it had three restaurants. Now New City has 11,200 people, so it should have seven restaurants.

$$\frac{4800 \text{ people}}{3 \text{ restaurants}} = \frac{11,200 \text{ people}}{7 \text{ restaurants}}$$

Determine which equations are proportions.

22. $\dfrac{8}{6} \stackrel{?}{=} \dfrac{20}{15}$

$8 \times 15 \stackrel{?}{=} 6 \times 20$
$120 = 120$
It is a proportion.

23. $\dfrac{10}{25} \stackrel{?}{=} \dfrac{6}{15}$

$10 \times 15 \stackrel{?}{=} 25 \times 6$
$150 = 150$
It is a proportion.

24. $\dfrac{14}{11} = \dfrac{12}{10}$

$14 \times 10 \stackrel{?}{=} 11 \times 12$
$140 \neq 132$
It is not a proportion.

25. $\dfrac{8}{10} = \dfrac{13}{15}$

$8 \times 15 \stackrel{?}{=} 10 \times 13$
$120 \neq 130$
It is not a proportion.

26. $\dfrac{99}{100} \stackrel{?}{=} \dfrac{49}{50}$

$99 \times 50 \stackrel{?}{=} 100 \times 49$
$4950 \neq 4900$
It is not a proportion.

27. $\dfrac{17}{75} \stackrel{?}{=} \dfrac{22}{100}$

$17 \times 100 \stackrel{?}{=} 75 \times 22$
$1700 \neq 1650$
It is not a proportion.

28. $\dfrac{315}{2100} \stackrel{?}{=} \dfrac{15}{100}$

$315 \times 100 \stackrel{?}{=} 2100 \times 15$
$31{,}500 = 31{,}500$
It is a proportion.

29. $\dfrac{102}{120} \stackrel{?}{=} \dfrac{85}{100}$

$102 \times 100 \stackrel{?}{=} 120 \times 85$
$10{,}200 = 10{,}200$
It is a proportion.

30. $\dfrac{6}{14} \stackrel{?}{=} \dfrac{4.5}{10.5}$

$6 \times 10.5 \stackrel{?}{=} 14 \times 4.5$
$63 = 63$
It is a proportion.

31. $\dfrac{2.5}{4} \stackrel{?}{=} \dfrac{7.5}{12}$

$2.5 \times 12 \stackrel{?}{=} 4 \times 7.5$
$30 = 30$
It is a proportion.

32. $\dfrac{11}{12} \stackrel{?}{=} \dfrac{9.5}{10}$

$11 \times 10 \stackrel{?}{=} 12 \times 9.5$
$110 \neq 114$
It is not a proportion.

33. $\dfrac{3}{17} \stackrel{?}{=} \dfrac{4.5}{24.5}$

$3 \times 24.5 \stackrel{?}{=} 17 \times 4.5$
$73.5 \neq 76.5$
It is not a proportion.

34. $\dfrac{7}{1\frac{1}{2}} = \dfrac{14}{3}$

$7 \times 3 \stackrel{?}{=} 1\frac{1}{2} \times 14$
$21 = 21$
It is a proportion.

35. $\dfrac{6}{2\frac{1}{2}} = \dfrac{12}{5}$

$6 \times 5 \stackrel{?}{=} 2\frac{1}{2} \times 12$
$30 = 30$
It is a proportion.

36. $\dfrac{2\frac{1}{3}}{3} \stackrel{?}{=} \dfrac{7}{15}$

$2\frac{1}{3} \times 15 \stackrel{?}{=} 3 \times 7$
$35 \neq 21$
It is not a proportion.

37. $\dfrac{7\frac{1}{3}}{3} \stackrel{?}{=} \dfrac{23}{9}$

$7\frac{1}{3} \times 9 \stackrel{?}{=} 3 \times 23$
$66 \neq 69$
It is not a proportion.

38. $\dfrac{2.5}{\frac{1}{2}} \stackrel{?}{=} \dfrac{21}{5}$

$2.5 \times 5 \stackrel{?}{=} \frac{1}{2} \times 21$
$12.5 = 10.5$
It is not a proportion.

39. $\dfrac{\frac{1}{4}}{2} \stackrel{?}{=} \dfrac{\frac{7}{20}}{2.8}$

$\frac{1}{4} \times 2.8 \stackrel{?}{=} 2 \times \frac{7}{20}$
$0.7 = 0.7$
It is a proportion.

40. $\dfrac{75 \text{ miles}}{5 \text{ hours}} \stackrel{?}{=} \dfrac{105 \text{ miles}}{7 \text{ hours}}$

$75 \times 7 \stackrel{?}{=} 5 \times 105$
$525 = 525$
It is a proportion.

41. $\dfrac{135 \text{ miles}}{3 \text{ hours}} \stackrel{?}{=} \dfrac{225 \text{ miles}}{5 \text{ hours}}$

$135 \times 5 \stackrel{?}{=} 3 \times 225$
$675 = 675$
It is a proportion.

42. $\dfrac{286 \text{ gallons}}{12 \text{ acres}} \stackrel{?}{=} \dfrac{429 \text{ gallons}}{18 \text{ acres}}$

$286 \times 18 \stackrel{?}{=} 12 \times 429$
$5148 = 5148$
It is a proportion.

43. $\dfrac{166 \text{ gallons}}{14 \text{ acres}} \stackrel{?}{=} \dfrac{249 \text{ gallons}}{21 \text{ acres}}$

$166 \times 21 \stackrel{?}{=} 14 \times 249$
$3486 = 3486$
It is a proportion.

44. $\dfrac{52 \text{ free throws}}{80 \text{ attempts}} \stackrel{?}{=} \dfrac{60 \text{ free throws}}{95 \text{ attempts}}$

$52 \times 95 \stackrel{?}{=} 80 \times 60$
$4940 \neq 4800$　It is not a proportion.

45. $\dfrac{21 \text{ home runs}}{96 \text{ games}} \stackrel{?}{=} \dfrac{18 \text{ home runs}}{81 \text{ games}}$

$21 \times 81 \stackrel{?}{=} 96 \times 18$
$1701 \neq 1728$　It is not a proportion.

46. *Concert Audiences* At the Michael W. Smith concert on Friday there were 9600 female fans and 8200 male fans. The concert on Saturday had 12,480 female fans and 10,660 male fans. Is the ratio of female fans to male fans the same for both nights of the concert?　yes

47. *Baseball Team Wins* Since Harding High School opened, they have won 132 baseball games and have lost 22. Derry High School has won 160 games and lost 32 since it opened. Is the ratio of lost games to won games the same for both schools?　no

48. *Machine Operating Rate* A machine folds 650 boxes in five hours. Another machine folds 580 boxes in 4 hours.

(a) Do they fold boxes at the same rate?　no
(b) Which machine folds more boxes in 24 hours?
The machine that folds 580 boxes in four hours will fold more.

49. *Speed of Vehicle* A bus traveled 675 miles in 18 hours. A passenger van traveled 820 miles in 20 hours.

(a) Did they travel at the same rate?　no
(b) Which vehicle traveled at a faster rate?
The van traveled at a faster rate.

▲**50.** *Television Screen Size* A common size for a color television screen is 22 inches wide by 16 inches tall. Does a smaller color television screen that is 11 inches wide by 8.5 inches tall have the same ratio of width to length? no

▲**51.** *Driveway Size* A common size for a driveway in suburban Wheaton, Illinois, is 75 feet long by 20 feet wide. Does a larger driveway that is 105 feet long by 28 feet wide have the same ratio of width to length? yes

75 feet

20 feet

To Think About

52. Determine whether $\dfrac{63}{161} = \dfrac{171}{437}$

(a) by reducing each side to lowest terms.
$\dfrac{63}{161} = \dfrac{9}{23}$ $\dfrac{171}{437} = \dfrac{9}{23}$ yes

(b) by using the equality test for fractions. (This is the cross-product method.)
$63 \times 437 \stackrel{?}{=} 161 \times 171$
$27{,}531 = 27{,}531$ yes

(c) Which method was faster? Why?
The equality test for fractions; for most students it is faster to multiply than to reduce fractions.

53. Determine whether $\dfrac{169}{221} = \dfrac{247}{323}$

(a) by reducing each side to lowest terms.
$\dfrac{169}{221} = \dfrac{13}{17}$ $\dfrac{247}{323} = \dfrac{13}{17}$ yes

(b) by using the equality test for fractions. (This is the cross-product method.)
$169 \times 323 \stackrel{?}{=} 221 \times 247$
$54{,}587 = 54{,}587$ yes

(c) Which method was faster? Why?
The equality test for fractions; for most students it is faster to multiply than to reduce fractions.

Cumulative Review *Calculate.*

54. [3.3.1] $9.6 + 7.8 + 2.56 + 3.004 + 0.1765$
23.1405

55. [3.4.1] 5.92×3.04
17.9968

56. [3.3.2] $29{,}366.215$
$-28{,}963.807$
402.408

57. [3.5.2] $7.03\overline{)181.374}$ 25.8

58. [2.9.1] *Walking Distance* Susan has a goal of walking 20 miles this week. On Monday she walked $3\frac{1}{4}$ miles and on Tuesday she walked $4\frac{3}{8}$ miles. How many more miles does she need to walk to reach her goal? $12\frac{3}{8}$ miles

Quick Quiz 4.2

1. Write as a proportion.
8 is to 18 as 28 is to 63. $\dfrac{8}{18} = \dfrac{28}{63}$

2. Write as a proportion.
13 is to 32 as $3\frac{1}{4}$ is to 8. $\dfrac{13}{32} = \dfrac{3\frac{1}{4}}{8}$

3. Determine if this equation is a proportion.
$\dfrac{15 \text{ shots}}{4 \text{ goals}} \stackrel{?}{=} \dfrac{75 \text{ shots}}{22 \text{ goals}}$ It is not a proportion.

4. **Concept Check** Explain how to determine if
$\dfrac{33 \text{ chairs}}{45 \text{ employees}} \stackrel{?}{=} \dfrac{165 \text{ chairs}}{225 \text{ employees}}$ is a proportion.
Answers may vary

Classroom Quiz 4.2 You may use these problems to quiz your students' mastery of Section 4.2.

1. Write as a proportion.
9 is to 15 as 6 is to 10. **Ans:** $\dfrac{9}{15} = \dfrac{6}{10}$

2. Write as a proportion.
$2\frac{1}{3}$ is to 7 as 7 is to 21. **Ans:** $\dfrac{2\frac{1}{3}}{7} = \dfrac{7}{21}$

3. Determine if this equation is a proportion.
$\dfrac{16 \text{ home runs}}{94 \text{ games}} \stackrel{?}{=} \dfrac{14 \text{ home runs}}{88 \text{ games}}$ **Ans:** It is not a proportion.

How are you doing with your homework assignments in Sections 4.1 to 4.2? Do you feel you have mastered the material so far? Do you understand the concepts you have covered? Before you go further in the textbook, take some time to do each of the following problems.

4.1 *In questions 1–4, write each ratio in simplest form.*

1. 13 to 18

2. 44 to 220

3. $72 to $16

4. 135 meters to 165 meters

5. Sam's take-home pay is $240 per week. $70 per week is withheld for federal taxes and $22 per week is withheld for state taxes.
  (a) Find the ratio of federal withholding to take-home pay.
  (b) Find the ratio of state withholding to take-home pay.

Write each rate in simplest form.

6. 9 flight attendants for 300 passengers

7. 620 gallons of water for each 840 square feet of lawn

Write as a unit rate. Round to the nearest tenth if necessary.

8. A professional bicyclist travels 65 miles in 4 hours. What is the rate in miles per hour?

9. 15 CD players are purchased for $435. What is the cost per CD player?

10. In a certain recipe, 2400 cookies are made with 15 pounds of cookie dough. How many cookies can be made with 1 pound of cookie dough?

4.2 *Write a proportion.*

11. 13 is to 40 as 39 is to 120

12. 116 is to 148 as 29 is to 37

13. If a speedboat can travel 33 nautical miles in 2 hours, then it can travel 49.5 nautical miles in 3 hours.

14. If the cost to manufacture 3000 athletic shoes is $370, then the cost to manufacture 7500 athletic shoes is $925.

Determine whether each equation is a proportion.

15. $\dfrac{14}{31} = \dfrac{42}{93}$

16. $\dfrac{17}{33} = \dfrac{19}{45}$

17. $\dfrac{6.5}{4.8} = \dfrac{120}{96}$

18. $\dfrac{15}{24} = \dfrac{1\frac{5}{8}}{2\frac{3}{5}}$

19. The Pine Street Inn can produce 670 servings of a chicken dinner for homeless people at a cost of $1541. It will therefore cost $1886 to produce 820 servings of the same chicken dinner.

20. For every 30 flights that arrive at Logan Airport in December approximately 4 of them are more than 15 minutes late. Therefore, for every 3000 flights that arrive at Logan Airport in December approximately 400 of them will be more than 15 minutes late.

Now turn to page SA-8 for the answer to each of these problems. Each answer also includes a reference to the objective in which the problem is first taught. If you missed any of these problems, you should stop and review the Examples and Practice Problems in the referenced objective. A little review now will help you master the material in the upcoming sections of the text.

1. $\dfrac{13}{18}$

2. $\dfrac{1}{5}$

3. $\dfrac{9}{2}$

4. $\dfrac{9}{11}$

5. (a) $\dfrac{7}{24}$ **(b)** $\dfrac{11}{120}$

6. $\dfrac{3 \text{ flight attendants}}{100 \text{ passengers}}$

7. $\dfrac{31 \text{ gallons}}{42 \text{ square feet}}$

8. 16.25 miles per hour

9. $29 per CD player

10. 160 cookies per pound of cookie dough

11. $\dfrac{13}{40} = \dfrac{39}{120}$

12. $\dfrac{116}{148} = \dfrac{29}{37}$

13. $\dfrac{33 \text{ nautical miles}}{2 \text{ hours}} = \dfrac{49.5 \text{ nautical miles}}{3 \text{ hours}}$

14. $\dfrac{3000 \text{ shoes}}{\$370} = \dfrac{7500 \text{ shoes}}{\$925}$

15. It is a proportion.

16. It is not a proportion.

17. It is not a proportion.

18. It is a proportion.

19. It is a proportion.

20. It is a proportion.

Student Learning Objectives

After studying this section, you will be able to:

 Solve for the variable *n* in an equation of the form *a* × *n* = *b*.

 Find the missing number in a proportion.

Teaching Tip Some instructors prefer to introduce the notation 3*n* to indicate a product instead of 3 × *n*. Either is acceptable, but students who have never had algebra before may have trouble with the notation.

 Solving for the Variable *n* in an Equation of the Form *a* × *n* = *b*

Consider this expression: "3 times a number yields 15. What is the number?" We could write this as

$$3 \times \boxed{?} = 15$$

and guess that the number $\boxed{?} = 5$. There is a better way of solving this problem, a way that eliminates the guesswork. We will begin by using a **variable.** That is, we will use a letter to represent a number we do not yet know. We briefly used variables in Chapters 1–3. Now we use them more extensively.

Let the letter *n* represent the unknown number. We write

$$3 \times n = 15.$$

This is called an **equation.** An equation has an equals sign. This indicates that the values on each side of it are equivalent. We want to find the number *n* in this equation without guessing. We will not change the value of *n* in the equation if we divide both sides of the equation by 3. Thus if

$$3 \times n = 15,$$

we can say
$$\frac{3 \times n}{3} = \frac{15}{3},$$

which is
$$\frac{3}{3} \times n = 5$$

or
$$1 \times n = 5.$$

Since 1 × any number is the same number, we know that *n* = 5. Any equation of the form *a* × *n* = *b* can be solved in this way. We divide both sides of an equation of the form *a* × *n* = *b* by the number that is multiplied by *n*. (We do this because division is the inverse operation of multiplication. This method will not work for 3 + *n* = 15, since here the 3 is added to *n* and not multiplied by *n*.)

Teaching Example 1 Solve for *n*.

(a) 15 × *n* = 45   **(b)** 32 × *n* = 128

Ans: (a) *n* = 3   **(b)** *n* = 4

EXAMPLE 1 Solve for *n*.

(a) 16 × *n* = 80   **(b)** 24 × *n* = 240

Solution

(a) 16 × *n* = 80

$$\frac{16 \times n}{16} = \frac{80}{16} \qquad \text{Divide each side by 16.}$$

$$n = 5 \qquad \text{because } 16 \div 16 = 1 \text{ and } 80 \div 16 = 5.$$

(b) 24 × *n* = 240

$$\frac{24 \times n}{24} = \frac{240}{24} \qquad \text{Divide each side by 24.}$$

$$n = 10 \qquad \text{because } 24 \div 24 = 1 \text{ and } 240 \div 24 = 10.$$

NOTE TO STUDENT: Fully worked-out solutions to all of the Practice Problems can be found at the back of the text starting at page SP-1

Practice Problem 1 Solve for *n*.

(a) 5 × *n* = 45   *n* = 9   **(b)** 7 × *n* = 84   *n* = 12

The same procedure is followed if the variable n is on the right side of the equation.

EXAMPLE 2 Solve for n.

(a) $66 = 11 \times n$

(b) $143 = 13 \times n$

Solution

(a) $66 = 11 \times n$

$$\frac{66}{11} = \frac{11 \times n}{11}$$ Divide each side by 11.

$$6 = n$$

(b) $143 = 13 \times n$

$$\frac{143}{13} = \frac{13 \times n}{13}$$ Divide each side by 13.

$$11 = n$$

Practice Problem 2 Solve for n.

(a) $108 = 9 \times n$ $12 = n$

(b) $210 = 14 \times n$ $15 = n$

Teaching Example 2 Solve for n.

(a) $420 = 21 \times n$ **(b)** $68 = 4 \times n$

Ans: (a) $20 = n$ **(b)** $17 = n$

Teaching Tip Remind students that $n = 6$ and $6 = n$ mean exactly the same thing. This is called the *symmetric property of equality*.

The numbers in the equations are not always whole numbers, and the answer to an equation is not always a whole number.

EXAMPLE 3 Solve for n.

(a) $16 \times n = 56$

(b) $18.2 = 2.6 \times n$

Solution

(a) $16 \times n = 56$

$$\frac{16 \times n}{16} = \frac{56}{16}$$ Divide each side by 16.

$$n = 3.5$$

$$\begin{array}{r} 3.5 \\ 16\overline{)56.0} \\ \underline{48} \\ 8\,0 \\ \underline{8\,0} \\ 0 \end{array}$$

(b) $18.2 = 2.6 \times n$

$$\frac{18.2}{2.6} = \frac{2.6 \times n}{2.6}$$ Divide each side by 2.6

$$7 = n$$

$$\begin{array}{r} 7. \\ 2.6_\wedge\overline{)18.2_\wedge} \\ \underline{18\,2} \\ 0 \end{array}$$

Teaching Example 3 Solve for n.

(a) $14 \times n = 21$ **(b)** $12.4 = 3.1 \times n$

Ans: (a) $n = 1.5$ **(b)** $4 = n$

Practice Problem 3 Solve for n.

(a) $15 \times n = 63$ $n = 4.2$

(b) $39.2 = 5.6 \times n$ $7 = n$

② Finding the Missing Number in a Proportion

Sometimes one of the pieces of a proportion is unknown. We can use an equation such as $a \times n = b$ and solve for n to find the unknown quantity. Suppose we want to know the value of n in the proportion

$$\frac{5}{12} = \frac{n}{144}.$$

Since this is a proportion, we know that $5 \times 144 = 12 \times n$. Simplifying, we have

$$720 = 12 \times n.$$

Next we divide both sides by 12.

$$\frac{720}{12} = \frac{12 \times n}{12}$$

$$60 = n$$

We check to see if this is correct. Do we have a true proportion?

$$\frac{5}{12} \stackrel{?}{=} \frac{60}{144}$$

$$\frac{5}{12} \diagdown \frac{60}{144} \quad \begin{array}{l} 12 \times 60 = 720 \\ 5 \times 144 = 720 \end{array} \quad \text{The cross products are equal.}$$

Thus $\dfrac{5}{12} = \dfrac{60}{144}$ is true. We have checked our answer.

TO SOLVE FOR A MISSING NUMBER IN A PROPORTION

1. Find the cross products.
2. Divide each side of the equation by the number multiplied by n.
3. Simplify the result.
4. Check your answer.

Teaching Example 4 Find the value of n in $\frac{13}{3} = \frac{26}{n}$.

Ans: $n = 6$

EXAMPLE 4 Find the value of n in $\dfrac{25}{4} = \dfrac{n}{12}$.

Solution $\quad\quad 25 \times 12 = 4 \times n \quad$ Find the cross products.

$$300 = 4 \times n$$

$$\frac{300}{4} = \frac{4 \times n}{4} \quad \text{Divide each side by 4.}$$

$$75 = n$$

Check.   Is this a proportion?

$$\frac{25}{4} \stackrel{?}{=} \frac{75}{12}$$

$$25 \times 12 \stackrel{?}{=} 4 \times 75$$

$$300 = 300 \quad\quad ✓$$

It is a proportion. The answer $n = 75$ is correct.

Practice Problem 4 Find the value of n in

$$\frac{24}{n} = \frac{3}{7}. \quad 56 = n$$

The answer to the next problem is not a whole number.

EXAMPLE 5 Find the value of n in $\dfrac{125}{2} = \dfrac{150}{n}$.

Teaching Example 5 Find the value of n in $\dfrac{6}{n} = \dfrac{30}{11}$.

Ans: $2.2 = n$

Solution

$125 \times n = 2 \times 150$ Find the cross products.   *Check.*

$125 \times n = 300$

$\dfrac{125 \times n}{125} = \dfrac{300}{125}$ Divide each side by 125.

$n = 2.4$

$\dfrac{125}{2} \overset{?}{=} \dfrac{150}{2.4}$

$125 \times 2.4 \overset{?}{=} 2 \times 150$

$300 = 300$ ✓

Practice Problem 5 Find the value of n in

$$\dfrac{176}{4} = \dfrac{286}{n}. \quad n = 6.5$$

EXAMPLE 6 Find the value of n in

$$\dfrac{n}{20} = \dfrac{\frac{3}{4}}{5}.$$

Teaching Example 6 Find the value of n in $\dfrac{\frac{5}{6}}{n} = \dfrac{5}{42}$.

Ans: $7 = n$

Solution $5 \times n = 20 \times \dfrac{3}{4}$ Find the cross products.

$5 \times n = 15$ Simplify.

$\dfrac{5 \times n}{5} = \dfrac{15}{5}$ Divide each side by 5.

$n = 3$

Check.   *Can you verify that this is a proportion?*

Practice Problem 6 Find the value of n in

$$\dfrac{n}{30} = \dfrac{\frac{2}{3}}{4}. \quad n = 5$$

Teaching Tip Stress the importance of labeling the answers with the proper units in real-life problems. Students tend to avoid this, so remind them of the need to establish good habits now.

In real-life situations it is helpful to write the units of measure in the proportion. Remember, order is important. The same units should be in the same position in the fractions.

EXAMPLE 7 If 5 grams of a non-icing additive are placed in 8 liters of diesel fuel, how many grams n should be added to 12 liters of diesel fuel?

Teaching Example 7 If the correct dose of an antibiotic for a 180-pound man is 15 milligrams, what is the correct dose for a 120-pound woman?

Ans: 10 milligrams

Solution   We need to find the value of n in $\dfrac{n \text{ grams}}{12 \text{ liters}} = \dfrac{5 \text{ grams}}{8 \text{ liters}}$.

$8 \times n = 12 \times 5$

$8 \times n = 60$

$\dfrac{8 \times n}{8} = \dfrac{60}{8}$

$n = 7.5$

The answer is 7.5 grams. 7.5 grams of the additive should be added to 12 liters of the diesel fuel.

Check.

$$\frac{7.5 \text{ grams}}{12 \text{ liters}} \stackrel{?}{=} \frac{5 \text{ grams}}{8 \text{ liters}}$$

$$7.5 \times 8 \stackrel{?}{=} 12 \times 5$$

$$60 = 60 \qquad ✓$$

Practice Problem 7 If 2.5 tablespoons of a lawn fertilizer is to be mixed with 3 gallons of water, how many tablespoons of fertilizer should be mixed with 24 gallons of water? $n = 20$

Some answers will be exact values. In other cases we will obtain answers that are rounded to a certain decimal place. Recall that we sometimes use the ≈ symbol, which means "is approximately equal to."

Teaching Example 8 Find the value of n in $\frac{102 \text{ dollars}}{25 \text{ square feet}} = \frac{n \text{ dollars}}{72 \text{ square feet}}$. Round to the nearest tenth.

Ans: $n \approx 293.8$

EXAMPLE 8 Find the value of n in $\frac{141 \text{ miles}}{4.5 \text{ hours}} = \frac{67 \text{ miles}}{n \text{ hours}}$. Round to the nearest tenth.

Solution

$$141 \times n = 67 \times 4.5$$

$$141 \times n = 301.5$$

$$\frac{141 \times n}{141} = \frac{301.5}{141}$$

If we calculate to four decimal places, we have $n = 2.1382$. Rounding to the nearest tenth, $n \approx 2.1$.

The answer to the nearest tenth is $n = 2.1$. The check is up to you.

Practice Problem 8

Find the value of n in $\frac{264 \text{ meters}}{3.5 \text{ seconds}} = \frac{n \text{ meters}}{2 \text{ seconds}}$. Round to the nearest tenth. $n \approx 150.9$

TO THINK ABOUT: Proportions with Mixed Numbers or Fractions Suppose that the proportion contains many fractions or mixed numbers. Could you still follow all the steps? For example, find n when

$$\frac{n}{3\frac{1}{4}} = \frac{5\frac{1}{6}}{2\frac{1}{3}}$$

We have $2\frac{1}{3} \times n = 5\frac{1}{6} \times 3\frac{1}{4}$.

This can be written as

$$\frac{7}{3} \times n = \frac{31}{6} \times \frac{13}{4}$$

$$\boxed{\frac{7}{3} \times n = \frac{403}{24}} \qquad \text{equation (1)}$$

Now we divide each side of equation (1) by $\frac{7}{3}$. Why?

$$\frac{\frac{7}{3} \times n}{\frac{7}{3}} = \frac{\frac{403}{24}}{\frac{7}{3}}$$

Be careful here. The right-hand side means $\frac{403}{24} \div \frac{7}{3}$, which we evaluate by *inverting* the second fraction and multiplying.

$$\frac{403}{\underset{8}{\cancel{24}}} \times \frac{\overset{1}{\cancel{3}}}{7} = \frac{403}{56}$$

Thus $n = \frac{403}{56}$ or $7\frac{11}{56}$. Think about all the steps to solving this problem. Can you follow them? There is another way to do the problem. We could multiply each side of equation (1) by $\frac{3}{7}$.

$$\frac{7}{3} \times n = \frac{403}{24} \qquad \text{equation (1)}$$

$$\frac{3}{7} \times \frac{7}{3} \times n = \frac{3}{7} \times \frac{403}{24}$$

$$n = \frac{403}{56}$$

Why does this work? Now try exercises 56–59 in 4.3 Exercises.

Accuracy is especially important in this section. When you do the 4.3 Exercises, be sure to verify your answers on page SA-8. Take a little extra time with those problems that result in fraction or decimal answers. It is very important that you learn how to do this type of problem.

PRACTICE WATCH DOWNLOAD READ REVIEW

Verbal and Writing Skills

1. Suppose you have an equation of the form $a \times n = b$, where the letters a and b represent whole numbers and $a \neq 0$. Explain in your own words how you would solve the equation.

Divide each side of the equation by the number a.
Calculate $\frac{b}{a}$. The value of n is $\frac{b}{a}$.

2. Suppose you have an equation of the form $\frac{n}{a} = \frac{b}{c}$, where a, b, and c represent whole numbers and $a, c \neq 0$. Explain in your own words how you would solve the equation.

Form the cross product $c \times n = a \times b$. Multiply $a \times b$.
Then use the steps explained in exercise 1.

Solve for n.

3. $8 \times n = 72$
$n = 9$

4. $6 \times n = 72$
$n = 12$

5. $3 \times n = 16.8$
$n = 5.6$

6. $2 \times n = 19.6$
$n = 9.8$

7. $n \times 6.7 = 134$
$n = 20$

8. $n \times 3.8 = 95$
$n = 25$

9. $50.4 = 6.3 \times n$
$n = 8$

10. $40.6 = 5.8 \times n$
$n = 7$

11. $\frac{4}{9} \times n = 22$ (*Hint:* Divide each side by $\frac{4}{9}$.)

$n = 49\frac{1}{2}$

12. $\frac{6}{7} \times n = 26$ (*Hint:* Divide each side by $\frac{6}{7}$.)

$n = 30\frac{1}{3}$

Find the value of n. Check your answer.

13. $\frac{n}{20} = \frac{3}{4}$
$n = 15$

14. $\frac{n}{28} = \frac{3}{7}$
$n = 12$

15. $\frac{6}{n} = \frac{3}{8}$
$n = 16$

16. $\frac{4}{n} = \frac{2}{7}$
$14 = n$

17. $\frac{12}{40} = \frac{n}{25}$
$7.5 = n$

18. $\frac{13}{30} = \frac{n}{15}$
$6.5 = n$

19. $\frac{50}{100} = \frac{2.5}{n}$
$n = 5$

20. $\frac{40}{160} = \frac{1.5}{n}$
$n = 6$

21. $\frac{n}{6} = \frac{150}{12}$
$n = 75$

22. $\frac{n}{22} = \frac{25}{11}$
$n = 50$

23. $\frac{15}{4} = \frac{n}{6}$
$22.5 = n$

24. $\frac{16}{10} = \frac{n}{9}$
$14.4 = n$

25. $\frac{240}{n} = \frac{5}{4}$
$n = 192$

26. $\frac{180}{n} = \frac{4}{3}$
$n = 135$

Find the value of n. Round your answer to the nearest tenth when necessary.

27. $\frac{21}{n} = \frac{2}{3}$
$n = 31.5$

28. $\frac{62}{n} = \frac{5}{4}$
$n = 49.6$

29. $\frac{9}{26} = \frac{n}{52}$
$n = 18$

30. $\frac{12}{8} = \frac{21}{n}$
$n = 14$

31. $\frac{15}{12} = \frac{10}{n}$
$n = 8$

32. $\frac{n}{18} = \frac{3.5}{1}$
$n = 63$

33. $\frac{n}{36} = \frac{4.5}{1}$
$n = 162$

34. $\frac{2.5}{n} = \frac{0.5}{10}$
$n = 50$

35. $\frac{1.8}{n} = \frac{0.7}{12}$
$n \approx 30.9$

36. $\frac{7}{16} = \frac{n}{26.2}$
$n \approx 11.5$

37. $\frac{11}{12} = \frac{n}{32.8}$
$n \approx 30.1$

38. $\frac{12.5}{16} = \frac{n}{12}$
$n \approx 9.4$

39. $\frac{13.8}{15} = \frac{n}{6}$
$n \approx 5.5$

40. $\frac{5}{n} = \frac{12\frac{1}{2}}{100}$
$n = 40$

41. $\frac{3}{n} = \frac{6\frac{1}{4}}{100}$
$n = 48$

Applications *Find the value of n. Round to the nearest hundredth when necessary.*

42. $\dfrac{n \text{ grams}}{10 \text{ liters}} = \dfrac{7 \text{ grams}}{25 \text{ liters}}$

$n = 2.8$

43. $\dfrac{n \text{ pounds}}{20 \text{ ounces}} = \dfrac{2 \text{ pounds}}{32 \text{ ounces}}$

$n = 1.25$

44. $\dfrac{190 \text{ kilometers}}{3 \text{ hours}} = \dfrac{n \text{ kilometers}}{5 \text{ hours}}$

$n \approx 316.67$

45. $\dfrac{145 \text{ kilometers}}{2 \text{ hours}} = \dfrac{220 \text{ kilometers}}{n \text{ hours}}$

$n \approx 3.03$

46. $\dfrac{50 \text{ gallons}}{12 \text{ acres}} = \dfrac{36 \text{ gallons}}{n \text{ acres}}$

$n = 8.64$

47. $\dfrac{32 \text{ meters}}{5 \text{ yards}} = \dfrac{24 \text{ meters}}{n \text{ yards}}$

$n = 3.75$

48. $\dfrac{3 \text{ kilograms}}{6.6 \text{ pounds}} = \dfrac{n \text{ kilograms}}{10 \text{ pounds}}$

$n \approx 4.55$

49. $\dfrac{36.4 \text{ feet}}{5 \text{ meters}} = \dfrac{n \text{ feet}}{12 \text{ meters}}$

$n = 87.36$

50. $\dfrac{12 \text{ quarters}}{3 \text{ dollars}} = \dfrac{87 \text{ quarters}}{n \text{ dollars}}$

$n = 21.75$

51. $\dfrac{35 \text{ dimes}}{3.5 \text{ dollars}} = \dfrac{n \text{ dimes}}{8 \text{ dollars}}$

$n = 80$

52. $\dfrac{2\frac{1}{2} \text{ acres}}{3 \text{ people}} = \dfrac{n \text{ acres}}{5 \text{ people}}$ $n = 4\frac{1}{6}$

53. $\dfrac{3\frac{1}{4} \text{ feet}}{8 \text{ pounds}} = \dfrac{n \text{ feet}}{12 \text{ pounds}}$ $n = 4\frac{7}{8}$

▲ **54.** *Photography* A photographic negative is 3.5 centimeters wide and 2.5 centimeters tall. If you want to make a color print that is 6 centimeters tall, how wide will the print be?

8.4 centimeters

▲ **55.** *Photography* A color photograph is 5 inches wide and 3 inches tall. If you want to make an enlargement of this photograph that is 6.6 inches tall, how wide will the enlargement be?

11 inches

To Think About *Study the "To Think About" example in the text on page 280. Then solve for n in exercises 56–59. Express n as a mixed number.*

56. $\dfrac{n}{7\frac{1}{4}} = \dfrac{2\frac{1}{5}}{4\frac{1}{8}}$ $n \times 4\frac{1}{8} = 15\frac{19}{20}$ $n = 3\frac{13}{15}$

57. $\dfrac{n}{2\frac{1}{3}} = \dfrac{4\frac{5}{6}}{3\frac{1}{9}}$ $n \times 3\frac{1}{9} = 11\frac{5}{18}$ $n = 3\frac{5}{8}$

58. $\dfrac{9\frac{3}{4}}{n} = \dfrac{8\frac{1}{2}}{4\frac{1}{3}}$ $n \times 8\frac{1}{2} = 42\frac{1}{4}$ $n = 4\frac{33}{34}$

59. $\dfrac{8\frac{1}{6}}{n} = \dfrac{5\frac{1}{2}}{7\frac{1}{3}}$ $n \times 5\frac{1}{2} = 59\frac{8}{9}$ $n = 10\frac{8}{9}$

Cumulative Review *Evaluate by doing each operation in the proper order.*

60. **[1.6.2]** $4^3 + 20 \div 5 + 6 \times 3 - 5 \times 2$
$64 + 4 + 18 - 10 = 76$

61. **[1.6.2]** $(3 + 1)^3 - 30 \div 6 - 144 \div 12$
47

62. **[3.1.1]** Write a word name for the decimal 0.563.
five hundred sixty-three thousandths

63. **[3.1.1]** Write thirty-four ten-thousandths in decimal notation. 0.0034

64. **[1.8.2]** *Profit on Cell Phones* If a man purchases 156 cell phones for $32 each and sells half of them for $45 and half of them for $39, how much profit will he make?
$1560

65. **[1.8.2]** *Soccer Team* The North Bend Women's Soccer League has eight teams. If each team plays all the others twice, how many games will have been played? (Think carefully. This is a challenging question.)
56 games

Quick Quiz 4.3

1. Solve.

$$\frac{n}{26} = \frac{9}{130} \quad n = 1.8$$

2. Solve.

$$\frac{8}{6} = \frac{2\frac{2}{3}}{n} \quad n = 2$$

3. Solve. Round to the nearest tenth.

$$\frac{17 \text{ hits}}{93 \text{ pitches}} = \frac{n \text{ hits}}{62 \text{ pitches}} \quad n \approx 11.3$$

4. **Concept Check** Explain how you would solve the proportion $\dfrac{2\frac{1}{2}}{3\frac{3}{4}} = \dfrac{16\frac{1}{2}}{n}$. Answers may vary

Classroom Quiz 4.3 You may use these problems to quiz your students' mastery of Section 4.3.

1. Solve.

$$\frac{n}{33} = \frac{28}{132} \quad \textbf{Ans: } n = 7$$

2. Solve.

$$\frac{3\frac{1}{4}}{2} = \frac{13}{n} \quad \textbf{Ans: } n = 8$$

3. Solve. Round to the nearest tenth.

$$\frac{7 \text{ inches of snow}}{56 \text{ inches of rain}} = \frac{n \text{ inches of snow}}{35 \text{ inches of rain}} \quad \textbf{Ans: } n \approx 4.4$$

Solving Applied Problems Using Proportions

Let us examine a variety of applied problems that can be solved by proportions.

Student Learning Objective

After studying this section, you will be able to:

 Solve applied problems using proportions.

EXAMPLE 1 A company that makes eyeglasses conducted a recent survey using a quality control test. It was discovered that 37 pairs of eyeglasses in a sample of 120 pairs of eyeglasses were defective. If this rate remains the same each year, how many of the 36,000 pairs of eyeglasses made by this company each year are defective?

Solution

Mathematics Blueprint for Problem Solving

Gather the Facts	What Am I Asked to Do?	How Do I Proceed?	Key Points to Remember
Sample: 37 defective pairs in a total of 120 pairs 36,000 pairs were made by the company.	Find how many of the 36,000 pairs of eyeglasses are defective.	Set up a proportion comparing defective eyeglasses to total eyeglasses.	Make sure one fraction represents the sample and one fraction represents the total number of eyeglasses made by the company.

We will use the letter n to represent the number of defective eyeglasses in the total.

$$\underbrace{\frac{37 \text{ defective pairs}}{120 \text{ total pairs of eyeglasses}}}_{\substack{\text{We compare} \\ \text{the sample}}} = \underbrace{\frac{n \text{ defective pairs}}{36,000 \text{ total pairs of eyeglasses}}}_{\text{the total number}}$$

$37 \times 36,000 = 120 \times n$ Find the cross products.

$1,332,000 = 120 \times n$ Simplify.

$\dfrac{1,332,000}{120} = \dfrac{120 \times n}{120}$ Divide each side by 120.

$11,100 = n$

Thus, if the rate of defective eyeglasses holds steady, there are about 11,100 defective pairs of eyeglasses made by the company each year.

Teaching Example 1 Students in a math class counted that 26 out of 234 cars in one of the school's parking lots were compact cars. If that rate is true for all of the parking lots, how many of the total of 1080 cars in those lots are compact?

Ans: 120 cars

Teaching Tip This is a good time to stress the helpfulness of making an estimate in solving a proportion problem. Usually, a mistake will be obvious when compared with an estimate.

Practice Problem 1 Yesterday an automobile assembly line produced 243 engines, of which 27 were defective. If the same rate is true each day, how many of the 4131 engines produced this month are defective?

about 459 engines

NOTE TO STUDENT: Fully worked-out solutions to all of the Practice Problems can be found at the back of the text starting at page SP-1

Looking back at Example 1, perhaps it occurred to you that the fractions in the proportion could be set up in an alternative way. **You can set up this problem in several different ways as long as the units are in correctly**

corresponding positions. It would be correct to set up the problem in the form

$$\frac{\text{defective pairs in sample}}{\text{total defective pairs}} = \frac{\text{total glasses in sample}}{\text{total glasses made by company}}$$

or

$$\frac{\text{total glasses in sample}}{\text{defective pairs in sample}} = \frac{\text{total glasses made by company}}{\text{total defective pairs}}$$

But we **cannot** set up the problem this way.

$$\frac{\text{defective pairs in sample}}{\text{total glasses made by company}} = \frac{\text{total defective pairs}}{\text{total glasses in sample}}$$

This is *not* correct. Do you see why?

EXAMPLE 2   Ted's car can go 245 miles on 7 gallons of gas. Ted wants to take a trip of 455 miles. Approximately how many gallons of gas will this take?

Solution   Let n = the unknown number of gallons.

$$\frac{245 \text{ miles}}{7 \text{ gallons}} = \frac{455 \text{ miles}}{n \text{ gallons}}$$

$245 \times n = 7 \times 455$ Find the cross products.

$245 \times n = 3185$ Simplify.

$\dfrac{245 \times n}{245} = \dfrac{3185}{245}$ Divide both sides by 245.

$n = 13$

Ted will need approximately 13 gallons of gas for the trip.

Practice Problem 2   Cindy's car travels 234 miles on 9 gallons of gas. How many gallons of gas will Cindy need to take a 312-mile trip?   12 gallons

EXAMPLE 3   In a certain gear, Alice's 18-speed bicycle has a gear ratio of three revolutions of the pedal for every two revolutions of the bicycle wheel. If her bicycle wheel is turning at 65 revolutions per minute, how many times must she pedal per minute?

Solution   Let n = the number of revolutions of the pedal.

$$\frac{3 \text{ revolutions of the pedal}}{2 \text{ revolutions of the wheel}} = \frac{n \text{ revolutions of the pedal}}{65 \text{ revolutions of the wheel}}$$

$3 \times 65 = 2 \times n$ Cross-multiply.

$195 = 2 \times n$ Simplify.

$\dfrac{195}{2} = \dfrac{2 \times n}{2}$ Divide both sides by 2.

$97.5 = n$

Alice will pedal at the rate of 97.5 revolutions per 1 minute.

Practice Problem 3 Alicia must pedal at 80 revolutions per minute to ride her bicycle at 16 miles per hour. If she pedals at 90 revolutions per minute, how fast will she be riding? 18 miles per hour

EXAMPLE 4 Tim operates a bicycle rental center during the summer months on the island of Martha's Vineyard. He discovered that when the ferryboats brought 8500 passengers a day to the island, his center rented 340 bicycles a day. Next summer the ferryboats plan to bring 10,300 passengers a day to the island. How many bicycles a day should Tim plan to rent?

Solution   Two important cautions are necessary before we solve the proportion. We need to be sure that the bicycle rentals are directly related to the number of people on the ferryboat. (Presumably, people who fly to the island or who take small pleasure boats to the island also rent bicycles.)

Next we need to be sure that the people who represent the increase in passengers per day would be as likely to rent bicycles as the present number of passengers do. For example, if the new visitors to the island are all senior citizens, they are not as likely to rent bicycles as younger people. If we assume those two conditions are satisfied, then we can solve the problem as follows.

$$\frac{8500 \text{ passengers per day now}}{340 \text{ bike rentals per day now}} = \frac{10,300 \text{ passengers per day later}}{n \text{ bike rentals per day later}}$$

$$8500 \times n = 340 \times 10,300$$
$$8500 \times n = 3{,}502{,}000$$
$$\frac{8500 \times n}{8500} = \frac{3{,}502{,}000}{8500}$$
$$n = 412$$

If the two conditions are satisfied, we would predict 412 bicycle rentals.

Practice Problem 4 For every 4050 people who walk into Tom's Souvenir Shop, 729 make a purchase. Assuming the same conditions, if 5500 people walk into Tom's Souvenir Shop, how many people may be expected to make a purchase? 990 people

Teaching Example 4 Cindi participates in an annual bird count for the local wildlife society. Over a one-hour period, she counted 84 birds in her backyard. 35 of those birds were sparrows. Assuming the same conditions throughout the day, how many of the 336 birds that she counted over the next few hours were sparrows?

Ans: 140 birds

Teaching Tip Gear ratio problems often confuse students. Ask them why a bicycle might have a gear ratio of 3 to 2 as well as 4 to 5. Ask them which ratio would help them to ride at a higher rate of speed. Usually, this type of topic will generate a good class discussion.

Wildlife Population Counting   Biologists and others who observe or protect wildlife sometimes use the capture-mark-recapture method to determine how many animals are in a certain region. In this approach some animals are caught and tagged in a way that does not harm them. They are then released into the wild, where they mix with their kind.

It is assumed (usually correctly) that the tagged animals will mix throughout the entire population in that region, so that when they are recaptured in a future sample, the biologists can use them to make reasonable estimates about the total population. We will employ the capture-mark-recapture method in the next example.

EXAMPLE 5 A biologist catches 42 fish in a lake and tags them. She then quickly returns them to the lake. In a few days she catches a new sample of 50 fish. Of those 50 fish, 7 have her tag. Approximately how many fish are in the lake?

Solution

$$\frac{42 \text{ fish tagged in 1st sample}}{n \text{ fish in lake}} = \frac{7 \text{ fish tagged in 2nd sample}}{50 \text{ fish caught in 2nd sample}}$$

$$42 \times 50 = 7 \times n$$
$$2100 = 7 \times n$$
$$\frac{2100}{7} = \frac{7 \times n}{7}$$
$$300 = n$$

Assuming that no tagged fish died and that the tagged fish mixed throughout the population of fish in the lake, we estimate that there are 300 fish in the lake.

Practice Problem 5 A park ranger in Alaska captures and tags 50 bears. He then releases them to range through the forest. Sometime later he captures 50 bears. Of the 50, 4 have tags from the previous capture. Estimate the number of bears in the forest.   625 bears

Developing Your Study Skills

Getting Help

Getting the right kind of help at the right time can be a key ingredient in being successful in mathematics. When you have gone to class on a regular basis, taken careful notes, methodically read your textbook, and diligently done your homework—in other words, when you have made every effort possible to learn the mathematics—you may still find that you are having difficulty. If this is the case, then you need to seek help. Make an appointment with your instructor to find out what help is available to you. The instructor, tutoring services, a mathematics lab, videotapes, and computer software may be among the resources you can draw on.

Verbal and Writing Skills

1. Dan saw 12 people on the beach on Friday night. He counted 5 dogs on the beach at that time. On Saturday night he saw 60 people on the same beach. He is trying to estimate how many dogs might be on the beach Saturday night. He started by writing the equation

$$\frac{12 \text{ people}}{5 \text{ dogs}} = \quad .$$

Explain how he should set up the rest of the proportion.

He should continue with people on the top of the fraction. That would be 60 people that he observed on Saturday night. He does not know the number of dogs, so this would be n. The proportion would be

$$\frac{12 \text{ people}}{5 \text{ dogs}} = \frac{60 \text{ people}}{n \text{ dogs}}.$$

2. Connie drove to the top of Mount Washington. As she drove up the roadway she observed 15 cars. On the same trip she observed 17 people walking. Later that afternoon she drove down the mountain. On that trip she observed 60 cars. She is trying to estimate how many people she might see walking. She started by writing the equation

$$\frac{15 \text{ cars}}{17 \text{ people}} = \quad .$$

Explain how she should set up the rest of the proportion.

She should continue with the number of cars observed on her trip down the mountain on the top of the fraction. That would be 60. She does not know the number of people, so this would be n. The proportion would be

$$\frac{15 \text{ cars}}{17 \text{ people}} = \frac{60 \text{ cars}}{n \text{ people}}.$$

Applications

3. *Car Repairs* An automobile dealership has found that for every 140 cars sold, 23 will be brought back to the dealer for major repairs. If the dealership sells 980 cars this year, approximately how many cars will be brought back for major repairs?   161 cars

4. *Hotel Management* The policy at the Colonnade Hotel is to have 19 desserts for every 16 people if a buffet is being served. If the Saturday buffet has 320 people, how many desserts must be available?   380 desserts

5. *Consumer Product Use* The directions on a bottle of bleach say to use $\frac{3}{4}$ cup bleach for every 1 gallon of soapy water. Ron needs a 4-gallon mixture to mop his floors. How many cups of bleach will he need?   3 cups

6. *Food Preparation* To make an 8-ounce serving of hot chocolate, $1\frac{1}{2}$ tablespoons of cocoa are needed. How much cocoa is needed to make 12 ounces of hot chocolate?

$2\frac{1}{4}$ tablespoons

7. *Measurement* There are approximately $1\frac{1}{2}$ kilometers in one mile. Approximately how many kilometers are in 5 miles?

$7\frac{1}{2}$ kilometers

8. *Measurement* There are approximately $2\frac{1}{2}$ centimeters in one inch. Approximately how many centimeters are in one foot?

30 centimeters

9. *Exchange Rate* When Julie flew to Hong Kong in 2007, the exchange rate was 39 Hong Kong dollars for every 5 U.S. dollars. If Julie brought 180 U.S. dollars for spending money, how many Hong Kong dollars did she receive?

1404 Hong Kong dollars

10. *Exchange Rate* One day in February 2007, one U.S. dollar was worth 0.511 British pounds. Frederich exchanged 220 U.S. dollars when he arrived in London. How many British pounds did he receive?

112.42 British pounds

Shadow Length *In exercises 11 and 12 two nearby objects cast shadows at the same time of day. The ratio of the height of one of the objects to the length of its shadow is equal to the ratio of the height of the other object to the length of its shadow.*

▲ **11.** A pro football offensive tackle who stands 6.5 feet tall casts a 5-foot shadow. At the same time, the football stadium, which he is standing next to, casts a shadow of 152 feet. How tall is the stadium? Round your answer to the nearest tenth.
197.6 feet

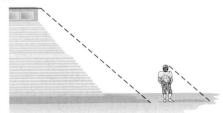

▲ **12.** In Copper Center, Alaska, at 2:00 P.M., a boulder that is 7 feet high casts a shadow that is 11 feet long. At that same time, Melinda Tobey is standing by a tree that is on the bank of the river. She measured the tree and found it is exactly 22 feet tall. The tree has a shadow that crosses the entire width of the river. How wide is the river? Round your answer to the nearest tenth.   34.6 feet

13. ***Map Scale*** On a tour guide map of Madagascar, the scale states that 3 inches represent 125 miles. Two beaches are 5.2 inches apart on the map. What is the approximate distance in miles between the two beaches? Round your answer to the nearest mile.   217 miles

14. ***Map Scale*** On Dr. Jennings's map of Antarctica, the scale states that 4 inches represent 250 miles of actual distance. Two Antarctic mountains are 5.7 inches apart on the map. What is the approximate distance in miles between the two mountains? Round your answer to the nearest mile.
356 miles

15. ***Food Preparation*** In his curry chicken recipe, Deepak uses 3 cups of curry sauce for every 8 people. How many cups of curry sauce will he need to make curry chicken for a dinner party of 34 people? Write your answer as a mixed number.
$12\frac{3}{4}$ cups

16. ***Food Preparation*** Bianca's chocolate fondue recipe uses 4 cups of chocolate chips to make fondue sauce for 6 people. How many cups of chocolate chips are needed to make fondue for 26 people? Write your answer as a mixed number.
$17\frac{1}{3}$ cups

17. ***Basketball*** During a basketball game against the Miami Heat, the Denver Nuggets made 17 out of 25 free throws attempted. If they attempt 150 free throws in the remaining games of the season, how many will they make if their success rate remains the same?
102 free throws

18. ***Baseball*** A baseball pitcher gave up 52 earned runs in 260 innings of pitching. At that rate, how many runs would he give up in a 9-inning game? (This decimal is called the pitcher's *earned run average*.)
1.8 runs

19. ***Fuel Efficiency*** In her Dodge Neon, Claire can drive 192 miles on 6 gallons of gas. During spring break, she plans to drive 600 miles. How many gallons of gas will she use?
18.75 gallons

20. ***Fuel Efficiency*** In her Nissan Pathfinder, Juanita can drive 75 miles on 5 gallons of gas. She drove 318 miles for a business trip. How many gallons of gas did she use?
21.2 gallons

21. ***Wildlife Population Counting*** An ornithologist is studying hawks in the Adirondack Mountains. She catches 24 hawks over a period of one month, tags them, and releases them back into the wild. The next month, she catches 20 hawks and finds that 12 are already tagged. Estimate the number of hawks in this part of the mountains.
40 hawks

22. ***Wildlife Population Counting*** In Kenya, a worker at the game preserve captures 26 giraffes, tags them, and then releases them back into the preserve. The next month, he captures 18 giraffes and finds that 6 of them have already been tagged. Estimate the number of giraffes on the preserve.
78 giraffes

23. *Farming* Bill and Shirley Grant are raising tomatoes to sell at the local co-op. The farm has a yield of 425 pounds of tomatoes for every 3 acres. The farm has 14 acres of good tomatoes. The crop this year should bring $1.80 per pound for each pound of tomatoes. How much will the Grants get from the sale of the tomato crop? $3570

▲ **24. *Painting*** A paint manufacturer suggests 2 gallons of flat latex paint for every 750 square feet of wall. A painter is going to paint 7875 square feet of wall in a Dallas office building with paint that costs $8.50 per gallon. How much will the painter spend for the paint?
$178.50

25. *Manufacturing Quality* A company that manufactures computer chips expects 5 out of every 100 made to be defective. In a shipment of 5400 chips, how many are expected to be defective?
270 chips

26. *Customer Satisfaction* The editor of a small-town newspaper conducted a survey to find out how many customers are satisfied with their delivery service. Of the 100 people surveyed, 88 customers said they were satisfied. If 1700 people receive the newspaper, how many are satisfied?
1496 people

To Think About

Cooking *The following chart is used for several brands of instant mashed potatoes. Use this chart in answering exercises 27–30.*

TO MAKE	WATER	MARGARINE OR BUTTER	SALT (optional)	MILK	FLAKES
2 servings	2/3 cup	1 tablespoon	1/8 teaspoon	1/4 cup	2/3 cup
4 servings	1-1/3 cups	2 tablespoons	1/4 teaspoon	1/2 cup	1-1/3 cups
6 servings	2 cups	3 tablespoons	1/2 teaspoon	3/4 cup	2 cups
Entire box	5 cups	1/2 cup	1 teaspoon	2-1/2 cups	Entire box

27. How many cups of water and how many cups of milk are needed to make enough mashed potatoes for three people?

1 cup of water and $\frac{3}{8}$ cup of milk

28. How many cups of water and how many cups of milk are needed to make enough mashed potatoes for five people?

$1\frac{2}{3}$ cups of water and $\frac{5}{8}$ cup of milk

29. Phil and Melissa live in Denver. They found that the instructions on the box say that at high altitudes (above 5000 feet) the amount of water should be reduced by $\frac{1}{4}$. If you had to make enough mashed potatoes for eight people in a high-altitude city, how many cups of water and how many cups of milk would be needed?

2 cups of water and 1 cup of milk

30. Noah and Olivia live in Denver. They found that the instructions on the box say that at high altitudes (above 5000 feet) the amount of water should be reduced by $\frac{1}{4}$.

(a) If you had to make two boxes of mashed potatoes at a high altitude, how many cups of water and how many cups of milk would be used?

(b) How many servings would be obtained?

(a) $7\frac{1}{2}$ cups of water and 5 cups of milk   **(b)** 30 servings

Baseball Salaries *During the 2006 playing season, Albert Pujols of the St. Louis Cardinals hit 49 home runs and was paid an annual salary of $14,000,000. During the same season, Alfonso Soriano of the Washington Nationals hit 46 home runs and was paid an annual salary of $10,000,000. (Source: www.usatoday.com)*

31. Express the salary of each player as a unit rate in terms of dollars paid to home runs hit.
Albert Pujols, approximately $285,714 for each home run; Alfonso Soriano, approximately $217,391 for each home run

32. Which player hit more home runs per dollar?
Alfonso Soriano

Basketball Salaries *During the 2005–2006 basketball season, Ray Allen of the Seattle Super Sonics made 269 three-point shots and was paid an annual salary of $13,220,000. During the same season, Gilbert Arenas of the Washington D.C. Wizards made 199 three-point shots and was paid an annual salary of $10,240,000. (Source: www.usatoday.com)*

33. Express the salary of each player as a unit rate in terms of dollars paid to three-point shots made.
Ray Allen, approximately $49,145 for each three-point shot; Gilbert Arenas, approximately $51,457 for each three-point shot

34. Which player made more three-point shots per dollar?
Ray Allen

Cumulative Review

35. **[1.7.1]** Round to the nearest hundred. 56,179
56,200

36. **[1.7.1]** Round to the nearest ten thousand. 196,379,910   196,380,000

37. **[3.2.3]** Round to the nearest tenth. 56.148
56.1

38. **[3.2.3]** Round to the nearest ten-thousandth. 2.74895
2.7490

▲ **39.** **[2.9.1]** ***Sunglass Production*** An eyewear company makes very expensive carbon fiber sunglasses. The material is made into long sheets, and the basic eyeglass frame is punched out by a machine. It takes a section measuring $1\frac{3}{16}$ feet by $\frac{4}{5}$ foot to make one pair of glasses.

(a) How many square feet of this material are needed for one frame? $\frac{19}{20}$ of a square foot

(b) How many square feet of this material are needed for 1500 frames?
1425 square feet

Quick Quiz 4.4 Solve using a proportion. Round your answer to the nearest hundredth when necessary.

1. A copper cable 36 feet long weighs 160 pounds. How much will 54 feet of this cable weigh?   240 pounds

2. If 11 inches on a map represent a distance of 64 miles, what distance does 5 inches represent?   29.09 miles

3. During the first few games of the basketball season, Caleb shot 16 free throws and made 7 of them. If Caleb shoots free throws at the same rate as the first few games, he expects to shoot 100 more during the rest of the season. How many of these additional free throws should he expect to make? Round to the nearest whole number.   44 free throws

4. **Concept Check** When Fred went to France he discovered that 70 euros were worth 104 American dollars. He brought 400 American dollars on his trip. Explain how he would find what that is worth in euros.   Answers may vary

Classroom Quiz 4.4 You may use these problems to quiz the students' mastery of Section 4.4.

Solve using a proportion. Round your answer to the nearest hundredth when necessary.

1. Sam's recipe for pancakes calls for 4 eggs and will serve 13 people. If he wants to feed 39 people, how many eggs will he need?   **Ans:** 12 eggs

2. Russ and Norma Camp found it would cost $320 per year to fertilize their lawn of 3000 square feet. How much would it cost to fertilize 5000 square feet?   **Ans:** $533.33

3. While playing softball, Olivia got 5 hits in 31 times at bat last week. During the summer she was at bat 120 times. How many hits would we expect she got over the summer if she gets hits at the same rate? Round your answer to the nearest whole number.
Ans: 19 hits

Putting Your Skills to Work: Use Math to Save Money

CHOOSING A CELL PHONE PLAN

Everyone wants to comparison shop in order to save money. Finding the best cell phone plan is one place where comparison shopping can really help, especially when you're trying to work within a budget. Consider the story of Jake.

Jake is interested in purchasing a new cell phone plan. He would like to call his family, friends, and co-workers, as well as send/receive text and picture messages. He hopes he can afford to access the Internet, too, so he can check his e-mail with his phone. He wants to find a cell phone plan that would allow him to do these things for between $69 and $89 per month.

Analyzing the Options

He did some research and he is considering these cell phone plans:

Calling Plan A: 300 minutes per month, $0.20 each additional minute: **$39.99 per month**

Calling Plan B: 450 minutes per month, $0.45 each additional minute: **$39.99 per month**

Calling Plan C: 600 minutes per month, $0.20 each additional minute: **$49.99 per month**

Calling Plan D: 900 minutes per month, $0.40 each additional minute: **$59.99 per month**

Which calling plan is the best choice if. . .

1. Jake plans to talk for 100 minutes per month?
   Either A or B

2. Jake plans to talk for 350 minutes per month?
   B

3. Jake plans to talk for 470 minutes per month?
   B

4. Jake plans to talk for 550 minutes per month?
   C

5. Jake plans to talk for 11 hours per month?
   D

6. Jake plans to talk for 16 hours, 40 minutes per month?   D

Making the Best Choice to Save Money

Jake expects he will talk between 500 and 750 minutes per month, and is therefore considering Calling Plan C or Calling Plan D. Remember, Jake would like to send/receive text and picture messages. He'd also like to check his e-mail on the Internet with his phone. Jake's research also shows that "message bundles" are available at an additional charge to the calling plans.

Bundle A: 200 messages (text, picture, video, and IM) per month, $0.10 each additional message: **$5.00 per month**

Bundle B: 1500 messages (text, picture, video, and IM) per month, $0.05 each additional message: **$15.00 per month**

Bundle C: Unlimited text, picture, video, and IM messages per month: **$20.00 per month**

Bundle D: Unlimited web access and unlimited text, picture, video, and IM messages per month: **$35.00 per month**

Bundle D is the only bundle that would provide Jake with web access to check his e-mail in addition to messaging. If Jake wants to keep the cost of his cell phone service between $69 and $89 per month, determine the following:

7. The cost of Calling Plan C and Bundle D per month. $84.99

8. The cost of Calling Plan D and Bundle D per month $94.99

Jake is excited that Calling Plan C and Bundle D fit his budget. However, he is nervous about what will happen if he goes over his minutes with Calling Plan C.

9. Determine the total cost for one month if Jake talks for 750 minutes with Calling Plan C and Bundle D. $114.99

10. If you were Jake, which calling plan and bundle would you choose? Why?
    Answers may vary

Topic	Procedure	Examples
Forming a ratio, p. 261.	A *ratio* is the comparison of two quantities that have the same units. A ratio is usually expressed as a fraction. The fraction should be in reduced form.	**1.** Find the ratio of 20 books to 35 books. $$\frac{20}{35} = \frac{4}{7}$$ **2.** Find the ratio in simplest form of $88 : 99$. $$\frac{88}{99} = \frac{8}{9}$$ **3.** Bob earns \$250 each week, but \$15 is withheld for medical insurance. Find the ratio of medical insurance to total pay. $$\frac{\$15}{\$250} = \frac{3}{50}$$
Forming a rate, p. 263.	A *rate* is a comparison of two quantities that have different units. A rate is usually expressed as a fraction in reduced form.	A college has 2520 students with 154 faculty. What is the rate of students to faculty? $$\frac{2520 \text{ students}}{154 \text{ faculty}} = \frac{180 \text{ students}}{11 \text{ faculty}}$$
Forming a unit rate, p. 263.	A *unit rate* is a rate with a denominator of 1. Divide the denominator into the numerator to obtain the unit rate.	A car traveled 416 miles in 8 hours. Find the unit rate. $$\frac{416 \text{ miles}}{8 \text{ hours}} = 52 \text{ miles/hour}$$ Bob spread 50 pounds of fertilizer over 1870 square feet of land. Find the unit rate of square feet per pound. $$\frac{1870 \text{ square feet}}{50 \text{ pounds}} = 37.4 \text{ square feet/pound}$$
Writing proportions, p. 269.	A *proportion* is a statement that two rates or two ratios are equal. The proportion statement a is to b as c is to d can be written $$\frac{a}{b} = \frac{c}{d}.$$	Write a proportion for 17 is to 34 as 13 is to 26. $$\frac{17}{34} = \frac{13}{26}$$
Determining whether a relationship is a proportion, p. 270.	For any two fractions where $b \neq 0$ and $d \neq 0$, $\frac{a}{b} = \frac{c}{d}$ if and only if $a \times d = b \times c$. A proportion is a statement that two rates or two ratios are equal.	**1.** Is this a proportion? $\frac{7}{56} \stackrel{?}{=} \frac{3}{24}$ $$7 \times 24 \stackrel{?}{=} 56 \times 3$$ $$168 = 168 \checkmark$$ It is a proportion. **2.** Is this a proportion? $$\frac{64 \text{ gallons}}{5 \text{ acres}} \stackrel{?}{=} \frac{89 \text{ gallons}}{7 \text{ acres}}$$ $$64 \times 7 \stackrel{?}{=} 5 \times 89$$ $$448 \neq 445$$ It is not a proportion.
Solving a proportion, p. 278.	To solve a proportion where the value n is not known: **1.** Cross-multiply. **2.** Divide both sides of the equation by the number multiplied by n.	Solve for n. $$\frac{17}{n} = \frac{51}{9}$$ $17 \times 9 = 51 \times n$ Cross-multiply. $153 = 51 \times n$ Simplify. $\dfrac{153}{51} = \dfrac{51 \times n}{51}$ Divide by 51. $3 = n$

Topic	Procedure	Examples
Solving applied problems, p. 285.	**1.** Write a proportion with n representing the unknown value. **2.** Solve the proportion.	Bob purchased eight notebooks for $19. How much would 14 notebooks cost? $$\frac{8 \text{ notebooks}}{\$19} = \frac{14 \text{ notebooks}}{n}$$ $$8 \times n = 19 \times 14$$ $$8 \times n = 266$$ $$\frac{8 \times n}{8} = \frac{266}{8}$$ $$n = 33.25$$ The 14 notebooks would cost $33.25.

Chapter 4 Review Problems

Section 4.1

Write in simplest form. Express your answer as a fraction.

1. 88 : 40

$\dfrac{11}{5}$

2. 65 : 39

$\dfrac{5}{3}$

3. 28 : 35

$\dfrac{4}{5}$

4. 250 : 475

$\dfrac{10}{19}$

5. $2\dfrac{1}{3}$ to $4\dfrac{1}{4}$

$\dfrac{28}{51}$

6. 27 to 81

$\dfrac{1}{3}$

7. 180 to 531

$\dfrac{20}{59}$

8. 168 to 300

$\dfrac{14}{25}$

9. 26 tons to 65 tons

$\dfrac{2}{5}$

Personal Finance *Bob earns $480 per week and has $60 withheld for federal taxes and $45 withheld for state taxes.*

10. Write the ratio of federal taxes withheld to earned income.

$\dfrac{1}{8}$

11. Write the ratio of total withholdings to earned income.

$\dfrac{7}{32}$

Write as a rate in simplest form.

12. $75 donated by every 6 people

$\dfrac{\$25}{2 \text{ people}}$

13. 44 revolutions every 121 minutes

$\dfrac{4 \text{ revolutions}}{11 \text{ minutes}}$

14. 75 heartbeats every 60 seconds

$\dfrac{5 \text{ heartbeats}}{4 \text{ seconds}}$

15. 12 cups of flour for every 27 cakes

$\dfrac{4 \text{ cups}}{9 \text{ cakes}}$

In exercises 16–21, write as a unit rate. Round to the nearest tenth when necessary.

16. $2125 was paid for 125 shares of stock. Find the cost per share.

$17/share

17. ***Education*** $1344 was paid for 12 credit-hours. Find the cost per credit-hour.

$112/credit-hour

▲ **18.** $742.50 was spent for 55 square yards of carpet. Find the cost per square yard.

$13.50/square yard

19. ***DVD Cost*** Larry spent $600 on 48 DVDs. Find the cost per DVD.

$12.50/DVD

20. ***Food Costs*** A 4-ounce jar of instant coffee costs $2.96. A 9-ounce jar of the same brand of instant coffee costs $5.22.

    (a) What is the cost per ounce of the 4-ounce jar? $0.74

    (b) What is the cost per ounce of the 9-ounce jar? $0.58

    (c) How much per ounce do you save by buying the larger jar? $0.16

21. ***Food Costs*** A 12.5-ounce can of white tuna costs $2.75. A 7.0-ounce can of the same brand of white tuna costs $1.75.

    (a) What is the cost per ounce of the large can? $0.22

    (b) What is the cost per ounce of the small can? $0.25

    (c) How much per ounce do you save by buying the larger can? $0.03

Section 4.2

Write as a proportion.

22. 12 is to 48 as 7 is to 28

$$\frac{12}{48} = \frac{7}{28}$$

23. $1\frac{1}{2}$ is to 5 as 4 is to $13\frac{1}{3}$

$$\frac{1\frac{1}{2}}{5} = \frac{4}{13\frac{1}{3}}$$

24. 7.5 is to 45 as 22.5 is to 135

$$\frac{7.5}{45} = \frac{22.5}{135}$$

25. ***Bus Capacity*** If three buses can transport 138 passengers, then five buses can transport 230 passengers. $\dfrac{3 \text{ buses}}{138 \text{ passengers}} = \dfrac{5 \text{ buses}}{230 \text{ passengers}}$

26. ***Cost of Products*** If 15 pounds cost $4.50, then 27 pounds will cost $8.10.

$$\frac{15 \text{ pounds}}{\$4.50} = \frac{27 \text{ pounds}}{\$8.10}$$

Determine whether each equation is a proportion.

27. $\dfrac{16}{48} \overset{?}{=} \dfrac{2}{12}$

It is not a proportion.

28. $\dfrac{20}{25} \overset{?}{=} \dfrac{8}{10}$

It is a proportion.

29. $\dfrac{36}{30} \overset{?}{=} \dfrac{60}{50}$

It is a proportion.

30. $\dfrac{28}{12} \overset{?}{=} \dfrac{84}{36}$

It is a proportion.

31. $\dfrac{37}{33} \overset{?}{=} \dfrac{22}{19}$

It is not a proportion.

32. $\dfrac{15}{18} \overset{?}{=} \dfrac{18}{22}$

It is not a proportion.

33. $\dfrac{84 \text{ miles}}{7 \text{ gallons}} \overset{?}{=} \dfrac{108 \text{ miles}}{9 \text{ gallons}}$

It is a proportion.

34. $\dfrac{156 \text{ revolutions}}{6 \text{ minutes}} \overset{?}{=} \dfrac{181 \text{ revolutions}}{7 \text{ minutes}}$

It is not a proportion.

Section 4.3

Solve for n.

35. $9 \times n = 162$

$n = 18$

36. $5 \times n = 38$

$n = 7\frac{3}{5}$ or 7.6

37. $442 = 20 \times n$

$n = 22\frac{1}{10}$ or 22.1

38. $663 = 39 \times n$

$n = 17$

Solve. Round to the nearest tenth when necessary.

39. $\dfrac{3}{11} = \dfrac{9}{n}$

$n = 33$

40. $\dfrac{2}{7} = \dfrac{12}{n}$

$n = 42$

41. $\dfrac{n}{28} = \dfrac{6}{24}$

$n = 7$

42. $\dfrac{n}{32} = \dfrac{15}{20}$

$n = 24$

43. $\dfrac{2\frac{1}{4}}{9} = \dfrac{4\frac{3}{4}}{n}$

$n = 19$

44. $\dfrac{3\frac{1}{3}}{2\frac{2}{3}} = \dfrac{7}{n}$

$n = 5\frac{3}{5}$ or 5.6

45. $\dfrac{42}{50} = \dfrac{n}{6}$

$n \approx 5.0$

46. $\dfrac{38}{45} = \dfrac{n}{8}$

$n \approx 6.8$

47. $\dfrac{2.25}{9} = \dfrac{4.75}{n}$
$n = 19$

48. $\dfrac{3.5}{5} = \dfrac{10.5}{n}$
$n = 15$

49. $\dfrac{20}{n} = \dfrac{43}{16}$
$n \approx 7.4$

50. $\dfrac{36}{n} = \dfrac{109}{18}$
$n \approx 5.9$

51. $\dfrac{35 \text{ miles}}{28 \text{ gallons}} = \dfrac{15 \text{ miles}}{n \text{ gallons}}$
$n = 12$

52. $\dfrac{8 \text{ defective parts}}{100 \text{ perfect parts}} = \dfrac{44 \text{ defective parts}}{n \text{ perfect parts}}$
$n = 550$

Section 4.4

Solve using a proportion. Round your answer to the nearest hundredth when necessary.

53. *Painting* The school volunteers used 3 gallons of paint to paint two rooms. How many gallons would they need to paint 10 rooms of the same size? 15 gallons

54. *Coffee Consumption* Several recent surveys show that 49 out of every 100 adults in America drink coffee. If a computer company employs 3450 people, how many of those employees would you expect would drink coffee? Round to the nearest whole number.
1691 employees

55. *Exchange Rate* When Marguerite traveled as a child, the rate of French francs to American dollars was 24 francs to 5 dollars. How many francs did Marguerite receive for 420 dollars?
2016 francs

56. *Exchange Rate* When John and Nancy traveled to Switzerland in 2007, the rate of Swiss francs to U.S. dollars was 6 francs to 4.8 dollars. How many Swiss francs would they receive for 125 U.S. dollars?
156.25 Swiss francs

57. *Map Scale* Two cities located 225 miles apart appear 3 inches apart on a map. If two other cities appear 8 inches apart on the map, how many miles apart are the cities?
600 miles

58. *Basketball* In the first three games of the basketball season, Kyle made 8 rebounds. There are 15 games left in the season. How many rebounds would you expect Kyle to make in these 15 games?
40 rebounds

▲ **59. *Shadow Length*** In the setting sun, a 6-foot man casts a shadow 16 feet long. At the same time a building casts a shadow of 320 feet. How tall is the building?
120 feet

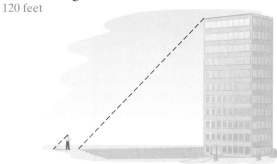

60. *Gasoline Consumption* During the first 680 miles of a trip, Johnny and Stephanie used 26 gallons of gas. They need to travel 200 more miles. Assume that the car will have the same rate of gas consumption.

   (a) How many more gallons of gas will they need? 7.65 gallons

   (b) If gas costs \$4.20 per gallon, what will fuel cost them for the last 200 miles? \$32.13

▲ **61.** *Photography* A film negative is 3.5 centimeters wide and 2.5 centimeters tall. If you want to make a color print that is 8 centimeters wide, how tall will the print be?
5.71 centimeters tall

62. The dosage of a certain medication is 3 grams for every 50 pounds of body weight. If a person weighs 125 pounds, how many grams of this medication should she take?
7.5 grams

63. Carl is designing a walkway around his swimming pool. He will need 22 pavers for each 3-foot section of walkway. How many pavers will he need to buy to make a walkway that is 65 feet long?
technically, 476.67 pavers, but in real life 477 pavers will be needed

64. Greta did a study at her community college for her sociology class. Of the 35 students she interviewed, 21 of them said they eat in the campus cafeteria at least once a week. If there are a total of 2800 students at the college, how many of them eat at least once a week in the cafeteria?
1680 students

▲ **65.** When Carlos was painting his apartment he found that he used 3 gallons of paint to cover 500 square feet of wall space. He is planning to paint his sister's apartment. She said there are 1400 square feet of wall space that need to be painted. How many gallons of paint will Carlos need?
technically 8.4 gallons, but in real life 9 gallons of paint will be needed

66. From previous experience, the directors of a large running race know they need 2 liters of water for every 3 runners. This year 1250 runners will participate in the race. How many liters of water do they need?
technically, 833.33 liters, but in real life, 834 liters will be needed

▲ **67.** A scale model of a new church sanctuary has a length of 14 centimeters. When the church is built, the actual length will be 145 feet. In the scale model, the width measures 11 centimeters. What will be the actual width of the church sanctuary?
approximately 113.93 feet

68. Jeff checked the time on his new watch. In 40 days his watch gained 3 minutes. How much time will the watch gain in a year? (Assume it is not a leap year.)
approximately 27.38 minutes

69. Jean, the top soccer player for the Springfield Comets, scored a total of 68 goals during the season. During the season the team played 32 games, but Jean played in only 27 of them due to a leg injury. The league has been expanded and next season the team will play 34 games. If Jean scores goals at the same rate and is able to play in every game, how many goals might she be expected to score? Round your answer to the nearest whole number.
86 goals

70. Hank found out there were 345 calories in the 10-ounce chocolate milkshake that he purchased yesterday. Today he decided to order the 16-ounce milkshake. How many calories would you expect to be in the 16-ounce milkshake?
552 calories

71. A recent survey showed that 3 out of every 10 people in Massachusetts read the *Boston Globe*. In a Massachusetts town of 45,600 people, how many people would you expect read the *Boston Globe?*
13,680 people

72. Greg and Marcia are managing a boat for Eastern Whale Watching Tours this summer. For every 16 trips out to the ocean, the passengers spotted at least one whale during 13 trips. If Greg and Marcia send out 240 trips this month, how many trips will have the passengers spotting at least one whale?
195 trips

Remember to use your Chapter Test Prep Video CD to see the worked-out solutions to the test problems you want to review.

Write as a ratio in simplest form.

1. 18 : 52

2. 70 to 185

Write as a rate in simplest form. Express your answer as a fraction.

3. 784 miles per 24 gallons

4. 2100 square feet per 45 pounds

Write as a unit rate. Round to the nearest hundredth when necessary.

5. 19 tons in five days

6. $57.96 for seven hours

7. 5400 feet per 22 telephone poles

8. $9373 for 110 shares of stock

Write as a proportion.

9. 17 is to 29 as 51 is to 87

10. $2\frac{1}{2}$ is to 10 as 6 is to 24

11. 490 miles is to 21 gallons as 280 miles is to 12 gallons

12. 3 hours is to 180 miles as 5 hours is to 300 miles

Determine whether each equation is a proportion.

13. $\dfrac{50}{24} = \dfrac{34}{16}$

14. $\dfrac{3\frac{1}{2}}{14} = \dfrac{5}{20}$

15. $\dfrac{32 \text{ smokers}}{46 \text{ nonsmokers}} = \dfrac{160 \text{ smokers}}{230 \text{ nonsmokers}}$

16. $\dfrac{\$0.74}{16 \text{ ounces}} = \dfrac{\$1.84}{40 \text{ ounces}}$

1. $\dfrac{9}{26}$

2. $\dfrac{14}{37}$

3. $\dfrac{98 \text{ miles}}{3 \text{ gallons}}$

4. $\dfrac{140 \text{ square feet}}{3 \text{ pounds}}$

5. 3.8 tons/day

6. $8.28/hour

7. 245.45 feet/pole

8. $85.21/share

9. $\dfrac{17}{29} = \dfrac{51}{87}$

10. $\dfrac{2\frac{1}{2}}{10} = \dfrac{6}{24}$

11. $\dfrac{490 \text{ miles}}{21 \text{ gallons}} = \dfrac{280 \text{ miles}}{12 \text{ gallons}}$

12. $\dfrac{3 \text{ hours}}{180 \text{ miles}} = \dfrac{5 \text{ hours}}{300 \text{ miles}}$

13. It is not a proportion.

14. It is a proportion.

15. It is a proportion.

16. It is not a proportion.

Chapter 4 Ratio and Proportion

17. *n* = 16

18. *n* = 22.5

19. *n* = 19

20. *n* = 29.4

21. *n* = 120

22. *n* = 70.4

23. *n* = 120

24. *n* = 52

25. 6 eggs

26. 80.95 pounds

27. 19 miles

28. $360

29. 136.6 miles

30. 696.67 kilometers

31. 88 free throws

32. 32 hits

Solve. Round to the nearest tenth when necessary.

17. $\dfrac{n}{20} = \dfrac{4}{5}$

18. $\dfrac{8}{3} = \dfrac{60}{n}$

19. $\dfrac{2\frac{2}{3}}{8} = \dfrac{6\frac{1}{3}}{n}$

20. $\dfrac{4.2}{11} = \dfrac{n}{77}$

21. $\dfrac{45 \text{ women}}{15 \text{ men}} = \dfrac{n \text{ women}}{40 \text{ men}}$

22. $\dfrac{5 \text{ kg}}{11 \text{ pounds}} = \dfrac{32 \text{ kg}}{n \text{ pounds}}$

23. $\dfrac{n \text{ inches of snow}}{14 \text{ inches of rain}} = \dfrac{12 \text{ inches of snow}}{1.4 \text{ inches of rain}}$

24. $\dfrac{5 \text{ pounds of coffee}}{\$n} = \dfrac{1/2 \text{ pound of coffee}}{\$5.20}$

Solve using a proportion. Round your answer to the nearest hundredth when necessary.

25. Bob's recipe for pancakes calls for three eggs and will serve 11 people. If he wants to feed 22 people, how many eggs will he need?

26. A steel cable 42 feet long weighs 170 pounds. How much will 20 feet of this cable weigh?

27. If 9 inches on a map represent 57 miles, what distance does 3 inches represent?

28. Dan and Connie found it would cost $240 per year to fertilize their front lawn of 4000 square feet. How much would it cost to fertilize 6000 square feet?

29. Tom Tobey knows that 1 mile is approximately 1.61 kilometers. While he is driving in Canada, a sign reads "Montreal 220 km." How many miles is Tom from Montreal? Round to the nearest tenth of a mile.

30. Stephen traveled 570 kilometers in 9 hours. At this rate, how far could he go in 11 hours?

31. During the first few games of the basketball season, Tyler shot 15 free throws and made 11 of them. If Tyler shoots free throws at the same rate as the first few games, he expects to shoot 120 more during the rest of the season. How many of these free throws should he expect to make?

32. On the tri-city softball league, Lexi got seven hits in 34 times at bat last week. During the entire playing season, she was at bat 155 times. If she gets hits at the same rate all season as during last week's game, how many hits would she have for the entire season? Round to the nearest whole number.

Approximately one-half of this test is based on Chapter 4 material. The remainder is based on material covered in Chapters 1–3.

1. Write in words. 26,597,089

2. Divide. $23\overline{)1564}$

3. Combine. $\dfrac{1}{4} + \dfrac{1}{8} \times \dfrac{3}{4}$

4. Subtract. $8\dfrac{1}{3} - 5\dfrac{3}{4}$

5. Multiply. $20 \times 3\dfrac{1}{4}$

6. Subtract. $\begin{array}{r} 12.1 \\ -\ 3.8416 \end{array}$

7. Multiply. $\begin{array}{r} 2.55 \\ \times 1.08 \end{array}$

8. Divide. $\dfrac{18}{25} \div \dfrac{14}{5}$

9. Multiply. 16.1455×10^3

10. Round to the nearest tenth. 56.8918

11. Add. $258.92 + 67.358$

12. Divide. $0.552 \div 0.15$

13. Change to a decimal. $\dfrac{5}{32}$

Determine whether each equation is a proportion.

14. $\dfrac{12}{17} = \dfrac{30}{42.5}$

15. $\dfrac{4\frac{1}{3}}{13} = \dfrac{2\frac{2}{3}}{8}$

Solve. Round to the nearest tenth when necessary.

16. $\dfrac{9}{2.1} = \dfrac{n}{0.7}$

17. $\dfrac{50}{20} = \dfrac{5}{n}$

18. $\dfrac{n}{70} = \dfrac{32}{51}$

19. $\dfrac{7}{n} = \dfrac{28}{36}$

20. $\dfrac{n}{11} = \dfrac{5}{16}$

21. $\dfrac{3\frac{1}{3}}{7} = \dfrac{10}{n}$

Solve. Round to the nearest hundredth when necessary.

22. Two cities that are located 300 miles apart appear 4 inches apart on a map. If two other cities are 625 miles apart, how far apart will they appear on the same map?

23. Jeanette has a new job as a hairstylist. During one week, she did 26 haircuts and earned $117 in tips. Next week she is scheduled to do 31 haircuts. Assuming the same ratio, how much will she earn in tips next week?

24. Emily Robinson's lasagna recipe feeds 14 people and calls for 3.5 pounds of sausage. If she wants to feed 20 people, how much sausage does she need?

25. Loring Kerr in Nova Scotia produces his own maple syrup. He has found that 39 gallons of maple sap produce 2 gallons of maple syrup. How much sap is needed to produce 11 gallons of syrup?

1. twenty-six million, five hundred ninety-seven thousand, eighty-nine

2. 68

3. $\dfrac{11}{32}$

4. $2\dfrac{7}{12}$

5. 65

6. 8.2584

7. 2.754

8. $\dfrac{9}{35}$

9. 16,145.5

10. 56.9

11. 326.278

12. 3.68

13. 0.15625

14. It is a proportion.

15. It is a proportion.

16. $n = 3$

17. $n = 2$

18. $n \approx 43.9$

19. $n = 9$

20. $n \approx 3.4$

21. $n = 21$

22. 8.33 inches

23. $139.50

24. 5 pounds

25. 214.5 gallons

Percent

CHAPTER

5

Compared to thirty years ago, today more high school students take a modern foreign language rather than an ancient language such as Latin.

This increase is partly due to a greater demand from businesses in our own country for more employees who are bilingual. However, there is a huge, growing need for companies that place employees overseas to have modern foreign language skills. Many companies require that a certain percent of their employees are bilingual. Many percent calculations require the knowledge of the mathematics of this chapter.

1 Writing a Fraction with a Denominator of 100 as a Percent

"My raise came through. I got a 6% increase!"

"The leading economic indicators show inflation rising at a rate of 1.3%."

"Mark McGwire and Babe Ruth each hit quite a few home runs. But I wonder who has the higher percentage of home runs per at-bat?"

We use percents often in our everyday lives. In business, in sports, in shopping, and in many areas of life, percentages play an important role. In this section we introduce the idea of percent, which means "*per centum*" or "per hundred." We then show how to use percentages.

In previous chapters, when we described parts of a whole, we used fractions or decimals. Using a percent is another way to describe a part of a whole. Percents can be described as ratios whose denominators are 100. The word **percent** means per 100. This sketch has 100 rectangles.

Of the 100 rectangles, 23 are shaded. We can say that 23 percent of the whole is shaded. We use the symbol % for percent. It means "parts per 100." When we write 23 percent as 23%, we understand that it means 23 parts per one hundred, or, as a fraction, $\frac{23}{100}$.

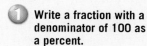
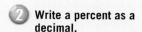
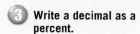

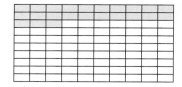

EXAMPLE 1 Recently 100 college students were surveyed about their intentions for voting in the next presidential election. 39 students intended to vote for the Republican candidate, 28 students intended to vote for the Democratic candidate, and 22 students were undecided about which candidate to vote for. The remaining 11 students admitted that they were not planning to vote.

(a) What percent of the students intended to vote for the Democratic candidate?

(b) What percent of the students intended to vote for the Republican candidate?

(c) What percent of the students were undecided as to which candidate they would vote for?

(d) What percent of the students were not planning to vote?

Solution

(a) $\frac{28}{100} = 28\%$ **(b)** $\frac{39}{100} = 39\%$

(c) $\frac{22}{100} = 22\%$ **(d)** $\frac{11}{100} = 11\%$

Percent notation is often used in circle graphs or pie charts.

Teaching Example 1 There are 100 vehicles parked in one of the college parking lots: 41 compact cars, 32 medium-size cars, 15 SUVs, and 12 pickup trucks.

(a) What percent of the vehicles are compact cars?

(b) What percent of the vehicles are medium-size cars?

(c) What percent of the vehicles are SUVs?

(d) What percent of the vehicles are pickup trucks?

Ans: **(a)** 41%   **(b)** 32%   **(c)** 15%
(d) 12%

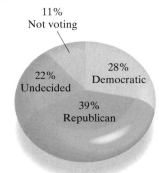

Practice Problem 1 Write as a percent.

(a) 51 out of 100 students in the class were women. 51%

(b) 68 out of 100 cars in the parking lot have front-wheel drive. 68%

(c) 7 out of 100 students in the dorm quit smoking. 7%

(d) 26 out of 100 students did not vote in class elections. 26%

NOTE TO STUDENT: Fully worked-out solutions to all of the Practice Problems can be found at the back of the text starting at page SP-1

Some percents are larger than 100%. When you see expressions like 140% or 400%, you need to understand what they represent. Consider the following situations.

Teaching Example 2
(a) Write $\frac{527}{100}$ as a percent.
(b) Last year 100 students competed on the math team. This year 113 students competed. Write this year's number as a percent of last year's number.

Ans: (a) 527%   **(b)** 113%

EXAMPLE 2

(a) Write $\dfrac{386}{100}$ as a percent.

(b) Twenty years ago, four car tires for a full-size car cost $100. Now the average price for four car tires for a full-size car is $270. Write the present cost as a percent of the cost 20 years ago.

Solution

(a) $\dfrac{386}{100} = 386\%$

(b) The ratio is $\dfrac{\$270 \text{ for four tires now}}{\$100 \text{ for four tires then}}$. $\dfrac{270}{100} = 270\%$

The present cost of four car tires for a full-size car is 270% of the cost 20 years ago.

Practice Problem 2

(a) Write $\dfrac{238}{100}$ as a percent. 238%

(b) Last year 100 students tried out for varsity baseball. This year 121 students tried out. Write this year's number as a percent of last year's number. 121%

Some percents are smaller than 1%.

$\dfrac{0.7}{100}$ can be written as 0.7%. $\dfrac{0.3}{100}$ can be written as 0.3%.

$\dfrac{0.04}{100}$ can be written as 0.04%.

Teaching Example 3 Write as a percent.
(a) $\dfrac{0.23}{100}$ (b) $\dfrac{0.01}{100}$ (c) $\dfrac{0.039}{100}$

Ans: (a) 0.23% **(b)** 0.01% **(c)** 0.039%

EXAMPLE 3 Write as a percent.

(a) $\dfrac{0.9}{100}$ (b) $\dfrac{0.002}{100}$ (c) $\dfrac{0.07}{100}$

Solution

(a) $\dfrac{0.9}{100} = 0.9\%$ (b) $\dfrac{0.002}{100} = 0.002\%$ (c) $\dfrac{0.07}{100} = 0.07\%$

Practice Problem 3 Write as a percent.

(a) $\dfrac{0.5}{100}$ 0.5%   (b) $\dfrac{0.06}{100}$ 0.06%   (c) $\dfrac{0.003}{100}$ 0.003%

Remember: Whenever the denominator of a fraction is 100, the numerator is the percent.

Writing a Percent as a Decimal

Suppose we have a percent such as 59%. What would be the equivalent in decimal form? Using our definition of percent, $59\% = \frac{59}{100}$. This fraction could be written in decimal form as 0.59. In a similar way, we could write 21% as $\frac{21}{100} = 0.21$. This pattern allows us to quickly change the form of a number from a percent to a fraction whose denominator is 100 to a decimal.

EXAMPLE 4   Write as a decimal.

(a) 38%   **(b)** 6%

Solution

(a) $38\% = \dfrac{38}{100} = 0.38$ **(b)** $6\% = \dfrac{6}{100} = 0.06$

Practice Problem 4   Write as a decimal.

(a) 47%   0.47   **(b)** 2%   0.02

Teaching Example 4 Write as a decimal.

(a) 49%   **(b)** 3%

Ans: (a) 0.49   **(b)** 0.03

NOTE TO STUDENT: Fully worked-out solutions to all of the Practice Problems can be found at the back of the text starting at page SP-1

The results of Example 4 suggest that **when you remove a percent sign (%) you are dividing by 100.** When we divide by 100 this moves the decimal point of a number two places to the left. Now that you understand this process we can abbreviate it with the following rule.

CHANGING A PERCENT TO A DECIMAL

1. Drop the % symbol.

2. Move the decimal point two places to the left.

EXAMPLE 5   Write as a decimal.

(a) 26.9%   **(b)** 7.2%   **(c)** 0.13%   **(d)** 158%

Solution   In each case, we drop the percent symbol and move the decimal point two places to the left.

(a) $26.9\% = 0.269 = 0.269$

(b) $7.2\% = 0.072 = 0.072$ Note that we need to add an extra zero to the left of the seven.

(c) $0.13\% = 0.0013 = 0.0013$ Here we added zeros to the left of the 1.

(d) $158\% = 1.58 = 1.58$

Teaching Example 5 Write as a decimal.

(a) 78.4%   **(b)** 6.1%

(c) 0.89%   **(d)** 394%

Ans: (a) 0.784   **(b)** 0.061

(c) 0.0089   **(d)** 3.94

Practice Problem 5   Write as a decimal.

(a) 80.6%   0.806   **(b)** 2.5%   0.025   **(c)** 0.29%   0.0029   **(d)** 231%   2.31

Writing a Decimal as a Percent

In Example 4(a) we changed 38% to $\frac{38}{100}$ to 0.38. We can start with 0.38 and reverse the process. We obtain $0.38 = \frac{38}{100} = 38\%$. Study all the parts of Examples 4 and 5. You will see that the steps are reversible. Thus $0.38 = 38\%$, $0.06 = 6\%$, $0.70 = 70\%$, $0.269 = 26.9\%$, $0.072 = 7.2\%$, $0.0013 = 0.13\%$, and $1.58 = 158\%$.

In each part we are multiplying by 100. **To change a decimal number to a percent we are multiplying the number by 100.** In each part the decimal point is moved two places to the right. Then the percent symbol is written after the number.

> **CHANGING A DECIMAL TO A PERCENT**
> 1. Move the decimal point two places to the right.
> 2. Then write the % symbol at the end of the number.

Teaching Example 6 Write as a percent.
(a) 0.16   (b) 0.03   (c) 4.05
(d) 0.017   (e) 0.002

Ans: (a) 16% **(b)** 3%   **(c)** 405%
(d) 1.7%   **(e)** 0.2%

EXAMPLE 6 Write as a percent.

(a) 0.47   (b) 0.08   (c) 6.31
(d) 0.055   (e) 0.001

Solution In each part we move the decimal point two places to the right and write the percent symbol at the end of the number.

(a) $0.47 = 47\%$ (b) $0.08 = 8\%$ (c) $6.31 = 631\%$
(d) $0.055 = 5.5\%$ (e) $0.001 = 0.1\%$

Practice Problem 6 Write as a percent.

(a) 0.78 78%   (b) 0.02 2%   (c) 5.07 507%
(d) 0.029 2.9%   (e) 0.006 0.6%

Calculator

 Percent to Decimal

You can use a calculator to change 52% to a decimal.
Enter

52 $\boxed{\%}$

The display should read

$\boxed{0.52}$

Try the following.

(a) 46%   (b) 137%
(c) 9.3%   (d) 6%

Note: The calculator divides by 100 when the percent key is pressed. If you do not have a $\boxed{\%}$ key then you can use the keystrokes $\boxed{\div}$ 100 $\boxed{=}$.

TO THINK ABOUT: The Meaning of Percent What is really happening when we change a decimal to a percent? Suppose that we wanted to change 0.59 to a percent.

$$0.59 = \frac{59}{100} \qquad \text{Definition of a decimal.}$$

$$= 59 \times \frac{1}{100} \qquad \text{Definition of multiplying fractions.}$$

$$= 59 \text{ percent} \qquad \text{Because "per 100" means percent.}$$

$$= 59\% \qquad \text{Writing the symbol for percent.}$$

Can you see why each step is valid? Since we know the reason behind each step, we know we can always move the decimal point two places to the right and write the percent symbol. See Exercises 5.1, exercises 81 and 82.

Verbal and Writing Skills

1. In this section we introduced percent, which means "per centum" or "per _____hundred_____."

2. The number 1 written as a percent is _____100%_____.

3. To change a percent to a decimal, move the decimal point _____two_____ places to the _____left_____. _____Drop_____ the % symbol.

4. To change a decimal to a percent, move the decimal point _____two_____ places to the _____right_____. _____Write_____ the % symbol at the end of the number.

Write as a percent.

5. $\dfrac{59}{100}$ 59%

6. $\dfrac{67}{100}$ 67%

7. $\dfrac{4}{100}$ 4%

8. $\dfrac{7}{100}$ 7%

9. $\dfrac{80}{100}$ 80%

10. $\dfrac{90}{100}$ 90%

11. $\dfrac{245}{100}$ 245%

12. $\dfrac{110}{100}$ 110%

13. $\dfrac{12.5}{100}$ 12.5%

14. $\dfrac{15.8}{100}$ 15.8%

15. $\dfrac{0.07}{100}$ 0.07%

16. $\dfrac{0.019}{100}$ 0.019%

Applications *Write a percent to express each of the following.*

17. 13 out of 100 loaves of bread had gone stale. 13%

18. 54 out of 100 dog owners have attended an obedience class. 54%

19. 9 out of 100 customers ordered black coffee. 9%

20. 7 out of 100 students majored in exercise science. 7%

Write as a decimal.

21. 51% 0.51

22. 42% 0.42

23. 7% 0.07

24. 6% 0.06

25. 20% 0.2

26. 40% 0.4

27. 43.6% 0.436

28. 81.5% 0.815

29. 0.03% 0.0003

30. 0.09% 0.0009

31. 0.72% 0.0072

32. 0.61% 0.0061

33. 1.25% 0.0125

34. 9.6% 0.096

35. 275% 2.75

36. 189% 1.89

Write as a percent.

37. 0.74 74%

38. 0.66 66%

39. 0.50 50%

40. 0.40 40%

41. 0.08 8%

42. 0.03 3%

43. 0.563 56.3%

44. 0.408 40.8%

45. 0.002 0.2%

46. 0.009 0.9%

47. 0.0057 0.57%

48. 0.0026 0.26%

49. 1.35 135%

50. 1.86 186%

51. 5.16 516%

52. 4.32 432%

53. *Income Taxes* Robert Tansill paid $\frac{27}{100}$ of his income for federal income taxes. This means that 0.27 of his income was paid for federal taxes. Express this as a percent. 27%

54. *Housing Costs* Sally LeBlanc spends $\frac{37}{100}$ of her income for housing. This means that 0.37 of her income was spent for housing. Express this as a percent. 37%

55. *Grade Distribution* Professor Harlin gave $\frac{2}{10}$ of his students a grade of A for the semester. This means that 0.2 of his students got an A. Express this as a percent. 20%

56. *Checking Account* Tomás Garcia puts $\frac{8}{10}$ of his income into his checking account each month. This means that 0.8 of his income goes into his checking account. Express this as a percent. 80%

Mixed Practice *Write as a percent.*

57. 0.94 94%

58. 0.25 25%

59. 2.31 231%

60. 1.48 148%

61. $\frac{10}{100}$ 10%

62. $\frac{40}{100}$ 40%

63. 0.089 8.9%

64. 0.055 5.5%

Write as a decimal.

65. 62% 0.62

66. 49% 0.49

67. 138% 1.38

68. 210% 2.1

69. $\frac{0.3}{100}$ 0.003

70. $\frac{0.8}{100}$ 0.008

71. $\frac{75}{100}$ 0.75

72. $\frac{35}{100}$ 0.35

Applications *Write as a percent.*

73. *Presidential Election* For every 100 people in Oregon who voted in the 2004 presidential election, 52 voted for John Kerry. 52%

74. *Presidential Election* For every 100 people in Montana who voted in the 2004 presidential election, 59 voted for George W. Bush. 59%

The following are statements found in newspapers. In each case write the percent as a decimal.

75. *House Value* The value of the Sanchez's home increased by 115 percent during the last five years. 1.15

76. *Charitable Donations* According to the Gallup Poll, 1 percent of Americans receiving money from a federal tax cut plan to donate the money. 0.01

77. *Home Value* 0.6 percent of homes in the United States are valued at more than $1 million. 0.006

78. *Vitamins* Americans get 30 percent of their vitamin A from carrots. 0.3

79. *College Education* In 1985, 16.5 percent of U.S. women ages 25 and older had bachelor's degrees. By 2005, that number had increased to 27 percent. (*Source:* U.S. Census Bureau) 0.165; 0.27

80. *Vehicle Sales* In February 2007, the number of vehicles sold by General Motors increased by 3.7 percent. The number of vehicles sold by Ford decreased by 13.4 percent. (*Source:* www. autodata.com) 0.037; 0.134

To Think About

81. Suppose that we want to change 36% to 0.36 by moving the decimal point two places to the left and dropping the % symbol. Explain the steps to show what is really involved in changing 36% to 0.36. Why does the rule work?

36% = 36 percent = 36 "per one hundred" = $36 \times \frac{1}{100} = \frac{36}{100} = 0.36$. The rule is using the fact that 36% means 36 per one hundred.

82. Suppose that we want to change 10.65 to 1065%. Give a complete explanation of the steps.

$10.65 = 1065 \times \frac{1}{100} = 1065$ "per one hundred" = 1065 percent = 1065%. We change 10.65 to $1065 \times \frac{1}{100}$ and use the idea that percent means "per one hundred."

Write the given value (a) as a decimal, (b) as a fraction with a denominator of 100, and (c) as a reduced fraction.

83. 1562%

(a) 15.62   (b) $\frac{1562}{100}$ (c) $\frac{781}{50}$

84. 3724%

(a) 37.24   (b) $\frac{3724}{100}$ (c) $\frac{931}{25}$

Cumulative Review *Write as a fraction in simplest form.*

85. **[3.1.3]** 0.56 $\frac{14}{25}$

86. **[3.1.3]** 0.78 $\frac{39}{50}$

Write as a decimal.

87. **[3.6.1]** $\frac{11}{16}$ 0.6875

88. **[3.6.1]** $\frac{7}{8}$ 0.875

89. **[1.8.2]** *Ceramics Studio* A very successful commercial ceramics studio makes beautiful vases for gift stores. In one corner of the warehouse, the storage area has 24 shelves. Three shelves have 246 vases each, seven shelves have 380 vases each, five shelves have 168 vases each, and nine shelves have 122 vases each. How many vases are there in this corner of the studio?

5336 vases

Quick Quiz 5.1   Write as a percent.

1. 0.007   0.7%

2. $\frac{4.5}{100}$ 4.5%

3. Write as a decimal. 1.25%   0.0125

4. **Concept Check** Explain how you would change 0.00072% to a decimal.

Answers may vary

Classroom Quiz 5.1   You may use these problems to quiz your students' mastery of Section 5.1.

Write as a percent.

1. 0.026   **Ans:** 2.6%

2. $\frac{3.7}{100}$ **Ans:** 3.7%

Write as a decimal.

3. 0.09%   **Ans:** 0.0009

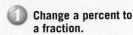
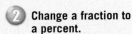
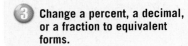
Teaching Example 1 Write as a fraction in simplest form.

(a) 49%   (b) 80%   (c) 12%

Ans: (a) $\frac{49}{100}$ (b) $\frac{4}{5}$ (c) $\frac{3}{25}$

Teaching Example 2 Write as a fraction in simplest form.

(a) 20.4%   (b) 68.26%

Ans: (a) $\frac{51}{250}$ (b) $\frac{3413}{5000}$

Teaching Tip Some students will forget to reduce after they change the percent to a mixed fraction. Remind them that if they write $375\% = 3\frac{75}{100}$ the answer is not completely finished (and thus not correct) until it is reduced to $3\frac{3}{4}$.

Teaching Example 3 Write as a mixed number.

(a) 540%   (b) 336%

Ans: (a) $5\frac{2}{5}$ (b) $3\frac{9}{25}$

---

1 Changing a Percent to a Fraction

By using the definition of percent, we can write any percent as a fraction whose denominator is 100. Thus when we change a percent to a fraction, we remove the percent symbol and write the number over 100. To write a number over 100 means that we are dividing by 100. If possible, we then simplify the fraction.

EXAMPLE 1 Write as a fraction in simplest form.

(a) 37%   (b) 75%   (c) 2%

Solution

(a) $37\% = \frac{37}{100}$ (b) $75\% = \frac{75}{100} = \frac{3}{4}$ (c) $2\% = \frac{2}{100} = \frac{1}{50}$

Practice Problem 1 Write as a fraction in simplest form.

(a) 71% $\frac{71}{100}$ (b) 25% $\frac{1}{4}$ (c) 8% $\frac{2}{25}$

In some cases, it may be helpful to write the percent as a decimal before you write it as a fraction in simplest form.

EXAMPLE 2 Write as a fraction in simplest form.

(a) 43.5%   (b) 36.75%

Solution

(a) $43.5\% = 0.435$ Change the percent to a decimal.

$= \frac{435}{1000}$ Change the decimal to a fraction.

$= \frac{87}{200}$ Reduce the fraction.

(b) $36.75\% = 0.3675 = \frac{3675}{10,000} = \frac{147}{400}$

Practice Problem 2 Write as a fraction in simplest form.

(a) 8.4% $\frac{21}{250}$ (b) 28.5% $\frac{57}{200}$

If the percent is greater than 100%, the simplified fraction is usually changed to a mixed number.

EXAMPLE 3 Write as a mixed number.

(a) 225%   (b) 138%

Solution

(a) $225\% = 2.25 = 2\frac{25}{100} = 2\frac{1}{4}$ (b) $138\% = 1.38 = 1\frac{38}{100} = 1\frac{19}{50}$

Practice Problem 3 Write as a mixed number.

(a) 170% $1\frac{7}{10}$ (b) 288% $2\frac{22}{25}$

Sometimes a percent is not a whole number, such as 9% or 10%. Instead, it contains a fraction, such as $9\frac{1}{12}\%$ or $9\frac{3}{8}\%$. Extra steps will be needed to write such a percent as a simplified fraction.

EXAMPLE 4   Convert $3\frac{3}{8}\%$ to a fraction in simplest form.

Solution

$$3\frac{3}{8}\% = \frac{3\frac{3}{8}}{100} \quad \text{Change the percent to a fraction.}$$

$$= 3\frac{3}{8} \div \frac{100}{1} \quad \text{Write the division horizontally. } \frac{3\frac{3}{8}}{100} \text{ means } 3\frac{3}{8} \text{ divided by 100.}$$

$$= \frac{27}{8} \div \frac{100}{1} \quad \text{Write } 3\frac{3}{8} \text{ as an improper fraction.}$$

$$= \frac{27}{8} \times \frac{1}{100} \quad \text{Use the definition of division of fractions.}$$

$$= \frac{27}{800} \quad \text{Simplify.}$$

Practice Problem 4   Convert $7\frac{5}{8}\%$ to a fraction in simplest form.

$$\frac{61}{800}$$

Teaching Example 4 Convert $5\frac{3}{4}\%$ to a fraction in simplest form.

Ans: $\frac{23}{400}$

EXAMPLE 5   In the fiscal 2007 budget of the United States, approximately $19\frac{5}{8}\%$ of the budget was designated for defense. (*Source:* U.S. Office of Management and Budget.) Write this percent as a fraction.

Solution

$$19\frac{5}{8}\% = \frac{19\frac{5}{8}}{100} = 19\frac{5}{8} \div 100 = \frac{157}{8} \times \frac{1}{100} = \frac{157}{800}$$

Thus we could say $\frac{157}{800}$ of the fiscal 2007 budget was designated for defense. That is, for every \$800 in the budget, \$157 was spent for defense.

Practice Problem 5   In the fiscal 2007 budget of the United States, approximately $20\frac{7}{8}\%$ was designated for social security. (*Source:* U.S. Office of Management and Budget.) Write this percent as a fraction.

$$\frac{167}{800}$$

Teaching Example 5 $13\frac{2}{3}\%$ of the residents of Lakeview have lived in that town for more than 20 years. Write this percent as a fraction.

Ans: $\frac{41}{300}$

Certain percents occur very often, especially in money matters. Here are some common equivalents that you may already know. If not, be sure to memorize them.

$$25\% = \frac{1}{4} \qquad 33\frac{1}{3}\% = \frac{1}{3} \qquad 10\% = \frac{1}{10}$$

$$50\% = \frac{1}{2} \qquad 66\frac{2}{3}\% = \frac{2}{3}$$

$$75\% = \frac{3}{4}$$

② Changing a Fraction to a Percent

A convenient way to change a fraction to a percent is to write the fraction in decimal form first and then convert the decimal to a percent.

Teaching Example 6 Write $\frac{9}{40}$ as a percent.

Ans: 22.5%

EXAMPLE 6 Write $\frac{3}{8}$ as a percent.

Solution We see that $\frac{3}{8} = 0.375$ by calculating $3 \div 8$.

$$
\begin{array}{r}
0.375 \\
8\overline{)3.000} \\
\underline{24} \\
60 \\
\underline{56} \\
40 \\
\underline{40} \\
0
\end{array}
$$

Thus $\frac{3}{8} = 0.375 = 37.5\%$.

Practice Problem 6 Write $\frac{5}{8}$ as a percent. 62.5%

Teaching Example 7 Write as a percent.

(a) $\frac{11}{20}$ (b) $\frac{9}{16}$

Ans: (a) 55%   (b) 56.25%

EXAMPLE 7 Write as a percent.

(a) $\frac{7}{40}$ (b) $\frac{39}{50}$

Solution

(a) $\frac{7}{40} = 0.175 = 17.5\%$ (b) $\frac{39}{50} = 0.78 = 78\%$

Practice Problem 7 Write as a percent.

(a) $\frac{21}{25}$ 84%   (b) $\frac{7}{16}$ 43.75%

Teaching Tip Remind students that if we want to write an exact value, we must say $\frac{2}{3} = 0.666666\ldots$ or use the bar notation over the 6. They must realize that $\frac{2}{3} \approx 0.67$ is a value rounded to the nearest hundredth. In contrast, $\frac{2}{3} \approx 0.66$ is a truncated value and is not an accurate approximation for our purposes.

Changing some fractions to decimal form results in infinitely repeating decimals. In such cases, we usually round to the nearest hundredth of a percent.

Teaching Example 8 Write as a percent. Round to the nearest hundredth of a percent.

(a) $\frac{5}{6}$ (b) $\frac{17}{60}$

Ans: (a) 83.33%   (b) 28.33%

EXAMPLE 8 Write as a percent. Round to the nearest hundredth of a percent.

(a) $\frac{1}{6}$ (b) $\frac{15}{33}$

Solution

(a) We find that $\frac{1}{6} = 0.1666\ldots$ by calculating $1 \div 6$.

$$
\begin{array}{r}
0.1666 \\
6{\overline{\smash{)}1.0000}} \\
\underline{6} \\
40 \\
\underline{36} \\
40 \\
\underline{36} \\
40 \\
\underline{36} \\
4
\end{array}
$$

We will need a four-place decimal so that we will obtain a percent to the nearest hundredth. If we round the decimal to the nearest ten-thousandth, we have $\frac{1}{6} \approx 0.1667$. If we change this to a percent, we have

$$\frac{1}{6} \approx 16.67\%.$$

This is correct to the nearest hundredth of a percent.

(b) By calculating $15 \div 33$, we see that $\frac{15}{33} = 0.45454545\ldots$. We will need a four-place decimal so that we will obtain a percent to the nearest hundredth. If we round to the nearest ten-thousandth, we have

$$\frac{15}{33} \approx 0.4545 = 45.45\%.$$

This rounded value is correct to the nearest hundredth of a percent.

Practice Problem 8 Write as a percent. Round to the nearest hundredth of a percent.

(a) $\frac{7}{9}$ 77.78%

(b) $\frac{19}{30}$ 63.33%

NOTE TO STUDENT: *Fully worked-out solutions to all of the Practice Problems can be found at the back of the text starting at page SP-1*

Recall that sometimes percents are written with fractions.

EXAMPLE 9 Express $\frac{11}{12}$ as a percent containing a fraction.

Solution We will stop the division after two steps and write the remainder in fraction form.

$$
\begin{array}{r}
0.91 \\
12{\overline{\smash{)}11.00}} \\
\underline{108} \\
20 \\
\underline{12} \\
8
\end{array}
$$

This division tells us that we can write

$$\frac{11}{12} \quad \text{as} \quad 0.91\frac{8}{12} \quad \text{or} \quad 0.91\frac{2}{3}.$$

Teaching Example 9 Express $\frac{3}{11}$ as a percent containing a fraction.

Ans: $27\frac{3}{11}\%$

We now have a decimal with a fraction. When we express this decimal as a percent, we move the decimal point two places to the right. We do not write the decimal point in front of the fraction.

$$0.91\frac{2}{3} = 91\frac{2}{3}\%$$

Note that our answer in Example 9 is an *exact answer*. We have not rounded off or approximated in any way.

Practice Problem 9 Express $\frac{7}{12}$ as a percent containing a fraction.

$$58\frac{1}{3}\%$$

③ Changing a Percent, a Decimal, or a Fraction to Equivalent Forms

We have seen so far that a fraction, a decimal, and a percent are three different forms (notations) for the same number. We can illustrate this in a chart.

Teaching Example 10 Complete the following table of equivalent notations. Round decimals to the nearest ten-thousandth. Round percents to the nearest hundredth of a percent.

Fraction	Decimal	Percent
$\frac{27}{35}$		
	0.425	
		$5\frac{3}{4}\%$

Ans:

Fraction	Decimal	Percent
$\frac{27}{35}$	0.7714	77.14%
$\frac{17}{40}$	0.425	42.5%
$\frac{23}{400}$	0.0575	$5\frac{3}{4}\%$

EXAMPLE 10 Complete the following table of equivalent notations. Round decimals to the nearest ten-thousandth. Round percents to the nearest hundredth of a percent.

Fraction	Decimal	Percent
$\frac{11}{16}$		
	0.265	
		$17\frac{1}{5}\%$

Solution Begin with the first row. The number is written as a fraction. We will change the fraction to a decimal and then to a percent.

The fraction is changed to a decimal is changed to a percent.

$$\frac{11}{16} \longrightarrow 16\overline{)11.0000} \xrightarrow{\ 0.6875\ } 68.75\%$$

In the second row the number is written as a decimal. This can easily be written as a percent.

$$0.265 \longrightarrow 26.5\%$$

Now write 0.265 as a fraction and simplify.

$$0.265$$
$$\downarrow$$
$$\frac{53}{200} \longleftarrow \frac{265}{1000}$$

In the third row the number is written as a percent. Proceed from right to left—that is, write the number as a decimal and then as a fraction.

$$\frac{17\frac{1}{5}}{100} \leftarrow 17\frac{1}{5}\%$$

$$\downarrow$$

$$\boxed{\frac{86}{5} \times \frac{1}{100}}$$

$$\downarrow$$

$$0.172 \leftarrow \frac{86}{500} \quad \text{Divide.} \quad \frac{0.172}{500)\overline{86.000}}$$

$$\text{and}$$

$$\frac{43}{250} \leftarrow \frac{86}{500}$$

Thus the completed table is as follows.

Fraction	Decimal	Percent
$\frac{11}{16}$	0.6875	68.75%
$\frac{53}{200}$	0.265	26.5%
$\frac{43}{250}$	0.172	$17\frac{1}{5}\%$

Practice Problem 10 Complete the following table of equivalent notations. Round decimals to the nearest ten-thousandth. Round percents to the nearest hundredth of a percent.

Fraction	Decimal	Percent
$\frac{23}{99}$	0.2323	23.23%
$\frac{129}{250}$	0.516	51.6%
$\frac{97}{250}$	0.388	$38\frac{4}{5}\%$

NOTE TO STUDENT: Fully worked-out solutions to all of the Practice Problems can be found at the back of the text starting at page SP-1

Calculator

Fraction to Decimal

You can use a calculator to change $\frac{3}{5}$ to a decimal. Enter

$$3 \;\boxed{\div}\; 5 \;\boxed{=}$$

The display should read

$$\boxed{0.6}$$

Try the following.

(a) $\frac{17}{25}$ (b) $\frac{2}{9}$

(c) $\frac{13}{10}$ (d) $\frac{15}{19}$

Note: 0.78947368 is an approximation for $\frac{15}{19}$. Some calculators round to only eight places.

ALTERNATIVE METHOD: Using Proportions to Convert from Fraction to Percent Another way to convert a fraction to a percent is to use a proportion. To change $\frac{7}{8}$ to a percent, write the proportion

$$\frac{7}{8} = \frac{n}{100}$$

$$7 \times 100 = 8 \times n \quad \text{Cross-multiply.}$$

$$700 = 8 \times n \quad \text{Simplify.}$$

$$\frac{700}{8} = \frac{8 \times n}{8} \quad \text{Divide each side by 8.}$$

$$87.5 = n \quad \text{Simplify.}$$

Thus $\frac{7}{8} = 87.5\%$. You will use this approach in Exercises 5.2, exercises 85 and 86.

Verbal and Writing Skills

1. Explain in your own words how to change a percent to a fraction.
Write the number in front of the percent symbol as the numerator of a fraction. Write the number 100 as the denominator of the fraction. Reduce the fraction if possible.

2. Explain in your own words how to change a fraction to a percent.
Change the fraction to a decimal by dividing the denominator into the numerator. Change the decimal to a percent by moving the decimal point two places to the right and adding the % symbol.

Write as a fraction or as a mixed number.

3. 6%
$\frac{3}{50}$

4. 8%
$\frac{2}{25}$

5. 33%
$\frac{33}{100}$

6. 47%
$\frac{47}{100}$

7. 55%
$\frac{11}{20}$

8. 35%
$\frac{7}{20}$

9. 75%
$\frac{3}{4}$

10. 25%
$\frac{1}{4}$

11. 20%
$\frac{1}{5}$

12. 40%
$\frac{2}{5}$

13. 9.5%
$\frac{19}{200}$

14. 6.5%
$\frac{13}{200}$

15. 22.5%
$\frac{9}{40}$

16. 92.5%
$\frac{37}{40}$

17. 64.8%
$\frac{81}{125}$

18. 12.2%
$\frac{61}{500}$

19. 71.25%
$\frac{57}{80}$

20. 38.75%
$\frac{31}{80}$

21. 168%
$\frac{168}{100} = 1\frac{17}{25}$

22. 256%
$\frac{256}{100} = 2\frac{14}{25}$

23. 340%
$\frac{340}{100} = 3\frac{2}{5}$

24. 420%
$\frac{420}{100} = 4\frac{1}{5}$

25. 1200%
$\frac{1200}{100} = 12$

26. 3600%
$\frac{3600}{100} = 36$

27. $3\frac{5}{8}$%
$\frac{\frac{29}{8}}{100} = \frac{29}{800}$

28. $4\frac{3}{5}$%
$\frac{\frac{23}{5}}{100} = \frac{23}{500}$

29. $12\frac{1}{2}$%
$\frac{\frac{25}{2}}{100} = \frac{1}{8}$

30. $37\frac{1}{2}$%
$\frac{\frac{75}{2}}{100} = \frac{3}{8}$

31. $8\frac{4}{5}$%
$\frac{\frac{44}{5}}{100} = \frac{11}{125}$

32. $9\frac{3}{5}$%
$\frac{\frac{48}{5}}{100} = \frac{12}{125}$

Applications

33. *Crime Rates* Between 1996 and 2005, the number of violent crimes in the United States decreased by 26.3%. Write the percent as a fraction. (*Source:* www.ojp.usdoj.gov)
$\frac{263}{1000}$

34. *Crime Rates* Between 1996 and 2005, the number of property crimes in the United States decreased by 22.9%. Write the percent as a fraction. (*Source:* www.ojp.usdoj.gov)
$\frac{229}{1000}$

35. *Gasoline Prices* On June 15, 2007, the average price in the United States for regular gasoline was $3.043 per gallon. This was a $2\frac{4}{5}$% decrease in the average price the previous week. Write this percent as a fraction. (*Source:* www.aaa.com)
$\frac{7}{250}$

36. *Gasoline Prices* On June 15, 2007, the average price in the United States for premium gasoline was $3.348. This was a $5\frac{2}{25}$% increase in the average price one year ago. Write this percent as a fraction. (*Source:* www.aaa.com)
$\frac{127}{2500}$

Write as a percent. Round to the nearest hundredth of a percent when necessary.

37. $\frac{3}{4}$ 75% **38.** $\frac{1}{4}$ 25% **39.** $\frac{7}{10}$ 70% **40.** $\frac{9}{10}$ 90% **41.** $\frac{7}{20}$ 35% **42.** $\frac{11}{20}$ 55%

43. $\frac{18}{25}$ 72% **44.** $\frac{22}{25}$ 88% **45.** $\frac{11}{40}$ 27.5% **46.** $\frac{13}{40}$ 32.5% **47.** $\frac{18}{5}$ 360% **48.** $\frac{7}{4}$ 175%

49. $2\frac{1}{2}$ 250% **50.** $3\frac{3}{4}$ 375% **51.** $4\frac{1}{8}$ 412.5% **52.** $2\frac{5}{8}$ 262.5% **53.** $\frac{1}{3}$ 33.33% **54.** $\frac{2}{3}$ 66.67%

55. $\frac{5}{12}$ 41.67% **56.** $\frac{8}{15}$ 53.33% **57.** $\frac{17}{4}$ 425% **58.** $\frac{12}{5}$ 240% **59.** $\frac{26}{50}$ 52% **60.** $\frac{43}{50}$ 86%

Applications *Round to the nearest hundredth of a percent.*

61. *Human Brain* The brain represents approximately $\frac{1}{40}$ of an average person's weight. Express this fraction as a percent. 2.5%

62. *Monthly House Payments* To calculate your maximum monthly house payment, a real estate agent multiplies your monthly income by $\frac{7}{25}$. Express this fraction as a percent. 28%

63. *Size of Africa* Africa is the second largest continent on Earth, measuring 30,301,596 sq km. However, it comprises only $\frac{119}{2000}$ of the earth's total surface area. Express this fraction as a percent. 5.95%

64. *Size of Antarctica* The continent of Antarctica takes up $\frac{11}{400}$ of the earth's total surface area. Express this fraction as a percent. 2.75%

Express as a percent containing a fraction. (See Example 9.)

65. $\frac{3}{8}$ $37\frac{1}{2}\%$ **66.** $\frac{5}{8}$ $62\frac{1}{2}\%$ **67.** $\frac{3}{40}$ $7\frac{1}{2}\%$ **68.** $\frac{11}{90}$ $12\frac{2}{9}\%$

69. $\frac{4}{15}$ $26\frac{2}{3}\%$ **70.** $\frac{11}{15}$ $73\frac{1}{3}\%$ **71.** $\frac{2}{9}$ $22\frac{2}{9}\%$ **72.** $\frac{8}{9}$ $88\frac{8}{9}\%$

Mixed Practice *In exercises 73–82, complete the table of equivalents. Round decimals to the nearest ten-thousandth. Round percents to the nearest hundredth of a percent.*

	Fraction	Decimal	Percent
73.	$\frac{11}{12}$	0.9167	91.67%
75.	$\frac{14}{25}$	0.56	56%
77.	$\frac{1}{200}$	0.005	0.5%
79.	$\frac{5}{9}$	0.5556	55.56%
81.	$\frac{1}{32}$	0.0313	$3\frac{1}{8}\%$

	Fraction	Decimal	Percent
74.	$\frac{1}{12}$	0.0833	8.33%
76.	$\frac{17}{20}$	0.85	85%
78.	$\frac{17}{200}$	0.085	8.5%
80.	$\frac{7}{9}$	0.7778	77.78%
82.	$\frac{21}{800}$	0.0263	$2\frac{5}{8}\%$

83. Write $28\frac{15}{16}$% as a fraction.

$$\frac{463}{16} \times \frac{1}{100} = \frac{463}{1600}$$

84. Write $18\frac{7}{12}$% as a fraction.

$$\frac{223}{12} \times \frac{1}{100} = \frac{223}{1200}$$

Change each fraction to a percent by using a proportion.

85. $\dfrac{123}{800}$ $\dfrac{123}{800} = \dfrac{n}{100}$ $n = 15.375$ 15.375%

86. $\dfrac{417}{600}$ $\dfrac{417}{600} = \dfrac{n}{100}$ $n = 69.5$ 69.5%

Cumulative Review *Solve for n.*

87. [4.3.2] $\dfrac{15}{n} = \dfrac{8}{3}$ $n = 5.625$

88. [4.3.2] $\dfrac{32}{24} = \dfrac{n}{3}$ $n = 4$

89. [1.8.1] *Law Firm* The law firm of Dewey, Cheatham, & Howe was required to review 54 years of documents of one of its clients. The first file contained 10,041 documents. The second file contained 986 documents. The third file contained 4,283 documents. The last file contained 533,855 documents. How many total documents were there?
549,165 documents

▲**90.** [2.9.1] *Restaurant Size* A small neighborhood café has an area of 1800 square feet. A new steak house across the street is $2\frac{1}{2}$ times the size of the café. How many square feet is the new steak house?
4500 square feet

Quick Quiz 5.2 Write as a fraction or as a mixed number in simplified form.

1. 45% $\dfrac{9}{20}$

2. $7\dfrac{3}{5}$% $\dfrac{19}{250}$

3. Change to a percent. $\dfrac{23}{25}$ 92%

4. Concept Check Explain how you would change $8\frac{3}{8}$% to a decimal.   Answers may vary

Classroom Quiz 5.2 You may use these problems to quiz your students' mastery of Section 5.2. Write as a fraction or as a mixed number in simplified form.

1. 62%   **Ans:** $\dfrac{31}{50}$

2. $8\dfrac{3}{4}$%   **Ans:** $\dfrac{7}{80}$

Change to a percent.

3. $\dfrac{17}{40}$ **Ans:** 42.5%

5.3A SOLVING PERCENT PROBLEMS USING EQUATIONS

 Translating a Percent Problem into an Equation

In word problems like the ones in this section, we can translate from words to mathematical symbols and back again. After we have the mathematical symbols arranged in an *equation,* we solve the equation. When we find the values that make the equation true, we have also found the answer to our word problem.

To solve a percent problem, we express it as an equation with an unknown quantity. We use the letter n to represent the number we do not know. The following table is helpful when translating from a percent problem to an equation.

Word	Mathematical Symbol
of	Any multiplication symbol: $\times$ or ( ) or $\cdot$
is	$=$
what	Any letter; for example, n
find	$n =$

In Examples 1–5 we show how to translate words into an equation. Please do **not** solve the problem. Translate into an equation only.

EXAMPLE 1 Translate into an equation.

What is 5% of 19.00?

Solution $n = 5\% \times 19.00$

Practice Problem 1 Translate into an equation. What is 26% of 35?

$n = 26\% \times 35$

EXAMPLE 2 Translate into an equation.

Find 0.6% of 400.

Solution Notice here that the words *what is* are missing. The word *find* is equivalent to *what is.*

Find 0.6% of 400.

$n = 0.6\% \times 400$

Practice Problem 2 Translate into an equation. Find 0.08% of 350.

$n = 0.08\% \times 350$

The unknown quantity, n, does not always stand alone in an equation.

Student Learning Objectives

After studying this section, you will be able to:

 Translate a percent problem into an equation.

 Solve a percent problem by solving an equation.

Teaching Example 1 Translate into an equation. What is 17% of 23?

Ans: $n = 17\% \times 23$

NOTE TO STUDENT: Fully worked-out solutions to all of the Practice Problems can be found at the back of the text starting at page SP-1

Teaching Example 2 Translate into an equation. Find 0.03% of 154.

Ans: $n = 0.03\% \times 154$

Teaching Tip Practicing translation is an excellent way for students to develop confidence with percent problems. Encourage them not to skip the step of practicing translation.

EXAMPLE 3 Translate into an equation.

(a) 35% of what is 60?   **(b)** 7.2 is 120% of what?

Solution

(a) 35% of what is 60?   **(b)** 7.2 is 120% of what?

$\downarrow \quad \downarrow \quad \downarrow \quad \downarrow \quad \downarrow$ $\downarrow \quad \downarrow \quad \downarrow \quad \quad \downarrow \quad \downarrow$

$35\% \times \quad n \quad = 60$ $7.2 = 120\% \times \quad n$

Practice Problem 3 Translate into an equation.

(a) 58% of what is 400?   **(b)** 9.1 is 135% of what?

(a) $58\% \times n = 400$ **(b)** $9.1 = 135\% \times n$

EXAMPLE 4 Translate into an equation.

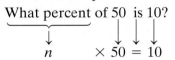

What percent of 50 is 10?

$n \quad \times 50 = 10$

Solution

We see here that the words *what percent* are represented by the letter n.

Practice Problem 4 Translate into an equation. What percent of 250 is 36? $n \times 250 = 36$

EXAMPLE 5 Translate into an equation.

(a) 30 is what percent of 16?   **(b)** What percent of 3000 is 2.6?

Solution

(a) 30 is what percent of 16?   **(b)** What percent of 3000 is 2.6?

$30 = \quad n \quad \times 16$ $n \quad \times 3000 = 2.6$

Practice Problem 5 Translate into an equation.

(a) 50 is what percent of 20?   **(b)** What percent of 2000 is 4.5?

(a) $50 = n \times 20$ **(b)** $n \times 2000 = 4.5$

2 Solving a Percent Problem by Solving an Equation

The percent problems we have translated are of three types. Consider the equation $60 = 20\% \times 300$. This problem has the form

$$\text{amount} = \text{percent} \times \text{base}$$

Any one of these quantities—amount, percent, or base—may be unknown.

1. When *we do not know the amount,* we have an equation like

$$n = 20\% \times 300.$$

2. When *we do not know the base,* we have an equation like

$$60 = 20\% \times n.$$

3. When *we do not know the percent,* we have an equation like

$$60 = n \times 300.$$

We will study each type separately. It is not necessary to memorize the three types, but it is helpful to look carefully at the examples we give of each. In each example, do the computation in a way that is easiest for you. This may be using a pencil and paper, using a calculator, or, in some cases, doing the problem mentally.

Solving Percent Problems When the Amount Is Unknown   In solving these equations we will need to change the percent number to decimal form.

EXAMPLE 6   What is 45% of 590?
$$\downarrow \quad \downarrow \quad \downarrow \quad \downarrow \quad \downarrow$$

Solution
$n = 45\% \times 590$ Translate into an equation.
$n = (0.45)(590)$ Change the percent to decimal form.
$n = 265.5$ Multiply 0.45×590.

Practice Problem 6   What is 82% of 350? $n = 287$

EXAMPLE 7   Find 160% of 500.

Find 160% of 500.   When you translate, remember that the
$$\downarrow \quad \downarrow \quad \downarrow \quad \downarrow$$ word *find* is equivalent to *what is*.

Solution
$n = 160\% \times 500$
$n = (1.60)(500)$ Change the percent to decimal form.
$n = 800$ Multiply 1.6 by 500.

Practice Problem 7   Find 230% of 400. $n = 920$

EXAMPLE 8   When Rick bought a new Toyota Yaris, he had to pay a sales tax of 5% on the cost of the car, which was $12,000. What was the sales tax?

Solution   This problem is asking
$$\text{What is 5\% of \$12,000?}$$
$$\downarrow \quad \downarrow\downarrow \quad \downarrow \quad \downarrow$$
$n = 5\% \times \$12,000$
$n = 0.05 \times 12,000$
$n = \$600$

The sales tax was $600.

Practice Problem 8   When Oprah bought an airplane ticket, she had to pay a tax of 8% on the cost of the ticket, which was $350. What was the tax?

$28

Solving Percent Problems When the Base Is Unknown   If a number is multiplied by the letter n, this can be indicated by a multiplication sign, parentheses, a dot, or placing the number in front of the letter. Thus $3 \times n = 3(n) = 3 \cdot n = 3n$.

In this section we use equations like $3n = 9$ and $0.5n = 20$. To solve these equations we use the procedures developed in Chapter 4. We divide each side by the number multiplied by n.

Teaching Example 6 What is 32% of 260?

Ans: $n = 83.2$

Teaching Example 7 Find 310% of 500.

Ans: $n = 1550$

Teaching Example 8 15% of the students at West Lake College are currently taking math courses. There are 3400 students at West Lake College. How many students are currently taking math courses?

Ans: 510 students

In solving these equations we will need to change the percent number to decimal form.

EXAMPLE 9 12 is 0.6% of what?

$$\downarrow \quad \downarrow \quad \downarrow \quad \quad \downarrow \quad \downarrow$$

Solution $\quad 12 = 0.6\% \times n \quad$ Translate into an equation.

$\qquad\qquad 12 = 0.006n \qquad$ Change 0.6% to a decimal.

$\qquad\qquad \dfrac{12}{0.006} = \dfrac{0.006n}{0.006} \qquad$ Divide each side of the equation by 0.006.

$\qquad\qquad 2000 = n \qquad$ Divide $12 \div 0.006$.

Practice Problem 9 32 is 0.4% of what? $\quad n = 8000$

EXAMPLE 10 Dave and Elsie went out to dinner. They gave the waiter a tip that was 15% of the total bill. The tip the waiter received was $6. What was the total bill (not including the tip)?

Solution This problem is asking

$$15\% \text{ of what is } \$6?$$

$$\downarrow \quad \downarrow \quad \downarrow \quad \downarrow \quad \downarrow$$

$$15\% \times \quad n \quad = \quad 6$$

$$0.15n = 6$$

$$\dfrac{0.15n}{0.15} = \dfrac{6}{0.15} \quad n = 40$$

The total bill for the meal (not including the tip) was $40.

Practice Problem 10 The coach of the university baseball team said that 30% of the players on his team are left-handed. Six people on the team are left-handed. How many people are on the team? 20 people

Solving Percent Problems When the Percent Is Unknown In solving these problems, we notice that there is no % symbol in the problem. The percent is what we are trying to find. Therefore, our answer for this type of problem will always have a percent symbol.

EXAMPLE 11 What percent of 5000 is 3.8?

$$\underbrace{\qquad\qquad}_{\downarrow} \quad \begin{array}{ccc} \downarrow & \downarrow & \downarrow \\ \end{array}$$

Solution $\qquad n \qquad \times 5000 = 3.8 \qquad$ Translate into an equation.

$\qquad\qquad 5000n = 3.8 \qquad$ Multiplication is commutative.
$\qquad\qquad\qquad\qquad\qquad n \times 5000 = 5000 \times n.$

$\qquad\qquad \dfrac{5000n}{5000} = \dfrac{3.8}{5000} \qquad$ Divide each side by 5000.

$\qquad\qquad\qquad n = 0.00076 \qquad$ Divide 3.8 by 5000.

$\qquad\qquad\qquad n = 0.076\% \qquad$ Express the decimal as a percent.

Practice Problem 11 What percent of 9000 is 4.5? $\quad n = 0.05\%$

EXAMPLE 12 90 is what percent of 20?

Solution   90 = n × 20   Translate into an equation.

$90 = 20n$ Multiplication is commutative. $n × 20 = 20 × n$.

$\dfrac{90}{20} = \dfrac{20n}{20}$ Divide each side by 20.

$4.5 = n$ Divide 90 by 20.

$450\% = n$ Express the decimal as a percent.

Practice Problem 12 198 is what percent of 33? $n = 600\%$

EXAMPLE 13 In a recent basketball game for the New York Knicks, Jamal Crawford made 10 of his 24 shots. What percent of his shots did he make? (Round to the nearest tenth of a percent.)

Solution This is equivalent to

10 is what percent of 24?

$10 = n$ × 24

$10 = 24n$

$\dfrac{10}{24} = \dfrac{24n}{24}$

$0.41666\ldots = n$

To the nearest tenth of a percent we have

$$n = 41.7\%$$

Jamal Crawford made 41.7% of his shots in this game.

Practice Problem 13 In a basketball game for the Los Angeles Lakers, Kobe Bryant made 5 of his 16 shots. What percent of his shots did he make? (Round to the nearest tenth of a percent.) 31.3%

Verbal and Writing Skills

1. Give an example of a percent problem when we do not know the amount.
What is 20% of $300?

2. Give an example of a percent problem when we do not know the base.
$500 is 30% of what number?

3. Give an example of a percent problem when we do not know the percent.
20 baskets out of 25 shots is what percent?

4. When you encounter a problem like "What is 65% of $600?" what type of percent problem is this? How would you solve such a problem?
This is a type called "a percent problem when we do not know the amount." We can translate this into an equation
$$n = 65\% \times 600$$
$$n = (0.65)(600)$$
$$n = 390$$

5. When you encounter a problem like "108 is 18% of what number?" what type of percent problem is this? How would you solve such a problem?
This is a type called "a percent problem when we do not know the base." We can translate this into an equation
$$108 = 18\% \times n$$
$$108 = 0.18n$$
$$\frac{108}{0.18} = \frac{0.18n}{0.18}$$
$$600 = n$$

6. When you encounter a problem like "What percent of 35 is 14?" what type of percent problem is this? How would you solve such a problem?
This is a type called "a percent problem when we do not know the percent." We can translate this into an equation
$$n \times 35 = 14$$
$$35n = 14$$
$$n = 0.4$$
The answer is 40%.

*Translate into a mathematical equation in exercises 7–12. Use the letter n for the unknown quantity. Do **not** solve, but rather just obtain the equation.*

7. What is 5% of 90?
$n = 5\% \times 90$

8. What is 9% of 65?
$n = 9\% \times 65$

9. 30% of what is 5?
$30\% \times n = 5$

10. 65% of what is 28?
$65\% \times n = 28$

11. 17 is what percent of 85?
$17 = n \times 85$

12. 24 is what percent of 144?
$24 = n \times 144$

Solve.

13. What is 20% of 140?
$n = 20\% \times 140 \quad n = 28$

14. What is 30% of 210?
$n = 30\% \times 210 \quad n = 63$

15. Find 40% of 140.
$n = 40\% \times 140 \quad n = 56$

16. Find 60% of 210.
$n = 60\% \times 210 \quad n = 126$

Applications

17. ***Sales Tax*** Malik bought a new flat-screen television. The price before the 6% sales tax was added on was $850. How much tax did Malik have to pay? $6\% \times 850 = \$51$

18. ***Coin-Counting Service*** At the local bank coins can be placed into a machine to be counted. You can then receive bills for the amount the coins are worth. However, the bank charges a fee that is 8% of the coins' value. How much would the service fee be if someone put $215 worth of coins into the machine? $8\% \times 215 = \$17.20$

Solve.

19. 2% of what is 26?
2% × *n* = 26   *n* = 1300

20. 3% of what is 18?
3% × *n* = 18   *n* = 600

21. 52 is 4% of what?
52 = 4% × *n*   *n* = 1300

22. 36 is 6% of what?
36 = 6% × *n*   *n* = 600

Applications

23. *Australia Tax* In Australia, all general sales (except for food) have a hidden tax of 22% built into the final price. Walter is planning to purchase a camera while in Australia. He wants to know the before-tax price that the dealer is charging before he adds on the hidden tax of $33. Can you determine the amount of the before-tax price? 22% × *n* = 33   *n* = $150

24. *Opinion Poll* A newspaper states that 522 of its residents are in favor of building a new high school. This is 12% of the town's population. What is the population of the town?
12% × *n* = 522
n = 4350

Solve.

25. What percent of 200 is 168?
n × 200 = 168   *n* = 84%

26. What percent of 300 is 135?
n × 300 = 135   *n* = 45%

27. 33 is what percent of 300?
33 = *n* × 300   11% = *n*

28. 78 is what percent of 200?
78 = *n* × 200   39% = *n*

Applications

29. *Basketball* The total number of points scored in a basketball game was 120. The winning team scored 78 of those points. What percent of the points were scored by the winning team?
78 = *n* × 120   *n* = 65%

30. *Car Repairs* Randy's bill for car repairs was $140. Of this amount, $28 was charged for labor and $112 was charged for parts. What percent of the bill was for labor?
28 = *n* × 140   *n* = 20%

Mixed Practice *Solve.*

31. 20% of 155 is what?
20% × 155 = *n*   *n* = 31

32. 60% of 215 is what?
60% × 215 = *n*   *n* = 129

33. 170% of what is 144.5?
170% × *n* = 144.5   *n* = 85

34. 160% of what is 152?
160% × *n* = 152   *n* = 95

35. 84 is what percent of 700?
84 = *n* × 700   12% = *n*

36. 72 is what percent of 900?
72 = *n* × 900   8% = *n*

37. Find 0.4% of 820.
n = 0.4% × 820   *n* = 3.28

38. Find 0.3% of 540.
n = 0.3% × 540   *n* = 1.62

39. What percent of 35 is 22.4?
n × 35 = 22.4   *n* = 64%

40. What percent of 45 is 16.2?
n × 45 = 16.2   *n* = 36%

41. 15 is 20% of what?
15 = 20% × *n*   75 = *n*

42. 10 is 25% of what?
10 = 25% × *n*   40 = *n*

43. 8 is what percent of 1000?
8 = *n* × 1000   0.8% = *n*

44. 6 is what percent of 800?
6 = *n* × 800   0.75% = *n*

45. What is 10.5% of 180?
n = 10.5% × 180   *n* = 18.9

46. What is 17.5% of 260?
n = 17.5% × 260   *n* = 45.5

47. Scoring 44 problems out of 55 problems correctly on a test is what percent? 80%

48. Scoring 27 problems out of 45 problems correctly on a test is what percent? 60%

Applications

49. *Computer Sales* It is projected that in 2008, 283.2 million computers will be sold worldwide. Of these, 171.8 million will be for commercial use. What percent of all computers sold in 2008 will be for commercial use? Round your answer to the nearest hundredth of a percent. (*Source:* www.usatoday.com, March 2007) 60.66%

50. *Equestrian Rider* An Olympic equestrian rider practiced jumping over a water hazard. In 400 attempts, she and her horse touched the water 15 times. What percent of her jump attempts were not perfect? 3.75%

51. *College Courses* At Monroe State College, 62% of the freshman class is enrolled in a composition course. There are 1070 freshmen this year. How many of them are taking a composition course? Round your answer to the nearest whole number. 663 students

52. *Student Health* A recent study indicates that 15% of all middle school students do not eat a proper breakfast. If Pineridge Middle School has 420 students, how many do not eat a proper breakfast? 63 students

53. *Swim Team* The swim team at Stonybrook College has gone on to the state championships 24 times over the years. If that translates to 60% of the time in which the team has qualified for the finals, how many years has the swim team qualified for the finals? 40 years

54. *Higher Education* North Shore Community College found that 60% of its graduates go on for further education. Last year 570 of the graduates went on for further education. How many students graduated from the college last year? 950 students

55. Find 12% of 30% of $1600. $57.60

56. Find 90% of 15% of 2700. 364.5

Cumulative Review *Multiply or divide.*

57. [3.4.1]
$$\begin{array}{r} 1.36 \\ \times\ 1.8 \\ \hline 2.448 \end{array}$$

58. [3.4.1]
$$\begin{array}{r} 5.06 \\ \times\ 0.82 \\ \hline 4.1492 \end{array}$$

59. [3.5.2] $0.06)\overline{170.04}$ 2834

60. [3.5.2] $0.9)\overline{2.124}$ 2.36

Quick Quiz 5.3A

1. What is 152% of 84? 127.68

2. 72 is 0.8% of what number? 9000

3. 68 is what percent of 400? 17%

4. Concept Check Explain how to solve the following problem using an equation. Jason found that 85% of all people who purchased a Mustang at Danvers Ford were previous Mustang owners. Last year 120 people purchased a Mustang at Danvers Ford. How many of them were previous Mustang owners? Answers may vary

Classroom Quiz 5.3A You may use these problems to quiz your students' mastery of Section 5.3A.

1. What is 0.06% of 27,000? **Ans:** 16.2

2. 115.2 is 72% of what number? **Ans:** 160

3. 39 is what percent of 300? **Ans:** 13%

Identifying the Parts of the Percent Proportion

In Section 5.3A we showed you how to use an equation to solve a percent problem. Some students find it easier to use proportions to solve percent problems. We will show you how to use proportions in this section. The two methods work equally well. Using percent proportions allows you to see another of the many uses of the proportions that we studied in Chapter 4.

Suppose your math class of 25 students has 19 right-handed students and 6 left-handed students. You could say that $\frac{19}{25}$ of the class or 76% is right-handed. Consider the following relationship.

$$\frac{19}{25} = 76\%$$

This can be written as

$$\frac{19}{25} = \frac{76}{100}$$

As a rule, we can write this relationship using the **percent proportion**

$$\frac{\text{amount}}{\text{base}} = \frac{\text{percent number}}{100}.$$

To use this equation effectively, we need to find the amount, base, and percent number in a word problem. The easiest of these three parts to find is the percent number. We use the letter p (a variable) to represent the **percent number.**

Student Learning Objectives

After studying this section, you will be able to:

 Identify the parts of the percent proportion.

 Use the percent proportion to solve percent problems.

6
Left-handed
students

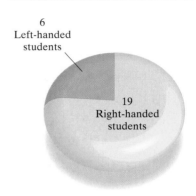

19
Right-handed
students

Teaching Tip Some students dislike solving percent problems by the proportion method; other students find it very helpful.

EXAMPLE 1 Identify the percent number p.

(a) Find 16% of 370
(b) 28% of what is 25?
(c) What percent of 18 is 4.5?

Solution

(a) Find 16% of 370.
The value of p is 16.

(b) 28% of what is 25?
The value of p is 28.

(c) What percent of 18 is 4.5?

$\overbrace{\qquad\qquad}$
$\downarrow$
p

We let p represent the unknown percent number.

Teaching Example 1 Identify the percent number p.

(a) Find 21% of 47.

(b) 13% of what is 89?

(c) What percent of 25 is 18.3?

Ans: (a) p is 21. **(b)** p is 13.

(c) p is the unknown percent number.

 Practice Problem 1 Identify the percent number p.

(a) Find 83% of 460. p is 83.
(b) 18% of what number is 90? p is 18.
(c) What percent of 64 is 8? p is the unknown percent number.

NOTE TO STUDENT: Fully worked-out solutions to all of the Practice Problems can be found at the back of the text starting at page SP-1

We use the letter b to represent the base number. The **base** is the entire quantity or the total involved. The number that is the base usually appears after the word *of*. The **amount,** which we represent by the letter a, is the part being compared to the whole.

EXAMPLE 2   Identify the base b and the amount a.

(a) 20% of 320 is 64.   **(b)** 12 is 60% of what?

Solution

(a) 20%   of   320   is   64.

> The base is the entire quantity. It follows the word *of*. Here $b = 320$.

> The amount is the part compared to the whole. Here $a = 64$.

(b) 12   is   60%   of   what?

> The amount 12 is the part of the base. Here $a = 12$.

> The base is unknown. We represent the base by the variable b.

Practice Problem 2   Identify the base b and the amount a.

(a) 30% of 52 is 15.6.   **(b)** 170 is 85% of what?

(a) $b = 52$; $a = 15.6$ (b) base $= b$; $a = 170$

When identifying p, b, and a in a problem, it is easiest to identify p and b first. The remaining quantity or variable is a.

EXAMPLE 3   Find p, b, and a.

(a) What is 52% of 300?   **(b)** What percent of 30 is 18?

> The value of p is 52.

Solution

(a) What   is   52%   of   300?

> The amount is unknown. We let $a =$ the amount.

> The base usually follows the word *of*. Here $b = 300$.

(b) > The value of p is not known. We let p represent the unknown percent.

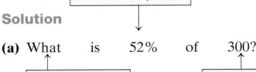

What percent of 30 is 18?

> The base usually follows the word *of*. Here $b = 30$.

> The amount is 18. Thus $a = 18$.

Practice Problem 3   Find p, b, and a.

(a) What is 18% of 240?   **(b)** What percent of 64 is 4?

(a) $p = 18$; $b = 240$; $a =$ the amount   (b) $p =$ the percent; $b = 64$; $a = 4$

Using the Percent Proportion to Solve Percent Problems

When we solve the percent proportion, we will have enough information to state the numerical value for two of the three variables a, b, p in the equation

$$\frac{a}{b} = \frac{p}{100}$$

We first identify those two values, and then substitute those values into the equation. Then we will use the skills that we acquired for solving proportions in Chapter 4 to find the value we do not know. Here and throughout the entire chapter we assume that $b \neq 0$.

When solving each problem it is a good idea to look at your answer and see if it is reasonable. Ask yourself, "Does my answer make sense?"

EXAMPLE 4   Find 260% of 40.

Solution   The percent $p = 260$. The number that is the base usually appears after the word *of*. The base $b = 40$. The amount is unknown. We use the variable a. Thus

$$\frac{a}{b} = \frac{p}{100} \qquad \text{becomes} \qquad \frac{a}{40} = \frac{260}{100}.$$

If we reduce the fraction on the right-hand side, we have

$$\frac{a}{40} = \frac{13}{5}$$

$5a = (40)(13)$ Cross-multiply.

$5a = 520$ Simplify.

$\dfrac{5a}{5} = \dfrac{520}{5}$ Divide each side of the equation by 5.

$a = 104$

Thus 260% of 40 is 104.

Practice Problem 4   Find 340% of 70.   238

EXAMPLE 5   85% of what is 221?

Solution   The percent $p = 85$. The base is unknown. We use the variable b. The amount a is 221. Thus

$$\frac{a}{b} = \frac{p}{100} \qquad \text{becomes} \qquad \frac{221}{b} = \frac{85}{100}.$$

If we reduce the fraction on the right-hand side, we have

$$\frac{221}{b} = \frac{17}{20}$$

$$(221)(20) = 17b \quad \text{Cross-multiply.}$$

$$4420 = 17b \quad \text{Simplify.}$$

$$\frac{4420}{17} = \frac{17b}{17} \quad \text{Divide each side by 17.}$$

$$260 = b. \quad \text{Divide 4420 by 17.}$$

Thus 85% of 260 is 221.

NOTE TO STUDENT: *Fully worked-out solutions to all of the Practice Problems can be found at the back of the text starting at page SP-1*

Practice Problem 5 68% of what is 476? 700

Teaching Example 6 A city planner estimates that 2.7% of the city's annual budget is used to pay her salary. If her salary is $60,912, what is the city's annual budget?

Ans: $2,256,000

EXAMPLE 6 George and Barbara purchased some no-load mutual funds. The account manager charged a service fee of 0.2% of the value of the mutual funds. George and Barbara paid this fee, which amounted to $53. When they got home they could not find the receipt that showed the exact value of the mutual funds that they purchased. Can you find the value of the mutual funds that they purchased?

Solution The basic situation here is that 0.2% of some number is $53. This is equivalent to saying $53 is 0.2% of what? If we want to answer the question "53 is 0.2% of what?", we need to identify a, b, and p.

The percent $p = 0.2$. The base is unknown. We use the variable b. The amount $a = 53$. Thus

$$\frac{a}{b} = \frac{p}{100} \quad \text{becomes} \quad \frac{53}{b} = \frac{0.2}{100}.$$

When we cross-multiply, we obtain

$$(53)(100) = 0.2b$$

$$5300 = 0.2b$$

$$\frac{5300}{0.2} = \frac{0.2b}{0.2}$$

$$26,500 = b.$$

Thus $53 is 0.2% of $26,500. Therefore the value of the mutual funds was $26,500.

Practice Problem 6 Everett Hatfield recently exchanged U.S. dollars to Canadian dollars for his company, Nova Scotia Central Trucking, Ltd. The bank charged a fee of 0.3% of the total U.S. dollars exchanged. The fee amounted to $216 in U.S. money. How many U.S. dollars were exchanged? $72,000

EXAMPLE 7 What percent of 4000 is 160?

Solution The percent is unknown. We use the variable p. The base $b = 4000$. The amount $a = 160$. Thus

$$\frac{a}{b} = \frac{p}{100} \quad \text{becomes} \quad \frac{160}{4000} = \frac{p}{100}.$$

If we reduce the fraction on the left-hand side, we have

$$\frac{1}{25} = \frac{p}{100}$$

$$100 = 25p \quad \text{Cross-multiply.}$$

$$\frac{100}{25} = \frac{25p}{25} \quad \text{Divide each side by 25.}$$

$$4 = p \quad \text{Divide 100 by 25.}$$

Thus 4% of 4000 is 160.

Practice Problem 7 What percent of 3500 is 105? 3%

Developing Your Study Skills

Reading the Textbook

Homework time each day should begin with the careful reading of the section(s) assigned in your textbook. Much time and effort have gone into the selection of a particular text, and your instructor has chosen a book that will help you become successful in this mathematics class. Expensive textbooks can be a wise investment if you take advantage of them by reading them.

Reading a mathematics textbook is unlike reading many other types of books that you may use in your literature, history, psychology, or sociology courses. Mathematics texts are technical books that provide you with exercises to practice on. Reading a mathematics text requires slow and careful reading of each word, which takes time and effort.

Begin reading your textbook with a paper and pencil in hand. As you come across a new definition, or concept, underline it in the text and/or write it down in your notebook. Whenever you encounter an unfamiliar term, look it up and make a note of it. When you come to an example, work through it step by step. Be sure to read each word and to follow directions carefully.

Notice the helpful hints the author provides to guide you to correct solutions and prevent you from making errors. Take advantage of these pieces of expert advice.

Be sure that you understand what you are reading. Make a note of any of those things that you do not understand and ask your instructor about them. Do not hurry through the material. Learning mathematics takes time.

PRACTICE   WATCH   DOWNLOAD   READ   REVIEW

Identify p, b, and a. Do not solve for the unknown.

	p	b	a
1. 75% of 660 is 495.	75	660	495
2. 65% of 820 is 532.	65	820	532
3. What is 22% of 60?	22	60	a
4. What is 35% of 95?	35	95	a
5. 49% of what is 2450?	49	b	2450
6. 38% of what is 2280?	38	b	2280
7. 30 is what percent of 50?	p	50	30
8. 50 is what percent of 250?	p	250	50

Solve using the percent proportion

$$\frac{a}{b} = \frac{p}{100}.$$

In exercises 9–14, the amount a is not known.

9. 40% of 70 is what?
$\frac{a}{70} = \frac{40}{100}$ $a = 28$

10. 80% of 90 is what?
$\frac{a}{90} = \frac{80}{100}$ $a = 72$

11. Find 210% of 40.
$\frac{a}{40} = \frac{210}{100}$ $a = 84$

12. Find 150% of 80.
$\frac{a}{80} = \frac{150}{100}$ $a = 120$

13. 0.7% of 8000 is what?
$\frac{a}{8000} = \frac{0.7}{100}$ $a = 56$

14. 0.8% of 9000 is what?
$\frac{a}{9000} = \frac{0.8}{100}$ $a = 72$

In exercises 15–20, the base b is not known.

15. 20 is 25% of what?
$\frac{20}{b} = \frac{25}{100}$ $b = 80$

16. 45 is 60% of what?
$\frac{45}{b} = \frac{60}{100}$ $b = 75$

17. 250% of what is 200?
$\frac{200}{b} = \frac{250}{100}$ $b = 80$

18. 120% of what is 90?
$\frac{90}{b} = \frac{120}{100}$ $b = 75$

19. 3000 is 0.5% of what?
$\frac{3000}{b} = \frac{0.5}{100}$ $b = 600,000$

20. 6000 is 0.4% of what?
$\frac{6000}{b} = \frac{0.4}{100}$ $b = 1,500,000$

In exercises 21–24, the percent p is not known.

21. 56 is what percent of 280?
$\frac{56}{280} = \frac{p}{100}$ $p = 20$

22. 70 is what percent of 1400?
$\frac{70}{1400} = \frac{p}{100}$ $p = 5$

23. What percent of 90 is 18?
$\frac{18}{90} = \frac{p}{100}$ $p = 20$

24. What percent of 120 is 18?
$\frac{18}{120} = \frac{p}{100}$ $p = 15$

Mixed Practice

25. 25% of 88 is what?

$\dfrac{a}{88} = \dfrac{25}{100}$ $a = 22$

26. 20% of 75 is what?

$\dfrac{a}{75} = \dfrac{20}{100}$ $a = 15$

27. 300% of what is 120?

$\dfrac{120}{b} = \dfrac{300}{100}$ $b = 40$

28. 200% of what is 120?

$\dfrac{120}{b} = \dfrac{200}{100}$ $b = 60$

29. 82 is what percent of 500?

$\dfrac{82}{500} = \dfrac{p}{100}$ $p = 16.4$ 16.4%

30. 75 is what percent of 600?

$\dfrac{75}{600} = \dfrac{p}{100}$ $p = 12.5$ 12.5%

31. Find 0.7% of 520.

$\dfrac{a}{520} = \dfrac{0.7}{100}$ $a = 3.64$

32. Find 0.4% of 650.

$\dfrac{a}{650} = \dfrac{0.4}{100}$ $a = 2.6$

33. What percent of 66 is 16.5?

$\dfrac{16.5}{66} = \dfrac{p}{100}$ $p = 25$ 25%

34. What percent of 49 is 34.3?

$\dfrac{34.3}{49} = \dfrac{p}{100}$ $p = 70$ 70%

35. 68 is 40% of what?

$\dfrac{68}{b} = \dfrac{40}{100}$ $b = 170$

36. 52 is 40% of what?

$\dfrac{40}{100} = \dfrac{52}{b}$ $b = 130$

Applications *When solving each applied problem, examine your answer and see if it is reasonable. Ask yourself, "Does my answer make sense?"*

37. *Paycheck Deposit* Each time Lowell gets paid, 5% of his paycheck is deposited in his retirement account. Last week, $48 was put into his retirement account. What was the amount of Lowell's paycheck? $960

38. *Income Tax* Last year, Rachel had 24% of her salary withheld for taxes. If the total amount withheld was $6300 for the year, what was Rachel's annual salary? $26,250

39. *Eating Out* Ed and Suzie went out to eat at Pizzeria Uno. The dinner check was $26.00. They left a tip of $3.90. What percent of the check was the tip? 15%

40. *Baseball* During the baseball season, Damon was up to bat 60 times. Of these at-bats, 12 were home runs. What percent of Damon's at-bats resulted in a home run? 20%

41. *Food Expiration Date* The Super Shop and Save store had 120 gallons of milk placed on the shelf one night. During the next morning's inspection, the manager found that 15% of the milk had passed the expiration date. How many gallons of milk had passed the expiration date? 18 gallons

42. *Police Arrests* During June the Wenham police stopped 250 drivers for speed violations. It was found that 8% of the people who were stopped had outstanding warrants for their arrest. How many people had outstanding warrants for their arrest? 20 people

43. *Car Purchase* Victor purchased a used car for $9500. He made a down payment of 24% of the purchase price. How much was his down payment? $2280

44. *Education* Trudy took a biology test with 40 problems. She got 8 of the problems wrong and 32 of the problems right. What percent of the test problems did she do incorrectly? 20%

To Think About

The Cost of Children *A study was conducted to determine the amount of money middle-income families in the United States spend per child, ages 6 to 14, for each of the following: housing, food, transportation, clothing, health care, child care and education, and miscellaneous (including personal care items, entertainment, and reading materials). The amount of money in each of these categories is given in the pie chart below. Use the pie chart to answer questions 45–48. Round all answers to the nearest tenth.*

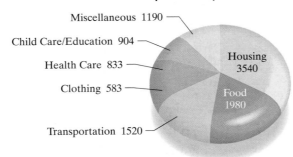

Middle-Income Families' Annual Expenditures per Child in 2005

Miscellaneous 1190
Child Care/Education 904
Health Care 833
Clothing 583
Transportation 1520
Housing 3540
Food 1980

Source: www.census.gov

45. What percent of the total expenditures in these seven categories was spent on housing?
33.5%

46. What percent of the total expenditures in these seven categories was spent on clothing?
5.5%

47. Suppose that compared to 2005, expenditures in 2010 for housing were 25% larger, expenditures for child care/education were 10% larger, and the other five categories remained at the same dollar amount. What percent of the total expenditures in 2010 in these seven categories would be used for food? 17.2%

48. Suppose that compared to 2005, expenditures in 2015 for health care were 25% larger, expenditures for food were 15% larger, and the other five categories remained at the same dollar amount. What percent of the total expenditures in 2015 in these seven categories would be used for miscellaneous items? 10.8%

Cumulative Review *Simplify.*

49. **[2.7.2]** $\frac{4}{5} + \frac{8}{9}$ $1\frac{31}{45}$

50. **[2.7.2]** $\frac{7}{13} - \frac{1}{2}$ $\frac{1}{26}$

51. **[2.4.3]** $\left(2\frac{4}{5}\right)\left(1\frac{1}{2}\right)$ $4\frac{1}{5}$

52. **[2.5.3]** $1\frac{2}{5} \div \frac{3}{4}$ $1\frac{13}{15}$

Quick Quiz 5.3B

1. What is 0.09% of 17,000? 1.53

2. 64.8 is 54% of what number?
120

3. 132 is what percent of 600?
22%

Explain how to solve the following problem using a proportion.

4. Concept Check Alice purchased some stock and was charged a service fee of 0.7% of the value of the stock. The fee she paid was $140. What was the value of the stock that she purchased? Answers may vary

Classroom Quiz 5.3B You may use these problems to quiz your students' mastery of Section 5.3B.

1. What is 145% of 96? **Ans:** 139.2

2. 88 is 0.4% of what number?
Ans: 22,000

3. 126 is what percent of 800?
Ans: 15.75%

How are you doing with your homework assignments in Sections 5.1 to 5.3? Do you feel you have mastered the material so far? Do you understand the concepts you have covered? Before you go further in the textbook, take some time to do each of the following problems.

5.1

Write as a percent.

1. 0.17   **2.** 0.387   **3.** 7.95   **4.** 5.18

5. 0.006   **6.** 0.0004   **7.** $\dfrac{17}{100}$ **8.** $\dfrac{89}{100}$

9. $\dfrac{13.4}{100}$ **10.** $\dfrac{19.8}{100}$ **11.** $\dfrac{6\frac{1}{2}}{100}$ **12.** $\dfrac{1\frac{3}{8}}{100}$

5.2

Change to a percent. Round to the nearest hundredth of a percent when necessary.

13. $\dfrac{8}{10}$ **14.** $\dfrac{15}{30}$ **15.** $\dfrac{52}{20}$ **16.** $\dfrac{17}{16}$

17. $\dfrac{5}{7}$ **18.** $\dfrac{2}{7}$ **19.** $\dfrac{18}{24}$ **20.** $\dfrac{9}{36}$

21. $4\dfrac{2}{5}$ **22.** $2\dfrac{3}{4}$ **23.** $\dfrac{1}{300}$ **24.** $\dfrac{1}{400}$

Write as a fraction in simplified form.

25. 22%   **26.** 53%   **27.** 150%   **28.** 160%

29. $6\dfrac{1}{3}\%$ **30.** $3\dfrac{1}{8}\%$ **31.** $51\dfrac{1}{4}\%$ **32.** $43\dfrac{3}{4}\%$

5.3

Solve. Round to the nearest hundredth when necessary.

33. What is 70% of 60?   **34.** Find 12% of 200.

35. 68 is what percent of 72?   **36.** What percent of 76 is 34?

37. 8% of what number is 240?   **38.** 354 is 40% of what number?

Now turn to page SA-10 for the answer to each of these problems. Each answer also includes a reference to the objective in which the problem is first taught. If you missed any of these problems, you should stop and review the Examples and Practice Problems in the referenced objective. A little review now will help you master the material in the upcoming sections of the text.

1.	17%
2.	38.7%
3.	795%
4.	518%
5.	0.6%
6.	0.04%
7.	17%
8.	89%
9.	13.4%
10.	19.8%
11.	$6\frac{1}{2}\%$
12.	$1\frac{3}{8}\%$
13.	80%
14.	50%
15.	260%
16.	106.25%
17.	71.43%
18.	28.57%
19.	75%
20.	25%
21.	440%
22.	275%
23.	0.33%
24.	0.25%
25.	$\frac{11}{50}$
26.	$\frac{53}{100}$
27.	$\frac{3}{2}$ or $1\frac{1}{2}$
28.	$\frac{8}{5}$ or $1\frac{3}{5}$
29.	$\frac{19}{300}$
30.	$\frac{1}{32}$
31.	$\frac{41}{80}$
32.	$\frac{7}{16}$
33.	42
34.	24
35.	94.44%
36.	44.74%
37.	3000
38.	885

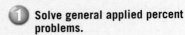
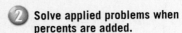

Teaching Example 1 117 cars passing through the turnpike tollbooth yesterday were over 15 years old. This number represents 4.5% of the cars that passed through the tollbooth. How many cars went through the tollbooth yesterday?

Ans: 2600 cars

 Solving General Applied Percent Problems

In Sections 5.3A and 5.3B, we learned the three types of percent problems. Some problems ask you to find a percent of a number. Some problems give you an amount and a percent and ask you to find the base (or whole). Other problems give an amount and a base and ask you to find the percent. We will now see how the three types of percent problems occur in real life.

EXAMPLE 1 Of all the computers manufactured last month, an inspector found 18 that were defective. This is 2.5% of all the computers manufactured last month. How many computers were manufactured last month?

Solution

Method A Translate to an equation.
The problem is equivalent to: 2.5% of the number of computers is 18.
Let n = the number of computers.

2.5% of the number of computers is 18

2.5% × n = 18

$$0.025n = 18$$
$$\frac{0.025n}{0.025} = \frac{18}{0.025}$$
$$n = 720$$

720 computers were manufactured last month.

Method B Use the percent proportion $\frac{a}{b} = \frac{p}{100}$.

The percent $p = 2.5$. The base is unknown. We will use the variable b. The amount $a = 18$. Thus

$$\frac{a}{b} = \frac{p}{100} \qquad \text{becomes} \qquad \frac{18}{b} = \frac{2.5}{100}.$$

Using cross multiplication, we have

$$(18)(100) = 2.5b$$
$$1800 = 2.5b$$
$$\frac{1800}{2.5} = \frac{2.5b}{2.5}$$
$$720 = b.$$

720 computers were manufactured last month.
By either Method A or Method B, we obtain the same number of computers, 720.

Substitute 720 into the original problem to check.

2.5% of 720 computers are defective.

$$(0.025)(720) = 18 \checkmark$$

Practice Problem 1 4800 people, or 12% of all passengers holding tickets for American Airlines flights in one month, did not show up for their flights. How many people held tickets that month? 40,000 people

NOTE TO STUDENT: Fully worked-out solutions to all of the Practice Problems can be found at the back of the text starting at page SP-1

EXAMPLE 2 How much sales tax will you pay on a color television priced at $299 if the sales tax is 5%?

Solution

Method A   Translate to an equation.

What is 5% of $299?

$\downarrow \quad \downarrow \quad \downarrow \quad \downarrow \quad \downarrow$

$$n = 5\% \times 299$$
$$n = (0.05)(299)$$
$$n = 14.95 \qquad \text{The tax is \$14.95.}$$

Method B   Use the percent proportion $\dfrac{a}{b} = \dfrac{p}{100}$.

The percent $p = 5$. The base $b = 299$. The amount is unknown. We use the variable a. Thus

$$\frac{a}{b} = \frac{p}{100} \qquad \text{becomes} \qquad \frac{a}{299} = \frac{5}{100}.$$

If we reduce the fraction on the right-hand side, we have

$$\frac{a}{299} = \frac{1}{20}.$$

We then cross-multiply to obtain

$$20a = 299$$
$$\frac{20a}{20} = \frac{299}{20}$$
$$a = 14.95 \quad \text{The tax is \$14.95.}$$

Thus, by either method, the amount of the sales tax is $14.95.

Let's see if our answer is reasonable. Is 5% of $299 really $14.95? If we round $299 to one nonzero digit, we have $300. Thus we have 5% of 300 = 15. Since 15 is quite close to our value of $14.95, our answer seems reasonable.

Practice Problem 2 A salesperson rented a hotel room for $62.30 per night. The tax in her state is 8%. What tax does she pay for one night at the hotel? Round to the nearest cent. $4.98

Teaching Example 2 Luz's lunch at the restaurant cost $10.50. The meal tax is 6%. How much meal tax will Luz pay?

Ans: $0.63

EXAMPLE 3 A failing student attended class 39 times out of the 45 times the class met last semester. What percent of the classes did he attend? Round to the nearest tenth of a percent.

Solution

Method A Translate to an equation.

This problem is equivalent to:

39 is what percent of 45?

$$39 = n \times 45$$

$$39 = 45n$$

$$\frac{39}{45} = \frac{45n}{45}$$

$$0.8666\ldots = n.$$

To the nearest tenth of a percent we have $n = 86.7\%$.

Method B Use the percent proportion $\dfrac{a}{b} = \dfrac{p}{100}$.

The percent is unknown. We use the variable p. The base b is 45. The amount a is 39. Thus

$$\frac{a}{b} = \frac{p}{100} \qquad \text{becomes} \qquad \frac{39}{45} = \frac{p}{100}.$$

When we cross-multiply, we get

$$(39)(100) = 45p$$

$$3900 = 45p$$

$$\frac{3900}{45} = \frac{45p}{45}$$

$$86.666\ldots = p.$$

To the nearest tenth, the answer is 86.7%.

By using either method, we discover that the failing student attended approximately 86.7% of the classes.

Verify by estimating that the answer is reasonable.

Practice Problem 3 Of the 130 flights at Orange County Airport yesterday, only 105 of them were on time. What percent of the flights were on time? (Round to the nearest tenth of a percent.) 80.8%

Now you have some experience solving the three types of percent problems in real-life applications. You can use either Method A or Method B to solve applied percent problems. In the following pages we will present more percent applications. We will not list all the steps of Method A or Method B. Most students will find after a careful study of Examples 1–3 that they do not need to write out all the steps of Method A or Method B when solving applied percent problems.

Solving Applied Problems When Percents Are Added

Percents can be added if the base (whole) is the same. For example, 50% of your salary added to 20% of your salary = 70% of your salary. 100% of your cost added to 15% of your cost = 115% of your cost. Problems like this are often called **markup problems.** If we add 15% of the cost of an item to the original cost, the markup is 15%. We will add percents in some applied situations.

The following example is interesting, but it is a little challenging. So please read it very carefully. A lot of students find it difficult at first.

EXAMPLE 4 Walter and Mary Ann are going out to a restaurant. They have a limit of $63.25 to spend for the evening. They want to tip the waitress 15% of the cost of the meal. How much money can they afford to spend on the meal itself? (Assume there is no tax.)

Solution In some of the problems in this section, it may help you to use the Mathematics Blueprint. We will use it here for Example 4.

Mathematics Blueprint for Problem Solving

Gather the Facts	What Am I Asked to Do?	How Do I Proceed?	Key Points to Remember
They have a spending limit of $63.25. They want to tip the waitress 15% of the cost of the meal.	Find the amount of money that the meal will cost.	Separate the $63.25 into two parts: the cost of the meal and the tip. Add these two parts to get $63.25.	We are not taking 15% of $63.25, but rather 15% of the cost of the meal.

Let n = the cost of the meal. 15% of the cost = the amount of the tip. We want to add the percents of the meal.

$$\boxed{\text{Cost of meal } n} + \boxed{\begin{array}{c}\text{tip of 15\%}\\\text{of the cost}\end{array}} = \boxed{\$63.25}$$

$$100\% \text{ of } n + 15\% \text{ of } n = \$63.25$$

Note that 100% of n added to 15% of n is 115% of n.

$$115\% \text{ of } n = \$63.25$$
$$1.15 \times n = 63.25$$
$$\frac{1.15 \times n}{1.15} = \frac{63.25}{1.15} \qquad \text{Divide both sides by 1.15.}$$
$$n = 55$$

They can spend up to $55.00 on the meal itself.

Does this answer seem reasonable?

Practice Problem 4 Sue and Sam have $46.00 to spend at a restaurant, including a 15% tip. How much can they spend on the meal itself? (Assume there is no tax.) $40

Solving Discount Problems

Frequently, we see signs urging us to buy during a sale when the list price is discounted by a certain percent.

The amount of a **discount** is the product of the discount rate and the list price.

Discount = discount rate × list price

Teaching Example 5 Will needs to buy a suit for a job interview. The price tag on the suit says $189. A nearby sign says that all suits are on sale for 30% off.

(a) What is the amount of the discount?

(b) What is the sale price of the suit?

Ans: (a) $56.70   **(b)** $132.30

EXAMPLE 5 Jeff purchased a large-screen color TV on sale at a 35% discount. The list price was $430.00.

(a) What was the amount of the discount?

(b) How much did Jeff pay for the large-screen color TV?

Solution

(a) Discount = discount rate × list price

= 35% × 430

= 0.35 × 430

= 150.5

The discount was $150.50.

(b) We subtract the discount from the list price to get the selling price.

$$\begin{array}{ll} \$430.00 & \text{list price} \\ - \$150.50 & \text{discount} \\ \hline \$279.50 & \text{selling price} \end{array}$$

Jeff paid $279.50 for the large-screen color TV.

Practice Problem 5 Betty bought a car that lists for $13,600 at a 7% discount.

(a) What was the discount? $952

(b) What did she pay for the car? $12,648

Applications *Exercises 1–18 present the three types of percent problems. They are similar to Examples 1–3. Take the time to master exercises 1–18 before going on to the next ones. Round to the nearest hundredth when necessary.*

1. **Education** No graphite was found in 4500 pencils shipped to Sureway School Supplies. This was 2.5% of the total number of pencils received by Sureway. How many pencils in total were in the order? 180,000 pencils

2. **Track and Field** A high-jumper on the track and field team hit the bar 58 times last week. This means that he did not succeed in 29% of his jump attempts. How many total attempts did he make last week? 200 attempts

3. **Cable Television Bill** Under Todd's new cable television plan, his bill averages $63 per month. This is 140% of his average monthly bill last year when he had the basic cable package. What was his average monthly cable bill last year? $45

4. **Salary Changes** Renata now earns $9.50 per hour. This is 125% of what she earned last year. What did she earn per hour last year? $7.60 per hour

5. **Square Footage of Home** A 2100-square-foot home is for sale. The finished basement has an area of 432 square feet. The basement accounts for what percent of the total square footage? 20.57%

6. **Coffee Bar** Every day this year, Sam ordered either cappuccino or espresso from the coffee bar downstairs. He had 85 espressos and 280 cappuccinos. What percent of the coffees were espressos? 23.29%

7. **Sales Tax** Elizabeth bought new towels and sheets for $65. How much tax did she pay if the sales tax is 6%? $3.90

8. **Sales Tax** Leon bought new clothes for his bank job. Before tax was added on, his total was $180. How much tax did he pay if the sales tax is 5%? $9

9. **Mountain Bike** Malia bought a new mountain bike. The sales tax in her state is 7%, and she paid $38.50 in tax. What was the price of the mountain bike before the tax? $550

10. **Sales Tax** Hiro bought some artwork and paid $10.75 in tax. The sales tax in his state is 5%. What was the price of the artwork? $215

11. **Mortgage Payment** Paul and Sue Yin together earn $4180 per month. Their mortgage payment is $1254 per month. What percent of their household income goes toward paying the mortgage? 30%

12. **Car Payment** Cora puts aside $52.50 per week for her monthly car payment. She earns $350 per week. What percent of her income is set aside for car payments? 15%

13. **Charities** The Children's Wish Charity raised 75% of its funds from sporting promotions. Last year the charity received $7,200,000 from its sporting promotions. What was the charity's total income last year? $9,600,000

14. **Taxes** Shannon paid $8400 in federal and state income taxes as a lab technician, which amounted to 28% of her annual income. What was her income last year? $30,000

15. **Baseball** 10,001 home runs have been hit in Boston's Fenway Park since it opened in 1912. Ted Williams of the Boston Red Sox hit 248 of these. What percent of the total home runs did Ted Williams hit? 2.48%

16. **Baseball** During the 2006 baseball season, Nomar Garciaparra was up to bat 469 times. Of these, 20 resulted in a home run. What percent of Garciaparra's at-bats resulted in a home run? 4.26%

17. **Pediatrics** In Flagstaff, Arizona, 0.9% of all babies walk before they reach the age of 11 months. If 24,000 babies were born in Flagstaff in the last 20 years, how many of them will have walked before they reached the age of 11 months? 216 babies

18. **Basketball** At a Boston Celtics basketball game scheduled for 8:30 P.M., 0.8% of the spectators were children under age 12. If 28,000 people showed up for the game, how many children under age 12 were in the stands? 224 children

Exercises 19–30 include percents that are added and discounts. Solve. Round to the nearest hundredth when necessary.

19. *Sales Tax* Henry has $800 total to spend on a new dining room table and chairs. If the sales tax is 5%, how much can he afford to spend on the table and chairs? $761.90

20. *Eating Out* Belinda asked Martin out to dinner. She has $47.50 to spend. She wants to tip the waitress 15% of the cost of their meal. How much money can she afford to spend on the meal itself? $41.30

21. *Building Costs* John and Chris Maney are building a new house. When finished, the house will cost $163,500. The price of the house is 9% higher than the price when the original plans were made. What was the price of the house when the original plans were made? $150,000

22. *SUV Purchase* Dan and Connie Lacorazza purchased a new Honda Pilot. The purchase price was $25,440. The price was 6% higher than the price of a similar Honda Pilot three years ago. What was the price of the Honda Pilot three years ago? $24,000

23. *Manufacturing* When a new computer case is made, there is some waste of the plastic material used for the front of the computer case. Approximately 3% of the plastic that is delivered is of poor quality and is thrown away. Furthermore, 8% of the plastic that is delivered is waste material that is thrown away as the front pieces are created by a giant stamping machine. If 20,000 pounds of plastic are delivered to make the fronts of computer cases each month, how many pounds are thrown away? 2200 pounds

24. *Airport Operations* In a recent survey of planes landing at Logan Airport in Boston, it was observed that 12% of the flights were delayed less than an hour and 7% of the flights were delayed an hour or more but less than two hours. If 9000 flights arrive at Logan Airport in a day, how many flights are delayed less than two hours? 1710 flights

25. *Political Parties* The Democratic National Committee has a budget of $33,000,000 to spend on the inauguration of the new president. 15% of the costs will be paid to personnel, 12% of the costs will go toward food, and 10% will go to decorations.

(a) How much money will go for personnel, food, and decorations?
$12,210,000 for personnel, food, and decorations

(b) How much will be left over to cover security, facility rental, and all other expenses?
$20,790,000 for security, facility rental, and all other expenses

26. *Medical Research* A major research facility has developed an experimental drug to treat Alzheimer's disease. Twenty percent of the research costs was paid to the staff. Sixteen percent of the research costs was paid to rent the building where the research was conducted. The rest of the money was used for research. The company has spent $6,000,000 in research on this new drug.

(a) How much was paid to cover the cost of staff and rental of the building?
$2,160,000

(b) How much was left over for research?
$3,840,000

27. *Clothes Purchase* Melinda purchased a new blouse, jeans, and a sweater in Naperville, Illinois. All of the clothes were discounted 35%. Before the sale, the total purchase price would have been $190 for these three items. How much did she pay for them with the discount? $123.50

28. *Tire Purchase* Juan went to purchase two new radial tires for his Honda Accord in Austin, Texas. The set of two tires normally costs $130. However, he bought them on sale at a discount of 30%. How much did he pay for the tires with the discount? $91

29. *Snowmobile Purchase* Jack bought his first Polaris snowmobile in Rice Lake, Wisconsin. The price was $8800, but the dealer gave him a discount of 15%.

(a) What was the discount? $1320

(b) How much did he pay for the snowmobile? $7480

30. *Appliance Purchase* Charlotte bought a stainless steel refrigerator and stove that had been used as floor models at Home Depot. The list price of the set was $1150, but the store manager gave her a discount of 25%.

(a) What was the discount? $287.50

(b) How much did she pay for the refrigerator and stove? $862.50

Cumulative Review

31. [1.7.1] Round to the nearest thousand. 1,698,481
1,698,000

32. [1.7.1] Round to the nearest hundred. 2,452,399
2,452,400

33. [3.2.3] Round to the nearest hundredth. 1.63474
1.63

34. [3.2.3] Round to the nearest thousandth. 0.7995
0.800

35. [3.2.3] Round to the nearest ten-thousandth. 0.055613   0.0556

36. [3.2.3] Round to the nearest ten-thousandth. 0.079152   0.0792

Quick Quiz 5.4

1. Chris Smith bought a new laptop computer. The list price was $596. He got a 28% discount.

(a) What was the discount? $166.88

(b) How much did Chris pay for the laptop? $429.12

2. Laurie left on a trip for a week. When she returned she had 87 e-mail messages. 56 of these messages were "spam" junk mail. What percent of her e-mail was spam? Round your answer to the nearest tenth if necessary.   64.4%

3. A total of 4500 people in the city bought take-out pizza at least once during the week. This was 30% of all the people who live in the city. How many people live in the city?   15,000 people

Explain how to solve the following problem.

4. Concept Check Sam works in sales for a pharmaceutical company. He can spend 23% of his budget for travel expenses. He can spend 14% for entertainment of clients. He can spend 17% of his budget for advertising. Last year he had a total budget of $80,000. Last year he spent a total of $48,000 for travel expenses, entertainment, and advertising. Did he stay within his budget allowance for those items?   Answers may vary

Classroom Quiz 5.4 You may use these problems to quiz your students' mastery of Section 5.4.

1. Melinda bought a new color television. The list price was $633. She got a 20% discount.

(a) What was the discount?   **Ans:** $126.60

(b) How much did Melinda pay for the television?   **Ans:** $506.40

2. The math instructor found that 19 out of 28 students in his class drove their cars to school. What percent of the students drove to school? Round your answer to the nearest tenth if necessary.   **Ans:** 67.9%

3. A total of 3380 customers visited Wal-Mart last week. This was 26% of all the people who live in the city where the store is located. How many people live in that city?   **Ans:** 13,000 people

1 Solving Commission Problems

If you work as a salesperson, your earnings may be in part or in total a certain percentage of the sales you make. The amount of money you get that is a percentage of the value of your sales is called your **commission.** It is calculated by multiplying the percentage (called the **commission rate**) by the value of the sales.

$$\text{Commission} = \text{commission rate} \times \text{value of sales}$$

EXAMPLE 1 A salesperson has a commission rate of 17%. She sells $32,500 worth of goods in a department store in two months. What is her commission?

Solution

$$\text{Commission} = \text{commission rate} \times \text{value of sales}$$
$$\text{Commission} = 17\% \times \$32,500$$
$$= 0.17 \times 32,500$$
$$= 5525$$

Her commission is $5525.00.

Does this answer seem reasonable? Check by estimating.

Practice Problem 1 A real estate salesperson earns a commission rate of 6% when he sells a $156,000 home. What is his commission? $9360

In some problems, the unknown quantity will be the commission rate or the value of sales. However, the same equation is used:

$$\text{Commission} = \text{commission rate} \times \text{value of sales}$$

2 Solving Percent-of-Increase or Percent-of-Decrease Problems

We sometimes need to find the percent by which a number increases or decreases. If a car costs $7000 and the price decreases $1750, we say that the percent of decrease is $\frac{1750}{7000} = 0.25 = 25\%$.

$$\textbf{Percent of decrease} = \frac{\text{amount of decrease}}{\text{original amount}}$$

Similarly, if a population of 12,000 people increases by 1920 people, we say that the percent of increase is $\frac{1920}{12,000} = 0.16 = 16\%$.

$$\textbf{Percent of increase} = \frac{\text{amount of increase}}{\text{original amount}}$$

Note that for these types of problems the base is always the *original amount*.

The most important thing to remember is that we must **first** find the amount of increase or decrease.

EXAMPLE 2 The population of Center City increased from 50,000 to 59,500. What was the percent of increase?

Solution For this problem as well as others in this section, you may find it helpful to use the Mathematics Blueprint.

Teaching Example 2 The number of a bank's customers who use online banking changed from 4550 to 4641 last month. What is the percent of increase?

Ans: 2%

Mathematics Blueprint for Problem Solving

Gather the Facts	What Am I Asked to Do?	How Do I Proceed?	Key Points to Remember
The population increased from 50,000 to 59,500.	We must find the percent of increase.	First subtract to find the amount of increase. Then divide the amount of increase by the original amount.	Always divide by the original amount.

Amount of increase
$$\begin{array}{r} 59,500 \\ -\ 50,000 \\ \hline 9500 \end{array}$$

$$\text{Percent of increase} = \frac{\text{amount of increase}}{\text{original amount}} = \frac{9500}{50,000}$$

$$= 0.19 = 19\%$$

The percent of increase is 19%.

Practice Problem 2 A new car is sold for $15,000. A year later its price had decreased to $10,500. What is the percent of decrease?   30%

3 Solving Simple Interest Problems

Interest is money paid for the use of money. If you deposit money in a bank, the bank uses that money and pays you interest. If you borrow money, you pay the bank interest for the use of that money.

The **principal** is the amount deposited or borrowed. Interest is usually expressed as a percent rate of the principal. The **interest rate** is assumed to be per year, unless otherwise stated. The formula used in business to compute simple interest is

$$\text{Interest} = \text{principal} \times \text{rate} \times \text{time}$$

$$I = P \times R \times T$$

If the interest rate is *per year,* the time *T must* be in *years.*

Teaching Tip If students have trouble with percent-of-increase and percent-of-decrease problems, remind them that the denominator is always the original amount. Usually, students make mistakes in this type of problem because they have forgotten this fact.

Calculator

 Interest

You can use a calculator to find simple interest. Find the interest on $450 invested at 6.5% for 15 months. Notice the time is in months. Since the interest formula $I = P \times R \times T$, is in years, you need to change 15 months to years by dividing 15 by 12.

Enter
15 ÷ 12 =

Display
1.25

Leave this on the display and multiply as follows:
1.25 × 450 ×
6.5 % =

The display should read
36.5625

which would round to $36.56.

Try the following.

(a) $9516 invested at 12% for 30 months

(b) $593 borrowed at 8% for 5 months

EXAMPLE 3 Find the simple interest on a loan of $7500 borrowed at 13% for one year.

Solution $I = P \times R \times T$

$P = \text{principal} = \$7500 \qquad R = \text{rate} = 13\% \qquad T = \text{time} = 1 \text{ year}$

$I = 7500 \times 13\% \times 1 = 7500 \times 0.13 = 975$

The interest is $975.

Practice Problem 3 Find the simple interest on a loan of $5600 borrowed at 12% for one year. $672

Our formula is based on a yearly interest rate. Time periods of more than one year or a fractional part of a year are sometimes needed.

EXAMPLE 4 Find the simple interest on a loan of $2500 that is borrowed at 9% for

(a) three years. **(b)** three months.

Solution

(a) $I = P \times R \times T$

$P = \$2500 \qquad R = 9\% \qquad T = 3 \text{ years}$

$I = 2500 \times 0.09 \times 3 = 225 \times 3 = 675$

The interest for three years is $675.

(b) Three months $= \dfrac{1}{4}$ year. The time period must be in years to use the formula.

Since $T = \dfrac{1}{4}$ year, we have

$$I = 2500 \times 0.09 \times \frac{1}{4}$$

$$= 225 \times \frac{1}{4}$$

$$= \frac{225}{4} = 56.25$$

The interest for three months is $56.25.

Practice Problem 4 Find the simple interest on a loan of $1800 that is borrowed at 11% for

(a) four years. $792 **(b)** six months. $99

Many loans today are based on **compound interest.** This topic is covered in more advanced mathematics courses. The calculations for compound interest are tedious to do by hand. Usually people use a computer or a compound interest table to do compound interest problems.

Applications

Exercises 1–18 are problems involving commissions, percent of increase or decrease, and simple interest.

1. *Appliance Sales* Walter works as an appliance salesman in a department store. Last month he sold $170,000 worth of appliances. His commission rate is 2%. How much money did he earn in commission last month? $3400

2. *Car Sales* Susan works at the Acura dealership in Winchester. Last month she had car sales totaling $230,000. Her commission rate is 3%. How much money did she earn in commission last month? $6900

3. *Mobile Phone Sales* Allison works in the local Verizon office selling mobile phones. She is paid $300 per month plus 4% of her total sales in mobile phones. Last month she sold $96,000 worth of mobile phones. What was her total income for the month? $4140

4. *Stockbroker* Matthew is a stockbroker. He is paid $500 per month plus 0.5% of the total sales of stocks that he sells. Last month he sold $340,000 worth of stock. What was his total income for the month? $2200

5. *Airline Tickets* Dawn is searching online for airline tickets. Two weeks ago the cost to fly from San Francisco to Minneapolis was $275. The price now is $330. What is the percent of increase? 20%

6. *Weight Loss* Tim weighed 267 pounds before starting an exercise routine. Two years later, he weighed a healthy 183 pounds. What was the percent of decrease in Tim's weight? Round your answer to the nearest tenth of a percent. 31.5%

7. *Computer Prices* In 2002, the average price for a notebook computer in the United States was $1496. In 2006, the average price was $948. What was the percent of decrease in average notebook computer price? Round your answer to the nearest tenth of a percent. (*Source:* www.npd.com) 36.6%

8. *Computer Prices* In 2002, the average price for a desktop computer in the United States was $807. In 2006, the average price was $635. What was the percent of decrease in average desktop computer price? (*Source:* www.npd.com) 21.3%

9. *CD Interest* Phil placed $2000 in a one-year CD at the bank. The bank is paying simple interest of 7% for one year on the CD. How much interest will Phil earn in one year? $140

10. *Checking Account Interest* Charlotte has a checking account that pays her simple interest of 1.2% on the average balance in her checking account. Last year her average balance was $450. How much interest did she earn in her checking account? $5.40

11. *Credit Card Expenses* Melinda has a MasterCard account with Centerville Bank. She has to pay a monthly interest rate of 1.5% on the average daily balance of the amount she owes on her credit card. Last month her average daily balance was $500. How much interest was she charged last month? (*Hint:* The formula $I = P \times R \times T$ can be used if the interest rate is *per month* and the time is in *months.*) $7.50

12. *Student Loan* Walter borrowed $3000 for a student loan to finish college this year. Next year he will need to pay 7% simple interest on the amount he borrowed. How much interest will he need to pay next year? $210

13. ***House Construction Loan*** James had to borrow $26,000 for a house construction loan for four months. The interest rate was 12% per year. How much interest did he have to pay for borrowing the money for four months? $1040

14. ***Small Business Loan*** Maya needed to borrow $9200 for three months to finance some renovations to her gift shop. The interest rate was 15% per year. How much interest did she have to pay for borrowing the money for three months? $345

15. ***Life Insurance*** Robert sells life insurance for a major insurance company for a commission. Last year he sold $12,000,000 worth of insurance. He earned $72,000 in commissions. What was his commission rate? 0.6%

16. ***Medical Supplies*** Hillary sells medical supplies to doctors' offices for a major medical supply company. Last year she sold $9,000,000 worth of medical supplies. She works on a commission basis and last year she earned $63,000 in commissions. What was her commission rate? 0.7%

17. ***Furniture Sales*** Jennifer sells furniture for a major department store. Last year she was paid $48,000 for commissions. If her commission rate is 3%, what was the sales total of the furniture that she sold last year? $1,600,000

18. ***Auto Sales*** Michael sells used cars for Beltway Motors. Last year he was paid $42,000 in commissions. Beltway Motors pays the salespeople a commission rate of 6%. What was the sales total of the cars that Michael sold last year? $700,000

Mixed Applications *Exercises 19–36 are a variety of percent problems. They involve commissions, percent of increase or decrease, and simple interest. There are also some of each kind of percent problem encountered in the chapter. Unless otherwise directed, round to the nearest hundredth.*

19. ***Entertainment Expenses*** Ted is trying to decrease his spending on entertainment. He earns $265 per week and is allowing himself to spend only 15% per week on movies, dining out, and so on. How much can Ted spend per week on entertainment? $39.75

20. ***Biology*** The maximum capacity of your lungs is 4.58 liters of air. In a typical breath, you breathe in 12% of the maximum capacity. How many liters of air do you breathe in a typical breath? approximately 0.55 liter

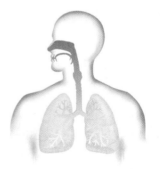

21. ***Girl Scout Cookies*** Of all the boxes of Girl Scout cookies sold, 25% are Thin Mints. If a Girl Scout troop sells 156 boxes of cookies, how many are Thin Mints? 39 boxes

22. ***Scotland*** 11% of the Scottish population have red hair. If the population of Scotland is 5,600,000, how many people have red hair? 616,000 people

23. ***Decrease in Crime*** In 1995, the number of non-fatal firearm incidents in the United States was 902,680. By 2005, the number had dropped to 419,640. What is the percent of decrease in the number of non-fatal firearm incidents? Round your answer to the nearest hundredth of a percent. (*Source:* www.usdoj.org) 53.51%

24. ***Carbon Dioxide Emissions*** In 2005, the amount of energy-related carbon dioxide emissions in the United States was 5955 million metric tons. In 2006, it is estimated that this number fell to 5877 million metric tons. What is the percent of decrease in carbon dioxide emissions? Round your answer to the nearest hundredth of a percent. (*Source:* www.energy.gov) 1.31%

25. ***Sporting Goods*** A sporting goods store buys cross-training shoes for $40, and sells them for $72. What is the percent of increase in the price of the shoes? 80%

26. ***Jewelry Costs*** A gift store buys earrings from an artist for $20 and sells them for $29. What is the percent of increase in the price of the earrings? 45%

27. ***Savings Account*** Adam deposited $3700 in his savings account for one year. His savings account earns 2.3% interest annually. He did not add any more money within the year, and at the end of that time, he withdrew all funds.

  (a) How much interest did he earn? $85.10
  (b) How much money did he withdraw from the bank? $3785.10

28. ***Credit Card*** Nikki had $1258 outstanding on her MasterCard, which charges 2% monthly interest. At the end of this month Nikki paid off the loan.

  (a) How much interest did Nikki pay for one month? $25.16
  (b) How much did it cost to pay off the loan totally? $1283.16

29. ***Shopping Trip*** Bryce went shopping and bought a pair of sandals for $52, swimming trunks for $38, and sunglasses for $26. The tax in Bryce's city is 6%.

  (a) What is the total sales tax? $6.96
  (b) What is the total of the purchases? $122.96

30. ***Automobile Purchase*** Jin bought a used Toyota Camry for $12,600. The tax in her state is 7%.

  (a) What is the sales tax? $882
  (b) What is the final price of the Camry? $13,482

31. ***Property Taxes*** Smithville Kitchen Cabinetry Inc. is late in paying $9500 in property taxes to the city of Springfield. It will be assessed 14% interest for being late in property tax payment. Find the one total amount it needs to pay off both the taxes and the interest charge. $10,830

32. ***Property Taxes*** Raymond and Elsie Ostram are late in paying $1600 in property taxes to the city of New Boston. They will be assessed 12% interest for being late in property tax payment. Find the one total amount they need to pay off both the taxes and the interest charge. $1792

33. ***Home Purchase*** Betty and Michael Bently purchased a new home for $349,000. They paid a down payment of 8% of the cost of the home. They took out a mortgage for the rest of the purchase price of the home.

  (a) What was the amount of their down payment? $27,920
  (b) What was the amount of their mortgage? $321,080

34. ***Home Purchase*** Marcia and Dan Perkins purchased a condominium for $188,000. They paid a down payment of 11% of the cost of the condominium. They took out a mortgage for the rest of the purchase price of the condominium.

  (a) What was the amount of their down payment? $20,680
  (b) What was the amount of their mortgage? $167,320

35. ***Interest Charges on a Mortgage*** Richard is making monthly mortgage payments of $840 for his home mortgage. He noticed on his monthly statement that $814 is used to pay off the interest charge. Only $26 is used to pay off the principal. What percent of his monthly mortgage payment is used to pay off the interest charge? Round to the nearest tenth of a percent.   96.9%

36. ***Interest Charges on a Mortgage*** Alicia is making monthly mortgage payments of $960 for her home mortgage. She noticed on her monthly statement that $917 is used to pay off the interest charge. Only $43 is used to pay off the principal. What percent of her monthly mortgage payment is used to pay off the interest charge? Round to the nearest tenth of a percent.   95.5%

Solve. Round to the nearest cent.

37. ***Sales Tax*** How much sales tax would you pay to purchase a new Honda Accord that costs $18,456.82 if the sales tax rate is 4.6%? $849.01

38. ***Living Room Set Purchase*** The Hartling family purchased a new living room set. The list price was $1249.95. However, they got a discount of 29%. How much did they pay for the new living room set? $887.46

Cumulative Review *Perform the following calculations using the correct order of operations.*

39. **[1.6.2]** $3(12 - 6) - 4(12 \div 3)$ 2

40. **[1.6.2]** $7 + 4^3 \times 2 - 15$ 120

41. **[2.8.3]** $\left(\dfrac{5}{2}\right)\left(\dfrac{1}{3}\right) - \left(\dfrac{2}{3} - \dfrac{1}{3}\right)^2$ $\dfrac{13}{18}$

42. **[3.6.2]** $(6.8 - 6.6)^2 + 2(1.8)$ 3.64

Quick Quiz 5.5

1. A construction contractor working for a real estate developer builds a house that sells for $325,000. He gets a commission of 8% on the house. What is his commission? $26,000

2. Susanne runs a bakery. Five years ago the bakery sold 160 loaves of bread each day. Today they sell 275 loaves each day. What is the percent of increase of the number of loaves sold each day?   71.875%

3. Find the simple interest on a loan of $4600 borrowed at 13% for six months. $299

4. **Concept Check** Explain how to find simple interest on a loan of $5800 borrowed at an annual rate of 16% for a period of three months.   Answers may vary

Classroom Quiz 5.5 You may use these problems to quiz your students' mastery of Section 5.5.

1. Barbara is a real estate agent who sells a house for $316,200. She gets a commission of 3% on the sale. What is her commission?   **Ans:** $9486

2. Ten years ago the Massachusetts Turnpike had 400 toll collectors. Today the turnpike has 225. What is the percent of decrease in the number of toll collectors?   **Ans:** 43.75%

3. Find the simple interest on a loan of $8500 borrowed at 12% for six months.   **Ans:** $510

Putting Your Skills to Work: Use Math to Save Money

AUTOMOBILE LEASING VS. PURCHASE

Louvy has his eye on a brand new car. He thinks he should lease the car because his best friend Tranh has a car lease and says he can get the same deal for Louvy. On the other hand, Louvy's girl-friend Allie says it is always better to buy the car and finance it by taking out a loan.

Louvy does some research and finds that it is not at all simple. While a lease offers lower monthly payments, at the end of the lease period you are left with nothing, except expenses.

Look at the following comparison for lease vs. buy for the car Louvy is considering.

	Lease	Purchase
Automobile price	$23,000.00	$23,000.00
Interest rate	6%	6%
Length of loan	36 months	36 months
Down payment	$1000.00	$1000.00
Residual (value of car you are turning in, amount you pay if you wish to purchase it)	$11,000.00	Not applicable
Monthly payment	$388.06	$669.28

1. How much would Louvy pay over the entire length of the loan? $1000 + 36 × $669.28 = $25,094.08

2. How much would Louvy pay over the entire length of the lease? $1000 + 36 × $388.06 = $14,970.16

3. If Louvy decided to buy the car at the end of the lease, he would have to pay $11,000.00 in addition to his lease cost. What would that bring the total cost of that car up to?
$14,970.16 + $11,000.00 = $25,970.16

Making It Personal for You

4. How much can you afford per month for pay-ments for a car? How much would you pay in insurance, taxes, and gas?   Answers may vary

5. Would you prefer to lease or buy? Why?
Answers may vary

Facts You Should Know

You may wish to lease a car if:

- You want a new vehicle every 2–3 years
- You don't drive an excessive number of miles each year

- You don't want major repairs risk
- You want a lower monthly payment

IF you lease a car:

- You pay only that portion of the vehicle you use
- There may be mileage restrictions, hidden fees, or security deposits
- You can sometimes find a lease with no down payment
- You pay sales tax only on monthly fees
- You usually must keep the car for the entire lease period, or pay a heavy penalty.

You may wish to buy a car if:

- You intend to keep it a long time
- You want to be debt-free after a time
- You qualify for a very low interest rate
- The long term cost is more important than the lower monthly payment

IF you buy a car:

- You pay for the entire vehicle
- You pay the sales tax on the entire price of the car
- There are no hidden mileage costs, except in wear and tear on the vehicle
- You can sell or trade the car during the period of the loan

Topic	Procedure	Examples
Converting a decimal to a percent, p. 306.	**1.** Move the decimal point two places to the right. **2.** Add the percent sign.	$0.19 = 19\%$ $0.516 = 51.6\%$ $0.04 = 4\%$ $1.53 = 153\%$ $0.006 = 0.6\%$
Converting a fraction with a denominator of 100 to a percent, p. 303.	**1.** Use the numerator only. **2.** Add the percent sign.	$\dfrac{29}{100} = 29\%$ $\dfrac{5.6}{100} = 5.6\%$ $\dfrac{3}{100} = 3\%$ $\dfrac{7\frac{1}{3}}{100} = 7\frac{1}{3}\%$ $\dfrac{231}{100} = 231\%$
Changing a fraction (whose denominator is not 100) to a percent, p. 312.	**1.** Divide the numerator by the denominator and obtain a decimal. **2.** Change the decimal to a percent.	$\dfrac{13}{50} = 0.26 = 26\%$ $\dfrac{1}{20} = 0.05 = 5\%$ $\dfrac{3}{800} = 0.00375 = 0.375\%$ $\dfrac{312}{200} = 1.56 = 156\%$
Changing a percent to a decimal, p. 305.	**1.** Drop the percent sign. **2.** Move the decimal point two places to the left.	$49\% = 0.49$ $2\% = 0.02$ $0.5\% = 0.005$ $196\% = 1.96$ $1.36\% = 0.0136$
Changing a percent to a fraction, p. 310.	**1.** If the percent does not contain a decimal point, remove the % and write the number over a denominator of 100. Reduce the fraction if possible. **2.** If the percent contains a decimal point, change the percent to a decimal by removing the % and moving the decimal point two places to the left. Then write the decimal as a fraction and reduce if possible. **3.** If the percent contains a fraction, remove the % and write the number over a denominator of 100. If the numerator is a mixed number, change the numerator to an improper fraction. Next simplify by the "invert and multiply" rule. Then reduce the fraction if possible.	$25\% = \dfrac{25}{100} = \dfrac{1}{4}$ $38\% = \dfrac{38}{100} = \dfrac{19}{50}$ $130\% = \dfrac{130}{100} = \dfrac{13}{10}$ $5.8\% = 0.058$ $\quad = \dfrac{58}{1000} = \dfrac{29}{500}$ $2.72\% = 0.0272$ $\quad = \dfrac{272}{10{,}000} = \dfrac{17}{625}$ $7\frac{1}{8}\% = \dfrac{7\frac{1}{8}}{100}$ $\quad = 7\frac{1}{8} \div \dfrac{100}{1}$ $\quad = \dfrac{57}{8} \times \dfrac{1}{100} = \dfrac{57}{800}$

Topic	Procedure	Examples
Changing a percent to a mixed number, p. 310.	**1.** Drop the percent sign and move the decimal point two places to the left. **2.** Write the decimal part of the number as a fraction. **3.** Reduce the fraction if possible.	$275\% = 2.75 = 2\dfrac{75}{100} = 2\dfrac{3}{4}$ $324\% = 3.24 = 3\dfrac{24}{100} = 3\dfrac{6}{25}$ $107\% = 1.07 = 1\dfrac{7}{100}$
Solving percent problems by translating to equations, p. 321.	**1.** Translate by replacing "of" with $\times$ "is" with $=$ "what" with n "find" with $n =$ **2.** Solve the resulting equation.	**(a)** What is 3% of 56? $n = 3\% \times 56$ $n = (0.03)(56)$ $n = 1.68$ **(b)** 16% of what is 208? $16\% \times n = 208$ $0.16n = 208$ $\dfrac{0.16n}{0.16} = \dfrac{208}{0.16}$ $n = 1300$ **(c)** What percent of 70 is 30? $n \times 70 = 30$ $70n = 30$ $\dfrac{70n}{70} = \dfrac{30}{70}$ $n = 0.428571\ldots$ $n \approx 42.86\%.$
Solving percent problems by using proportions, p. 329.	**1.** Identify the parts of the percent proportion. $a =$ the amount $b =$ the base (the whole; it usually appears after the word "of") $p =$ the percent number **2.** Write the percent proportion $\dfrac{a}{b} = \dfrac{p}{100}$ using the values obtained in step 1 and solve.	**(a)** What is 28% of 420? The percent $p = 28$. The base $b = 420$. The amount a is unknown. We use the variable a. $\dfrac{a}{b} = \dfrac{p}{100}$ becomes $\dfrac{a}{420} = \dfrac{28}{100}$ If we reduce the fraction on the right-hand side, we have $\dfrac{a}{420} = \dfrac{7}{25}$ $25a = (7)(420)$ $25a = 2940$ $\dfrac{25a}{25} = \dfrac{2940}{25}$ $a = 117.6$ Thus, 28% of 420 is 117.6. **(b)** 64% of what is 320? The percent $p = 64$. The base is unknown. We use the variable b. The amount $a = 320$. $\dfrac{a}{b} = \dfrac{p}{100}$ becomes $\dfrac{320}{b} = \dfrac{64}{100}$ If we reduce the fraction on the right-hand side, we have $\dfrac{320}{b} = \dfrac{16}{25}$ $(320)(25) = 16b$ $8000 = 16b$ $\dfrac{8000}{16} = \dfrac{16b}{16}$ $500 = b$ Thus, 64% of 500 is 320.

(Continued on next page)

Topic	Procedure	Examples
Solving percent problems by using proportions, p. 331.	**1.** Identify the parts of the percent proportion. a = the amount b = the base (the whole; it usually appears after the word "of") p = the percent number **2.** Write the percent proportion $\dfrac{a}{b} = \dfrac{p}{100}$ using the values obtained in step 1 and solve.	**(c)** What percent of 140 is 105? The percent is unknown. The base $b = 140$, the amount $a = 105$. $\dfrac{a}{b} = \dfrac{p}{100}$ becomes $\dfrac{105}{140} = \dfrac{p}{100}$ If we reduce the fraction on the left side, we have $$\dfrac{3}{4} = \dfrac{p}{100}$$ $$(100)(3) = 4p$$ $$300 = 4p$$ $$\dfrac{300}{4} = \dfrac{4p}{4}$$ $$75 = p$$ Thus 105 is 75% of 140.
Solving discount problems, p. 340.	Discount = discount rate × list price	Carla purchased a color TV set that lists for $350 at an 18% discount. **(a)** How much was the discount? **(b)** How much did she pay for the TV set? **(a)** Discount = $(0.18)(350) = \$63$ **(b)** $350 - 63 = 287$ She paid $287 for the color TV set.
Solving commission problems, p. 344.	Commission = commission rate × value of sales	A housewares salesperson gets a 16% commission on sales he makes. How much commission does he earn if he sells $12,000 in housewares? Commission = $(0.16)(12{,}000)$ $= 1920$ He earns a commission of $1920.
Solving simple-interest problems, p. 345.	Interest = principal × rate × time $$I = P \times R \times T$$ I = interest P = principal R = rate T = time	Hector borrowed $3000 for 4 years at a simple interest rate of 12%. How much interest did he owe after 4 years? $$I = P \times R \times T$$ $$I = (3000)(0.12)(4)$$ $$= (360)(4)$$ $$= 1440$$ Hector owed $1440 in interest.
Percent-of-increase or percent-of-decrease problems, p. 344.	Percent of increase or decrease = amount of increase or decrease ÷ base	A car that costs $16,500 now cost only $15,000 last year. What is the percent of increase? $\begin{array}{r} 16{,}500 \\ -\,15{,}000 \\ \hline 1500 \ \text{increase} \end{array}$ $\dfrac{1500}{15{,}000} = 0.10$ Percent of increase = 10%

Chapter 5 Review Problems

Section 5.1

Write as a percent. Round to the nearest hundredth of a percent when necessary.

1. 0.62 62% **2.** 0.43 43% **3.** 0.372 37.2% **4.** 0.529 52.9% **5.** 2.2 220% **6.** 1.8 180%

7. 2.52 252% **8.** 4.37 437% **9.** 1.036 103.6% **10.** 1.052 105.2% **11.** 0.006 0.6% **12.** 0.002 0.2%

13. $\dfrac{62.5}{100}$ 62.5% **14.** $\dfrac{37.5}{100}$ 37.5% **15.** $\dfrac{4\frac{1}{12}}{100}$ $4\frac{1}{12}$% **16.** $\dfrac{3\frac{5}{12}}{100}$ $3\frac{5}{12}$% **17.** $\dfrac{317}{100}$ 317% **18.** $\dfrac{225}{100}$ 225%

Section 5.2

Change to a percent. Round to the nearest hundredth of a percent when necessary.

19. $\dfrac{19}{25}$ 76% **20.** $\dfrac{13}{25}$ 52% **21.** $\dfrac{11}{20}$ 55% **22.** $\dfrac{9}{40}$ 22.5% **23.** $\dfrac{7}{12}$ 58.33%

24. $\dfrac{14}{15}$ 93.33% **25.** $2\dfrac{1}{4}$ 225% **26.** $3\dfrac{3}{4}$ 375% **27.** $2\dfrac{7}{9}$ 277.78% **28.** $5\dfrac{5}{9}$ 555.56%

29. $\dfrac{152}{80}$ 190% **30.** $\dfrac{200}{80}$ 250% **31.** $\dfrac{3}{800}$ 0.38% **32.** $\dfrac{5}{800}$ 0.63%

Change to decimal form.

33. 32% 0.32 **34.** 68% 0.68 **35.** 15.75% 0.1575 **36.** 12.35% 0.1235

37. 236% 2.36 **38.** 177% 1.77 **39.** $32\dfrac{1}{8}$% 0.32125 **40.** $26\dfrac{3}{8}$% 0.26375

Change to fractional form.

41. 72% $\dfrac{18}{25}$ **42.** 92% $\dfrac{23}{25}$ **43.** 175% $\dfrac{7}{4}$ **44.** 260% $\dfrac{13}{5}$ **45.** 16.4% $\dfrac{41}{250}$

46. 30.5% $\dfrac{61}{200}$ **47.** $31\dfrac{1}{4}$% $\dfrac{5}{16}$ **48.** $43\dfrac{3}{4}$% $\dfrac{7}{16}$ **49.** 0.08% $\dfrac{1}{1250}$ **50.** 0.04% $\dfrac{1}{2500}$

Complete the following chart.

	Fraction	Decimal	Percent
51.	$\dfrac{3}{5}$	0.6	60%
52.	$\dfrac{7}{10}$	0.7	70%
53.	$\dfrac{3}{8}$	0.375	37.5%

	Fraction	Decimal	Percent
54.	$\dfrac{9}{16}$	0.5625	56.25%
55.	$\dfrac{1}{125}$	0.008	0.8%
56.	$\dfrac{9}{20}$	0.45	45%

Section 5.3

Solve. Round to the nearest hundredth when necessary.

57. What is 20% of 85?
17

58. What is 25% of 92?
23

59. 15 is 25% of what number?
60

60. 30 is 75% of what number?
40

61. 50 is what percent of 130?
38.46%

62. 70 is what percent of 180?
38.89%

63. Find 162% of 60.
97.2

64. Find 124% of 80.
99.2

65. 92% of what number is 147.2?
160

66. 68% of what number is 95.2?
140

67. What percent of 70 is 14?
20%

68. What percent of 60 is 6?
10%

Sections 5.4 and 5.5

Solve. Round your answer to the nearest hundredth when necessary.

69. *Education* Professor Padron found that 35% of his World History course is sophomores. He has 140 students in his class. How many are sophomores? 49 students

70. *Truck Dealer* A Vermont truck dealer found that 64% of all the trucks he sold had four-wheel drive. If he sold 150 trucks, how many had four-wheel drive? 96 trucks

71. *Car Depreciation* Today Yvonne's car has 61% of the value that it had two years ago. Today it is worth $6832. What was it worth two years ago? $11,200

72. *Administrative Expenses* A charity organization spent 12% of its budget for administrative expenses. It spent $9624 on administrative expenses. What was the total budget? $80,200

73. *Rain in Seattle* In Seattle it rained 20 days in February, 18 days in March, and 16 days in April. What percent of those three months did it rain? (Assume it was a leap year.) 60%

74. *Job Applications* Moorehouse Industries received 600 applications and hired 45 of the applicants. What percent of the applicants obtained a job? 7.5%

75. *Appliance Purchase* Nathan bought new appliances for $3670. The sales tax in his state is 5%. What did he pay in sales tax? $183.50

76. *Boat Purchase* Chris and Annette bought a boat for $12,600. The sales tax is 6% in their state. What did they pay in sales tax? $756

77. *Budgets* Joan and Michael budget 38% of their income for housing. They spend $684 per month for housing. What is their monthly income? $1800

78. *Real Estate Sales* Beachfront property is very expensive. A real estate agent in South Carolina earned $26,000 in commissions. The property she sold was worth $650,000. What was her commission rate? 4%

79. *Encyclopedia Sales* Adam sold encyclopedias last summer to raise tuition money. He sold $83,500 worth of encyclopedias and was paid $5010 in commissions. What commission rate did he earn? 6%

80. *Commission Sales* Roberta earns a commission at the rate of 7.5%. Last month she sold $16,000 worth of goods. How much commission did she make last month? $1200

81. *Furniture Set* Amy purchased a table and chairs set for her new patio at a 25% discount. The list price was $1450.
(a) What was the discount? $362.50
(b) What did she pay for the set? $1087.50

82. *Laptop Computer* A Dell laptop computer listed for $2125. This week, Lisa heard that the manufacturer is offering a rebate of 12%.
(a) What is the rebate? $255
(b) How much will Lisa pay for the computer? $1870

83. *Medical School Applications* In 2002, the number of students who applied to medical school in the United States was 33,625. In 2006, the number who applied was 39,109. What was the percent of increase in the number of medical school applications? (*Source:* www.aamc.org) 16.31%

84. *Cost of Education* For the 1996–97 academic year, the average cost of tuition, fees, and room and board at a four-year public college in the United States was $9258. By the 2006–07 academic year, the cost had risen to $12,796. Find the percent of increase. Round to the nearest whole percent. (*Source:* www.collegeboard.com) 38%

85. *Log Cabin* Mark and Julie wanted to buy a pre-fabricated log cabin to put on their property in the Colorado Rockies. The price of the kit is listed at $24,000. At the after-holiday cabin sale, a discount of 14% was offered.
(a) What was the discount? $3360
(b) How much did they pay for the cabin? $20,640

86. *Mutual Funds* Sally invested $6000 in mutual funds earning 11% simple interest in one year. How much interest will she earn in
(a) six months? $330
(b) two years? $1320

87. *College Loan* Reed took out a college loan of $3000. He will be charged 8% simple interest on the loan.
(a) How much interest will be due on the loan in three months? $60
(b) How much interest will be due on the loan in three years? $720

Note to Instructor: The Chapter 5 Test file in the TestGen program provides algorithms specifically matched to these problems so you can easily replicate this test for additional practice or assessment purposes.

1.	57%
2.	1%
3.	0.8%
4.	1280%
5.	356%
6.	71%
7.	1.8%
8.	$3\frac{1}{7}\%$
9.	47.5%
10.	75%
11.	300%
12.	175%
13.	8.25%
14.	302.4%
15.	$1\frac{13}{25}$
16.	$\frac{31}{400}$
17.	20
18.	130
19.	55.56%
20.	200

How Am I Doing? Chapter 5 Test

Remember to use your Chapter Test Prep Video CD to see the worked-out solutions to the test problems you want to review.

Write as a percent. Round to the nearest hundredth of a percent when necessary.

1. 0.57

2. 0.01

3. 0.008

4. 12.8

5. 3.56

6. $\frac{71}{100}$

7. $\frac{1.8}{100}$

8. $\frac{3\frac{1}{7}}{100}$

Change to a percent. Round to the nearest hundredth of a percent when necessary.

9. $\frac{19}{40}$

10. $\frac{27}{36}$

11. $\frac{225}{75}$

12. $1\frac{3}{4}$

Write as a percent.

13. 0.0825

14. 3.024

Write as a fraction in simplified form.

15. 152%

16. $7\frac{3}{4}\%$

Solve. Round to the nearest hundredth if necessary.

17. What is 40% of 50?

18. 33.8 is 26% of what number?

19. What percent of 72 is 40?

20. Find 0.8% of 25,000.

21. 16% of what number is 800?

22. 92 is what percent of 200?

23. 132% of 530 is what number?

24. What percent is 15 of 75?

Solve. Round to the nearest hundredth if necessary.

25. A real estate agent sells a house for $152,300. She gets a commission of 4% on the sale. What is her commission?

26. Julia and Charles bought a new dishwasher at a 33% discount. The list price was $457.
   (a) What was the discount?
   (b) How much did they pay for the dishwasher?

27. An inspector found that 75 out of 84 parts were not defective. What percent of the parts were not defective?

28. Last year Charlotte was the top player on the basketball team, scoring 185 points. This year, she scored 228 points. What is the percentage of increase in the number of points?

29. A total of 5160 people voted in the city election. This was 43% of the registered voters. How many registered voters are in the city?

30. Wanda borrowed $3000 at a simple interest rate of 16%.
   (a) How much interest did she pay in six months?
   (b) How much interest did she pay in two years?

Cumulative Test for Chapters 1–5

Approximately one-half of this test is based on Chapter 5 material. The remainder is based on material covered in Chapters 1–4.

Solve. Simplify your answer.

1. Add.   38
          196
       + 2007

2. Subtract.   23,007
              − 14,563

3. Multiply.   126
             × 42

4. Divide. $36\overline{)3204}$

5. Add. $2\frac{1}{4} + 3\frac{1}{3}$

6. Subtract. $5\frac{2}{5} - 2\frac{7}{10}$

7. Multiply. $3\frac{1}{8} \times \frac{12}{5}$

8. Divide. $\frac{21}{4} \div 1\frac{3}{4}$

9. Round to the nearest thousandth. 77.1832

10. Add.   5.6
          3.21
         18.3
       + 7.008

11. Multiply.   3.16
               × 2.8

12. Divide. $1.4\overline{)0.5152}$

▲ **13.** Write as a unit rate. 36 tiles in 9 square feet

14. Is this equation a proportion? $\frac{20}{25} = \frac{300}{375}$

15. Solve the proportion. $\frac{8}{2.5} = \frac{n}{7.5}$

16. A college has a ratio of 3 faculty members for every 19 students. The student body presently has 4263 students. How many faculty members are there? Round to the nearest whole number.

1. 2241
2. 8444
3. 5292
4. 89
5. $\frac{67}{12}$ or $5\frac{7}{12}$
6. $2\frac{7}{10}$
7. $\frac{15}{2}$ or $7\frac{1}{2}$
8. 3
9. 77.183
10. 34.118
11. 8.848
12. 0.368
13. 4 tiles/square foot
14. yes
15. $n = 24$
16. 673 faculty members

In questions 17–30, round to the nearest hundredth when necessary.

Write as a percent.

17. 0.355

18. $\dfrac{46.8}{100}$

19. 1.98

20. $\dfrac{3}{80}$

In questions 21 and 22, write as a decimal.

21. 243%

22. $6\dfrac{3}{4}\%$

23. What percent of 214 is 38?

24. Find 1.7% of 6740.

25. 40 is 25% of what number?

26. 95% of 200 is what number?

27. While shopping for appliances, Cecilia sees a new washer and dryer for $680. A sign in the store reads "20% off." How much would she pay for the appliances?

28. At Waldoch Auto Sales, 58 Volkswagens were sold last year. This made up 25% of all vehicles sold. How many vehicles were sold last year?

29. The air pollution level in Centerville is 8.86 parts per million. Ten years ago it was 7.96 parts per million. What is the percent of increase of the air pollution level?

30. Carlo borrowed $3400 for two years. He was charged simple interest at a rate of 9%. How much interest did he pay?

17.	35.5%
18.	46.8%
19.	198%
20.	3.75%
21.	2.43
22.	0.0675
23.	17.76%
24.	114.58
25.	160
26.	190
27.	$544
28.	232 vehicles
29.	11.31%
30.	$612

Practice Final Examination

This examination is based on Chapters 1–5 of the book. There are 10 questions covering the content of each chapter.

Chapter 1

1. Write in words. 82,367

2. Add.   13,428
         + 16,905

3. Add.   19
            23
            16
            45
          + 70

4. Subtract.   89,071
              − 54,968

Multiply.

5.   78
       × 54

6.   2035
        × 107

In questions 7 and 8, divide. (Be sure to indicate the remainder if one exists.)

7. $7\overline{)1106}$

8. $26\overline{)15,756}$

9. Evaluate. Perform operations in the proper order. $3^4 + 20 \div 4 \times 2 + 5^2$

10. Melinda traveled 512 miles in her car. The car used 16 gallons of gas on the entire trip. How many miles per gallon did the car achieve?

Chapter 2

11. Reduce the fraction. $\dfrac{14}{30}$

12. Change to an improper fraction. $3\dfrac{9}{11}$

13. Add. $\dfrac{1}{10} + \dfrac{3}{4} + \dfrac{4}{5}$

14. Add. $2\dfrac{1}{3} + 3\dfrac{3}{5}$

15. Subtract. $4\dfrac{5}{7} - 2\dfrac{1}{2}$

16. Multiply. $1\dfrac{1}{4} \times 3\dfrac{1}{5}$

17. Divide. $\dfrac{7}{9} \div \dfrac{5}{18}$

18. Divide. $\dfrac{5\frac{1}{2}}{3\frac{1}{4}}$

19. Lucinda jogged $1\frac{1}{2}$ miles on Monday, $3\frac{1}{4}$ miles on Tuesday, and $2\frac{1}{10}$ miles on Wednesday. How many miles in all did she jog over the three-day period?

20. A butcher has $11\frac{2}{3}$ pounds of steak. She wishes to place them in several equal-size packages. Each package will hold $2\frac{1}{3}$ pounds of steak. How many packages can be made?

---

1. eighty-two thousand, three hundred sixty-seven

2. 30,333

3. 173

4. 34,103

5. 4212

6. 217,745

7. 158

8. 606

9. 116

10. 32 mi/gal

11. $\dfrac{7}{15}$

12. $\dfrac{42}{11}$

13. $\dfrac{33}{20}$ or $1\dfrac{13}{20}$

14. $\dfrac{89}{15}$ or $5\dfrac{14}{15}$

15. $\dfrac{31}{14}$ or $2\dfrac{3}{14}$

16. 4

17. $\dfrac{14}{5}$ or $2\dfrac{4}{5}$

18. $\dfrac{22}{13}$ or $1\dfrac{9}{13}$

19. $6\dfrac{17}{20}$ mi

20. 5 packages

Chapter 3

21. Express as a decimal. $\dfrac{719}{1000}$

22. Write in reduced fractional notation. 0.86

23. Fill in the blank with $<$, $=$, or $>$. 0.315 _____ 0.309

24. Round to the nearest hundredth. 506.3782

25. Add.   9.6
3.82
1.05
+ 7.3

26. Subtract.   3.61
− 2.853

27. Multiply.   1.23
× 0.4

28. Divide. $0.24\overline{)0.8856}$

29. Write as a decimal. $\dfrac{13}{16}$

30. Evaluate by performing operations in proper order.
$0.7 + (0.2)^3 − 0.08(0.03)$

Chapter 4

31. Write a rate in simplest form to compare 7000 students to 215 faculty.

32. Is this a proportion? $\dfrac{12}{15} = \dfrac{17}{21}$

Solve the proportion. Round to the nearest tenth when necessary.

33. $\dfrac{5}{9} = \dfrac{n}{17}$

34. $\dfrac{3}{n} = \dfrac{7}{18}$

35. $\dfrac{n}{12} = \dfrac{5}{4}$

36. $\dfrac{n}{7} = \dfrac{36}{28}$

Solve using a proportion. Round to the nearest hundredth when necessary.

37. Bob earned $2000 for painting three houses. How much would he earn for painting five houses?

38. Two cities that are actually 200 miles apart appear 6 inches apart on the map. Two other cities are 325 miles apart. How far apart will they appear on the same map?

39. Roberta earned $68 last week on her part-time job. She had $5 withheld for federal income tax. Last year she earned $4000 on her part-time job. Assuming the same rate, how much was withheld for federal income tax last year?

21.	0.719
22.	$\dfrac{43}{50}$
23.	$>$
24.	506.38
25.	21.77
26.	0.757
27.	0.492
28.	3.69
29.	0.8125
30.	0.7056
31.	$\dfrac{1400 \text{ students}}{43 \text{ faculty}}$
32.	no
33.	$n \approx 9.4$
34.	$n \approx 7.7$
35.	$n = 15$
36.	$n = 9$
37.	$3333.33
38.	9.75 in.
39.	$294.12

40.	1.6 lb
41.	0.63%
42.	21.25%
43.	1.64
44.	17.33%
45.	302.4
46.	250
47.	4284
48.	$10,856
49.	4500 students
50.	34.3%

40. Malaga's recipe feeds 18 people and calls for 1.2 pounds of butter. If she wants to feed 24 people, how many pounds of butter does she need?

Chapter 5

Round to the nearest hundredth when necessary in problems 41–44.

41. Write as a percent. 0.0063

42. Change $\dfrac{17}{80}$ to a percent.

43. Write as a decimal. 164%

44. What percent of 300 is 52?

Round to the nearest tenth when necessary in problems 45–50.

45. Find 6.3% of 4800.

46. 145 is 58% of what number?

47. 126% of 3400 is what number?

48. Pauline bought a new car. She got an 8% discount. The car listed for $11,800. How much did she pay for the car?

49. A total of 1260 freshmen were admitted to Central College. This is 28% of the student body. How big is the student body?

50. There are 11.28 centimeters of water in the rain gauge this week. Last week the rain gauge held 8.40 centimeters of water. What is the percentage of increase from last week to this week?

Appendix A Consumer Finance Applications

A.1 BALANCING A CHECKING ACCOUNT

 Calculating a Checkbook Balance

If you have a checking account, you should keep records of the checks written, ATM withdrawals, deposits, and other transactions on a check register. To find the amount of money in a checking account you subtract debits and add credits to the balance in the account. Debits are checks written, withdrawals made, or any other amount charged to a checking account. Credits include deposits made, as well as any other money credited to the account.

Student Learning Objectives

After studying this section, you will be able to:

1. Calculate a checkbook balance.

2. Balance a checkbook.

EXAMPLE 1 Jesse Holm had a balance of $1254.32 in his checking account before writing five checks and making a deposit. On 9/2 Jesse wrote check #243 to the Manor Apartments for $575, check #244 to the Electric Company for $23.41, and check #245 to the Gas Company for $15.67. Then on 9/3, he wrote check #246 to Jack's Market for $125.57, check #247 to Clothing Mart for $35.85, and made a $634.51 deposit. Record the checks and deposit in Jesse's check register and then find Jesse's ending balance.

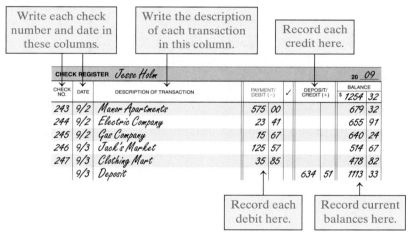

| Write each check number and date in these columns. | | Write the description of each transaction in this column. | | | Record each credit here. | | |

CHECK REGISTER *Jesse Holm*   20 _09_

CHECK NO.	DATE	DESCRIPTION OF TRANSACTION	PAYMENT/ DEBIT (−)		✓	DEPOSIT/ CREDIT (+)		BALANCE $ 1254	32
243	9/2	Manor Apartments	575	00				679	32
244	9/2	Electric Company	23	41				655	91
245	9/2	Gas Company	15	67				640	24
246	9/3	Jack's Market	125	57				514	67
247	9/3	Clothing Mart	35	85				478	82
	9/3	Deposit				634	51	1113	33

| Record each debit here. | Record current balances here. |

Solution To find the ending balance, we subtract each check written and add the deposit to the current balance. Then we record these amounts in the check register.

$$
\begin{array}{cccccc}
1254.32 & 679.32 & 655.91 & 640.24 & 514.67 & 478.82 \\
- \ 575.00 & - \ 23.41 & - \ 15.67 & - \ 125.57 & - \ 35.85 & + \ 634.51 \\
\hline
679.32 & 655.91 & 640.24 & 514.67 & 478.82 & 1113.33
\end{array}
$$

Jesse's balance is 1113.33.

Practice Problem 1 My Chung Nguyen had a balance of $1434.52 in her checking account before writing three checks and making a deposit. On 3/1/2009 My Chung wrote check #144 to the Leland Mortgage Company for $908 and check #145 to the Phone Company for $33.21. Then on 3/2/2009 she wrote check #146 to Sam's Food Market for $102.37 and made a $524.41 deposit. Record the checks and deposit in My Chung's check register, and then find My Chung's ending balance.

NOTE TO STUDENT: Fully worked-out solutions to all of the Practice Problems can be found at the back of the text starting at page SP-1

CHECK REGISTER	*My Chung Nguyen*					20 _09_
CHECK NO.	DATE	DESCRIPTION OF TRANSACTION	PAYMENT/ DEBIT (–)	✓	DEPOSIT/ CREDIT (+)	BALANCE $

Balancing a Checkbook

The bank provides customers with bank statements each month. This statement lists the checks the bank paid, ATM withdrawals, deposits made, and all other debits and credits made to a checking account. It is very important to verify that these bank records match ours. We must make sure that we deducted all debits and added all credits in our check register. This is called **balancing a checkbook.** Balancing our checkbook allows us to make sure that the balance we think we have in our checkbook is correct. If our checkbook does not balance, we must look for any mistakes.

> To balance a checkbook, proceed as follows.
>
> 1. First, *adjust the check register balance* so that it includes all credits and debits listed on the bank statement.
>
> 2. Then, *adjust the bank statement balance* so that it includes all credits and debits that may not have been received by the bank when the statement was printed. Checks that were written, but not received by the bank, are called **checks outstanding.**
>
> 3. Finally, *compare both balances* to verify that they are equal. If they are equal, the checking account balances. If they are not equal, we must find the error and make adjustments.

There are several ways to balance a checkbook. Most banks include a form you can fill out to assist you in this process.

EXAMPLE 2 Balance Jesse's checkbook using his check register and bank statement.

CHECK REGISTER	*Jesse Holm*					20 _09_
CHECK NO.	DATE	DESCRIPTION OF TRANSACTION	PAYMENT/ DEBIT (–)	✓	DEPOSIT/ CREDIT (+)	BALANCE $ 1254 32
243	9/2	Manor Apartments	575 00	✓		679 32
244	9/2	Electric Company	23 41	✓		655 91
245	9/2	Gas Company	15 67	✓		640 24
246	9/3	Jack's Market	125 57	✓		514 67
247	9/3	Clothing Mart	35 85			478 82
	9/3	Deposit		✓	634 51	1113 33
248	9/12	College Bookstore	168 96	✓		944 37
	9/18	ATM	100 00	✓		844 37
249	9/25	Telephone Company	43 29	✓		801 08
250	9/30	Sports Emporium	40 00			761 08
	10/1	Deposit			530 90	1291 98

Bank Statement: JESSE HOLM   9/1/2009 to 9/30/2009

Beginning Balance $1254.32
Ending Balance $831.68

Checks cleared by the bank

#243	$575.00	#245	$15.67	#248	$168.96
#244	$23.41	#246*	$125.57	#249*	$43.29

*Indicates that the next check in the sequence is outstanding (hasn't cleared).

Deposits

9/3 $634.51

Other withdrawals

9/18 ATM $100.00   Service charge $5.25

Solution   Follow steps 1–6 on the Checking Account Balance Form below.

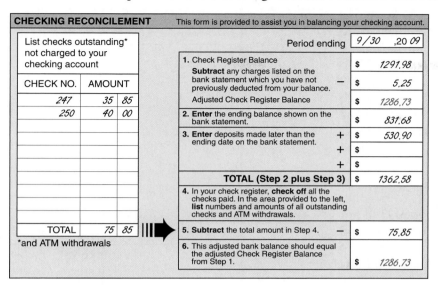

The balances in steps **1** and **6** are equal, so Jesse's checkbook is balanced.

Practice Problem 2   Balance Anthony's checkbook using his check register, his bank statement, and the given Checking Account Balance Form.

NOTE TO STUDENT: Fully worked-out solutions to all of the Practice Problems can be found at the back of the text starting at page SP-1

CHECK REGISTER *Anthony Maida*   20 _09_

CHECK NO.	DATE	DESCRIPTION OF TRANSACTION	PAYMENT/ DEBIT (−)	✓	DEPOSIT/ CREDIT (+)	BALANCE
						$ 1823 00
211	7/2	Apple Apartments	985 00			838 00
	7/9	ATM	101 50			736 50
212	7/10	Leland Groceries	98 87			637 63
213	7/21	The Gas Company	45 56			592 07
	7/21	Deposit			687 10	1279 17
214	7/28	Cellular for Less	59 98			1279 19
215	7/28	The Electric Company	89 75			1129 44
216	7/28	Sports World	129 99			999 45
217	7/28	Leland Groceries	205 99			793 46
	7/30	ATM	141 50			651 96
	8/1	Deposit			398 50	1050 46

Bank Statement: ANTHONY MAIDA   7/1/2009 to 8/1/2009

Beginning Balance $1823.00
Ending Balance $934.95

Checks cleared by the bank

#211	$985.00	#213	$45.56	#216	$129.99
#212	$98.87	#214*	$59.98		

*Indicates that the next check in the sequence is outstanding (hasn't cleared).

Deposits

7/21 $687.10

Other withdrawals

7/9 ATM	$101.50	Service charge	$3.50
7/30 ATM	$141.50	Check purchase	$9.25

CHECKING RECONCILEMENT   This form is provided to assist you in balancing your checking account.

List checks outstanding* not charged to your checking account

CHECK NO.	AMOUNT
TOTAL	

*and ATM withdrawals

Period ending ____ ,20 ____

1. Check Register Balance $ ____
   Subtract any charges listed on the bank statement which you have not previously deducted from your balance.   − $ ____
   Adjusted Check Register Balance $ ____

2. **Enter** the ending balance shown on the bank statement. $ ____

3. **Enter** deposits made later than the ending date on the bank statement.   + $ ____
    + $ ____
    + $ ____

   TOTAL (Step 2 plus Step 3) $ ____

4. In your check register, **check off** all the checks paid. In the area provided to the left, **list** numbers and amounts of all outstanding checks and ATM withdrawals.

5. **Subtract** the total amount in Step 4.   − $ ____

6. This adjusted bank balance should equal the adjusted Check Register Balance from Step 1. $ ____

Note: To calculate the service charges, we add 3.50 + 9.25 = 12.75.

1. Shin Karasuda had a balance of $532 in his checking account before writing four checks and making a deposit. On 6/1 he wrote check #122 to the Mini Market for $124.95 and check #123 to Better Be Dry Cleaners for $41.50. Then on 6/9 he made a $384.10 deposit, wrote check #124 to Macy's Department Store for $72.98, and check #125 to Costco for $121.55. Record the checks and deposit in Shin's check register, and then find the ending balance.

CHECK REGISTER	*Shin Karasuda*			20 09				
CHECK NO.	DATE	DESCRIPTION OF TRANSACTION	PAYMENT/ DEBIT (−)	✓	DEPOSIT/ CREDIT (+)	BALANCE $ 532 00		
122	6/1	MiniMarket	124 95				407	05
123	6/1	BetterBe Dry Cleaners	41 50				365	55
	6/9	Deposit			384 10		749	65
124	6/9	Macy's Dept. Store	72 98				676	67
125	6/9	Costco	121 55				555	12

2. Mary Beth O'Brian had a balance of $493 in her checking account before writing four checks and making a deposit. On 9/4 she wrote check #311 to Ben's Garage for $213.45 and #312 to Food Mart for $132.50. Then on 9/5 she made a $387.50 check deposit, wrote check #313 to the Shoe Pavilion for $69.98, and check #314 to the Electric Company for $92.45. Record the checks and deposit in Mary Beth's check register, and then find the ending balance.

CHECK REGISTER	*Mary Beth O'Brian*			20 09				
CHECK NO.	DATE	DESCRIPTION OF TRANSACTION	PAYMENT/ DEBIT (−)	✓	DEPOSIT/ CREDIT (+)	BALANCE $ 493 00		
311	9/4	Ben's Garage	213 45				279	55
312	9/4	Food Mart	132 50				147	05
	9/5	Deposit			387 50		534	55
313	9/5	Shoe Pavilion	69 98				464	57
314	9/5	Electric Company	92 45				372	12

3. The Harbor Beauty Salon had a balance of $2498.90 in its business checking account on 3/4. The manager made a deposit on the same day for $786 and wrote check #734 to the Beauty Supply Factory for $980. Then on 3/9 he wrote check #735 to the Water Department for $131.85 and check #736 to the Electric Company for $251.50. On 3/19 he made a $2614.10 deposit and wrote two payroll checks: #737 to Ranik Ghandi for $873 and #738 to Eduardo Gomez for $750. Record the checks and deposits in the Harbor Beauty Salon's check register, and then find the ending balance.

CHECK REGISTER	*Harbor Beauty Salon*			20 09				
CHECK NO.	DATE	DESCRIPTION OF TRANSACTION	PAYMENT/ DEBIT (−)	✓	DEPOSIT/ CREDIT (+)	BALANCE $ 2498 90		
	3/4	Deposit			786 00		3284	90
734	3/4	Beauty Supply Factory	980 00				2304	90
735	3/9	Water Department	131 85				2173	05
736	3/9	Electric Company	251 50				1921	55
	3/19	Deposit			2614 10		4535	65
737	3/19	Ranik Ghandi	873 00				3662	65
738	3/19	Eduardo Gomez	750 00				2912	65

4. Joanna's Coffee Shop had a balance of $1108.50 in its business checking account on 1/7. The owner made a deposit on the same day for $963 and wrote check #527 to the Restaurant Supply Company for $492. Then on 1/11 she wrote check #528 to the Gas Company for $122.45 and check #529 to the Electric Company for $321.20. Then on 1/12 she made a $1518.20 deposit and wrote two payroll checks: #530 to Sara O'Conner for $579, and #531 to Nlegan Raskin for $466. Record the checks and deposits in Joanna's Coffee Shop's check register, and then find the ending balance.

CHECK REGISTER	*Joanna's Coffee Shop*			20 09				
CHECK NO.	DATE	DESCRIPTION OF TRANSACTION	PAYMENT/ DEBIT (−)	✓	DEPOSIT/ CREDIT (+)	BALANCE $ 1108 50		
	1/7	Deposit			963 00		2071	50
527	1/7	Restaurant Supply Company	492 00				1579	50
528	1/11	Gas Company	122 45				1457	05
529	1/11	Electric Company	321 20				1135	85
	1/12	Deposit			1518 20		2654	05
530	1/12	Sara O'Conner	579 00				2075	05
531	1/12	Nlegan Raskin	466 00				1609	05

5. On 3/3 Justin Larkin had $321.94 in his checking account before he withdrew $101.50 at the ATM. On 3/7 he made a $601.90 deposit and wrote checks to pay the following bills: check #114 to the Third Street Apartments for $550, check #115 to the Cable Company for $59.50, check #116 to the Electric Company for $43.50, and check #117 to the Gas Company for $15.90. Does Justin have enough money left in his checking account to pay $99 for his car insurance?

CHECK REGISTER	Justin Larkin				20 09	
CHECK NO.	DATE	DESCRIPTION OF TRANSACTION	PAYMENT/DEBIT (−)	✓	DEPOSIT/CREDIT (+)	BALANCE $ 321 94
	3/3	ATM	101 50			220 44
	3/7	Deposit			601 90	822 34
114	3/7	Third Street Apartments	550 00			272 34
115	3/7	Cable Company	59 50			212 84
116	3/7	Electric Company	43 50			169 34
117	3/7	Gas Company	15 90			153 44

Yes, Justin can pay his car insurance.

6. On 5/2 Leon Jones had $423.54 in his checking account when he withdrew $51.50 at the ATM. On 5/9 he made a $601.80 deposit and wrote checks to pay the following bills: check #334 to Fasco Car Finance for $150.25, check #335 to A-1 Car Insurance for $89.20, check #336 to the Telephone Company for $33.40, and check #337 to the Apple Apartments for $615. Does Leon have enough money left in his checking account to pay $59 for his electric bill?

CHECK REGISTER	Leon Jones				20 09	
CHECK NO.	DATE	DESCRIPTION OF TRANSACTION	PAYMENT/DEBIT (−)	✓	DEPOSIT/CREDIT (+)	BALANCE $ 423 54
	5/2	ATM	51 50			372 04
	5/9	Deposit			601 80	973 84
334	5/9	Fasco Car Finance	150 25			823 59
335	5/9	A-1 Car Insurance	89 20			734 39
336	5/9	Telephone Company	33 40			700 99
337	5/9	Apple Apartments	615 00			85 99

Yes, Leon can pay his electric bill.

7. Balance the monthly statement for the Carson Maid Service on the Checking Account Balance Form shown below.

CHECK REGISTER	Carson Maid Service				20 09	
CHECK NO.	DATE	DESCRIPTION OF TRANSACTION	PAYMENT/DEBIT (−)	✓	DEPOSIT/CREDIT (+)	BALANCE $ 1721 50
102	4/3	A&R Cleaning Supplies	422 33			1299 17
103	4/9	Allison De Julio	510 50			788 67
104	4/9	Mai Vu	320 00			468 67
105	4/9	Jon Veldez	320 00			148 67
	4/11	Deposit			1890 00	2038 67
106	4/20	Mobil Gas Company	355 35			1683 32
107	4/25	Peterson Property Mangt. warehouse rent	525 00			1158 32
108	4/29	Jack's Garage	450 10			708 22
	5/2	Deposit			540 00	1248 22

Bank Statement: CARSON MAID SERVICE   4/1/2009 to 4/30/2009

Beginning Balance $1721.50
Ending Balance $1564.82

Checks cleared by the bank

#102	$422.33	#104	$320.00	#108	$450.10
#103	$510.50	#105*	$320.00		

*Indicates that the next check in the sequence is outstanding (hasn't cleared).

Deposits

4/11 $1890.00

Other withdrawals

Service charge $4.50
Check purchase $19.25

CHECKING RECONCILEMENT	This form is provided to assist you in balancing your checking account.

List checks outstanding* not charged to your checking account

Period ending 4 / 30 ,20 09

CHECK NO.	AMOUNT
106	355 35
107	525 00
TOTAL	880 35

*and ATM withdrawals

1. Check Register Balance — $ 1248.22
   Subtract any charges listed on the bank statement which you have not previously deducted from your balance. — $ 23.75
   Adjusted Check Register Balance — $ 1224.47

2. **Enter** the ending balance shown on the bank statement. — $ 1564.82

3. **Enter** deposits made later than the ending date on the bank statement. + $ 540.00
   + $
   + $

   TOTAL (Step 2 plus Step 3) $ 2104.82

4. In your check register, **check off** all the checks paid. In the area provided to the left, **list** numbers and amounts of all outstanding checks and ATM withdrawals.

5. **Subtract** the total amount in Step 4. − $ 880.35

6. This adjusted bank balance should equal the adjusted Check Register Balance from Step 1. $ 1224.47

The account balances.

8. Balance the monthly statement for The Flower Shop on the Checking Account Balance Form shown below.

CHECK NO.	DATE	DESCRIPTION OF TRANSACTION	PAYMENT/ DEBIT (−)		✓	DEPOSIT/ CREDIT (+)		BALANCE $ 3459 40	
502	9/7	Whole Sale Flower Company	733	67				2725	73
503	9/7	Alexsandra Kruse	580	20				2145	53
504	9/7	Jose Sanchez	430	50				1715	03
505	9/7	Kamir Kosedag	601	90				1113	13
	9/11	Deposit				2654	00	3767	13
506	9/21	L&S Pottery	466	84				3300	29
507	9/24	Peterson Property Mangt. (warehouse rent)	985	00				2315	29
508	9/28	Barton Electric Company	525	60				1789	69
	10/2	Deposit				540	00	2329	69

CHECK REGISTER *The Flower Shop*   20 09

Bank Statement: THE FLOWER SHOP   9/1/2009 to 9/30/2009

Beginning Balance $3459.40
Ending Balance $3220.28

Checks cleared by the bank

#502	$733.67	#504	$430.50	#508	$525.60
#503	$580.20	#505*	$601.90		

*Indicates that the next check in the sequence is outstanding (hasn't cleared).

Deposits

9/11 $2654.00

Other withdrawals

Service charge $4.75
Check purchase $16.50

CHECKING RECONCILEMENT   This form is provided to assist you in balancing your checking account.

List checks outstanding* not charged to your checking account

CHECK NO.	AMOUNT	
506	466	84
507	985	00
TOTAL	1451	84

*and ATM withdrawals

Period ending 9/30 ,20 09

1. Check Register Balance	$	2329.69
Subtract any charges listed on the bank statement which you have not previously deducted from your balance.	− $	21.25
Adjusted Check Register Balance	$	2308.44
2. **Enter** the ending balance shown on the bank statement.	$	3220.28
3. **Enter** deposits made later than the ending date on the bank statement.	+ $	540.00
	+ $	
	+ $	
TOTAL (Step 2 plus Step 3)	$	3760.28
4. In your check register, **check off** all the checks paid. In the area provided to the left, **list** numbers and amounts of all outstanding checks and ATM withdrawals.		
5. **Subtract** the total amount in Step 4.	− $	1451.84
6. This adjusted bank balance should equal the adjusted Check Register Balance from Step 1.	$	2308.44

The account balances.

9. On 2/1 Jeremy Sirk had a balance of $672.10 in his checking account. On the same day he deposited $735 of his paycheck into his checking account and wrote check #233 to Stanton Sporting Goods for $92.99 and check #234 to the Garden Apartments for $680. Then on 2/20 he wrote check #235 to the Gas Company for $31.85, check #236 to the Cable Company for $51.50, check #237 to Ralph's Market for $173.98, and also made an ATM withdrawal for $101.50. Then on 3/1 he made an $814.10 deposit, wrote check #238 to State Farm Insurance for $98, and made another ATM withdrawal for $41.50. Record the checks and deposits in Jeremy's check register, and then find the ending balance.

CHECK REGISTER *Jeremy Sirk*   20 09

CHECK NO.	DATE	DESCRIPTION OF TRANSACTION	PAYMENT/ DEBIT (−)		✓	DEPOSIT/ CREDIT (+)		BALANCE $ 672 10	
	2/1	Deposit				735	00	1407	10
233	2/1	Stanton Sporting Goods	92	99				1314	11
234	2/1	Garden Apartments	680	00				634	11
235	2/20	Gas Company	31	85				602	26
236	2/20	Cable Company	51	50				550	76
237	2/20	Ralph's Market	173	98				376	78
	2/20	ATM	101	50				275	28
	3/1	Deposit				814	10	1089	38
238	3/1	State Farm Insurance	98	00				991	38
	3/1	ATM	41	50				949	88

10. On 8/1 Shannon Mending had a balance of $525.90 in her checking account. Later that same day she deposited $588.23 into her checking account and wrote check #333 to Verizon Telephone Company for $33.20 and check #334 to Discount Car Insurance for $332.50. Then on 8/8 she wrote check #335 to the Walden Market for $21.35, made an ATM withdrawal for $81.50, and wrote check #336 to the Cable Company for $41.50. Then on 9/1 she made a $904.10 deposit, wrote check #337 to Next Day Dry Cleaners for $33.50, and check #338 to Marty's Dress Shop for $87.99. Record the checks and deposits in Shannon's check register, and then find the ending balance.

CHECK NO.	DATE	DESCRIPTION OF TRANSACTION	PAYMENT/ DEBIT (−)	✓	DEPOSIT/ CREDIT (+)	BALANCE $ 525 90
	8/1	Deposit			588 23	1114 13
333	8/1	Verizon Telephone Company	33 20			1080 93
334	8/1	Discount Car Insurance	332 50			748 43
335	8/8	Walden Market	21 35			727 08
	8/8	ATM	81 50			645 58
336	8/8	Cable Company	41 50			604 08
	9/1	Deposit			904 10	1508 18
337	9/1	Next Day Dry Cleaners	33 50			1474 68
338	9/1	Marty's Dress Shop	87 99			1386 69

CHECK REGISTER *Shannon Mending* 20 _09_

11. Refer to exercise 9 and use the bank statement and Checking Account Balance Form below to balance Jeremy's checking account.

Bank Statement: JEREMY SIRK 2/1/2009 to 2/28/2009

Beginning Balance $672.10
Ending Balance $293.88

Checks cleared by the bank

| #233 | $92.99 | #236 | $51.50 |
| #234* | $680.00 | #237* | $173.98 |

*Indicates that the next check in the sequence is outstanding (hasn't cleared).

Deposits
2/1 $735.00

Other withdrawals
2/20 ATM $101.50   Service charge $3.50
Check purchase $9.75

Jeremy's account balances.

CHECKING RECONCILEMENT   This form is provided to assist you in balancing your checking account.

List checks outstanding* not charged to your checking account

Period ending 2/28 ,20 09

CHECK NO.	AMOUNT
235	31 85
238	98 00
ATM	41 50

1. Check Register Balance $ 949.88
   Subtract any charges listed on the bank statement which you have not previously deducted from your balance. − $ 13.25
   Adjusted Check Register Balance $ 936.63
2. Enter the ending balance shown on the bank statement. $ 293.88
3. Enter deposits made later than the ending date on the bank statement. + $ 814.10
   + $
   + $
   TOTAL (Step 2 plus Step 3) $ 1107.98
4. In your check register, **check off** all the checks paid. In the area provided to the left, **list** numbers and amounts of all outstanding checks and ATM withdrawals.

| TOTAL | 171 35 |

*and ATM withdrawals

5. Subtract the total amount in Step 4. − $ 171.35
6. This adjusted bank balance should equal the adjusted Check Register Balance from Step 1. $ 936.63

12. Refer to exercise 10 and use the bank statement and Checking Account Balance Form below to balance Shannon's checking account.

Bank Statement: SHANNON MENDING 8/1/2009 to 8/31/2009

Beginning Balance $525.90
Ending Balance $923.83

Checks cleared by the bank

| #333* | $33.20 | #336* | $41.50 |
| #335 | $21.35 | | |

*Indicates that the next check in the sequence is outstanding (hasn't cleared).

Deposits
8/1 $588.23

Other withdrawals
8/8 ATM $81.50   Service charge $3.25
Check purchase $9.50

Shannon's account balances.

CHECKING RECONCILEMENT   This form is provided to assist you in balancing your checking account.

List checks outstanding* not charged to your checking account

Period ending 8/31 ,20 09

CHECK NO.	AMOUNT
334	332 50
337	33 50
338	87 99

1. Check Register Balance $ 1386.69
   Subtract any charges listed on the bank statement which you have not previously deducted from your balance. − $ 12.75
   Adjusted Check Register Balance $ 1373.94
2. Enter the ending balance shown on the bank statement. $ 923.83
3. Enter deposits made later than the ending date on the bank statement. + $ 904.10
   + $
   + $
   TOTAL (Step 2 plus Step 3) $ 1827.93
4. In your check register, **check off** all the checks paid. In the area provided to the left, **list** numbers and amounts of all outstanding checks and ATM withdrawals.

| TOTAL | 453 99 |

*and ATM withdrawals

5. Subtract the total amount in Step 4. − $ 453.99
6. This adjusted bank balance should equal the adjusted Check Register Balance from Step 1. $ 1373.94

Student Learning Objectives

After studying this section, you will be able to:

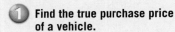 Find the true purchase price of a vehicle.

 Find the total cost of a vehicle to determine the best deal.

When we buy a car there are several facts to consider in order to determine which car has the best price. The sale price offered by a car dealer or seller is only one factor we must consider—others include the interest rate on the loan, sales tax, license fee, and sale promotions such as cash rebates or 0% interest. We must also consider the cost of options we choose such as extended warranties, sun roof, tinted glass, etc. In this section we will see how to calculate the total purchase price and determine the best deal when buying a vehicle.

① Finding the True Purchase Price of a Vehicle

In most states there is a sales tax and a license or title fee that must be paid on all vehicles purchased. To find the true purchase price for a vehicle, we *add* these extra costs to the sale price. In addition, we must add to the sale price the cost of any extended warranties and extra options or accessories we buy.

purchase price = sale price + sales tax + license fee + extended warranty (and other accessories)

Sometimes lenders (banks and finance companies) require a **down payment.** The amount of the down payment is usually a percent of the purchase price. We subtract the down payment from the purchase price to find the amount we must finance.

down payment = percent × purchase price

amount financed = purchase price − down payment

EXAMPLE 1 Daniel bought a truck that was on sale for $28,999 in a city that has a 6% sales tax and a 2% license fee.

(a) Find the sales tax and license fee Daniel paid.

(b) Daniel also bought an extended warranty for $1550. Find the purchase price of the truck.

Solution

(a) sales tax = 6% of sale price   license fee = 2% of sale price
          = 0.06 × 28,999   = 0.02 × 28,999
   sales tax = $1739.94   license fee = $579.98

(b) purchase price
      = sales price + sales tax + license fee + extended warranty
      = 28,999 + 1739.94 + 579.98 + 1550
   purchase price = $32,868.92

Practice Problem 1 Huy Nguyen bought a van that was on sale for $24,999 in a city that has a 7% sales tax and a 2% license fee.

(a) Find the sales tax and license fee Huy paid.
(b) Huy also bought an extended warranty for $1275. Find the purchase price of the van.

NOTE TO STUDENT: Fully worked-out solutions to all of the Practice Problems can be found at the back of the text starting at page SP-1

EXAMPLE 2 The purchase price of a car Jerome plans to buy is $19,999. In order to qualify for the loan on the car, Jerome must make a down payment of 20% of the purchase price.

(a) Find the down payment.
(b) Find the amount financed.

Solution

(a) down payment = percent × purchase price

down payment = 20%   × 19,999
            = 0.20   × 19,999
down payment = $3999.80

(b) amount financed = purchase price − down payment

            =   19,999   −   3999.80
amount financed = $15,999.20

Practice Problem 2 The purchase price of a Jeep Cheryl plans to buy is $32,499. In order to qualify for the loan on the Jeep, Cheryl must make a down payment of 15% of the purchase price.

(a) Find the down payment.   (b) Find the amount financed.

Finding the Total Cost of a Vehicle to Determine the Best Deal

When we borrow money to buy a car, we often pay interest on the loan and we must consider this extra cost when we calculate the **total cost** of the vehicle. If we know the amount of the car payment and the number of months it will take to pay off the loan, we can find the total payments on the car (the amount we borrowed plus interest) by multiplying the monthly payment amount times the number of months of the loan. Then we must add the down payment to that amount.

total cost = (monthly payment × number of months in loan) + down payment

EXAMPLE 3 Marvin went to two dealerships to find the best deal on the truck he plans to purchase. From which dealership should Marvin buy the truck so that the *total cost* of the truck is the least expensive?

Dealership 1	Dealership 2
• Purchase price: $39,999	• Purchase price: $36,499
• Financing option: 0% financing with $5000 down payment	• Financing option: 4% financing with no down payment
• Monthly payments: $838.52 per month for 48 months	• Monthly payments: $763.71 per month for 60 months

Solution First, we find the total cost of the truck at Dealership 1.

total cost = (monthly payment × number of months in loan) + down payment

   = (838.52 × 48)   + 5000   We multiply, then add.

   = 45,248.96

The total cost of the truck at Dealership 1 is $45,248.96.

Next, we find the total cost of the truck at Dealership 2.

total cost = (monthly payment × number of months in loan) + down payment

   = (763.71 × 60)   + 0   There is no down payment.

   = 45,822.60

The total cost of the truck at Dealership 2 is $45,822.60.

We see that the best deal on the truck Marvin plans to buy is at Dealership 1.

Practice Problem 3 Phoebe went to two dealerships to find the best deal on the minivan she plans to purchase. From which dealership should Phoebe buy the minivan so that the *total cost* of the minivan is the least expensive?

Dealership 1	Dealership 2
• Purchase price: $22,999	• Purchase price: $24,299
• Financing option: 4% financing with no down payment	• Financing option: 0% financing with $3000 down payment
• Monthly payments: $453.21 per month for 60 months	• Monthly payments: $479.17 per month for 48 months

1. A college student buys a car and pays a 6% sales tax on the $21,599 sale price. How much sales tax did the student pay? $1259.94

2. A high school teacher buys a minivan and pays a 5% sales tax on the $26,800 sale price. How much sales tax did the teacher pay? $1340

3. Francis is planning to buy a four-door sedan that is on sale for $18,999. She must pay a 2% license fee. Find the license fee. $379.98

4. Mai Vu saw an ad for a short-bed truck that is on sale for $17,599. If she buys the truck she must pay a 2% license fee. Find the license fee. $351.98

5. John must make a 10% down payment on the purchase price of the $42,450 sports car he is planning to buy. Find the down payment. $4245

6. Kamir must make a 15% down payment on the purchase price of the $31,500 extended cab truck he is planning to buy. Find the down payment. $4725

7. Tabatha bought a minivan that was on sale for $24,899 in a city that has a 5% sales tax and a 2% license fee.
   (a) Find the sales tax and license fee Tabatha paid. $1244.95; $497.98
   (b) Tabatha also bought an extended warranty for $1100. Find the purchase price of the minivan. $27,741.93

8. Dante bought a truck that was on sale for $32,499 in a city that has a 7% sales tax and a 2% license fee.
   (a) Find the sales tax and license fee Dante paid. $2274.93; $649.98
   (b) Dante also bought an extended warranty for $1600. Find the purchase price of the truck. $37,023.91

9. Jeremiah bought a sports car that was on sale for $44,799 in a city that has a 7% sales tax and a 2% license fee.
   (a) Find the sales tax and license fee Jeremiah paid. $3135.93; $895.98
   (b) Jeremiah also bought an extended warranty for $2100. Find the purchase price of the sports car. $50,930.91

10. Dawn bought a four-door sedan that was on sale for $31,899 in a city that has a 6% sales tax and a 2% license fee.
    (a) Find the sales tax and license fee Dawn paid. $1913.94; $637.98
    (b) Dawn also bought an extended warranty for $1300. Find the purchase price of the sedan. $35,750.92

11. The purchase price of an SUV that a soccer coach plans to buy is $49,999. In order to qualify for the loan on the SUV, the coach must make a down payment of 15% of the purchase price.
    (a) Find the down payment. $7499.85
    (b) Find the amount financed. $42,499.15

12. The purchase price of a flat-bed truck a contractor plans to buy is $39,999. In order to qualify for the loan on the truck, the contractor must make a down payment of 10% of the purchase price.
    (a) Find the down payment. $3999.90
    (b) Find the amount financed. $35,999.10

13. Tammy went to two dealerships to find the best deal on the truck she plans to purchase. From which dealership should Tammy buy the truck so that the *total cost* of the truck is the least expensive? Dealership 1

Dealership 1	Dealership 2
• Purchase price: $35,999	• Purchase price: $32,499
• Financing option: 0% financing with $4000 down payment	• Financing option: 4% financing with $2000 down payment
• Monthly payments: $696.65 per month for 48 months	• Monthly payments: $603.58 per month for 60 months

14. John went to two dealerships to find the best deal on a luxury SUV for his company to purchase. From which dealership should John buy the SUV so that the *total cost* of the SUV is the least expensive? Dealership 1

Dealership 1	Dealership 2
• Purchase price: $49,999	• Purchase price: $46,499
• Financing option: 0% financing with $5000 down payment	• Financing option: 4% financing with no down payment
• Monthly payments: $1010.39 per month for 48 months	• Monthly payments: $916.29 per month for 60 months

To Think About

Natasha went to three car dealerships to check the prices of the same Ford two-door coupe. The city where the dealerships are located has a sales tax of 5% and a license fee of 2%. All dealerships offer extended warranties that are 3 years/70,000 miles. Use the following information gathered by Natasha to answer Exercises 15 and 16.

Dealership 1—Ford Coupe	Dealership 2—Ford Coupe	Dealership 3—Ford Coupe
• $24,999 plus $2000 rebate	• $23,799; dealer pays sales tax	• $23,999
• extended warranty $1350	• extended warranty $1450	• free extended warranty

15. **(a)** Which dealership offers the least expensive *purchase price?* State this amount.   Dealership 3; $25,678.93

(b) Each dealership offers a *different interest rate* on a 60-month loan without a down payment, resulting in the following monthly payments:

Dealership 1	Dealership 2	Dealership 3
$480.65/month	$485.46/month	$496.44/month

From which dealership should Natasha buy the Ford coupe so that the *total cost* of the car is the least expensive? State this amount.   Dealership 1; $28,839

(c) Compare the results of parts **(a)** and **(b)**. What conclusion can you make?
The least expensive purchase price does not guarantee the least expensive total cost. Many factors need to be considered to determine the best deal.

16. **(a)** Which dealership offers the most expensive *purchase price?* State this amount.   Dealership 1; $26,098.93

(b) Each dealership offers a *different interest rate* on a 48-month loan without a down payment, resulting in the following monthly payments:

Dealership 1	Dealership 2	Dealership 3
$589.27/month	$594.07/month	$603.07/month

From which dealership should Natasha buy the Ford coupe so that the *total cost* of the car is the most expensive? State this amount.   Dealership 3; $28,947.36

(c) Compare the results of parts **(a)** and **(b)**. What conclusion can you make?
The most expensive purchase price does not guarantee the most expensive total cost. Many factors need to be considered to determine the best deal.

Appendix B Tables

Table of Basic Addition Facts

+	0	1	2	3	4	5	6	7	8	9
0	0	1	2	3	4	5	6	7	8	9
1	1	2	3	4	5	6	7	8	9	10
2	2	3	4	5	6	7	8	9	10	11
3	3	4	5	6	7	8	9	10	11	12
4	4	5	6	7	8	9	10	11	12	13
5	5	6	7	8	9	10	11	12	13	14
6	6	7	8	9	10	11	12	13	14	15
7	7	8	9	10	11	12	13	14	15	16
8	8	9	10	11	12	13	14	15	16	17
9	9	10	11	12	13	14	15	16	17	18

Table of Basic Multiplication Facts

×	0	1	2	3	4	5	6	7	8	9	10	11	12
0	0	0	0	0	0	0	0	0	0	0	0	0	0
1	0	1	2	3	4	5	6	7	8	9	10	11	12
2	0	2	4	6	8	10	12	14	16	18	20	22	24
3	0	3	6	9	12	15	18	21	24	27	30	33	36
4	0	4	8	12	16	20	24	28	32	36	40	44	48
5	0	5	10	15	20	25	30	35	40	45	50	55	60
6	0	6	12	18	24	30	36	42	48	54	60	66	72
7	0	7	14	21	28	35	42	49	56	63	70	77	84
8	0	8	16	24	32	40	48	56	64	72	80	88	96
9	0	9	18	27	36	45	54	63	72	81	90	99	108
10	0	10	20	30	40	50	60	70	80	90	100	110	120
11	0	11	22	33	44	55	66	77	88	99	110	121	132
12	0	12	24	36	48	60	72	84	96	108	120	132	144

Table of Prime Factors

Number	Prime Factors	Number	Prime Factors	Number	Prime Factors	Number	Prime Factors
2	prime	52	$2^2 \times 13$	102	$2 \times 3 \times 17$	152	$2^3 \times 19$
3	prime	53	prime	103	prime	153	$3^2 \times 17$
4	2^2	54	2×3^3	104	$2^3 \times 13$	154	$2 \times 7 \times 11$
5	prime	55	5×11	105	$3 \times 5 \times 7$	155	5×31
6	2×3	56	$2^3 \times 7$	106	2×53	156	$2^2 \times 3 \times 13$
7	prime	57	3×19	107	prime	157	prime
8	2^3	58	2×29	108	$2^2 \times 3^3$	158	2×79
9	3^2	59	prime	109	prime	159	3×53
10	2×5	60	$2^2 \times 3 \times 5$	110	$2 \times 5 \times 11$	160	$2^5 \times 5$
11	prime	61	prime	111	3×37	161	7×23
12	$2^2 \times 3$	62	2×31	112	$2^4 \times 7$	162	2×3^4
13	prime	63	$3^2 \times 7$	113	prime	163	prime
14	2×7	64	2^6	114	$2 \times 3 \times 19$	164	$2^2 \times 41$
15	3×5	65	5×13	115	5×23	165	$3 \times 5 \times 11$
16	2^4	66	$2 \times 3 \times 11$	116	$2^2 \times 29$	166	2×83
17	prime	67	prime	117	$3^2 \times 13$	167	prime
18	2×3^2	68	$2^2 \times 17$	118	2×59	168	$2^3 \times 3 \times 7$
19	prime	69	3×23	119	7×17	169	13^2
20	$2^2 \times 5$	70	$2 \times 5 \times 7$	120	$2^3 \times 3 \times 5$	170	$2 \times 5 \times 17$
21	3×7	71	prime	121	11^2	171	$3^2 \times 19$
22	2×11	72	$2^3 \times 3^2$	122	2×61	172	$2^2 \times 43$
23	prime	73	prime	123	3×41	173	prime
24	$2^3 \times 3$	74	2×37	124	$2^2 \times 31$	174	$2 \times 3 \times 29$
25	5^2	75	3×5^2	125	5^3	175	$5^2 \times 7$
26	2×13	76	$2^2 \times 19$	126	$2 \times 3^2 \times 7$	176	$2^4 \times 11$
27	3^3	77	7×11	127	prime	177	3×59
28	$2^2 \times 7$	78	$2 \times 3 \times 13$	128	2^7	178	2×89
29	prime	79	prime	129	3×43	179	prime
30	$2 \times 3 \times 5$	80	$2^4 \times 5$	130	$2 \times 5 \times 13$	180	$2^2 \times 3^2 \times 5$
31	prime	81	3^4	131	prime	181	prime
32	2^5	82	2×41	132	$2^2 \times 3 \times 11$	182	$2 \times 7 \times 13$
33	3×11	83	prime	133	7×19	183	3×61
34	2×17	84	$2^2 \times 3 \times 7$	134	2×67	184	$2^3 \times 23$
35	5×7	85	5×17	135	$3^3 \times 5$	185	5×37
36	$2^2 \times 3^2$	86	2×43	136	$2^3 \times 17$	186	$2 \times 3 \times 31$
37	prime	87	3×29	137	prime	187	11×17
38	2×19	88	$2^3 \times 11$	138	$2 \times 3 \times 23$	188	$2^2 \times 47$
39	3×13	89	prime	139	prime	189	$3^3 \times 7$
40	$2^3 \times 5$	90	$2 \times 3^2 \times 5$	140	$2^2 \times 5 \times 7$	190	$2 \times 5 \times 19$
41	prime	91	7×13	141	3×47	191	prime
42	$2 \times 3 \times 7$	92	$2^2 \times 23$	142	2×71	192	$2^6 \times 3$
43	prime	93	3×31	143	11×13	193	prime
44	$2^2 \times 11$	94	2×47	144	$2^4 \times 3^2$	194	2×97
45	$3^2 \times 5$	95	5×19	145	5×29	195	$3 \times 5 \times 13$
46	2×23	96	$2^5 \times 3$	146	2×73	196	$2^2 \times 7^2$
47	prime	97	prime	147	3×7^2	197	prime
48	$2^4 \times 3$	98	2×7^2	148	$2^2 \times 37$	198	$2 \times 3^2 \times 11$
49	7^2	99	$3^2 \times 11$	149	prime	199	prime
50	2×5^2	100	$2^2 \times 5^2$	150	$2 \times 3 \times 5^2$	200	$2^3 \times 5^2$
51	3×17	101	prime	151	prime		

Table of Square Roots

Square Root Values Are Rounded to the Nearest Thousandth Unless the Answer Ends in .000

n	$\sqrt{n}$	n	$\sqrt{n}$	n	$\sqrt{n}$	n	$\sqrt{n}$	n	$\sqrt{n}$
1	1.000	41	6.403	81	9.000	121	11.000	161	12.689
2	1.414	42	6.481	82	9.055	122	11.045	162	12.728
3	1.732	43	6.557	83	9.110	123	11.091	163	12.767
4	2.000	44	6.633	84	9.165	124	11.136	164	12.806
5	2.236	45	6.708	85	9.220	125	11.180	165	12.845
6	2.449	46	6.782	86	9.274	126	11.225	166	12.884
7	2.646	47	6.856	87	9.327	127	11.269	167	12.923
8	2.828	48	6.928	88	9.381	128	11.314	168	12.961
9	3.000	49	7.000	89	9.434	129	11.358	169	13.000
10	3.162	50	7.071	90	9.487	130	11.402	170	13.038
11	3.317	51	7.141	91	9.539	131	11.446	171	13.077
12	3.464	52	7.211	92	9.592	132	11.489	172	13.115
13	3.606	53	7.280	93	9.644	133	11.533	173	13.153
14	3.742	54	7.348	94	9.695	134	11.576	174	13.191
15	3.873	55	7.416	95	9.747	135	11.619	175	13.229
16	4.000	56	7.483	96	9.798	136	11.662	176	13.266
17	4.123	57	7.550	97	9.849	137	11.705	177	13.304
18	4.243	58	7.616	98	9.899	138	11.747	178	13.342
19	4.359	59	7.681	99	9.950	139	11.790	179	13.379
20	4.472	60	7.746	100	10.000	140	11.832	180	13.416
21	4.583	61	7.810	101	10.050	141	11.874	181	13.454
22	4.690	62	7.874	102	10.100	142	11.916	182	13.491
23	4.796	63	7.937	103	10.149	143	11.958	183	13.528
24	4.899	64	8.000	104	10.198	144	12.000	184	13.565
25	5.000	65	8.062	105	10.247	145	12.042	185	13.601
26	5.099	66	8.124	106	10.296	146	12.083	186	13.638
27	5.196	67	8.185	107	10.344	147	12.124	187	13.675
28	5.292	68	8.246	108	10.392	148	12.166	188	13.711
29	5.385	69	8.307	109	10.440	149	12.207	189	13.748
30	5.477	70	8.367	110	10.488	150	12.247	190	13.784
31	5.568	71	8.426	111	10.536	151	12.288	191	13.820
32	5.657	72	8.485	112	10.583	152	12.329	192	13.856
33	5.745	73	8.544	113	10.630	153	12.369	193	13.892
34	5.831	74	8.602	114	10.677	154	12.410	194	13.928
35	5.916	75	8.660	115	10.724	155	12.450	195	13.964
36	6.000	76	8.718	116	10.770	156	12.490	196	14.000
37	6.083	77	8.775	117	10.817	157	12.530	197	14.036
38	6.164	78	8.832	118	10.863	158	12.570	198	14.071
39	6.245	79	8.888	119	10.909	159	12.610	199	14.107
40	6.325	80	8.944	120	10.954	160	12.649	200	14.142

Appendix C Scientific Calculators

This text *does not require* the use of a calculator. However, you may want to consider the purchase of an inexpensive scientific calculator. It is wise to ask your instructor for advice before you purchase any calculator for this course. It should be stressed that students are asked to avoid using a calculator for any of the exercises in which the calculations can be readily done by hand. The only problems in the text that really demand the use of a scientific cal culator are marked with the ▪ symbol. Dependence on the use of the scientific calculator for regular exercises in the text will only hurt the student in the long run.

The Two Types of Logic Used in Scientific Calculators

Two major types of scientific calculators are popular today. The most common type employs a type of logic known as **algebraic** logic. The calculators manufactured by Casio, Sharp, and Texas Instruments as well as many other companies employ this type of logic. An example of calculation on such a calculator would be the following. To add 14 + 26 on an algebraic logic calculator, the sequence of buttons would be:

$$14 \boxed{+} 26 \boxed{=}$$

The second type of scientific calculator requires the entry of data in **Reverse Polish Notation (RPN).** Calculators manufactured by Hewlett-Packard and a few other specialized calculators use RPN. To add 14 + 26 on an RPN calculator, the sequence of buttons would be:

$$14 \boxed{\text{enter}} 26 \boxed{+}$$

Graphing scientific calculators such as the TI-83 and TI-84 have a large display for viewing graphs. To perform the calculation on most graphing calculators, the sequence of buttons would be:

$$14 \boxed{+} 26 \boxed{\text{enter}}$$

Mathematicians and scientists do not agree on which type of scientific calculator is superior. However, the clear majority of college students own calculators that employ *algebraic* logic. Therefore this section of the text is explained with reference to the sequence of steps employed by an *algebraic* logic calculator. If you already own or intend to purchase a scientific calculator that uses RPN or a graphing calculator, you are encouraged to study the instruction booklet that comes with the calculator and practice the problems shown in the booklet. After this practice you will be able to solve the calculator problems discussed in this section.

Performing Simple Calculations

The following example will illustrate the use of a scientific calculator in doing basic arithmetic calculations.

EXAMPLE 1 Add. 156 + 298

Solution We first enter the number 156, then press the $\boxed{+}$ key, then enter the number 298, and finally press the $\boxed{=}$ key.

$$156 \boxed{+} 298 \boxed{=} 454$$

Practice Problem 1 Add. 3792 + 5896

NOTE TO STUDENT: Fully worked-out solutions to all of the Practice Problems can be found at the back of the text starting at page SP-1

EXAMPLE 2 Subtract. 1508 − 963

Solution We first enter the number 1508, then press the $\boxed{-}$ key, then enter the number 963, and finally press the $\boxed{=}$ key.

$$1508 \boxed{-} 963 \boxed{=} 545$$

Practice Problem 2 Subtract. 7930 − 5096

EXAMPLE 3 Multiply. 196 × 358

Solution $196 \boxed{\times} 358 \boxed{=} 70168$

Practice Problem 3 Multiply. 896 × 273

EXAMPLE 4 Divide. 2054 ÷ 13

Solution $2054 \boxed{\div} 13 \boxed{=} 158$

Practice Problem 4 Divide. 2352 ÷ 16

Decimal Problems

Problems involving decimals can be readily done on a calculator. Entering numbers with a decimal point is done by pressing the decimal point key, the $\boxed{\cdot}$ key, at the appropriate time.

EXAMPLE 5 Calculate. 4.56 × 283

Solution To enter 4.56, we press the $\boxed{4}$ key, the decimal point key, then the $\boxed{5}$ key, and finally the $\boxed{6}$ key.

$$4.56 \boxed{\times} 283 \boxed{=} 1290.48$$

The answer is 1290.48. Observe how your calculator displays the decimal point.

Practice Problem 5 Calculate. 72.8 × 197

EXAMPLE 6   Add. 128.6 + 343.7 + 103.4 + 207.5

Solution   128.6 $\boxed{+}$ 343.7 $\boxed{+}$ 103.4 $\boxed{+}$ 207.5 $\boxed{=}$ 783.2

The answer is 783.2. Observe how your calculator displays the answer.

Practice Problem 6   Add. 52.98 + 31.74 + 40.37 + 99.82

Combined Operations

You must use extra caution concerning the order of mathematical operations when you are doing a problem on the calculator that involves two or more different operations.

Any scientific calculator with algebraic logic uses a priority system that has a clearly defined order of operations. It is the same order we use in performing arithmetic operations by hand. In either situation, calculations are performed in the following order:

1. First calculations within parentheses are completed.
2. Then numbers are raised to a power or a square root is calculated.
3. Then multiplication and division operations are performed from left to right.
4. Then addition and subtraction operations are performed from left to right.

This order is carefully followed on *scientific calculators* and *graphing calculators*. Small inexpensive calculators that do not have scientific functions often do not follow this order of operations.

The number of digits displayed in the answer varies from calculator to calculator. In the following examples, your calculator may display more or fewer digits than the answer we have listed.

EXAMPLE 7   Evaluate. 5.3 × 1.62 + 1.78 ÷ 3.51

Solution   This problem requires that we multiply 5.3 by 1.62 and divide 1.78 by 3.51 first and then add the two results. If the numbers are entered directly into the calculator exactly as the problem is written, the calculator will perform the calculations in the correct order.

5.3 $\boxed{\times}$ 1.62 $\boxed{+}$ 1.78 $\boxed{\div}$ 3.51 $\boxed{=}$ 9.09312251

Practice Problem 7   Evaluate. 0.0618 × 19.22 − 59.38 ÷ 166.3

The Use of Parentheses

In order to perform some calculations on a calculator, the use of parentheses is helpful. These parentheses may or may not appear in the original problem.

EXAMPLE 8 Evaluate. $5 \times (2.123 + 5.786 - 12.063)$

Solution The problem requires that the numbers in the parentheses be combined first. By entering the parentheses on the calculator this will be accomplished.

5 $\boxed{\times}$ $\boxed{(}$ 2.123 $\boxed{+}$ 5.786 $\boxed{-}$ 12.063 $\boxed{)}$ $\boxed{=}$ -20.77

Note: The result is a negative number.

Practice Problem 8 Evaluate. $3.152 \times (0.1628 + 3.715 - 4.985)$

NOTE TO STUDENT: *Fully worked-out solutions to all of the Practice Problems can be found at the back of the text starting at page SP-1*

Negative Numbers

To enter a negative number, enter the number followed by the $\boxed{+/-}$ button. Some calculators require a different order. So on some calculators you first use the $\boxed{+/-}$ button and then enter the number.

EXAMPLE 9 Evaluate. $(-8.634)(5.821) + (1.634)(-16.082)$

Solution The products will be evaluated first by the calculator. Therefore, parentheses are not needed as we enter the data.

8.634 $\boxed{+/-}$ $\boxed{\times}$ 5.821 $\boxed{+}$ 1.634 $\boxed{\times}$ 16.082 $\boxed{+/-}$ $\boxed{=}$ -76.536502

Note: The result is negative.

Practice Problem 9 Evaluate. $(0.5618)(-98.3) - (76.31)(-2.98)$

Scientific Notation

If you wish to enter a number in scientific notation, you should use the special scientific notation button. On most calculators it is denoted as $\boxed{\text{EXP}}$ or $\boxed{\text{EE}}$.

EXAMPLE 10 Multiply. $(9.32 \times 10^6)(3.52 \times 10^8)$

Solution 9.32 $\boxed{\text{EXP}}$ 6 $\boxed{\times}$ 3.52 $\boxed{\text{EXP}}$ 8 $\boxed{=}$ $3.28064 \quad 15$

This notation means the answer is 3.28064×10^{15}.

Practice Problem 10 Divide. $(3.76 \times 10^{15}) \div (7.76 \times 10^7)$

Raising a Number to a Power

All scientific calculators have a key for finding powers of numbers. It is usually labeled $\boxed{y^x}$. (On a few calculators the notation is $\boxed{x^y}$ or sometimes $\boxed{\wedge}$.) To raise a number to a power on most scientific calculators, first you enter the base, then push the $\boxed{y^x}$ key. Then you enter the exponent, then finally the $\boxed{=}$ button.

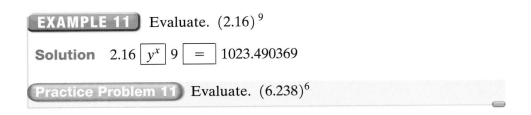

EXAMPLE 11 Evaluate. $(2.16)^9$

Solution   2.16 $\boxed{y^x}$ 9 $\boxed{=}$ 1023.490369

Practice Problem 11 Evaluate. $(6.238)^6$

There is a special key to square a number. It is usually labeled $\boxed{x^2}$.

EXAMPLE 12 Evaluate. $(76.04)^2$

Solution   76.04 $\boxed{x^2}$ 5782.0816

Practice Problem 12 Evaluate. $(132.56)^2$

Finding Square Roots of Numbers

To approximate square roots on a scientific calculator, use the key labeled $\boxed{\sqrt{}}$. In this example we will need to use parentheses.

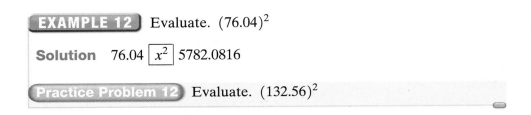

EXAMPLE 13 Evaluate. $\sqrt{5618 + 2734 + 3913}$

Solution $\boxed{(}$ 5618 $\boxed{+}$ 2734 $\boxed{+}$ 3913 $\boxed{)}$ $\boxed{\sqrt{}}$ 110.7474605

Practice Problem 13 Evaluate. $\sqrt{0.0782 - 0.0132 + 0.1364}$

On some calculators, you enter the square root key first and then enter the number. You will need to try this on your own calculator.

Use your calculator to complete each of the following. Your answers may vary slightly because of the characteristics of individual calculators.

Complete the table.

To Do This Operation	Use These Keystrokes	Record Your Answer Here
1. 8963 + 2784	8963 $+$ 2784 $=$	11,747
2. 15,308 − 7980	15308 $-$ 7980 $=$	7328
3. 2631 × 134	2631 $\times$ 134 $=$	352,554
4. 70,221 ÷ 89	70221 $\div$ 89 $=$	789
5. 5.325 − 4.031	5.325 $-$ 4.031 $=$	1.294
6. 184.68 + 73.98	184.68 $+$ 73.98 $=$	258.66
7. 2004.06 ÷ 7.89	2004.06 $\div$ 7.89 $=$	254
8. 1.34 × 0.763	1.34 $\times$ 0.763 $=$	1.02242

Write down the answer and then show what problem you have solved.

9. 123.45 $+$ 45.9876 $+$ 8765.3 $=$ 8934.7376
123.45 + 45.9876 + 8765.3

10. 0.0897 $\times$ 234.56 $\times$ 2.5428 $=$
53.50059337
0.0897 × 234.56 × 2.5428

11. 34 $\div$ 8 $+$ 12.56 $=$ 16.81
$\frac{34}{8}$ + 12.56

12. 458 $\div$ 4 $-$ 16.897 $=$ 97.603
$\frac{458}{4}$ − 16.897

Perform each calculation using your calculator.

13. 9.467 + 0.563   10.03

14. 0.347 + 23.457   23.804

15. 34.89 + 39.6 + 214.897   289.387

16. 12.567 + 48.31 + 189.38   250.257

17. 412,899 − 34,675   378,224

18. 87,456 − 2876   84,580

19. 3,567,089 − 2,876,805   690,284

20. 8,345,802 − 4,985,004   3,360,798

21. 234 × 4.567   1068.678

22. 1.9876 × 347   689.6972

23. 0.456 × 3.48   1.58688

24. 67,876 × 0.0946   6421.0696

25. 3458 ÷ 2.5   1383.2

26. 9764 ÷ 8   1220.5

27. 12.107524 ÷ 15.86   0.7634

28. 16.06513 ÷ 17.98   0.8935

Perform each calculation using your calculator.

29.
1.98
6.34
+ 7.71
16.03

30.
8.92
9.31
+ 7.79
26.02

31.
$103.91
$2653.82
+ $9804.61
$12,562.34

32.
$3986.21
$4502.89
+ $989.30
$9478.40

33.
368,781.5
− 283,617.8
85,163.7

34.
571,809.6
− 539,376.8
32,432.8

35.
$1,393,271.86
− $1,289,663.21
$103,608.65

36.
$8,571,300.76
− $4,098,789.39
$4,472,511.37

37.
345.34
× 45.7
15,782.038

38.
8954.34
× 425.4
3,809,176.236

39.
0.6314
× 3.96
2.500344

40.
0.0789
× 12.38
0.976782

41. $40.36\overline{)36202.92}$ 897

42. $52.98\overline{)172,608.84}$ 3258

43. $0.7613\overline{)17.12925}$ 22.5

44. $0.9854\overline{)3.59671}$ 3.65

Perform the following operations in the proper order using your calculator.

45. $4.567 + 87.89 - 2.45 \times 3.3$ 84.372

46. $4.891 + 234.5 - 0.98 \times 23.4$ 216.459

47. $7 \div 8 + 3.56$ 4.435

48. $9 \div 4.5 + 0.6754$ 2.6754

49. $(9.34)(0.345) + 98.345$ 101.5673

50. $(0.628)(398) + 34.4581$ 284.4021

51. $\dfrac{(95.34)(0.9874)}{381.36}$ 0.24685

52. $\dfrac{(0.8759)(45.87)}{183.48}$ 0.218975

53. $2.56 + 8.98 \times 3.14$ 30.7572

54. $1.62 + 3.81 - 5.23 \times 6.18$ −26.8914

55. $(-4.23)(1.863) - 5.998$ −13.87849

56. $12.34 - (26.314)(-1.856)$ 61.178784

57. $5.62(5 \times 3.16 - 18.12)$ −13.0384

58. $9.356(4.8 - 7.2 - 15.94)$ −171.58904

59. $(3.42 \times 10^8)(0.97 \times 10^{10})$ 3.3174×10^{18}

60. $(6.27 \times 10^{20})(1.35 \times 10^3)$ 8.4645×10^{23}

61. $\dfrac{(2.16 \times 10^3)(1.37 \times 10^{14})}{6.39 \times 10^5}$ $4.630985915 \times 10^{11}$

62. $\dfrac{(3.84 \times 10^{12})(1.62 \times 10^5)}{7.78 \times 10^8}$ 7.9958869×10^8

63. $\dfrac{2.3 + 5.8 - 2.6 - 3.9}{5.3 - 8.2}$ -0.5517241379

64. $\dfrac{(2.6)(-3.2) + (5.8)(-0.9)}{2.614 + 5.832}$ -1.60312574

65. $\sqrt{253.12}$ 15.90974544

66. $\sqrt{0.0713}$ 0.2670205985

67. $\sqrt{5.6213 - 3.7214}$ 1.378368601

68. $\sqrt{3417.2 - 2216.3}$ 34.6540041

69. $(1.78)^3 + 6.342$ 11.981752

70. $(2.26)^8 - 3.1413$ 677.4204134

71. $\sqrt{(6.13)^2 + (5.28)^2}$ 8.090444982

72. $\sqrt{(0.3614)^2 + (0.9217)^2}$ 0.9900206311

73. $\sqrt{56 + 83} - \sqrt{12}$ 8.325724507

74. $\sqrt{98 + 33} - \sqrt{17}$ 7.322417517

Find an approximate value. Round to five decimal places.

75. $\dfrac{7}{18} + \dfrac{9}{13}$ 1.08120

76. $\dfrac{5}{22} + \dfrac{1}{31}$ 0.25953

77. $\dfrac{7}{8} + \dfrac{3}{11}$ 1.14773

78. $\dfrac{9}{14} + \dfrac{5}{19}$ 0.90602

Solutions to Practice Problems

Chapter 1

1.1 Practice Problems

1. (a) $3182 = 3000 + 100 + 80 + 2$
   (b) $520,890 = 500,000 + 20,000 + 800 + 90$
   (c) $709,680,059 = 700,000,000 + 9,000,000 + 600,000$
   $+ 80,000 + 50 + 9$

2. (a) 492   **(b)** 80,427
3. (a) 7   **(b)** 9   **(c)** 4000
   (d) 900,000 for the first 9; 9 for the last 9
4. two hundred sixty-seven million, three hundred fifty-eight thousand, nine hundred eighty-one
5. (a) two thousand, seven hundred thirty-six
   (b) nine hundred eighty thousand, three hundred six
   (c) twelve million, twenty-one
6. The world population on January 1, 2004 was six billion, three hundred ninety-three million, six hundred forty-six thousand, five hundred twenty-five.
7. (a) 803   **(b)** 30,229
8. (a) 13,000   **(b)** 88,000   **(c)** 10,000

1.2 Practice Problems

1. (a) $\begin{array}{r} 7 \\ + 5 \\ \hline 12 \end{array}$ **(b)** $\begin{array}{r} 9 \\ + 4 \\ \hline 13 \end{array}$ **(c)** $\begin{array}{r} 3 \\ + 0 \\ \hline 3 \end{array}$

2. $\begin{array}{r} 7 \\ 6 \\ 5 \\ 8 \\ +2 \\ \hline 28 \end{array}$
  $7 + 6 = 13$
  $13 + 5 = 18$
  $18 + 8 = 26$
  $26 + 2 = 28$

3. $\begin{array}{r} 1 \\ 7 \\ 2 \\ 9 \\ +3 \\ \hline 22 \end{array}$ 10 10

4. $\begin{array}{r} 8246 \\ + 1702 \\ \hline 9948 \end{array}$

5. $\begin{array}{r} \overset{1}{5}6 \\ + 36 \\ \hline 92 \end{array}$

6. $\begin{array}{r} \overset{2\,1}{7}89 \\ 63 \\ + 297 \\ \hline 1149 \end{array}$

7. (a) $\begin{array}{r} \overset{1\,1\,1}{1}27 \\ 9876 \\ + 342 \\ \hline 10,345 \end{array}$ **(b)** Check by adding in opposite order.   same $\begin{array}{r} \overset{1\,1\,1}{3}42 \\ 9876 \\ + 127 \\ \hline 10,345 \end{array}$

8. $\begin{array}{r} \overset{1\,2\,1\,2}{1}8,316 \\ 24,789 \\ + 22,965 \\ \hline 66,070 \end{array}$ total women

9. $\begin{array}{r} 1000 \\ 2000 \\ 1000 \\ + 2000 \\ \hline 6000 \text{ ft} \end{array}$

1.3 Practice Problems

1. (a) $\begin{array}{r} 9 \\ - 6 \\ \hline 3 \end{array}$ **(b)** $\begin{array}{r} 12 \\ - 5 \\ \hline 7 \end{array}$ **(c)** $\begin{array}{r} 17 \\ - 8 \\ \hline 9 \end{array}$ **(d)** $\begin{array}{r} 14 \\ - 0 \\ \hline 14 \end{array}$ **(e)** $\begin{array}{r} 18 \\ - 9 \\ \hline 9 \end{array}$

2. $\begin{array}{r} 7695 \\ - 3481 \\ \hline 4214 \end{array}$

3. $\begin{array}{r} \overset{2\;\,14}{3}\,\cancel{4} \\ - 1\;6 \\ \hline 1\;8 \end{array}$

4. $\begin{array}{r} 6\,\overset{8\;\;13}{\cancel{9}}\,\cancel{3} \\ - 4\;2\;6 \\ \hline 2\;6\;7 \end{array}$

5. $\begin{array}{r} \overset{8\;\;10\;\;6\;\;10}{\cancel{9}\;\cancel{0}\;\cancel{7}\;\cancel{0}} \\ - 5\;8\;8\;6 \\ \hline 3\;1\;8\;4 \end{array}$

6. (a) $\begin{array}{r} 8964 \\ - 985 \\ \hline 7979 \end{array}$ **(b)** $\begin{array}{r} 50,000 \\ - 32,508 \\ \hline 17,492 \end{array}$

7. Subtraction   Checking by addition
$\begin{array}{r} 9763 \\ - 5732 \\ \hline 4031 \end{array}$ | IT CHECKS | $\begin{array}{r} 5732 \\ + 4031 \\ \hline 9763 \end{array}$

8. (a) $\begin{array}{r} 284,000 \\ - 96,327 \\ \hline 187,673 \end{array}$ Checking by addition | IT CHECKS | $\begin{array}{r} 96,327 \\ + 187,673 \\ \hline 284,000 \end{array}$

(b) $\begin{array}{r} 8,526,024 \\ - 6,397,518 \\ \hline 2,128,506 \end{array}$ Checking by addition | IT CHECKS | $\begin{array}{r} 6,397,518 \\ + 2,128,506 \\ \hline 8,526,024 \end{array}$

9. (a) $17 = 12 + x$ **(b)** $22 = 10 + x$
       $17 - 12 = x$ $22 - 10 = x$
              $5 = x$ $12 = x$
5 vessels left in the afternoon.   12 hikers were still on the mountain.

10. (a) $\begin{array}{r} 23,667,947 \\ - 14,227,799 \\ \hline 9,440,148 \end{array}$ **(b)** $\begin{array}{r} 11,198,655 \\ - 9,579,677 \\ \hline 1,618,978 \end{array}$

11. (a) From the bar graph: **(b)** From the bar graph:
2008 sales   114   Springfield   91
$\begin{array}{r} \text{2007 sales} \quad - 78 \\ \hline \text{Sales increase} \quad 36 \end{array}$ $\begin{array}{r} \text{Riverside} \quad - 78 \\ \hline 13 \text{ more homes} \end{array}$

(c) 2008 sales   271   2009 sales   284
$\begin{array}{r} \text{2007 sales} \quad - 240 \\ \hline 31 \end{array}$ $\begin{array}{r} \text{2008 sales} \quad - 271 \\ \hline 13 \end{array}$

Therefore, the greatest increase in sales occurred from 2007 to 2008.

1.4 Practice Problems

1. (a)
$$\begin{array}{r} 8 \\ \times\ 8 \\ \hline 64 \end{array}$$
(b)
$$\begin{array}{r} 7 \\ \times\ 6 \\ \hline 42 \end{array}$$
(c)
$$\begin{array}{r} 5 \\ \times\ 8 \\ \hline 40 \end{array}$$
(d)
$$\begin{array}{r} 9 \\ \times\ 7 \\ \hline 63 \end{array}$$
(e)
$$\begin{array}{r} 9 \\ \times\ 9 \\ \hline 81 \end{array}$$

2.
$$\begin{array}{r} 3021 \\ \times\ 3 \\ \hline 9063 \end{array}$$

3.
$$\begin{array}{r} {}^{2}\ \\ 43 \\ \times 8 \\ \hline 344 \end{array}$$

4.
$$\begin{array}{r} {}^{5\,6}\ \\ 579 \\ \times\ 7 \\ \hline 4053 \end{array}$$

5. (a) $1267 \times 10 = 12{,}670$ (one zero)
  (b) $1267 \times 1000 = 1{,}267{,}000$ (three zeros)
  (c) $1267 \times 10{,}000 = 12{,}670{,}000$ (four zeros)
  (d) $1267 \times 1{,}000{,}000 = 1{,}267{,}000{,}000$ (six zeros)
6. (a) $9 \times 60{,}000 = 9 \times 6 \times 10{,}000 = 54 \times 10{,}000 = 540{,}000$
  (b) $15 \times 400 = 15 \times 4 \times 100 = 60 \times 100 = 6000$
  (c) $270 \times 800 = 27 \times 8 \times 10 \times 100 = 216 \times 1000 = 216{,}000$

7.
$$\begin{array}{r} 323 \\ \times\ 32 \\ \hline 646\ \\ 9690\ \\ \hline 10{,}336 \end{array}$$

8.
$$\begin{array}{r} 385 \\ \times\ 69 \\ \hline 3465\ \\ 23100\ \\ \hline 26{,}565 \end{array}$$

9.
$$\begin{array}{r} 34 \\ \times\ 20 \\ \hline 0\ \\ 680\ \\ \hline 680 \end{array}$$

10.
$$\begin{array}{r} 130 \\ \times\ 50 \\ \hline 0\ \\ 6500\ \\ \hline 6500 \end{array}$$

11.
$$\begin{array}{r} 923 \\ \times\ 675 \\ \hline 4615\ \\ 6461\ \\ 5538\ \\ \hline 623{,}025 \end{array}$$

12. $25 \times 4 \times 17 = (25 \times 4) \times 17 = 100 \times 17 = 1700$
13. $8 \times 4 \times 3 \times 25 = 8 \times 3 \times 4 \times 25$
                   $= 8 \times 3 \times (4 \times 25)$
                   $= 24 \times 100$
                   $= 2400$

14.
$$\begin{array}{r} 17348 \\ \times\ 378 \\ \hline 138784\ \\ 121436\ \\ 52044\ \\ \hline 6{,}557{,}544 \end{array}$$
The total sales of cars was $6,557,544.
15. Area $= 5$ yards $\times 7$ yards $= 35$ square yards.

1.5 Practice Problems

1. (a) $4\overline{)36}$ → 9
  (b) $5\overline{)25}$ → 5
  (c) $9\overline{)72}$ → 8
  (d) $6\overline{)30}$ → 5
2. (a) $\dfrac{7}{1} = 7$
  (b) $\dfrac{9}{9} = 1$
  (c) $\dfrac{0}{5} = 0$
  (d) $\dfrac{12}{0}$ cannot be done

3.
$$\begin{array}{r} 7\ \text{R}\ 3 \\ 6\overline{)45} \\ \underline{42}\ \\ 3 \end{array}$$
Check
$$\begin{array}{r} 6 \\ \times\ 7 \\ \hline 42 \\ +\ 3 \\ \hline 45 \end{array}$$

4.
$$\begin{array}{r} 21\ \text{R}\ 3 \\ 6\overline{)129} \\ \underline{12}\ \ \\ 9 \\ \underline{6} \\ 3 \end{array}$$
Check
$$\begin{array}{r} 21 \\ \times\ 6 \\ \hline 126 \\ +\ 3 \\ \hline 129 \end{array}$$

5.
$$\begin{array}{r} 529\ \text{R}\ 5 \\ 8\overline{)4237} \\ \underline{40}\ \ \\ 23 \\ \underline{16} \\ 77 \\ \underline{72} \\ 5 \end{array}$$

6.
$$\begin{array}{r} 7\ \text{R}\ 19 \\ 32\overline{)243} \\ \underline{224} \\ 19 \end{array}$$

7.
$$\begin{array}{r} 1278\ \text{R}\ 9 \\ 33\overline{)42183} \\ \underline{33}\ \ \ \ \\ 91 \\ \underline{66} \\ 258 \\ \underline{231} \\ 273 \\ \underline{264} \\ 9 \end{array}$$

8.
$$\begin{array}{r} 25\ \text{R}\ 27 \\ 128\overline{)3227} \\ \underline{256}\ \ \\ 667 \\ \underline{640} \\ 27 \end{array}$$

9.
$$\begin{array}{r} 16{,}852 \\ 7\overline{)117{,}964} \end{array}$$
Check
$$\begin{array}{r} 16{,}852 \\ \times\ \ 7 \\ \hline 117{,}964 \end{array}$$
The cost of one car is $16,852.

10. $14\overline{)5138}$ → 367 The average speed was 367 mph.

1.6 Practice Problems

1. (a) $12 \times 12 \times 12 \times 12 = 12^4$
  (b) $2 \times 2 \times 2 \times 2 \times 2 \times 2 = 2^6$
2. (a) $12^2 = 12 \times 12 = 144$
  (b) $6^3 = 6 \times 6 \times 6 = 216$
  (c) $2^6 = 2 \times 2 \times 2 \times 2 \times 2 \times 2 = 64$
  (d) $1^{10} = 1 \times 1 \times 1 \times 1 \times 1 \times 1 \times 1 \times 1 \times 1 \times 1 = 1$

3. (a) $7^3 + 8^2 = (7)(7)(7) + (8)(8) = 343 + 64 = 407$
(b) $9^2 + 6^0 = (9)(9) + 1 = 81 + 1 = 82$
(c) $5^4 + 5 = (5)(5)(5)(5) + 5 = 625 + 5 = 630$

4. $7 + 4^3 \times 3 = 7 + 64 \times 3$ Exponents
$ = 7 + 192$ Multiply
$ = 199$ Add

5. $37 - 20 \div 5 + 2 - 3 \times 4$
$ = 37 - 4 + 2 - 3 \times 4$ Divide
$ = 37 - 4 + 2 - 12$ Multiply
$ = 33 + 2 - 12$ Subtract
$ = 35 - 12$ Add
$ = 23$ Subtract

6. $4^3 - 2 + 3^2$
$ = 4 \times 4 \times 4 - 2 + 3 \times 3$ Evaluate exponents.
$ = 64 - 2 + 9$ $4^3 = 64$ and $3^2 = 9$.
$ = 62 + 9$ Subtract
$ = 71$ Add

7. $(17 + 7) \div 6 \times 2 + 7 \times 3 - 4$
$ = 24 \div 6 \times 2 + 7 \times 3 - 4$ Combine inside parentheses
$ = 4 \times 2 + 7 \times 3 - 4$ Divide
$ = 8 + 7 \times 3 - 4$ Multiply
$ = 8 + 21 - 4$ Multiply
$ = 29 - 4$ Add
$ = 25$ Subtract

8. $5^2 - 6 \div 2 + 3^4 + 7 \times (12 - 10)$
$ = 5^2 - 6 \div 2 + 3^4 + 7 \times 2$ Combine inside parentheses
$ = 25 - 6 \div 2 + 81 + 7 \times 2$ Exponents
$ = 25 - 3 + 81 + 7 \times 2$ Divide
$ = 25 - 3 + 81 + 14$ Multiply
$ = 22 + 81 + 14$ Subtract
$ = 103 + 14$ Add
$ = 117$ Add

1.7 Practice Problems

1. $6\,5\,,5\,2\,8$ ↓   Locate the thousands round-off place.

$6\,5\,,\!\widehat{5}\,2\,8$ ↓   The first digit to the right is 5 or more. We will increase the thousands digit by 1.

$6\,6\,,0\,0\,0$ All digits to the right of the thousands place are replaced by zero.

2. $1\,7\,\widehat{2}\,,9\,6\,3 = 170{,}000$ to the nearest ten thousand.

3. (a) $5\,3\,,2\,\overset{\downarrow}{8}\,2 = 53{,}280$ to the nearest ten. The digit to the right of the tens place was less than 5.

(b) $1\,6\,4\,,4\,\overset{\downarrow}{8}\,5 = 164{,}000$ to the nearest thousand. The digit to the right of the thousands place was less than 5.

(c) $1\,,3\,6\,5\,,\overset{\downarrow}{2}\,7\,3 = 1{,}400{,}000$ to the nearest hundred thousand. The digit to the right of the hundred thousands place was greater than 5.

1.8 Practice Problems

Practice Problem 1

1. Understand the problem.

2. Solve and state the answer:
$135 + 28 + 13 + 34 = 210$
The total amount taken out of Diane's paycheck is $210.

3. *Check.* Estimate to see if the answer is reasonable.

4. (a) $9\,3\,\overset{\downarrow}{5}\,,6\,8\,2 = 936{,}000$ to the nearest thousand. The digit to the right of the thousands place is greater than 5.

(b) $9\,3\,5\,,\overset{\downarrow}{6}\,8\,2 = 900{,}000$ to the nearest hundred thousand. The digit to the right of the hundred thousands place is less than 5.

(c) $9\,3\,5\,,6\,8\,2 = 1{,}000{,}000$ to the nearest million. The digit to the right of the millions place is greater than 5.

5. 9,460,000,000,000,000 meters $= 9{,}500{,}000{,}000{,}000{,}000$ meters to the nearest hundred trillion meters.

6.

Actual Sum	Estimated Sum	
3456	3000	
9876	10000	
5421	5000	
+ 1278	+ 1000	
20,031	19,000	Close to the actual sum

7.

$697	$700	
35	40	
+ 19	+ 20	
	$760	

We estimate that the total cost is $760. (The exact answer is $751, so we can see that our answer is quite close.)

8. Estimate: $10{,}000 + 10{,}000 + 20{,}000 + 60{,}000 = 100{,}000$
This is significantly different from 81,358, so we would suspect that an error has been made. In fact, Ming did make an error. The exact sum is actually 101,358!

9. Estimate: $30{,}000{,}000 - 20{,}000{,}000 = 10{,}000{,}000$
We estimate that 10,000,000 more people lived in California than in Florida.

10. Estimate: $9000 \times 7000 = 63{,}000{,}000$
We estimate the product to be 63,000,000.

11. $40\overline{)80{,}000}$ gives 2000 Our estimate is 2000.

12. $60\overline{)2{,}000{,}000}$ gives $33{,}333$ R 20   Our estimate is $33,333 for one truck.

Mathematics Blueprint for Problem Solving

Gather the Facts	What Am I Asked to Do?	How Do I Proceed?	Key Points to Remember
The deductions are $135, $28, $13, and $34.	Find out the total amount of deductions.	I must add the four deductions to obtain the total.	Watch out! Gross pay of $1352 is not needed to solve the problem.

Practice Problem 2

1. Understand the problem.

Mathematics Blueprint for Problem Solving			
Gather the Facts	What Am I Asked to Do?	How Do I Proceed?	Key Points to Remember
Gore had 50,999,897 votes. Bush had 50,456,002 votes.	Find out by how many votes Gore beat Bush.	I must subtract the amounts.	Gore is a Democrat and Bush is a Republican.

2. Solve and state the answer:

    50,999,897
  − 50,456,002
      543,895

Gore beat Bush by 543,895 votes. Bush had more electoral college votes.

3. *Check.* Estimate to see if the answer is reasonable.

Practice Problem 3

1. Understand the problem.

Mathematics Blueprint for Problem Solving			
Gather the Facts	What Am I Asked to Do?	How Do I Proceed?	Key Points to Remember
1 gallon is 1024 fluid drams.	Find out how many fluid drams in 9 gallons.	I need to multiply 1024 by 9.	I must use fluid drams as the measure in my answer.

2. Solve and state the answer:

    1024
  ×   9
   9216

There are 9216 fluid drams in 9 gallons.

3. *Check.* Estimate to see if the answer is reasonable.

Practice Problem 4

1. Understand the problem.

Mathematics Blueprint for Problem Solving			
Gather the Facts	What Am I Asked to Do?	How Do I Proceed?	Key Points to Remember
Donna bought 45 shares of stock. She paid $1620 for them.	Find out the cost per share of stock.	I need to divide 1620 by 45.	Use dollars as the unit in the answer.

2. Solve and state the answer:

$$
\begin{array}{r}
36 \\
45\overline{)1620} \\
135 \\
\hline
270 \\
270 \\
\hline
0
\end{array}
$$

Donna paid $36 per share for the stock.

3. *Check.* Estimate to see if the answer is reasonable.

Practice Problem 5

1. Understand the problem. We will make an imaginary bill of sale.

2. Solve and state the answer. We do the calculation and enter the results in the bill of sale.

Customer: Anderson Dining Commons			
Quantity	**Item**	**Cost per Item**	**Amount for This Item**
50	Tables	$200	$10,000 (50 × $200 = $10,000)
180	Chairs	$ 40	$ 7200 (180 × $40 = $7200)
6	Moving Carts	$ 65	$ 390 (6 × $65 = $390)
		Total	$17,590 (sum of the three amounts)

The total cost of purchase was $17,590.

3. **Check.** Estimate to see if the answer is reasonable.

Practice Problem 6

1. Understand the problem.

Mathematics Blueprint for Problem Solving			
Gather the Facts	**What Am I Asked to Do?**	**How Do I Proceed?**	**Key Points to Remember**

Old balance: $498
New deposits: $607
                $163
Interest: $36
Withdrawals: $ 19
                $158
                $582
                $ 74
	Find her new balance after the transactions.	(a) Add the new deposits and interest to the old balance.
		(b) Add the withdrawals.
		(c) Subtract the results from steps (a) and (b).
			Deposits and interest are added and withdrawals are subtracted from savings accounts.

2. Solve and state the answer:

(a)   498   (b)   19   (c)   1304
       607   158   − 833
       163   582   471
     + 36   + 74
     1304   833

Her balance this month is $471.

3. **Check.** Estimate to see if the answer is reasonable.

Practice Problem 7

1. Understand the problem.

Mathematics Blueprint for Problem Solving			
Gather the Facts	**What Am I Asked to Do?**	**How Do I Proceed?**	**Key Points to Remember**
Odometer reading at end of trip: 51,118 miles Odometer reading at start of trip: 50,698 miles Used on trip: 12 gallons of gas	Find the number of miles per gallon that the car obtained on the trip.	(a) Subtract the two odometer readings. (b) Divide that number by 12.	The gas tank was full at the beginning of the trip. 12 gallons fills the tank at the end of the trip.

2. Solve and state the answer:

    51,118   odometer at end of trip
  − 50,698   odometer at start of trip
      420   miles traveled on trip

$$\frac{420 \text{ miles}}{12 \text{ gallons of gas used}}$$

$$= 12\overline{)420} \quad \begin{array}{c} 35 \\ \underline{36} \\ 60 \\ \underline{60} \\ 0 \end{array}$$

35 miles per gallon on the trip

3. **Check.** Estimate to see if the answer is reasonable.

Chapter 2 2.1 Practice Problems

1. (a) Four parts of twelve are shaded. The fraction is $\dfrac{4}{12}$.

(b) Three parts out of six are shaded. The fraction is $\dfrac{3}{6}$.

(c) Two parts of three are shaded. The fraction is $\dfrac{2}{3}$.

2. (a) Shade $\dfrac{4}{5}$ of the object.

(b) Shade $\dfrac{3}{7}$ of the group.

3. (a) $\dfrac{9}{17}$ represents 9 players out of 17.

(b) The total class is $382 + 351 = 733$.

The fractional part that is men is $\dfrac{382}{733}$.

(c) $\dfrac{7}{8}$ of a yard of material.

4. Total number of defective items $1 + 2 = 3$. Total number of items $7 + 9 = 16$. A fraction that represents the portion of the items that were defective is $\dfrac{3}{16}$.

2.2 Practice Problems

1. (a) $18 = 2 \times 9$
$= 2 \times 3 \times 3$
$= 2 \times 3^2$

(b) $72 = 8 \times 9$
$= 2 \times 2 \times 2 \times 3 \times 3$
$= 2^3 \times 3^2$

(c) $400 = 10 \times 40$
$= 5 \times 2 \times 5 \times 8$
$= 5 \times 2 \times 5 \times 2 \times 2 \times 2$
$= 2^4 \times 5^2$

2. (a) $\dfrac{30}{42} = \dfrac{30 \div 6}{42 \div 6} = \dfrac{5}{7}$

(b) $\dfrac{60}{132} = \dfrac{60 \div 12}{132 \div 12} = \dfrac{5}{11}$

3. (a) $\dfrac{120}{135} = \dfrac{2 \times 2 \times 2 \times \cancel{3} \times \cancel{5}}{3 \times 3 \times \cancel{3} \times \cancel{5}} = \dfrac{8}{9}$

(b) $\dfrac{715}{880} = \dfrac{\cancel{5} \times \cancel{11} \times 13}{2 \times 2 \times 2 \times 2 \times \cancel{5} \times \cancel{11}} = \dfrac{13}{16}$

4. (a) $\dfrac{84}{108} \overset{?}{=} \dfrac{7}{9}$

$84 \times 9 \overset{?}{=} 108 \times 7$
$756 = 756$ Yes

(b) $\dfrac{3}{7} \overset{?}{=} \dfrac{79}{182}$

$3 \times 182 \overset{?}{=} 7 \times 79$
$546 \neq 553$ No

2.3 Practice Problems

1. (a) $4\dfrac{3}{7} = \dfrac{4 \times 7 + 3}{7} = \dfrac{28 + 3}{7} = \dfrac{31}{7}$

(b) $6\dfrac{2}{3} = \dfrac{6 \times 3 + 2}{3} = \dfrac{18 + 2}{3} = \dfrac{20}{3}$

(c) $19\dfrac{4}{7} = \dfrac{19 \times 7 + 4}{7} = \dfrac{133 + 4}{7} = \dfrac{137}{7}$

2. (a) $4\overline{)17}$ so $\dfrac{17}{4} = 4\dfrac{1}{4}$
$\underline{16}$
1

(b) $5\overline{)36}$ so $\dfrac{36}{5} = 7\dfrac{1}{5}$
$\underline{35}$
1

(c) $27\overline{)116}$ so $\dfrac{116}{27} = 4\dfrac{8}{27}$
$\underline{108}$
8

(d) $13\overline{)91}$ so $\dfrac{91}{13} = 7$
$\underline{91}$
0

3. $\dfrac{51}{15} = \dfrac{\cancel{3} \times 17}{\cancel{3} \times 5} = \dfrac{17}{5}$

4. $\dfrac{16}{80} = \dfrac{1}{5}$ so

$3\dfrac{16}{80} = 3\dfrac{1}{5}$.

5. $\dfrac{1001}{572} = 1\dfrac{429}{572}$

Now the fraction $\dfrac{429}{572} = \dfrac{3 \times \cancel{11} \times \cancel{13}}{2 \times 2 \times \cancel{11} \times \cancel{13}} = \dfrac{3}{4}$.

Thus $\dfrac{1001}{572} = 1\dfrac{429}{572} = 1\dfrac{3}{4}$.

2.4 Practice Problems

1. (a) $\dfrac{6}{7} \times \dfrac{3}{13} = \dfrac{6 \times 3}{7 \times 13} = \dfrac{18}{91}$

(b) $\dfrac{1}{5} \times \dfrac{11}{12} = \dfrac{1 \times 11}{5 \times 12} = \dfrac{11}{60}$

2. $\dfrac{55}{72} \times \dfrac{16}{33} = \dfrac{5 \cdot 11}{2 \cdot 2 \cdot 2 \cdot 3 \cdot 3} \times \dfrac{2 \cdot 2 \cdot 2 \cdot 2}{3 \cdot 11}$
$= \dfrac{\cancel{2} \cdot \cancel{2} \cdot \cancel{2} \cdot 2 \cdot 5 \cdot \cancel{11}}{\cancel{2} \cdot \cancel{2} \cdot \cancel{2} \cdot 3 \cdot 3 \cdot 3 \cdot \cancel{11}}$
$= \dfrac{10}{27}$

3. (a) $7 \times \dfrac{5}{13} = \dfrac{7}{1} \times \dfrac{5}{13} = \dfrac{35}{13}$ or $2\dfrac{9}{13}$

(b) $\dfrac{13}{4} \times 8 = \dfrac{13}{\cancel{4}} \times \dfrac{\overset{2}{\cancel{8}}}{1} = \dfrac{26}{1} = 26$

4. $\dfrac{3}{\cancel{8}} \times \overset{12,300}{\cancel{98,400}} = \dfrac{3}{1} \times 12,300 = 36,900$

There are 36,900 square feet in the wetland area.

5. (a) $2\dfrac{1}{6} \times \dfrac{4}{7} = \dfrac{13}{\underset{3}{\cancel{6}}} \times \dfrac{\overset{2}{\cancel{4}}}{7} = \dfrac{26}{21}$ or $1\dfrac{5}{21}$

(b) $10\dfrac{2}{3} \times 13\dfrac{1}{2} = \dfrac{\overset{16}{\cancel{32}}}{\cancel{3}} \times \dfrac{\overset{9}{\cancel{27}}}{\cancel{2}} = \dfrac{144}{1} = 144$

(c) $\dfrac{3}{5} \times 1\dfrac{1}{3} \times \dfrac{5}{8} = \dfrac{\cancel{3}}{\cancel{5}} \times \dfrac{\overset{1}{\cancel{4}}}{\cancel{3}} \times \dfrac{\cancel{5}}{\underset{2}{\cancel{8}}} = \dfrac{1}{2}$

(d) $3\dfrac{1}{5} \times 2\dfrac{1}{2} = \dfrac{\overset{8}{\cancel{16}}}{\cancel{5}} \times \dfrac{\cancel{5}}{\cancel{2}} = \dfrac{8}{1} = 8$

6. Area $= 1\dfrac{1}{5} \times 4\dfrac{5}{6} = \dfrac{\overset{1}{\cancel{6}}}{5} \times \dfrac{29}{\cancel{6}} = \dfrac{29}{5} = 5\dfrac{4}{5}$

The area is $5\dfrac{4}{5}$ square meters.

7. Since $8 \cdot 10 = 80$ and $9 \cdot 9 = 81$,

we know that $\dfrac{8}{9} \cdot \dfrac{10}{9} = \dfrac{80}{81}$.

Therefore $x = \dfrac{10}{9}$.

2.5 Practice Problems

1. (a) $\dfrac{7}{13} \div \dfrac{3}{4} = \dfrac{7}{13} \times \dfrac{4}{3} = \dfrac{28}{39}$

(b) $\dfrac{16}{35} \div \dfrac{24}{25} = \dfrac{\overset{2}{\cancel{16}}}{\underset{7}{\cancel{35}}} \times \dfrac{\overset{5}{\cancel{25}}}{\underset{3}{\cancel{24}}} = \dfrac{10}{21}$

2. (a) $\dfrac{3}{17} \div 6 = \dfrac{3}{17} \div \dfrac{6}{1} = \dfrac{\overset{1}{\cancel{3}}}{17} \times \dfrac{1}{\underset{2}{\cancel{6}}} = \dfrac{1}{34}$

(b) $14 \div \dfrac{7}{15} = \dfrac{14}{1} \div \dfrac{7}{15} = \dfrac{\overset{2}{\cancel{14}}}{1} \times \dfrac{15}{\underset{1}{\cancel{7}}} = 30$

3. (a) $1 \div \dfrac{11}{13} = \dfrac{1}{1} \times \dfrac{13}{11} = \dfrac{13}{11} \text{ or } 1\dfrac{2}{11}$

(b) $\dfrac{14}{17} \div 1 = \dfrac{14}{17} \times \dfrac{1}{1} = \dfrac{14}{17}$

(c) $\dfrac{3}{11} \div 0$ Division by zero is undefined.

(d) $0 \div \dfrac{9}{16} = \dfrac{0}{1} \times \dfrac{16}{9} = \dfrac{0}{9} = 0$

4. (a) $1\dfrac{1}{5} \div \dfrac{7}{10} = \dfrac{6}{5} \div \dfrac{7}{10} = \dfrac{6}{\underset{1}{\cancel{5}}} \times \dfrac{\overset{2}{\cancel{10}}}{7} = \dfrac{12}{7} \text{ or } 1\dfrac{5}{7}$

(b) $2\dfrac{1}{4} \div 1\dfrac{7}{8} = \dfrac{9}{4} \div \dfrac{15}{8} = \dfrac{\overset{3}{\cancel{9}}}{\underset{1}{\cancel{4}}} \times \dfrac{\overset{2}{\cancel{8}}}{\underset{5}{\cancel{15}}} = \dfrac{6}{5} \text{ or } 1\dfrac{1}{5}$

5. (a) $\dfrac{5\dfrac{2}{3}}{7} = 5\dfrac{2}{3} \div 7 = \dfrac{17}{3} \times \dfrac{1}{7} = \dfrac{17}{21}$

(b) $\dfrac{1\dfrac{2}{5}}{2\dfrac{1}{3}} = 1\dfrac{2}{5} \div 2\dfrac{1}{3} = \dfrac{7}{5} \div \dfrac{7}{3} = \dfrac{\overset{1}{\cancel{7}}}{5} \times \dfrac{3}{\underset{1}{\cancel{7}}} = \dfrac{3}{5}$

6. $x \div \dfrac{3}{2} = \dfrac{22}{36}$

$x \cdot \dfrac{2}{3} = \dfrac{22}{36}$

$\dfrac{11}{12} \cdot \dfrac{2}{3} = \dfrac{22}{36}$ Thus $x = \dfrac{11}{12}$.

7. $19\dfrac{1}{4} \div 14 = \dfrac{\overset{11}{\cancel{77}}}{4} \times \dfrac{1}{\underset{2}{\cancel{14}}} = \dfrac{11}{8} \text{ or } 1\dfrac{3}{8}$

Each piece will be $1\dfrac{3}{8}$ feet long.

2.6 Practice Problems

1. The multiples of 14 are 14, 28, 42, 56, 70, 84, . . .
The multiples of 21 are 21, 42, 63, 84, 105, 126, . . .
42 is the least common multiple of 14 and 21.

2. The multiples of 10 are 10, 20, 30, 40 . . .
The multiples of 15 are 15, 30, 45 . . .
30 is the least common multiple of 10 and 15.

3. 54 is a multiple of 6. We know that $6 \times 9 = 54$.
The least common multiple of 6 and 54 is 54.

4. (a) The LCD of $\dfrac{3}{4}$ and $\dfrac{11}{12}$ is 12.

12 can be divided by 4 and 12.

(b) The LCD of $\dfrac{1}{7}$ and $\dfrac{8}{35}$ is 35.

35 can be divided by 7 and 35.

5. The LCD of $\dfrac{3}{7}$ and $\dfrac{5}{6}$ is 42.

42 can be divided by 7 and 6.

6. (a) $14 = 2 \times 7$
$10 = 2 \times 5$
LCD $= 2 \times 5 \times 7 = 70$

(b) $15 = 3 \times 5$
$50 = 2 \times 5 \times 5$
LCD $= 2 \times 3 \times 5 \times 5 = 150$

(c) $16 = 2 \times 2 \times 2 \times 2$
$12 = 2 \times 2 \times 3$
LCD $= 2 \times 2 \times 2 \times 2 \times 3 = 48$

7. $49 = 7 \times 7$
$21 = 7 \times 3$
$7 = 7 \times 1$
LCD $= 7 \times 7 \times 3 = 147$

8. (a) $\dfrac{3}{5} = \dfrac{3}{5} \times \dfrac{8}{8} = \dfrac{24}{40}$ **(c)** $\dfrac{2}{7} = \dfrac{2}{7} \times \dfrac{4}{4} = \dfrac{8}{28}$

(b) $\dfrac{7}{11} = \dfrac{7}{11} \times \dfrac{4}{4} = \dfrac{28}{44}$ $\dfrac{3}{4} = \dfrac{3}{4} \times \dfrac{7}{7} = \dfrac{21}{28}$

9. (a) $20 = 2 \times 2 \times 5$
$15 = 3 \times 5$
LCD $= 2 \times 2 \times 3 \times 5 = 60$

(b) $\dfrac{3}{20} = \dfrac{3}{20} \times \dfrac{3}{3} = \dfrac{9}{60}$ $\dfrac{11}{15} = \dfrac{11}{15} \times \dfrac{4}{4} = \dfrac{44}{60}$

10. (a) $64 = 2 \times 2 \times 2 \times 2 \times 2 \times 2$
$80 = 2 \times 2 \times 2 \times 2 \times 5$
LCD $= 2 \times 2 \times 2 \times 2 \times 2 \times 2 \times 5 = 320$

(b) $\dfrac{5}{64} = \dfrac{5}{64} \times \dfrac{5}{5} = \dfrac{25}{320}$

$\dfrac{3}{80} = \dfrac{3}{80} \times \dfrac{4}{4} = \dfrac{12}{320}$

2.7 Practice Problems

1. $\dfrac{3}{17} + \dfrac{12}{17} = \dfrac{15}{17}$

2. (a) $\dfrac{1}{12} + \dfrac{5}{12} = \dfrac{6}{12} = \dfrac{1}{2}$

(b) $\dfrac{13}{15} + \dfrac{7}{15} = \dfrac{20}{15} = \dfrac{4}{3} \text{ or } 1\dfrac{1}{3}$

3. (a) $\dfrac{5}{19} - \dfrac{2}{19} = \dfrac{3}{19}$ **(b)** $\dfrac{21}{25} - \dfrac{6}{25} = \dfrac{15}{25} = \dfrac{3}{5}$

4.
$\begin{array}{r} \dfrac{2}{15} = \dfrac{2}{15} \\ + \dfrac{1}{5} \times \dfrac{3}{3} = + \dfrac{3}{15} \\ \hline \dfrac{5}{15} = \dfrac{1}{3} \end{array}$

5. LCD $= 48$ $\dfrac{5}{12} \times \dfrac{4}{4} = \dfrac{20}{48}$ $\dfrac{5}{16} \times \dfrac{3}{3} = \dfrac{15}{48}$

$\dfrac{5}{12} + \dfrac{5}{16} = \dfrac{20}{48} + \dfrac{15}{48} = \dfrac{35}{48}$

6. LCD $= 48$

$\dfrac{3}{16} \times \dfrac{3}{3} = \dfrac{9}{48}$ $\dfrac{1}{8} \times \dfrac{6}{6} = \dfrac{6}{48}$ $\dfrac{1}{12} \times \dfrac{4}{4} = \dfrac{4}{48}$

$\dfrac{3}{16} + \dfrac{1}{8} + \dfrac{1}{12} = \dfrac{9}{48} + \dfrac{6}{48} + \dfrac{4}{48} = \dfrac{19}{48}$

7. LCD $= 96$ $\dfrac{9}{48} \times \dfrac{2}{2} = \dfrac{18}{96}$ $\dfrac{5}{32} \times \dfrac{3}{3} = \dfrac{15}{96}$

$\dfrac{9}{48} - \dfrac{5}{32} = \dfrac{18}{96} - \dfrac{15}{96} = \dfrac{3}{96} = \dfrac{1}{32}$

8. $\dfrac{9}{10} \times \dfrac{2}{2} = \dfrac{18}{20}$ $\dfrac{1}{4} \times \dfrac{5}{5} = \dfrac{5}{20}$

$\dfrac{9}{10} - \dfrac{1}{4} = \dfrac{18}{20} - \dfrac{5}{20} = \dfrac{13}{20}$

There is $\dfrac{13}{20}$ gallon left.

9. The LCD of $\dfrac{3}{10}$ and $\dfrac{23}{25}$ is 50.

$\dfrac{3}{10} \times \dfrac{5}{5} = \dfrac{15}{50}$ Now rewriting: $x + \dfrac{15}{50} = \dfrac{46}{50}$

$\dfrac{23}{25} \times \dfrac{2}{2} = \dfrac{46}{50}$ $\dfrac{31}{50} + \dfrac{15}{50} = \dfrac{46}{50}$

So, $x = \dfrac{31}{50}$

10. $\dfrac{15}{16} + \dfrac{3}{40}$

$\dfrac{15}{16} \times \dfrac{40}{40} = \dfrac{600}{640}$ $\dfrac{3}{40} \times \dfrac{16}{16} = \dfrac{48}{640}$

Thus $\dfrac{15}{16} + \dfrac{3}{40} = \dfrac{600}{640} + \dfrac{48}{640} = \dfrac{648}{640} = \dfrac{81}{80}$ or $1\dfrac{1}{80}$

2.8 Practice Problems

1. $5\dfrac{1}{12}$

$\underline{+9\dfrac{5}{12}}$

$14\dfrac{6}{12} = 14\dfrac{1}{2}$

2. The LCD is 20.

$\dfrac{1}{4} \times \dfrac{5}{5} = \dfrac{5}{20}$ $\dfrac{2}{5} \times \dfrac{4}{4} = \dfrac{8}{20}$

$6\dfrac{1}{4} = 6\dfrac{5}{20}$

$\underline{+2\dfrac{2}{5} = +2\dfrac{8}{20}}$

$8\dfrac{13}{20}$

3. LCD = 12 $7\boxed{\dfrac{1}{4} \times \dfrac{3}{3}} = 7\dfrac{3}{12}$

$\underline{+3\boxed{\dfrac{5}{6} \times \dfrac{2}{2}} = +3\dfrac{10}{12}}$

$10\dfrac{13}{12} = 10 + 1\dfrac{1}{12} = 11\dfrac{1}{12}$

4. LCD = 12 $12\dfrac{5}{6} = 12\dfrac{10}{12}$

$\underline{-7\dfrac{5}{12} = -7\dfrac{5}{12}}$

$5\dfrac{5}{12}$

5. (a) LCD = 24 $9\boxed{\dfrac{1}{8} \times \dfrac{3}{3}} = 9\dfrac{3}{24} = 8\dfrac{27}{24}$

$\underline{-3\boxed{\dfrac{2}{3} \times \dfrac{8}{8}} = -3\dfrac{16}{24} = -3\dfrac{16}{24}}$

$5\dfrac{11}{24}$

Borrow 1 from 9:

$9\dfrac{3}{24} = 8 + 1\dfrac{3}{24} = 8\dfrac{27}{24}$

(b) $18 = 17\dfrac{18}{18}$

$\underline{-6\dfrac{7}{18} = -6\dfrac{7}{18}}$

$11\dfrac{11}{18}$

6. $6\dfrac{1}{4} = 6\dfrac{3}{12} = 5\dfrac{15}{12}$

$\underline{-4\dfrac{2}{3} = -4\dfrac{8}{12} = -4\dfrac{8}{12}}$

$1\dfrac{7}{12}$

They had $1\dfrac{7}{12}$ gallons left over.

7. $\dfrac{3}{5} - \dfrac{1}{15} \times \dfrac{10}{13}$

$= \dfrac{3}{5} - \dfrac{2}{39}$ LCD = $5 \cdot 39 = 195$

$= \dfrac{117}{195} - \dfrac{10}{195}$

$= \dfrac{107}{195}$

8. $\dfrac{1}{7} \times \dfrac{5}{6} + \dfrac{5}{3} \div \dfrac{7}{6} = \dfrac{1}{7} \times \dfrac{5}{6} + \dfrac{5}{3} \times \dfrac{6}{7}$

$= \dfrac{5}{42} + \dfrac{10}{7}$ LCD = 42

$= \dfrac{5}{42} + \dfrac{60}{42}$

$= \dfrac{65}{42}$ or $1\dfrac{23}{42}$

2.9 Practice Problems

Practice Problem 1

1. Understand the problem.

Mathematics Blueprint for Problem Solving			
Gather the Facts	What Am I Asked to Do?	How Do I Proceed?	Key Points to Remember
Gas amounts: $18\dfrac{7}{10}$ gal $15\dfrac{2}{5}$ gal $14\dfrac{1}{2}$ gal	Find out how many gallons of gas she bought altogether.	Add the three amounts.	When adding mixed numbers, the LCD is needed for the fractions.

2. Solve and state the answer:

$$\text{LCD} = 10 \qquad 18\frac{7}{10} = \quad 18\frac{7}{10}$$

$$15\frac{2}{5} = \quad 15\frac{4}{10}$$

$$14\frac{1}{2} = \ +14\frac{5}{10}$$

$$47\frac{16}{10} = 48\frac{6}{10}$$

$$= 48\frac{3}{5}$$

The total is $48\frac{3}{5}$ gallons.

3. *Check.* Estimate to see if the answer is reasonable.

Practice Problem 2

1. Understand the problem.

Mathematics Blueprint for Problem Solving			
Gather the Facts	What Am I Asked to Do?	How Do I Proceed?	Key Points to Remember
Poster: $12\frac{1}{4}$ in. Top border: $1\frac{3}{8}$ in. Bottom border: 2 in.	Find the length of the inside portion of the poster.	**(a)** Add the two border lengths. **(b)** Subtract this total from the poster length.	When adding mixed numbers, the LCD is needed for the fractions.

2. Solve and state the answer:

$$\textbf{(a)} \qquad 1\frac{3}{8}$$

$$+2$$

$$3\frac{3}{8}$$

$$\textbf{(b)} \qquad 12\frac{1}{4} = \quad 12\frac{2}{8} = \quad 11\frac{10}{8}$$

$$-3\frac{3}{8} = \ -3\frac{3}{8} = \ -3\frac{3}{8}$$

$$8\frac{7}{8}$$

The length of the inside portion is $8\frac{7}{8}$ inches.

3. *Check.* Estimate to see if the answer is reasonable or work backward to check.

Practice Problem 3

1. Understand the problem.

Mathematics Blueprint for Problem Solving			
Gather the Facts	What Am I Asked to Do?	How Do I Proceed?	Key Points to Remember
Regular tent uses $8\frac{1}{4}$ yards. Large tent uses $1\frac{1}{2}$ times the regular. She makes 6 regular and 16 large tents.	Find out how many yards of cloth will be needed to make the tents.	Find the amount used for regular tents, and the amount used for large tents. Then add the two.	Large tents use $1\frac{1}{2}$ times the regular amount.

2. Solve and state the answer:

We multiply $6 \times 8\frac{1}{4}$ for regular tents and $16 \times 1\frac{1}{2} \times 8\frac{1}{4}$ for large tents. Then add total yardage.

Regular tents: $6 \times 8\frac{1}{4} = \overset{3}{\cancel{6}} \times \frac{33}{\underset{2}{\cancel{4}}} = \frac{99}{2} = 49\frac{1}{2}$

Large tents: $16 \times 1\frac{1}{2} \times 8\frac{1}{4} = \overset{\overset{2}{\cancel{8}}}{\cancel{16}} \times \frac{3}{\underset{1}{\cancel{2}}} \times \frac{33}{\underset{1}{\cancel{4}}} = \frac{198}{1} = 198$

Total yardage for all tents is $198 + 49\frac{1}{2} = 247\frac{1}{2}$ yards.

3. ***Check.*** Estimate to see if the answer is reasonable.

Practice Problem 4

1. Understand the problem.

Mathematics Blueprint for Problem Solving

Gather the Facts	What Am I Asked to Do?	How Do I Proceed?	Key Points to Remember
He purchases 12-foot boards. Each shelf is $2\frac{3}{4}$ ft. He needs four shelves for each bookcase and he is making two bookcases.	**(a)** Find out how many boards he needs to buy. **(b)** Find out how many feet of shelving are actually needed. **(c)** Find out how many feet will be left over.	Find out how many $2\frac{3}{4}$-ft shelves he can get from one board. Then see how many boards he needs to make all eight shelves.	There will be three answers to this problem. Don't forget to calculate the leftover wood.

2. Solve and state the answer:

We want to know how many $2\frac{3}{4}$-ft shelves are in a 12-ft board.

$$12 \div 2\frac{3}{4} = \frac{12}{1} \div \frac{11}{4} = \frac{12}{1} \times \frac{4}{11} = \frac{48}{11} = 4\frac{4}{11}$$

He will get 4 shelves from each board with some left over.

(a) For two bookcases, he needs eight shelves. He gets four shelves out of each board. $8 \div 4 = 2$. He will need two 12-ft boards.

(b) He needs 8 shelves at $2\frac{3}{4}$ feet.

$$8 \times 2\frac{3}{4} = 8 \times \frac{11}{4} = 22$$

He actually needs 22 feet of shelving.

(c)   24 feet of shelving bought
     $-$ 22 feet of shelving used
          2 feet of shelving left over.

3. ***Check.*** Work backward to check the answer.

Practice Problem 5

1. Understand the problem.

Mathematics Blueprint for Problem Solving

Gather the Facts	What Am I Asked to Do?	How Do I Proceed?	Key Points to Remember
Distance is $199\frac{3}{4}$ miles. He uses $8\frac{1}{2}$ gallons of gas.	Find out how many miles per gallon he gets.	Divide the distance by the number of gallons.	Change mixed numbers to improper fractions before dividing.

2. Solve and state the answer:

$$199\frac{3}{4} \div 8\frac{1}{2} = \frac{799}{4} \div \frac{17}{2}$$

$$= \frac{\overset{47}{\cancel{799}}}{\underset{2}{\cancel{4}}} \times \frac{\overset{1}{\cancel{2}}}{\underset{1}{\cancel{17}}}$$

$$= \frac{47}{2} = 23\frac{1}{2}$$

He gets $23\frac{1}{2}$ miles per gallon.

3. ***Check.*** Estimate to see if the answer is reasonable.

Chapter 3   3.1 Practice Problems

1. (a) 0.073   seventy-three thousandths
(b) 4.68   four and sixty-eight hundredths
(c) 0.0017   seventeen ten-thousandths
(d) 561.78   five hundred sixty-one and seventy-eight hundredths

2. seven thousand, eight hundred sixty-three and $\frac{4}{100}$ dollars

3. (a) $\frac{9}{10} = 0.9$ **(b)** $\frac{136}{1000} = 0.136$

(c) $2\frac{56}{100} = 2.56$ **(d)** $34\frac{86}{1000} = 34.086$

4. (a) $0.37 = \frac{37}{100}$ **(b)** $182.3 = 182\frac{3}{10}$

(c) $0.7131 = \frac{7131}{10,000}$ **(d)** $42.019 = 42\frac{19}{1000}$

5. (a) $8.5 = 8\frac{5}{10} = 8\frac{1}{2}$ **(b)** $0.58 = \frac{58}{100} = \frac{29}{50}$

(c) $36.25 = 36\frac{25}{100} = 36\frac{1}{4}$ **(d)** $106.013 = 106\frac{13}{1000}$

6. $\frac{2}{1,000,000,000} = \frac{1}{500,000,000}$

The concentration of PCBs is $\frac{1}{500,000,000}$.

3.2 Practice Problems

1. Since $4 < 5$, therefore $5.74 < 5.75$.

5.74   **5.75**

2. $0.894 > 0.890$, so $0.894 > 0.89$
3. 2.45, 2.543, 2.46, 2.54, 2.5
It is helpful to add extra zeros and to place the decimals that begin with 2.4 in a group and the decimals that begin with 2.5 in the other.
    2.450, 2.460, 2.543, 2.540, 2.500
In order, we have from smallest to largest
    2.450, 2.460, 2.500, 2.540, 2.543.
It is OK to leave the extra terminal zeros in the answer.
4. 723.88
723.9   Since the digit to right of tenths is greater than 5, we round up.

5. (a) 12.92 6 47
12.926   Since the digit to right of thousandths is less than 5, we drop the digits 4 and 7.

(b) 0.00 7 892
0.008   Since the digit to right of thousandths is greater than 5, we round up.

6. 15,699.953
15,700.0   Since the digit to right of tenths is five, we round up.

7.

		Rounded to Nearest Dollar
Medical bills	375.50	376
Taxes	971.39	971
Retirement	980.49	980
Charity	817.65	818

3.3 Practice Problems

1. (a)
9.8
3.6
+ 5.4
18.8

(b)
300.72
163.75
+ 291.08
755.55

(c)
8.9000
37.0560
0.0023
+ 945.0000
990.9583

2.
93,521.8
+ 1634.8
95,156.6

The odometer reading was 95,156.6 miles.

3.
$ 80.95
133.91
256.47
53.08
+ 381.32
$905.73

4. (a)
38.8
− 26.9
11.9

(b)
2 0 3 4 . 9 0 8
− 1 9 8 6 . 3 2 5
4 8 . 5 8 3

5. (a)
1 9 . 0 0 0
− 1 2 . 5 7 9
6 . 4 2 1

(b)
2 8 3 . 0 7 6
− 9 6 . 3 8 0
1 8 6 . 6 9 6

6.
8 7 , 1 6 0 . 1
− 8 2 , 3 7 0 . 9
4 7 8 9 . 2

He had driven 4789.2 miles.

7.
1 5 . 3
− 1 0 . 8
4 . 5
$x = 4.5$

3.4 Practice Problems

1.
0.09   2 decimal places
× 0.6   1 decimal place
0.054   3 decimal places in product

2. (a)
0.47   2 decimal places
× 0.28   2 decimal places
376
94
0.1316   4 decimal places in product

(b)
0.436   3 decimal places
× 18.39   2 decimal places
3924
1308
3488
436
8.01804   5 decimal places in product

3.
0.4264   4 decimal places
×   38   0 decimal places
34112
12792
16.2032   4 decimal places in product

4. Area = length × width
1.26
× 2.3
378
252
2.898

The area is 2.898 square millimeters.

5. (a) $0.0561 \times 1\underline{0} = 0.561$ Decimal point moved one place to the right.
  (b) $1462.37 \times 1\underline{00} = 146{,}237.$ Decimal point moved two places to the right.
6. (a) $0.26 \times 1\underline{000} = 260.$ Decimal point moved three places to the right. One extra zero needed.
  (b) $5862.89 \times 1\underline{0{,}000} = 58{,}628{,}900.$ Decimal point moved four places to the right. Two extra zeros needed.
7. $7.684 \times 10^4 = 76{,}840.$ Decimal point moved four places to the right. One extra zero needed.
8. $156.2 \times 1000 = 156{,}200$
  156.2 kilometers is equal to 156,200 meters.

3.5 Practice Problems

1. (a)
$$
\begin{array}{r}
0.258 \\
7\overline{)1.806} \\
14 \\ \hline
40 \\
35 \\ \hline
56 \\
56 \\ \hline
0
\end{array}
$$

(b)
$$
\begin{array}{r}
0.0058 \\
16\overline{)0.0928} \\
80 \\ \hline
128 \\
128 \\ \hline
0
\end{array}
$$

2. $0.517 = 0.52$ to the nearest hundredth
$$
\begin{array}{r}
0.517 \\
46\overline{)23.820} \\
230 \\ \hline
82 \\
46 \\ \hline
360 \\
322 \\ \hline
38
\end{array}
$$

3.
$$
\begin{array}{r}
186.25 \\
19\overline{)3538.75} \\
19 \\ \hline
163 \\
152 \\ \hline
118 \\
114 \\ \hline
4\,7 \\
3\,8 \\ \hline
95 \\
95 \\ \hline
0
\end{array}
$$
He pays \$186.25 per month.

4. (a)
$$
\begin{array}{r}
1.12 \\
0.09_\wedge\overline{)0.10_\wedge08} \\
9 \\ \hline
1\,0 \\
9 \\ \hline
18 \\
18 \\ \hline
0
\end{array}
$$

(b)
$$
\begin{array}{r}
46. \\
0.037_\wedge\overline{)1.702_\wedge} \\
1.48 \\ \hline
222 \\
222 \\ \hline
0
\end{array}
$$

5. (a)
$$
\begin{array}{r}
0.023 \\
1.8_\wedge\overline{)0.0_\wedge414} \\
36 \\ \hline
54 \\
54 \\ \hline
0
\end{array}
$$

(b)
$$
\begin{array}{r}
2310. \\
0.0036_\wedge\overline{)8.3160_\wedge} \\
72 \\ \hline
111 \\
108 \\ \hline
36 \\
36 \\ \hline
0
\end{array}
$$

6. (a)
$$
\begin{array}{r}
137.26 \\
3.8_\wedge\overline{)521.6_\wedge00} \\
38 \\ \hline
141 \\
114 \\ \hline
27\,6 \\
26\,6 \\ \hline
1\,0\,0 \\
7\,6 \\ \hline
2\,40 \\
2\,28 \\ \hline
12
\end{array}
$$
The answer rounded to the nearest tenth is 137.3.

(b)
$$
\begin{array}{r}
0.0211 \\
8.05_\wedge\overline{)0.17_\wedge0000} \\
16\,10 \\ \hline
900 \\
805 \\ \hline
950 \\
805 \\ \hline
145
\end{array}
$$
The answer rounded to the nearest thousandth is 0.021.

7.
$$
\begin{array}{r}
1\,5.94 \\
28.5_\wedge\overline{)454.4_\wedge00} \\
285 \\ \hline
169\,4 \\
142\,5 \\ \hline
26\,9\,0 \\
25\,6\,5 \\ \hline
1\,2\,50 \\
1\,1\,40 \\ \hline
1\,10
\end{array}
$$
The truck got approximately 15.9 miles per gallon.

8.
$$
\begin{array}{r}
5.8 \\
0.12_\wedge\overline{)0.69_\wedge6} \\
60 \\ \hline
9\,6 \\
9\,6 \\ \hline
0
\end{array}
$$
n is 5.8.

9. Find the sum of levels for the years 1985, 1990, and 1995.
$$
\begin{array}{r}
9.30 \\
8.68 \\
+\,7.37 \\ \hline
25.35
\end{array}
$$
Then divide by three to obtain the average.
$$
\begin{array}{r}
8.45 \\
3\overline{)25.35} \\
24 \\ \hline
1\,3 \\
1\,2 \\ \hline
15 \\
15 \\ \hline
0
\end{array}
$$
The three-year average is 8.45 million tons. The five-year average was found to be 8.156 in Example 9. Find the difference between the averages.
$$
\begin{array}{r}
8.450 \\
-\,8.156 \\ \hline
0.294
\end{array}
$$
The three-year average differs from the five-year average by 0.294 million tons.

3.6 Practice Problems

1. (a)
$$
\begin{array}{r}
0.3125 \\
16\overline{)5.0000} \\
\underline{48} \\
20 \\
\underline{16} \\
40 \\
\underline{32} \\
80 \\
\underline{80} \\
0
\end{array}
$$

$\dfrac{5}{16} = 0.3125$

(b)
$$
\begin{array}{r}
0.1375 \\
80\overline{)11.0000} \\
\underline{80} \\
300 \\
\underline{240} \\
600 \\
\underline{560} \\
400 \\
\underline{400} \\
0
\end{array}
$$

$\dfrac{11}{80} = 0.1375$

2. (a)
$$
\begin{array}{r}
0.6363 \\
11\overline{)7.0000} \\
\underline{66} \\
40 \\
\underline{33} \\
70 \\
\underline{66} \\
40 \\
\underline{33} \\
7
\end{array}
$$

$\dfrac{7}{11} = 0.\overline{63}$

(b)
$$
\begin{array}{r}
0.533 \\
15\overline{)8.000} \\
\underline{75} \\
50 \\
\underline{45} \\
50 \\
\underline{45} \\
5
\end{array}
$$

$\dfrac{8}{15} = 0.5\overline{3}$

(c)
$$
\begin{array}{r}
0.29545 \\
44\overline{)13.000000} \\
\underline{88} \\
420 \\
\underline{396} \\
240 \\
\underline{220} \\
200 \\
\underline{176} \\
240 \\
\underline{220} \\
20
\end{array}
$$

$\dfrac{13}{44} = 0.29\overline{54}$

3. (a) $2\dfrac{11}{18} = 2 + \dfrac{11}{18}$

$$
\begin{array}{r}
0.611 = 0.6\overline{1} \\
18\overline{)11.000} \\
\underline{108} \\
20 \\
\underline{18} \\
20 \\
\underline{18} \\
2
\end{array}
$$

$2\dfrac{11}{18} = 2.6\overline{1}$

(b)
$$
\begin{array}{r}
1.03703 \\
27\overline{)28.00000} \\
\underline{27} \\
1\ 00 \\
\underline{81} \\
190 \\
\underline{189} \\
100 \\
\underline{81} \\
19
\end{array}
$$

$\dfrac{28}{27} = 1.\overline{037}$

4.
$$
\begin{array}{r}
0.7916 \\
24\overline{)19.0000} \\
\underline{168} \\
220 \\
\underline{216} \\
40 \\
\underline{24} \\
160 \\
\underline{144} \\
16
\end{array}
$$

$\dfrac{19}{24} = 0.792$ rounded to the nearest thousandth.

5. Divide to find the decimal equivalent of $\dfrac{5}{8}$.

$$
\begin{array}{r}
0.625 \\
8\overline{)5.000} \\
\underline{48} \\
20 \\
\underline{16} \\
40 \\
\underline{40} \\
0
\end{array}
$$

In the hundredths place $2 < 3$, so we know
$$0.6\underline{2}5 < 0.6\underline{3}0.$$

Therefore, $\dfrac{5}{8} < 0.63$.

6. $0.3 \times 0.5 + (0.4)^3 - 0.036 = 0.3 \times 0.5 + 0.064 - 0.036$
$$
\begin{aligned}
&= 0.15 + 0.064 - 0.036 \\
&= 0.214 - 0.036 \\
&= 0.178
\end{aligned}
$$

7. $6.56 \div (2 - 0.36) + (8.5 - 8.3)^2$
$$
\begin{aligned}
&= 6.56 \div (1.64) + (0.2)^2 &&\text{Parentheses} \\
&= 6.56 \div 1.64 + 0.04 &&\text{Exponents} \\
&= 4 + 0.04 &&\text{Divide} \\
&= 4.04 &&\text{Add}
\end{aligned}
$$

3.7 Practice Problems

Practice Problem 1

(a) $385.98 + 875.34 \approx 400 + 900 = 1300$

(b) $0.0932 - 0.0579 \approx 0.09 - 0.06 = 0.03$

(c) $5876.34 \times 0.087 \approx$
$$
\begin{array}{r}
6000 \\
\times\ 0.09 \\
\hline
540.00
\end{array}
$$

(d)
$$
46{,}873 \div 8.456 \approx
\begin{array}{r}
6250 \\
8\overline{)50{,}000} \\
\underline{48} \\
2\ 0 \\
\underline{1\ 6} \\
40 \\
\underline{40} \\
0
\end{array}
$$

Practice Problem 2

1. Understand the problem.

Mathematics Blueprint for Problem Solving

Gather the Facts	What Am I Asked to Do?	How Do I Proceed?	Key Points to Remember
She worked 51 hours. She gets paid $9.36 per hour for 40 hours. She gets paid time-and-a-half for 11 hours.	Find the amount Melinda earned working 51 hours last week.	Add the earnings of 40 hours at $9.36 per hour to the earnings of 11 hours at overtime pay.	Overtime pay is time-and-a-half, which is $1.5 \times \$9.36$.

2. Solve and state the answer:

(a) Calculate regular earnings for 40 hours.

$$\begin{array}{r} \$9.36 \\ \times \quad 40 \\ \hline \$374.40 \end{array}$$

(b) Calculate overtime pay rate.

$$\begin{array}{r} \$9.36 \\ \times \quad 1.5 \\ \hline 4680 \\ 936 \quad \\ \hline \$14.040 \end{array}$$

(c) Calculate overtime earnings for 11 hours.

$$\begin{array}{r} \$14.04 \\ \times \quad 11 \\ \hline 1404 \\ 1404 \quad \\ \hline \$154.44 \end{array}$$

(d) Add the two amounts.

$$\begin{array}{r} 1 \quad\quad \\ \$374.40 \quad \text{Regular earnings} \\ + \quad 154.44 \quad \text{Overtime earnings} \\ \hline \$528.84 \quad \text{Total earnings} \end{array}$$

Melinda earned $528.84 last week.

3. Check. Regular pay: $40 \times \$9 = \$360$
Overtime pay: $2 \times \$9 = \$18$
$10 \times \$20 = \$200$

$$\begin{array}{r} \$360 \\ + \ 200 \\ \hline \$560 \end{array}$$ The answer is reasonable.

Practice Problem 3

1. Understand the problem.

Mathematics Blueprint for Problem Solving

Gather the Facts	What Am I Asked to Do?	How Do I Proceed?	Key Points to Remember
The total amount of steak is 17.4 pounds. Each package contains 1.45 pounds. Prime steak costs $4.60 per pound.	**(a)** Find out how many packages of steak the butcher will have. **(b)** Find the cost of each package.	**(a)** Divide the total, 17.4, by the amount in each package, 1.45, to find the number of packages. **(b)** Multiply the cost of one pound, $4.60, by the amount in one package, 1.45.	There will be two answers to this problem.

2. Solve and state the answer.

(a)
$$\begin{array}{r} 12. \quad\quad\;\; \\ 1.45_\wedge \overline{)17.40_\wedge} \\ \underline{14\ 5} \quad\;\; \\ 2\ 90 \\ \underline{2\ 90} \\ 0 \end{array}$$
The butcher will have 12 packages of steak.

(b)
$$\begin{array}{r} \$4.60 \\ \times \ 1.45 \\ \hline 2300 \\ 1840 \\ 460 \quad \\ \hline \$6.6700 \end{array}$$
Each package will cost $6.67.

3. Check.

(a)
$$\begin{array}{r} 1.45 \\ \times \ 12 \\ \hline 290 \\ 145 \quad \\ \hline 17.40 \end{array}$$

(b) $\$5 \times 1 = \$5$

The answers are reasonable.

Chapter 4 4.1 Practice Problems

1. (a) $\dfrac{36}{40} = \dfrac{9}{10}$ **(b)** $\dfrac{18}{15} = \dfrac{6}{5}$ **(c)** $\dfrac{220}{270} = \dfrac{22}{27}$

2. (a) $\dfrac{200}{450} = \dfrac{4}{9}$

(b) The total number of students surveyed is
$200 + 450 + 300 + 150 + 100 = 1200.$ $\dfrac{300}{1200} = \dfrac{1}{4}$

3. $\dfrac{44 \text{ dollars}}{900 \text{ tons}} = \dfrac{11 \text{ dollars}}{225 \text{ tons}}$

4. $\dfrac{212 \text{ miles}}{4 \text{ hours}} = \dfrac{53 \text{ miles}}{1 \text{ hour}}$ 53 miles/hour

5.
selling price	$170.40
− purchase price	− 129.60
profit	$ 40.80

She made a profit of $40.80 on 120 batteries.

$$120\overline{)40.80} \quad \begin{array}{r} 0.34 \\ \hline \end{array}$$

$$\begin{array}{r} 0.34 \\ 120\overline{)40.80} \\ \underline{360} \\ 480 \\ \underline{480} \\ 0 \end{array}$$

Her profit was $0.34 per battery.

6. (a) $\dfrac{\$2.04}{12 \text{ ounces}} = \$0.17/\text{ounce}$ $\dfrac{\$2.80}{20 \text{ ounces}} = \$0.14/\text{ounce}$

(b) Fred saves $0.03/ounce by buying the larger size.

4.2 Practice Problems

1. 6 is to 8 as 9 is to 12.
$$\dfrac{6}{8} = \dfrac{9}{12}$$

2. $\dfrac{2 \text{ hours}}{72 \text{ miles}} = \dfrac{3 \text{ hours}}{108 \text{ miles}}$

3. (a) $\dfrac{10}{18} \overset{?}{=} \dfrac{25}{45}$

$$18 \times 25 = 450$$

$\dfrac{10}{18} \bowtie \dfrac{25}{45}$ The cross products are equal.

$$10 \times 45 = 450$$

Thus $\dfrac{10}{18} = \dfrac{25}{45}$. This is a proportion.

(b) $\dfrac{42}{100} \overset{?}{=} \dfrac{22}{55}$

$$100 \times 22 = 2200$$

$\dfrac{42}{100} \bowtie \dfrac{22}{55}$ The cross products are not equal.

$$42 \times 55 = 2310$$

Thus $\dfrac{42}{100} \neq \dfrac{22}{55}$. This is not a proportion.

4. (a) $\dfrac{2.4}{3} \overset{?}{=} \dfrac{12}{15}$

$$3 \times 12 = 36$$

$\dfrac{2.4}{3} \bowtie \dfrac{12}{15}$ The cross products are equal.

$$2.4 \times 15 = 36$$

Thus $\dfrac{2.4}{3} = \dfrac{12}{15}$. This is a proportion.

(b) $\dfrac{2\frac{1}{3}}{6} \overset{?}{=} \dfrac{14}{38}$

$$2\dfrac{1}{3} \times 38 = \dfrac{7}{3} \times \dfrac{38}{1} = \dfrac{266}{3} = 88\dfrac{2}{3}$$

$\dfrac{2\frac{1}{3}}{6} \bowtie \dfrac{14}{38}$

$$6 \times 14 = 84$$

The cross products are not equal.

$$2\dfrac{1}{3} \times 38 = 88\dfrac{2}{3}$$

Thus $\dfrac{2\frac{1}{3}}{6} \neq \dfrac{14}{38}$. This is not a proportion.

5. (a) $\dfrac{1260}{7} \overset{?}{=} \dfrac{3530}{20}$

$$7 \times 3530 = 24{,}710$$

$\dfrac{1260}{7} \bowtie \dfrac{3530}{20}$ The cross products are not equal.

$$1260 \times 20 = 25{,}200$$

The rates are not equal. This is not a proportion.

(b) $\dfrac{2}{11} \overset{?}{=} \dfrac{16}{88}$

$$11 \times 16 = 176$$

$\dfrac{2}{11} \bowtie \dfrac{16}{88}$ The cross products are equal.

$$2 \times 88 = 176$$

The rates are equal. This is a proportion.

4.3 Practice Problems

1. (a) $5 \times n = 45$
$$\dfrac{5 \times n}{5} = \dfrac{45}{5}$$
$$n = 9$$

(b) $7 \times n = 84$
$$\dfrac{7 \times n}{7} = \dfrac{84}{7}$$
$$n = 12$$

2. (a) $108 = 9 \times n$
$$\dfrac{108}{9} = \dfrac{9 \times n}{9}$$
$$12 = n$$

(b) $210 = 14 \times n$
$$\dfrac{210}{14} = \dfrac{14 \times n}{14}$$
$$15 = n$$

3. (a) $15 \times n = 63$
$$\dfrac{15 \times n}{15} = \dfrac{63}{15}$$
$$n = 4.2$$

$$\begin{array}{r} 4.2 \\ 15\overline{)63.0} \\ \underline{60} \\ 3\,0 \\ \underline{3\,0} \\ 0 \end{array}$$

(b) $39.2 = 5.6 \times n$
$$\dfrac{39.2}{5.6} = \dfrac{5.6 \times n}{5.6}$$
$$7 = n$$

$$\begin{array}{r} 7. \\ 5.6_\wedge\overline{)39.2_\wedge} \\ \underline{39\,2} \\ 0 \end{array}$$

4. $\dfrac{24}{n} = \dfrac{3}{7}$
$$24 \times 7 = n \times 3$$
$$168 = n \times 3$$
$$\dfrac{168}{3} = \dfrac{n \times 3}{3}$$
$$56 = n$$

5. $\dfrac{176}{4} = \dfrac{286}{n}$

$\quad\quad\quad\quad\quad\quad\quad\quad\quad$ 176$\overline{)1144.0}$

$176 \times n = 286 \times 4$ $\quad\quad\quad\quad$ $\underline{1056}$

$176 \times n = 1144$ $\quad\quad\quad\quad\quad\quad$ 88 0

$\dfrac{176 \times n}{176} = \dfrac{1144}{176}$ $\quad\quad\quad\quad$ $\underline{88\ 0}$

$\quad\quad\quad\quad\quad\quad\quad\quad\quad\quad\quad$ 0

$\quad\quad n = 6.5$

6. $\dfrac{n}{30} = \dfrac{\frac{2}{3}}{4}$

$4 \times n = 30 \times \dfrac{2}{3}$

$4 \times n = 20$

$\dfrac{4 \times n}{4} = \dfrac{20}{4}$

$\quad\quad n = 5$

7. $\dfrac{n \text{ tablespoons}}{24 \text{ gallons}} = \dfrac{2.5 \text{ tablespoons}}{3 \text{ gallons}}$

$3 \times n = 24 \times 2.5$

$3 \times n = 60$

$\dfrac{3 \times n}{3} = \dfrac{60}{3}$

$\quad\quad n = 20$

The answer is 20.

8. $264 \times 2 = 3.5 \times n$

$\quad\quad\ 528 = 3.5 \times n$

$\quad\ \dfrac{528}{3.5} = \dfrac{3.5 \times n}{3.5}$

$150.9 \approx n$

The answer to the nearest tenth is 150.9.

4.4 Practice Problems

1. $\dfrac{27 \text{ defective engines}}{243 \text{ engines produced}} = \dfrac{n \text{ defective engines}}{4131 \text{ engines produced}}$

$27 \times 4131 = 243 \times n$

$111{,}537 = 243 \times n$

$\dfrac{111{,}537}{243} = \dfrac{243 \times n}{243}$

$459 = n$

Thus we estimate that 459 engines are defective.

2. $\dfrac{9 \text{ gallons of gas}}{234 \text{ miles traveled}} = \dfrac{n \text{ gallons of gas}}{312 \text{ miles traveled}}$

$9 \times 312 = 234 \times n$

$2808 = 234 \times n$

$\dfrac{2808}{234} = \dfrac{234 \times n}{234}$

$12 = n$

She will need 12 gallons of gas.

3. $\dfrac{80 \text{ revolutions per minute}}{16 \text{ miles per hour}} = \dfrac{90 \text{ revolutions per minute}}{n \text{ miles per hour}}$

$80 \times n = 16 \times 90$

$80 \times n = 1440$

$\dfrac{80 \times n}{80} = \dfrac{1440}{80}$

$n = 18$

Alicia will be riding 18 miles per hour.

4. $\dfrac{4050 \text{ walk in}}{729 \text{ purchase}} = \dfrac{5500 \text{ walk in}}{n \text{ purchase}}$

$4050 \times n = 729 \times 5500$

$4050 \times n = 4{,}009{,}500$

$\dfrac{4050 \times n}{4050} = \dfrac{4{,}009{,}500}{4050}$

$n = 990$

Tom will expect 990 people to make a purchase in his store.

5. $\dfrac{50 \text{ bears tagged in 1st sample}}{n \text{ bears in forest}} = \dfrac{4 \text{ bears tagged in 2nd sample}}{50 \text{ bears caught in 2nd sample}}$

$50 \times 50 = n \times 4$

$2500 = n \times 4$

$\dfrac{2500}{4} = \dfrac{n \times 4}{4}$

$625 = n$

We estimate that there are 625 bears in the forest.

Chapter 5 5.1 Practice Problems

1. (a) $\dfrac{51}{100} = 51\%$ $\quad\quad\quad$ **(b)** $\dfrac{68}{100} = 68\%$

(c) $\dfrac{7}{100} = 7\%$ $\quad\quad\quad$ **(d)** $\dfrac{26}{100} = 26\%$

2. (a) $\dfrac{238}{100} = 238\%$ $\quad\quad$ **(b)** $\dfrac{121}{100} = 121\%$

3. (a) $\dfrac{0.5}{100} = 0.5\%$ $\quad\quad$ **(b)** $\dfrac{0.06}{100} = 0.06\%$

(c) $\dfrac{0.003}{100} = 0.003\%$

4. (a) $47\% = \dfrac{47}{100} = 0.47$ $\quad$ **(b)** $2\% = \dfrac{2}{100} = 0.02$

5. (a) $80.6\% = 0.806$ $\quad\quad$ **(b)** $2.5\% = 0.025$

(c) $0.29\% = 0.0029$ $\quad\quad$ **(d)** $231\% = 2.31$

6. (a) $0.78 = 78\%$ $\quad\quad\quad$ **(b)** $0.02 = 2\%$

(c) $5.07 = 507\%$ $\quad\quad\quad$ **(d)** $0.029 = 2.9\%$

(e) $0.006 = 0.6\%$

5.2 Practice Problems

1. (a) $71\% = \dfrac{71}{100}$ $\quad\quad\quad$ **(b)** $25\% = \dfrac{25}{100} = \dfrac{1}{4}$

(c) $8\% = \dfrac{8}{100} = \dfrac{2}{25}$

2. (a) $8.4\% = 0.084 = \dfrac{84}{1000} = \dfrac{21}{250}$

(b) $28.5\% = 0.285 = \dfrac{285}{1000} = \dfrac{57}{200}$

3. (a) $170\% = 1.70 = 1\dfrac{7}{10}$ $\quad$ **(b)** $288\% = 2.88 = 2\dfrac{88}{100} = 2\dfrac{22}{25}$

4. $7\dfrac{5}{8}\% = 7\dfrac{5}{8} \div 100$

$= \dfrac{61}{8} \times \dfrac{1}{100}$

$= \dfrac{61}{800}$

5. $20\dfrac{7}{8}\% = 20\dfrac{7}{8} \div 100$

$= 20\dfrac{7}{8} \times \dfrac{1}{100}$

$= \dfrac{167}{8} \times \dfrac{1}{100}$

$= \dfrac{167}{800}$

6. $\dfrac{5}{8}$ $\quad$ 8$\overline{)5.000}$ $\ \dfrac{0.625}{}$ $\quad$ 62.5%

7. (a) $\dfrac{21}{25} = 0.84 = 84\%$ $\quad$ **(b)** $\dfrac{7}{16} = 0.4375 = 43.75\%$

8. (a) $\dfrac{7}{9} = 0.77777\overline{7} \approx 0.7778 = 77.78\%$

(b) $\dfrac{19}{30} = 0.63333\overline{3} \approx 0.6333 = 63.33\%$

9. $\frac{7}{12}$ If we divide

$$12\overline{)7.00} \quad \begin{array}{r} 0.58 \\ \hline \end{array}$$

$$\frac{60}{100}$$ Thus $\frac{7}{12} = 0.58\frac{4}{12} = 58\frac{1}{3}\%.$

$$\frac{96}{4}$$

10.

Fraction	Decimal	Percent
$\frac{23}{99}$	0.2323	23.23%
$\frac{129}{250}$	0.516	51.6%
$\frac{97}{250}$	0.388	$38\frac{4}{5}\%$

5.3A Practice Problems

1. What is 26% of 35?

$$n \ = 26\% \times 35$$

2. Find 0.08% of 350.

$$n = 0.08\% \times 350$$

3. (a) 58% of what is 400?

$$58\% \times n = 400$$

(b) 9.1 is 135% of what?

$$9.1 = 135\% \times n$$

4. What percent of 250 is 36?

$$n \times 250 = 36$$

5. (a) 50 is what percent of 20?

$$50 = n \times 20$$

(b) What percent of 2000 is 4.5?

$$n \times 2000 = 4.5$$

6. What is 82% of 350?

$$n = 82\% \times 350$$
$$n = 0.82(350)$$
$$n = 287$$

7. $n = 230\% \times 400$
$n = (2.30)(400)$
$n = 920$

8. The problem asks: What is 8% of \$350?
$n = 8\% \times \$350$
$n = \$28$
The tax was \$28.

9. $32 = 0.4\% \times n$
$32 = 0.004n$
$\dfrac{32}{0.004} = \dfrac{0.004n}{0.004}$
$8000 = n$

10. The problem asks: 30% of what is 6?
$30\% \times n = 6$
$0.30n = 6$
$\dfrac{0.30n}{0.30} = \dfrac{6}{0.30}$
$n = 20$
There are 20 people on the team.

11. What percent of 9000 is 4.5?

$$n \quad \times 9000 = 4.5$$

$$9000n = 4.5$$
$$\frac{9000n}{9000} = \frac{4.5}{9000}$$
$$n = 0.0005$$
$$n = 0.05\%$$

12. $198 = n \times 33$
$\dfrac{198}{33} = \dfrac{33n}{33}$
$6 = n$
Now express n as a percent: 600%

13. The problem asks: 5 is what percent of 16?
$5 = n \times 16$
$\dfrac{5}{16} = \dfrac{16n}{16}$
$0.3125 = n$
Now express n as a percent rounded to the nearest tenth: 31.3%

5.3B Practice Problems

1. (a) Find 83% of 460.
   $p = 83$

(b) 18% of what number is 90?
   $p = 18$

(c) What percent of 64 is 8?
   The percent is unknown. Use the variable p.

2. (a) 30% of 52 is 15.6
   $b = 52, a = 15.6$

(b) 170 is 85% of what? Base $= b, a = 170$

3. (a) What is 18% of 240?
   Percent $p = 18$
   Base $b = 240$
   Amount is unknown; use the variable a.

(b) What percent of 64 is 4?
   Percent is unknown; use the variable p.
   Base $b = 64$
   Amount $a = 4$

4. Find 340% of 70.
   Percent $p = 340$
   Base $b = 70$
   Amount is unknown; use amount $= a$.
   $\dfrac{a}{b} = \dfrac{p}{100}$ becomes $\dfrac{a}{70} = \dfrac{340}{100}$
   $\dfrac{a}{70} = \dfrac{17}{5}$
   $5a = (70)(17)$
   $5a = 1190$
   $\dfrac{5a}{5} = \dfrac{1190}{5}$
   $a = 238$
   Thus 340% of 70 is 238.

5. 68% of what is 476?
   Percent $p = 68$
   Base is unknown; use base $= b$.
   Amount $a = 476$
   $\dfrac{a}{b} = \dfrac{p}{100}$ becomes $\dfrac{476}{b} = \dfrac{68}{100}$
   $\dfrac{476}{b} = \dfrac{17}{25}$
   $(476)(25) = 17b$
   $11,900 = 17b$
   $\dfrac{11,900}{17} = \dfrac{17b}{17}$
   $700 = b$
   Thus 68% of 700 is 476.

6. 216 is 0.3% of what?

Percent $p = 0.3$

Base is unknown; use base $= b$.

Amount $a = 216$

$\dfrac{a}{b} = \dfrac{p}{100}$ becomes $\dfrac{216}{b} = \dfrac{0.3}{100}$

$(216)(100) = 0.3b$

$21{,}600 = 0.3b$

$\dfrac{21{,}600}{0.3} = \dfrac{0.3b}{0.3}$

$72{,}000 = b$

Thus $72,000 was exchanged.

7. What percent of 3500 is 105?

Percent is unknown; use percent $= p$.

Base $b = 3500$

Amount $a = 105$

$\dfrac{a}{b} = \dfrac{p}{100}$ becomes $\dfrac{105}{3500} = \dfrac{p}{100}$

$\dfrac{3}{100} = \dfrac{p}{100}$

$300 = 100p$

$\dfrac{300}{100} = \dfrac{100p}{100}$

$3 = p$

Thus 3% of 3500 is 105.

5.4 Practice Problems

1. Method A Let $n =$ number of people with reserved airline tickets.

12% of $n = 4800$

$0.12 \times n = 4800$

$\dfrac{0.12 \times n}{0.12} = \dfrac{4800}{0.12}$

$n = 40{,}000$

Method B The percent $p = 12$. Use b for the unknown base. The amount $a = 4800$.

$\dfrac{a}{b} = \dfrac{p}{100}$ becomes $\dfrac{4800}{b} = \dfrac{12}{100}$.

$(4800)(100) = 12b$

$480{,}000 = 12b$

$\dfrac{480{,}000}{12} = b$

$40{,}000 = b$

40,000 people held airline tickets that month.

2. Method A The problem asks: What is 8% of $62.30?

$n = 0.08 \times 62.30$

$n = 4.984$

Method B The percent $p = 8$. The base $b = 62.30$. Use a for the unknown amount.

$\dfrac{a}{b} = \dfrac{p}{100}$ becomes $\dfrac{a}{62.30} = \dfrac{8}{100}$.

$\dfrac{a}{62.3} = \dfrac{2}{25}$

$25a = (2)(62.3)$

$25a = 124.60$

$\dfrac{25a}{25} = \dfrac{124.60}{25}$

$a = 4.984$

The tax is $4.98.

3. Method A The problem asks: 105 is what percent of 130?

$105 = n \times 130$

$\dfrac{105}{130} = n$

$0.8077 \approx n$

Method B Use p for the unknown percent. The base $b = 130$. The amount $a = 105$.

$\dfrac{a}{b} = \dfrac{p}{100}$ becomes $\dfrac{105}{130} = \dfrac{p}{100}$.

$\dfrac{21}{26} = \dfrac{p}{100}$

$(21)(100) = 26p$

$2100 = 26p$

$\dfrac{2100}{26} = \dfrac{26p}{26}$

$80.769230\ldots = p$

Thus 80.8% of the flights were on time.

4. 100% Cost of meal + tip of 15% $= \$46.00$

Let $n =$ Cost of meal

100% of n + 15% of $n = \$46.00$

115% of $n = 46.00$

$1.15 \times n = 46.00$

$\dfrac{1.15 \times n}{1.15} = \dfrac{46.00}{1.15}$

$n = 40.00$

They can spend $40.00 on the meal itself.

5. (a) 7% of $13,600 is the discount.

$0.07 \times 13{,}600 =$ the discount

$952 is the discount.

(b) $13,600   list price

$\underline{-\quad 952}$ discount

$12,648   Amount Betty paid for the car.

5.5 Practice Problems

1. Commission $=$ commission rate $\times$ value of sales

Commission $= 6\% \times \$156{,}000$

$= 0.06 \times 156{,}000$

$= 9360$

His commission is $9360.

2.   15,000

$\underline{-10{,}500}$

4500   the amount of decrease

Percent of decrease $= \dfrac{\text{amount of decrease}}{\text{original amount}} = \dfrac{4500}{15{,}000}$

$= 0.30 = 30\%$

The percent of decrease is 30%.

3. $I = P \times R \times T$

$P = \$5600 \qquad R = 12\% \qquad T = 1 \text{ year}$

$I = 5600 \times 12\% \times 1$

$= 5600 \times 0.12$

$= 672$

The interest is $672.

4. (a) $I = P \times R \times T$

$= 1800 \times 0.11 \times 4$

$= 198 \times 4$

$= 792$

The interest for four years is $792.

(b) $I = P \times R \times T$

$= 1800 \times 0.11 \times \dfrac{1}{2}$

$= 198 \times \dfrac{1}{2}$

$= 99$

The interest for six months is $99.

Appendix A.1 Balancing a Checking Account

Practice Problems

1.

CHECK REGISTER	*My Chung Nguyen*								20 *09*	
CHECK NO.	DATE	DESCRIPTION OF TRANSACTION	PAYMENT/ DEBIT (−)		✓	DEPOSIT/ CREDIT (+)		BALANCE $ *1434*	*52*
144	*3/1*	*Leland Mortgage Company*	*908*	*00*				*526*	*52*
145	*3/1*	*Phone Company*	*33*	*21*				*493*	*31*
146	*3/2*	*Sam's Food Market*	*102*	*37*				*390*	*94*
	3/2	*Deposit*				*524*	*41*	*915*	*35*

To find the ending balance we subtract each check written and add the deposit to the current balance. Then we record these amounts in the check register.

$$
\begin{array}{cccc}
1434.52 & 526.52 & 493.31 & 390.94 \\
-\ 908.00 & -\ 33.21 & -\ 102.37 & +\ 524.41 \\
\hline
526.52 & 493.31 & 390.94 & 915.35
\end{array}
$$

My Chung's balance is $915.35

2.

CHECKING RECONCILEMENT	This form is provided to assist you in balancing your checking account.

List checks outstanding* not charged to your checking account		Period ending *8/1* ,20 *09*

CHECK NO.	AMOUNT
215	*89 75*
217	*205 99*
TOTAL	*295 74*

* and ATM withdrawals

1. Check Register Balance — $ *1050.46*
   Subtract any charges listed on the bank statement which you have not previously deducted from your balance. − $ *12.75*
   Adjusted Check Register Balance $ *1037.71*
2. **Enter** the ending balance shown on the bank statement. $ *934.95*
3. **Enter** deposits made later than the ending date on the bank statement. + $ *398.50*
   + $
   + $
   TOTAL (Step 2 plus Step 3) $ *1333.45*
4. In your check register, **check off** all the checks paid. In the area provided to the left, **list** numbers and amounts of all outstanding checks and ATM withdrawals.
5. **Subtract** the total amount in Step 4. − $ *295.74*
6. This adjusted bank balance should equal the adjusted Check Register Balance from Step 1. $ *1037.71*

The balances in steps **1** and **6** are equal, so Anthony's checkbook is balanced.

Appendix A.2 Determining the Best Value When Purchasing a Vehicle

Practice Problems

1. (a) sales tax = 7% of sale price
= 0.07 × 24,999
sales tax = $1749.93
license fee = 2% of sale price
= 0.02 × 24,999
license fee = $499.98

(b) purchase price = sales price + sales tax + license fee + extended warranty
= 24,999 + 1749.93 + 499.98 + 1275
purchase price = $28,523.91

2. (a) down payment = percent × purchase price.
down payment = 15% × 32,499
= 0.15 × 32,499
down payment = $4874.85

(b) amount financed = purchase price − down payment
= 32,499 − 4874.85
amount financed = $27,624.15

3. First, we find the total cost of the minivan at Dealer 1.
Total Cost = (monthly payment × number of months in loan) + down payment
= (453.21 × 60) + 0   There is no down payment.
= 27,192.60
The total cost of the minivan at Dealership 1 is $27,192.60.

Next, we find the cost of the minivan at Dealer 2.
Total Cost = (monthly payment × number of months in loan) + down payment
= (479.17 × 48) + 3000   We multiply, then add.
= 26,000.16
The total cost of the minivan at Dealership 2 is $26,000.16.

We see that the best deal on the minivan Phoebe plans to buy is at Dealership 2.

Appendix C Practice Problems

1. 3792 $\boxed{+}$ 5896 $\boxed{=}$ 9688

2. 7930 $\boxed{-}$ 5096 $\boxed{=}$ 2834

3. 896 $\boxed{\times}$ 273 $\boxed{=}$ 244608

4. 2352 $\boxed{\div}$ 16 $\boxed{=}$ 147

5. 72.8 $\boxed{\times}$ 197 $\boxed{=}$ 14341.6

6. 52.98 $\boxed{+}$ 31.74 $\boxed{+}$ 40.37 $\boxed{+}$ 99.82 $\boxed{=}$ 224.91

7. 0.0618 $\boxed{\times}$ 19.22 $\boxed{-}$ 59.38 $\boxed{\div}$ 166.3 $\boxed{=}$ 0.830730456

8. 3.152 $\boxed{\times}$ $\boxed{(}$ 0.1628 $\boxed{+}$ 3.715 $\boxed{-}$ 4.985 $\boxed{)}$ $\boxed{=}$ −3.4898944

9. 0.5618 $\boxed{\times}$ 98.3 $\boxed{+/-}$ $\boxed{-}$ 76.31 $\boxed{\times}$ 2.98 $\boxed{+/-}$ $\boxed{=}$ 172.17886

10. 3.76 $\boxed{\text{EXP}}$ 15 $\boxed{\div}$ 7.76 $\boxed{\text{EXP}}$ 7 $\boxed{=}$ 48453608.25

11. 6.238 $\boxed{y^x}$ 6 $\boxed{=}$ 58921.28674

12. 132.56 $\boxed{x^2}$ 17572.1536

13. $\boxed{(}$ 0.0782 $\boxed{-}$ 0.0132 $\boxed{+}$ 0.1364 $\boxed{)}$ $\boxed{\sqrt{}}$ 0.448776113

Answers to Selected Exercises

Chapter 1

1.1 Exercises **1.** 6000 + 700 + 30 + 1   **3.** 100,000 + 8000 + 200 + 70 + 6
5. 20,000,000 + 3,000,000 + 700,000 + 60,000 + 1000 + 300 + 40 + 5   **7.** 100,000,000 + 3,000,000 + 200,000 + 60,000 + 700 + 60 + 8
9. 671   **11.** 9863   **13.** 40,885   **15.** 706,200   **17.** (a) 7 (b) 30,000   **19.** (a) 2 (b) 200,000   **21.** one hundred forty-two
23. nine thousand, three hundred four   **25.** thirty-six thousand, one hundred eighteen   **27.** one hundred five thousand, two hundred sixty-one
29. fourteen million, two hundred three thousand, three hundred twenty-six   **31.** four billion, three hundred two million, one hundred fifty-six
thousand, two hundred   **33.** 1561   **35.** 33,809   **37.** 100,079,826   **39.** one thousand, nine hundred sixty-five   **41.** 9 million or 9,000,000
43. 38 million or 38,000,000   **45.** 930,000   **47.** 52,566,000   **49.** (a) 5 (b) 2   **51.** (a) 2 (b) 1   **53.** 613,001,033,208,003   **55.** three
quintillion, six hundred eighty-two quadrillion, nine hundred sixty-eight trillion, nine billion, nine hundred thirty-one million, nine hundred sixty thousand, seven hundred forty-seven   **57.** You would obtain 2 E 20. This is 200,000,000,000,000,000,000 in standard form.

Quick Quiz 1.1 **1.** 70,000 + 3000 + 900 + 50 + 2   **2.** eight million, nine hundred thirty-two thousand, four hundred seventy-five
3. 964,257   **4.** See Student Solutions Manual

1.2 Exercises **1.** (a) You can change the order of the addends without changing the sum.   (b) You can group the addends in any way without changing the sum.

3.

+	3	5	4	8	0	6	7	2	9	1
2	5	7	6	10	2	8	9	4	11	3
7	10	12	11	15	7	13	14	9	16	8
5	8	10	9	13	5	11	12	7	14	6
3	6	8	7	11	3	9	10	5	12	4
0	3	5	4	8	0	6	7	2	9	1
4	7	9	8	12	4	10	11	6	13	5
1	4	6	5	9	1	7	8	3	10	2
8	11	13	12	16	8	14	15	10	17	9
6	9	11	10	14	6	12	13	8	15	7
9	12	14	13	17	9	15	16	11	18	10

5. 23   **7.** 26   **9.** 57   **11.** 99   **13.** 4125   **15.** 9994   **17.** 13,861   **19.** 117,240   **21.** 121   **23.** 1143   **25.** 10,130
27. 11,579,426   **29.** 1,135,280,240   **31.** 2,303,820   **33.** 300   **35.** 335   **37.** $723   **39.** $5549   **41.** 468 feet   **43.** 121,100,000 square
miles   **45.** 16,934,720 yards   **47.** (a) 1134 students (b) 1392 students   **49.** 202 miles   **51.** 434 feet   **53.** (a) $9553 (b) $7319
(c) $13,047   **55.** 1161   **57.** Answers may vary. A sample is: You could not group the addends in groups that sum to 10s to make column addition
easier.   **58.** seventy-six million, two hundred eight thousand, nine hundred forty-one.   **59.** one hundred twenty-one million, three hundred
seventy-four   **60.** 8,724,396   **61.** 9,051,719   **62.** 28,387,018

Quick Quiz 1.2 **1.** 212   **2.** 1615   **3.** 1,004,811   **4.** See Student Solutions Manual

1.3 Exercises **1.** In subtraction the minuend minus the subtrahead equals the difference. To check the problem we add the subtrahead and the difference to see if we get the minuend. If we do, the answer is correct.   **3.** We know that 1683 + 1592 = 32?5. Therefore if we add 8 tens and 9 tens we get 17 tens, which is 1 hundred and 7 tens. Thus the ? should be replaced by 7.   **5.** 5   **7.** 6   **9.** 16   **11.** 9   **13.** 7   **15.** 6   **17.** 3   **19.** 9

21.
21
+ 21
———
47

23.
12   73
+ 12
———
85

25.
343   36
+ 343
———
379

27.
321   548
+ 321
———
869

29.
4203   596
+ 4203
———
4799

31.
143,235   12,600
+ 143,235
————
155,835

33.
553,101   433,201
+ 553,101
————
986,302

35.
19
+ 110
———
129
Correct

37.
3215
+ 5781
———
8996
Incorrect
Correct answer: 5381

39.
5020
+ 1020
———
6040
Incorrect
Correct answer: 1010

41.
33,846
+ 13,023
————
46,869
Incorrect
Correct answer: 14,023

43. 46   **45.** 92   **47.** 384

49. 718   **51.** 10,715   **53.** 34,092   **55.** 7447   **57.** 908,930   **59.** $x = 5$ **61.** $x = 8$ **63.** $x = 27$ **65.** 54,892 votes
67. 6,543,635   **69.** $762   **71.** 1,416,920 people   **73.** 2,004,796 people   **75.** 320,317 people   **77.** 93,154 people   **79.** 93 homes
81. 13 homes   **83.** between 2006 and 2007   **85.** Willow Creek and Harvey   **87.** It is true if a and b represent the same number, for example,
if $a = 10$ and $b = 10$.   **89.** $550   **91.** 8,466,084   **92.** two hundred ninety-six thousand, three hundred eight   **93.** 218   **94.** 1,174,750

Quick Quiz 1.3 **1.** 4454   **2.** 222,933   **3.** 5,638,122   **4.** See Student Solutions Manual

1.4 Exercises
1. (a) You can change the order of the factors without changing the product.   **(b)** You can group the factors in any way without changing the product.

3.

×	6	2	3	8	0	5	7	9	12	4
5	30	10	15	40	0	25	35	45	60	20
7	42	14	21	56	0	35	49	63	84	28
1	6	2	3	8	0	5	7	9	12	4
0	0	0	0	0	0	0	0	0	0	0
6	36	12	18	48	0	30	42	54	72	24
2	12	4	6	16	0	10	14	18	24	8
3	18	6	9	24	0	15	21	27	36	12
8	48	16	24	64	0	40	56	72	96	32
4	24	8	12	32	0	20	28	36	48	16
9	54	18	27	72	0	45	63	81	108	36

5. 96   **7.** 70   **9.** 522   **11.** 693   **13.** 1932   **15.** 18,306   **17.** 36,609   **19.** 31,308   **21.** 100,208   **23.** 3,101,409   **25.** 1560
27. 2,715,800   **29.** 482,000   **31.** 372,560,000   **33.** 8460   **35.** 63,600   **37.** 56,000,000   **39.** 6168   **41.** 7884   **43.** 5696
45. 15,175   **47.** 20,672   **49.** 69,312   **51.** 148,567   **53.** 823,823   **55.** 1,881,810   **57.** 89,496   **59.** 217,980   **61.** 2,653,296
63. 720,000   **65.** 10,000   **67.** 90,600   **69.** 70   **71.** 308   **73.** 13,596   **75.** 1600   **77.** $x = 0$ **79.** 384 square feet
81. 195 square feet   **83.** $1200   **85.** $3192   **87.** 612 miles   **89.** $5040   **91.** $7,372,300,000   **93.** 198   **95.** 62   **97.** $x = 8$
99. $x = 9$ **101.** No, it would not always be true. In our number system $62 = 60 + 2$. But in Roman numerals IV ≠ I + V. The digit system in Roman numerals involves subtraction. Thus (XII) × (IV) ≠ (XII × I) + (XII × V).   **103.** 6756   **104.** 1249   **105.** $805   **106.** $86
107. 2137 people   **108.** $392,739,000,000

Quick Quiz 1.4   **1.** 174,930   **2.** 5056   **3.** 207,306   **4.** See Student Solutions Manual

1.5 Exercises
1. (a) When you divide a nonzero number by itself, the result is one.   **(b)** When you divide a number by 1, the result is that number.   **(c)** When you divide zero by a nonzero number, the result is zero.   **(d)** You cannot divide a number by zero. Division by zero is undefined.   **3.** 7   **5.** 3   **7.** 5   **9.** 4   **11.** 3   **13.** 6   **15.** 9   **17.** 9   **19.** 6   **21.** 9   **23.** 0   **25.** undefined   **27.** 0
29. 1   **31.** 4 R 5   **33.** 9 R 4   **35.** 25 R 3   **37.** 21 R 7   **39.** 32   **41.** 37   **43.** 322 R 1   **45.** 127 R 1   **47.** 563   **49.** 1122 R 1
51. 2056 R 2   **53.** 2562 R 3   **55.** 30 R 5   **57.** 5 R 7   **59.** 7   **61.** 418 R 8   **63.** 48 R 12   **65.** 845   **67.** 210 R 8   **69.** 14 R 2
71. 4 R 4   **73.** 125   **75.** 37   **77.** 61,693 runs per day   **79.** $288   **81.** $21,053   **83.** $245   **85.** 165 sandwiches
87. (a) 41,808 km   **(b)** 8192 km   **89.** a and b must represent the same number. For example, if $a = 12$, then $b = 12$.   **91.** 5400
92. 1,038,490   **93.** 406,195   **94.** 66,844

Quick Quiz 1.5   **1.** 467   **2.** 3287 R 3   **3.** 328   **4.** See Student Solutions Manual

How Am I Doing? Sections 1.1–1.5
1. seventy-eight million, three hundred ten thousand, four hundred thirty-six. (obj. 1.1.3)   **2.** $30,000 + 8000 + 200 + 40 + 7$ (obj. 1.1.1)
3. 5,064,122 (obj. 1.1.2)   **4.** 2,747,000 (obj. 1.1.4)   **5.** 2,802,000 (obj. 1.1.4)   **6.** 244 (obj. 1.2.4)   **7.** 50,570 (obj. 1.2.4)
8. 1,351,461 (obj. 1.2.4)   **9.** 3993 (obj. 1.3.3)   **10.** 76,311 (obj. 1.3.3)   **11.** 1,981,652 (obj. 1.3.3)   **12.** 108 (obj. 1.4.1)
13. 100,000 (obj. 1.4.4)   **14.** 18,606 (obj. 1.4.2)   **15.** 3740 (obj. 1.4.4)   **16.** 331,420 (obj. 1.4.4)   **17.** 10,605 (obj. 1.5.2)
18. 7376 R 1 (obj. 1.5.2)   **19.** 26 R 8 (obj. 1.5.3)   **20.** 139 (obj. 1.5.3)

1.6 Exercises
1. 5^3 means $5 \times 5 \times 5$. $5^3 = 125$.   **3.** base
5. To ensure consistency we
   1. perform operations inside parentheses
   2. simplify any expressions with exponents
   3. multiply or divide from left to right
   4. add or subtract from left to right
7. 6^4 **9.** 5^6 **11.** 9^4 **13.** 9^1 **15.** 16   **17.** 64   **19.** 36   **21.** 10,000   **23.** 1   **25.** 64   **27.** 243   **29.** 225   **31.** 343
33. 256   **35.** 1   **37.** 625   **39.** 1,000,000   **41.** 169   **43.** 9   **45.** 64   **47.** 10   **49.** 108   **51.** 520   **53.** $90 - 35 = 55$
55. $27 - 5 = 22$ **57.** $48 \div 8 + 4 = 6 + 4 = 10$ **59.** $3 \times 36 - 50 = 108 - 50 = 58$ **61.** $100 + 3 \times 5 = 100 + 15 = 115$
63. $20 \div 20 = 1$ **65.** $950 \div 5 = 190$ **67.** $60 - 17 = 43$ **69.** $9 + 16 \div 4 = 9 + 4 = 13$ **71.** $42 - 4 \div 4 = 42 - 1 = 41$
73. $100 - 9 \times 4 = 100 - 36 = 64$ **75.** $25 + 4 + 27 = 56$ **77.** $8 \times 3 \times 1 \div 2 = 24 \div 2 = 12$ **79.** $144 - 0 = 144$
81. $16 \times 6 \div 3 = 96 \div 3 = 32$ **83.** $60 - 40 + 10 = 20 + 10 = 30$ **85.** $3 + 9 \times 6 + 4 = 3 + 54 + 4 = 61$
87. $32 \div 2 \times 16 = 16 \times 16 = 256$ **89.** $9 \times 6 \div 9 + 4 \times 3 = 6 + 12 = 18$ **91.** $36 + 1 + 8 = 45$ **93.** $1200 - 8(3) \div 6 = 1200 - 4 = 1196$
95. $120 \div 40 - 1 = 3 - 1 = 2$ **97.** $4 + 10 - 1 = 13$ **99.** $5 \times 2 + (3)^3 + 2^0 = 10 + 27 + 1 = 38$ **101.** 86,164 seconds
103. (a) 3   **(b)** 2,000,000   **104.** 200,765,909   **105.** two hundred sixty-one million, seven hundred sixty-three thousand, two
106. 1460 feet of fencing will be needed. 120,000 square feet of grass must be planted

Quick Quiz 1.6   **1.** 12^5 **2.** 1296   **3.** 91   **4.** See Student Solutions Manual

1.7 Exercises
1. Locate the rounding place. If the digit to the right of the rounding place is 5 or greater than 5, round up. If the digit to the right of the rounding place is less than 5, round down.   **3.** 80   **5.** 70   **7.** 170   **9.** 7440   **11.** 2960   **13.** 200   **15.** 2800   **17.** 7700
19. 8000   **21.** 1000   **23.** 28,000   **25.** 800,000   **27.** 15,000,000 stars   **29. (a)** 2,400,000 **(b)** 2,358,000   **31. (a)** 3,700,000 square miles; 9,600,000 square kilometers   **(b)** 3,710,000 square miles; 9,600,000 square kilometers

33.
$$\begin{array}{r} 800 \\ 300 \\ +\ 200 \\ \hline 1300 \end{array}$$

35.
$$\begin{array}{r} 40 \\ 70 \\ 100 \\ +\ 20 \\ \hline 230 \end{array}$$

37.
$$\begin{array}{r} 200,000 \\ 50,000 \\ +\ 9,000 \\ \hline 259,000 \end{array}$$

39.
$$\begin{array}{r} 300,000 \\ -\ 70,000 \\ \hline 230,000 \end{array}$$

41.
$$\begin{array}{r} 800,000 \\ -\ 80,000 \\ \hline 720,000 \end{array}$$

43.
$$\begin{array}{r} 30,000,000 \\ -\ 20,000,000 \\ \hline 10,000,000 \end{array}$$

45.
$$\begin{array}{r} 50 \\ \times\ 60 \\ \hline 3000 \end{array}$$

47.
$$\begin{array}{r} 1000 \\ \times\ 8 \\ \hline 8000 \end{array}$$

49.
$$\begin{array}{r} 600,000 \\ \times\ 300 \\ \hline 180,000,000 \end{array}$$

51. $40\overline{)6000}$ = 150

53. $40\overline{)400,000}$ = 10,000

55. $800\overline{)4,000,000}$ = 5000

57. Incorrect
$$\begin{array}{r} 400 \\ 500 \\ 900 \\ +\ 200 \\ \hline 2000 \end{array}$$

59. Incorrect
$$\begin{array}{r} 100,000 \\ 50,000 \\ +\ 40,000 \\ \hline 190,000 \end{array}$$

61. Correct
$$\begin{array}{r} 300,000 \\ -\ 90,000 \\ \hline 210,000 \end{array}$$

63. Incorrect
$$\begin{array}{r} 80,000,000 \\ -\ 50,000,000 \\ \hline 30,000,000 \end{array}$$

65. Incorrect
$$\begin{array}{r} 400 \\ \times\ 30 \\ \hline 12,000 \end{array}$$

67. Correct
$$\begin{array}{r} 6000 \\ \times\ 70 \\ \hline 420,000 \end{array}$$

69. $40\overline{)80,000}$ = 2000 Correct

71. $400\overline{)200,000}$ = 500 Correct

73. 400 square feet   **75.** 11,000,000 people   **77.** 30,000 pizzas   **79.** 360,000 flights   **81.** 590,000 − 270,000 = 320,000 square miles
83. (a) 400,000 hours   **(b)** 20,000 days   **85.** 83   **86.** 27   **87.** 28   **88.** 66   **89.** 367,763   **90.** 87

Quick Quiz 1.7 **1.** 92,400   **2.** 2,340,000   **3.** 2,400,000,000   **4.** See Student Solutions Manual

1.8 Exercises 1. $8800   **3.** 1560 bagels   **5.** 7¢ per ounce   **7.** $64   **9.** 2,980,000 people   **11.** $20,382   **13.** 25,231; 466
15. 800,000 people   **17.** $192   **19.** $1360   **21.** $16,405   **23.** 25 miles per gallon   **25.** There are 54 oak trees, 108 maple trees, and 756
pine trees. In total there are 936 trees.   **27.** 118   **29.** 217   **31.** $14,734,000,000   **33.** $69,603,000,000   **35.** 343   **36.** 21   **37.** 4788
38. 258   **39.** 802   **40.** 23,285   **41.** 526,196,000   **42.** 3,400,603,025

Quick Quiz 1.8 **1.** $269   **2.** $858   **3.** $126   **4.** See Student Solutions Manual

Putting Your Skills to Work

1. $100, $300, $2000, $8000, $8000, $8000, $12,000   **2.** $365   **3.** 7 months

Chapter 1 Review Problems

1. eight hundred ninety-two   **2.** fifteen thousand, eight hundred two   **3.** one hundred nine thousand, two hundred seventy-six
4. four hundred twenty-three million, five hundred seventy-six thousand, fifty-five   **5.** 4000 + 300 + 60 + 4   **6.** 30,000 + 5000 + 400 + 10 + 4
7. 40,000,000 + 2,000,000 + 100,000 + 60,000 + 6000 + 30 + 7   **8.** 1,000,000 + 300,000 + 5000 + 100 + 20 + 8   **9.** 924   **10.** 5302
11. 1,328,828   **12.** 24,705,112   **13.** 115   **14.** 300   **15.** 400   **16.** 150   **17.** 400   **18.** 953   **19.** 1007   **20.** 60,100   **21.** 14,703
22. 10,582   **23.** 17   **24.** 6   **25.** 27   **26.** 171   **27.** 6155   **28.** 3167   **29.** 80,722   **30.** 105,818   **31.** 6,236,011   **32.** 5,332,991
33. 144   **34.** 0   **35.** 800   **36.** 1500   **37.** 62,100   **38.** 84,312,000   **39.** 780,000   **40.** 536,000,000   **41.** 1856   **42.** 1752
43. 4050   **44.** 13,680   **45.** 25,524   **46.** 24,096   **47.** 87,822   **48.** 268,513   **49.** 543,510   **50.** 255,068   **51.** 111,370   **52.** 113,946
53. 7,200,000   **54.** 7,500,000   **55.** 2,000,000,000   **56.** 12,000,000,000   **57.** 2   **58.** 5   **59.** 0   **60.** 12   **61.** 7   **62.** 0   **63.** 9
64. 7   **65.** undefined   **66.** 4   **67.** 7   **68.** 9   **69.** 125   **70.** 125   **71.** 207   **72.** 309   **73.** 2504   **74.** 3064   **75.** 36,958
76. 36,921   **77.** 15,046 R 3   **78.** 35,783 R 4   **79.** 7 R 21   **80.** 4 R 37   **81.** 31 R 15   **82.** 14 R 11   **83.** 38 R 30   **84.** 60 R 22
85. 195   **86.** 258   **87.** 54   **88.** 19   **89.** 13^2 **90.** 21^3 **91.** 8^5 **92.** 10^6 **93.** 64   **94.** 81   **95.** 128   **96.** 125   **97.** 49
98. 81   **99.** 216   **100.** 64   **101.** 8   **102.** 11   **103.** 22   **104.** 66   **105.** 22   **106.** 78   **107.** 17   **108.** 26   **109.** 86
110. 3360   **111.** 5900   **112.** 15,310   **113.** 42,640   **114.** 12,000   **115.** 23,000   **116.** 676,000   **117.** 202,000   **118.** 4,600,000
119. 10,000,000   **120.**
$$\begin{array}{r} 300 \\ 700 \\ 200 \\ +\ 200 \\ \hline 1400 \end{array}$$
121.
$$\begin{array}{r} 20,000 \\ 8000 \\ +\ 40,000 \\ \hline 68,000 \end{array}$$
122.
$$\begin{array}{r} 4,000,000 \\ -\ 3,000,000 \\ \hline 1,000,000 \end{array}$$
123.
$$\begin{array}{r} 30,000 \\ -\ 20,000 \\ \hline 10,000 \end{array}$$
124.
$$\begin{array}{r} 1000 \\ \times\ 6000 \\ \hline 6,000,000 \end{array}$$
125.
$$\begin{array}{r} 3,000,000 \\ \times\ 900 \\ \hline 2,700,000,000 \end{array}$$

126. $20\overline{)80,000}$ = 4000   **127.** $300\overline{)900,000}$ = 3000   **128.** 240 donut holes   **129.** 175 words   **130.** 7020 people   **131.** $59,470   **132.** 10,301 feet
133. $3348   **134.** $1356   **135.** $74   **136.** $278   **137.** 25 miles per gallon   **138.** $5041   **139.** $2031   **140.** 40,500,000 tons
141. 21,400,000 tons from 1990 to 1995   **142.** 93,400,000 tons   **143.** 2284   **144.** 7867   **145.** 11,088   **146.** 129   **147.** 29   **148.** $747
149. (a) 330 square feet   **(b)** 74 feet

How Am I Doing? Chapter 1 Test

1. forty-four million, seven thousand, six hundred thirty-five (obj. 1.1.3)   **2.** 20,000 + 6000 + 800 + 50 + 9 (obj. 1.1.1)   **3.** 3,581,076 (obj. 1.1.2)
4. 831 (obj. 1.2.4)   **5.** 1491 (obj. 1.2.4)   **6.** 318,977 (obj. 1.2.4)   **7.** 8067 (obj. 1.3.3)   **8.** 172,858 (obj. 1.3.3)   **9.** 5,225,768 (obj. 1.3.3)
10. 378 (obj. 1.4.1)   **11.** 4320 (obj. 1.4.4)   **12.** 192,992 (obj. 1.4.4)   **13.** 129,437 (obj. 1.4.2)   **14.** 3014 R 1 (obj. 1.5.2)   **15.** 2358 (obj. 1.5.2)
16. 352 (obj. 1.5.3)   **17.** 14^3 (obj. 1.6.1)   **18.** 64 (obj. 1.6.1)   **19.** 23 (obj. 1.6.2)   **20.** 50 (obj. 1.6.2)   **21.** 79 (obj. 1.6.2)
22. 94,800 (obj. 1.7.1)   **23.** 6,460,000 (obj. 1.7.1)   **24.** 5,300,000 (obj. 1.7.1)   **25.** 150,000,000,000 (obj. 1.7.2)   **26.** 16,000 (obj. 1.7.2)
27. $2148 (obj. 1.8.1)   **28.** 467 feet (obj. 1.8.1)   **29.** $127 (obj. 1.8.2)   **30.** $292 (obj. 1.8.2)   **31.** 748,000 square feet (obj. 1.8.2)
32. 46 feet (obj. 1.8.2)

Chapter 2

2.1 Exercises **1.** fraction **3.** denominator **5.** N: 3; D: 5 **7.** N: 7; D: 8 **9.** N: 1; D: 17 **11.** $\frac{1}{3}$ **13.** $\frac{7}{9}$

15. $\frac{3}{4}$ **17.** $\frac{3}{7}$ **19.** $\frac{2}{5}$ **21.** $\frac{7}{10}$ **23.** $\frac{5}{8}$ **25.** $\frac{4}{7}$ **27.** $\frac{7}{8}$ **29.** $\frac{9}{15}$ **31.** ▨☐☐☐ **33.** ▨▨☐☐☐☐☐☐

35. ▨▨▨▨▨▨☐☐☐ **37.** $\frac{42}{83}$ **39.** $\frac{209}{750}$ **41.** $\frac{89}{211}$ **43.** $\frac{9}{26}$ **45.** $\frac{24}{40}$ **47. (a)** $\frac{90}{195}$ **(b)** $\frac{22}{195}$

49. The amount of money each of six business owners gets if the business has a profit of $0. **51.** 241 **52.** 13,216 **53.** 146,188 **54.** 1258 R 4

Quick Quiz 2.1 **1.** $\frac{4}{7}$ **2.** $\frac{204}{371}$ **3.** $\frac{13}{33}$ **4.** See Student Solutions Manual

2.2 Exercises **1.** 11, 19, 41, 5 **3.** composite number **5.** $56 = 2 \times 2 \times 2 \times 7$ **7.** 3×5 **9.** 5×7 **11.** 7^2 **13.** 2^4
15. 5×11 **17.** $3^2 \times 7$ **19.** $2^2 \times 3 \times 7$ **21.** 2×3^3 **23.** $2^3 \times 3 \times 5$ **25.** $2^3 \times 23$ **27.** prime **29.** 3×19 **31.** prime
33. 2×31 **35.** prime **37.** prime **39.** 11×11 **41.** 3×43 **43.** $\frac{18 \div 9}{27 \div 9} = \frac{2}{3}$ **45.** $\frac{36 \div 12}{48 \div 12} = \frac{3}{4}$ **47.** $\frac{63 \div 9}{90 \div 9} = \frac{7}{10}$
49. $\frac{210 \div 10}{310 \div 10} = \frac{21}{31}$ **51.** $\frac{3 \times 1}{3 \times 5} = \frac{1}{5}$ **53.** $\frac{2 \times 3 \times 11}{2 \times 2 \times 2 \times 11} = \frac{3}{4}$ **55.** $\frac{2 \times 3 \times 5}{3 \times 3 \times 5} = \frac{2}{3}$ **57.** $\frac{2 \times 2 \times 3 \times 5}{3 \times 5 \times 5} = \frac{4}{5}$ **59.** $\frac{3 \times 11}{3 \times 12} = \frac{11}{12}$
61. $\frac{9 \times 7}{9 \times 12} = \frac{7}{12}$ **63.** $\frac{11 \times 8}{11 \times 11} = \frac{8}{11}$ **65.** $\frac{40 \times 3}{40 \times 5} = \frac{3}{5}$ **67.** $\frac{11 \times 20}{13 \times 20} = \frac{11}{13}$ **69.** $4 \times 28 \stackrel{?}{=} 16 \times 7$ **71.** no **73.** no **75.** yes
$112 = 112$
yes

77. yes **79.** $\frac{3}{4}$ **81.** $\frac{1}{8}$ failed; $\frac{7}{8}$ passed **83.** $\frac{5}{7}$ **85.** $\frac{17}{45}$ **87.** $\frac{8}{45}$ **89.** 164,050 **90.** 1296 **91.** 960,000 **92.** $571,600,000

Quick Quiz 2.2 **1.** $\frac{5}{7}$ **2.** $\frac{1}{6}$ **3.** $\frac{21}{8}$ **4.** See Student Solutions Manual

2.3 Exercises **1. (a)** Multiply the whole number by the denominator of the fraction. **(b)** Add the numerator of the fraction to the product
formed in step (a). **(c)** Write the sum found in step (b) over the denominator of the fraction. **3.** $\frac{7}{3}$ **5.** $\frac{17}{7}$ **7.** $\frac{83}{9}$ **9.** $\frac{32}{3}$ **11.** $\frac{58}{5}$
13. $\frac{55}{6}$ **15.** $\frac{121}{6}$ **17.** $\frac{131}{12}$ **19.** $\frac{79}{10}$ **21.** $\frac{201}{25}$ **23.** $\frac{65}{12}$ **25.** $\frac{494}{3}$ **27.** $\frac{131}{15}$ **29.** $\frac{113}{25}$ **31.** $1\frac{1}{3}$ **33.** $2\frac{3}{4}$ **35.** $2\frac{1}{2}$ **37.** $3\frac{3}{8}$
39. 25 **41.** $9\frac{5}{9}$ **43.** $23\frac{1}{3}$ **45.** $6\frac{1}{4}$ **47.** $5\frac{7}{10}$ **49.** $17\frac{1}{2}$ **51.** 13 **53.** 14 **55.** 6 **57.** $5\frac{15}{32}$ **59.** $5\frac{1}{2}$ **61.** $4\frac{1}{6}$ **63.** $15\frac{1}{4}$
65. 4 **67.** $\frac{12}{5}$ **69.** $\frac{15}{4}$ **71.** $2\frac{88}{126} = 2\frac{44}{63}$ **73.** $2\frac{20}{280} = 2\frac{1}{14}$ **75.** $1\frac{212}{296} = 1\frac{53}{74}$ **77.** $\frac{1082}{3}$ yards **79.** $50\frac{1}{3}$ acres
81. $141\frac{3}{8}$ pounds **83.** No, 101 is prime and is not a factor of 5687. **85.** 260,247 **86.** 16,000,000,000 **87.** 300 **88.** 37 full cartons are
needed. There are 5 books in the carton that is not full.

Quick Quiz 2.3 **1.** $\frac{59}{13}$ **2.** $7\frac{5}{12}$ **3.** 3 **4.** See Student Solutions Manual

2.4 Exercises **1.** $\frac{21}{55}$ **3.** $\frac{15}{52}$ **5.** 1 **7.** $\frac{1}{16}$ **9.** $\frac{12}{55}$ **11.** $\frac{21}{8}$ or $2\frac{5}{8}$ **13.** $\frac{24}{7}$ or $3\frac{3}{7}$ **15.** $\frac{10}{3}$ or $3\frac{1}{3}$ **17.** $\frac{1}{6}$ **19.** 3 **21.** $\frac{1}{2}$
23. 31 **25.** 0 **27.** $3\frac{7}{8}$ **29.** $\frac{55}{12}$ or $4\frac{7}{12}$ **31.** $\frac{69}{50}$ or $1\frac{19}{50}$ **33.** 35 **35.** $\frac{8}{5}$ or $1\frac{3}{5}$ **37.** $\frac{7}{9}$ **39.** $\frac{38}{3}$ or $12\frac{2}{3}$ **41.** $x = \frac{7}{9}$ **43.** $x = \frac{8}{9}$
45. $37\frac{11}{12}$ square miles **47.** 1560 miles **49.** 1629 grams **51.** 5332 students **53.** 377 companies **55.** $1\frac{8}{9}$ miles **57.** The step of
dividing the numerator and denominator by the same number allows us to work with smaller numbers when we do the multiplication. Also, this
allows us to avoid the step of having to simplify the fraction in the final answer. **59.** 529 cars **60.** 368 calls **61.** 5040 miles
62. 173,040 gallons

Quick Quiz 2.4 **1.** 10 **2.** $\frac{44}{65}$ **3.** $\frac{143}{12}$ or $11\frac{11}{12}$ **4.** See Student Solutions Manual

2.5 Exercises **1.** Think of a simple problem like $3 \div \frac{1}{2}$. One way to think of it is how many $\frac{1}{2}$'s can be placed in 3? For example, how many
$\frac{1}{2}$-pound rocks could be put in a bag that holds 3 pounds of rocks? The answer is 6. If we inverted the first fraction by mistake, we would have
$\frac{1}{3} \times \frac{1}{2} = \frac{1}{6}$. We know this is wrong since there are obviously several $\frac{1}{2}$-pound rocks in a bag that holds 3 pounds of rocks. The answer $\frac{1}{6}$ would make
no sense. **3.** $\frac{7}{12}$ **5.** $\frac{9}{2}$ or $4\frac{1}{2}$ **7.** $\frac{1}{9}$ **9.** $\frac{25}{9}$ or $2\frac{7}{9}$ **11.** 1 **13.** $\frac{9}{49}$ **15.** $\frac{4}{5}$ **17.** $\frac{3}{44}$ **19.** $\frac{27}{7}$ or $3\frac{6}{7}$ **21.** 0 **23.** undefined
25. 10 **27.** $\frac{7}{32}$ **29.** $\frac{3}{4}$ **31.** $\frac{13}{9}$ or $1\frac{4}{9}$ **33.** 2 **35.** 5000 **37.** $\frac{1}{250}$ **39.** $\frac{7}{40}$ **41.** 16 **43.** $\frac{7}{18}$ **45.** 2 **47.** 4
49. $\frac{68}{27}$ or $2\frac{14}{27}$ **51.** 1 **53.** $\frac{91}{75}$ or $1\frac{16}{75}$ **55.** $\frac{5}{12}$ **57.** $\frac{30}{19}$ or $1\frac{11}{19}$ **59.** 0 **61.** $\frac{7}{44}$ **63.** 12 **65.** $x = \frac{7}{5}$ **67.** $x = \frac{3}{10}$
69. $2\frac{1}{4}$ gallons **71.** $37\frac{1}{2}$ miles per hour **73.** 58 students **75.** 100 large Styrofoam cups **77.** It took six drill attempts.
79. We estimate by dividing $15 \div 5$, which is 3. The exact value is $2\frac{26}{31}$, which is very close. Our answer is off by only $\frac{5}{31}$. **81.** thirty-nine million,
five hundred seventy-six thousand, three hundred four **82.** $500,000 + 9000 + 200 + 70$ **83.** 1099 **84.** 87,595,631

Quick Quiz 2.5 **1.** $\frac{3}{4}$ **2.** $\frac{76}{29}$ or $2\frac{18}{29}$ **3.** $\frac{31}{16}$ or $1\frac{15}{16}$ **4.** See Student Solutions Manual

How Am I Doing? Sections 2.1–2.5 **1.** $\frac{3}{8}$ (obj. 2.1.1) **2.** $\frac{8}{69}$ (obj. 2.1.3) **3.** $\frac{5}{124}$ (obj. 2.1.3) **4.** $\frac{1}{6}$ (obj. 2.2.2) **5.** $\frac{1}{3}$ (obj. 2.2.2)
6. $\frac{1}{7}$ (obj. 2.2.2) **7.** $\frac{7}{8}$ (obj. 2.2.2) **8.** $\frac{4}{11}$ (obj. 2.2.2) **9.** $\frac{11}{3}$ (obj. 2.3.1) **10.** $\frac{46}{3}$ (obj. 2.3.1) **11.** $20\frac{1}{4}$ (obj. 2.3.2) **12.** $5\frac{4}{5}$ (obj. 2.3.2)
13. $2\frac{2}{17}$ (obj. 2.3.2) **14.** $\frac{5}{44}$ (obj. 2.4.1) **15.** $\frac{2}{3}$ (obj. 2.4.1) **16.** $\frac{160}{9}$ or $17\frac{7}{9}$ (obj. 2.4.3) **17.** 1 (obj. 2.5.1) **18.** $\frac{1}{2}$ (obj. 2.5.1)
19. $\frac{69}{13}$ or $5\frac{4}{13}$ (obj. 2.5.3) **20.** 21 (obj. 2.5.2)

How Am I Doing? Test on Sections 2.1–2.5 **1.** $\frac{33}{40}$ **2.** $\frac{85}{113}$ **3.** $\frac{1}{2}$ **4.** $\frac{5}{7}$ **5.** $\frac{4}{11}$ **6.** $\frac{25}{31}$ **7.** $\frac{5}{14}$ **8.** $\frac{7}{3}$ or $2\frac{1}{3}$ **9.** $\frac{38}{3}$
10. $\frac{33}{8}$ **11.** $6\frac{3}{7}$ **12.** $8\frac{1}{3}$ **13.** $\frac{21}{88}$ **14.** $\frac{7}{4}$ or $1\frac{3}{4}$ **15.** 15 **16.** $\frac{33}{2}$ or $16\frac{1}{2}$ **17.** $\frac{161}{12}$ or $13\frac{5}{12}$ **18.** 80 **19.** $\frac{16}{21}$ **20.** $\frac{16}{3}$ or $5\frac{1}{3}$
21. 7 **22.** $\frac{12}{5}$ or $2\frac{2}{5}$ **23.** $\frac{63}{8}$ or $7\frac{7}{8}$ **24.** 14 **25.** $\frac{8}{3}$ or $2\frac{2}{3}$ **26.** $\frac{23}{8}$ or $2\frac{7}{8}$ **27.** $\frac{13}{16}$ **28.** $\frac{1}{14}$ **29.** $\frac{9}{32}$ **30.** $\frac{13}{15}$
31. $45\frac{15}{16}$ square feet **32.** 4 cups **33.** $46\frac{7}{8}$ miles **34.** 16 full packages; $\frac{3}{8}$ lb left over **35.** 51 computers **36.** 24,600 gallons
37. 16 hours **38.** 6 tents; 7 yards left over **39.** 41 days

2.6 Exercises **1.** 24 **3.** 100 **5.** 60 **7.** 30 **9.** 147 **11.** 10 **13.** 28 **15.** 35 **17.** 18 **19.** 60 **21.** 32 **23.** 90
25. 80 **27.** 105 **29.** 120 **31.** 6 **33.** 12 **35.** 132 **37.** 84 **39.** 120 **41.** 3 **43.** 35 **45.** 20 **47.** 40 **49.** 96
51. 63 **53.** $\frac{21}{36}$ and $\frac{20}{36}$ **55.** $\frac{25}{80}$ and $\frac{68}{80}$ **57.** $\frac{18}{20}$ and $\frac{19}{20}$ **59.** LCD = 35; $\frac{14}{35}$ and $\frac{9}{35}$ **61.** LCD = 24; $\frac{5}{24}$ and $\frac{9}{24}$
63. LCD = 30; $\frac{16}{30}$ and $\frac{5}{30}$ **65.** LCD = 60; $\frac{16}{60}$ and $\frac{25}{60}$ **67.** LCD = 36; $\frac{10}{36}, \frac{11}{36}, \frac{21}{36}$ **69.** LCD = 56; $\frac{3}{56}, \frac{49}{56}, \frac{40}{56}$ **71.** LCD = 63; $\frac{5}{63}, \frac{12}{63}, \frac{56}{63}$
73. (a) LCD = 16 **(b)** $\frac{3}{16}, \frac{12}{16}, \frac{6}{16}$ **75.** 208 R 13 **76.** 76,980 **77.** 25

Quick Quiz 2.6 **1.** 42 **2.** 140 **3.** $\frac{21}{78}$ **4.** See Student Solutions Manual

2.7 Exercises **1.** $\frac{7}{9}$ **3.** $\frac{11}{9}$ or $1\frac{2}{9}$ **5.** $\frac{2}{5}$ **7.** $\frac{17}{44}$ **9.** $\frac{5}{6}$ **11.** $\frac{9}{20}$ **13.** $\frac{7}{8}$ **15.** $\frac{23}{20}$ or $1\frac{3}{20}$ **17.** $\frac{37}{100}$ **19.** $\frac{7}{15}$ **21.** $\frac{31}{24}$ or $1\frac{7}{24}$
23. $\frac{27}{40}$ **25.** $\frac{19}{18}$ or $1\frac{1}{18}$ **27.** 0 **29.** $\frac{5}{12}$ **31.** $\frac{11}{60}$ **33.** $\frac{1}{4}$ **35.** $\frac{2}{3}$ **37.** $\frac{1}{36}$ **39.** 0 **41.** $\frac{5}{12}$ **43.** 1 **45.** $\frac{11}{30}$ **47.** $1\frac{7}{15}$
49. $x = \frac{3}{14}$ **51.** $x = \frac{5}{33}$ **53.** $x = \frac{17}{30}$ **55.** $1\frac{5}{12}$ cups **57.** $\frac{17}{12}$ or $1\frac{5}{12}$ pounds of nuts; $\frac{7}{8}$ pound of dried fruit **59.** $\frac{19}{60}$ of the book report
61. 16 chocolates **63.** $\frac{7}{40}$ of the membership **64.** $\frac{3}{17}$ **65.** $\frac{3}{23}$ **66.** $8\frac{13}{14}$ **67.** $\frac{101}{7}$ **68.** $2\frac{8}{9}$ **69.** 7

Quick Quiz 2.7 **1.** $\frac{19}{16}$ or $1\frac{3}{16}$ **2.** $\frac{32}{21}$ or $1\frac{11}{21}$ **3.** $\frac{19}{45}$ **4.** See Student Solutions Manual

2.8 Exercises **1.** $9\frac{3}{4}$ **3.** $4\frac{1}{7}$ **5.** $17\frac{1}{2}$ **7.** 13 **9.** $\frac{4}{7}$ **11.** $2\frac{1}{16}$ **13.** $9\frac{4}{9}$ **15.** 0 **17.** $4\frac{14}{15}$ **19.** $14\frac{4}{7}$ **21.** $7\frac{2}{5}$
23. $10\frac{3}{10}$ **25.** $41\frac{4}{5}$ **27.** $8\frac{5}{12}$ **29.** $8\frac{1}{6}$ **31.** $73\frac{37}{40}$ **33.** $5\frac{1}{2}$ **35.** $\frac{2}{3}$ **37.** $4\frac{41}{60}$ **39.** $8\frac{8}{15}$ **41.** $102\frac{5}{8}$ **43.** $14\frac{1}{24}$
45. $43\frac{1}{8}$ miles **47.** $6\frac{9}{10}$ miles **49.** $2\frac{3}{4}$ inches **51. (a)** $3\frac{11}{12}$ pounds **(b)** $4\frac{1}{12}$ pounds **53.** $\frac{2607}{40}$ or $65\frac{7}{40}$
55. We estimate by adding $35 + 24$ to obtain 59. The exact answer is $59\frac{7}{12}$. Our estimate is very close. We are off by only $\frac{7}{12}$. **57.** $\frac{2}{3}$
59. 1 **61.** $\frac{3}{2}$ or $1\frac{1}{2}$ **63.** $\frac{3}{5}$ **65.** $\frac{9}{25}$ **67.** $\frac{1}{4}$ **69.** $\frac{1}{9}$ **71.** $\frac{6}{5}$ or $1\frac{1}{5}$ **73.** 480,000 **74.** 8,529,300

Quick Quiz 2.8 **1.** $9\frac{7}{40}$ **2.** $1\frac{43}{60}$ **3.** $\frac{41}{55}$ **4.** See Student Solutions Manual

2.9 Exercises **1.** $23\frac{13}{30}$ inches **3.** 385 gorillas **5.** $1\frac{9}{16}$ inches **7.** $9\frac{19}{20}$ miles **9.** $147 **11.** $275\frac{5}{8}$ gallons
13. $106\frac{7}{8}$ nautical miles **15.** $451 per week **17. (a)** 33 bracelets **(b)** $\frac{1}{5}$ foot **(c)** $313\frac{1}{2}$ **19. (a)** $14\frac{1}{8}$ ounces of bread **(b)** $\frac{5}{8}$ ounce
21. (a) $30\frac{1}{2}$ knots **(b)** 7 hours **23. (a)** 5485 bushels **(b)** $11,998\frac{7}{16}$ cubic feet **(c)** $9598\frac{3}{4}$ bushels **25.** 44,245 **26.** 22,437
27. 45,441 **28.** 356

Quick Quiz 2.9 **1.** 168 square feet **2.** 16 packets **3.** $5\frac{3}{8}$ miles **4.** See Student Solutions Manual

Putting Your Skills to Work **1. (a)** $360 **(b)** $4320 **2. (a)** Yes **(b)** Yes, there would be $520 left over for the celebration dinner.

3. If the cost of the television is $\frac{3}{4}$ of $2000, then the total would only be $1500. Thus $1020 would be left over for the birthday dinner.

4. (a) $600 **(b)** $7200 **(c)** $2880 **5.** Answers may vary

Chapter 2 Review Problems **1.** $\frac{3}{8}$ **2.** $\frac{5}{12}$ **3.** answers will vary **4.** answers will vary **5.** $\frac{9}{80}$ **6.** $\frac{87}{100}$ **7.** 2×3^3

8. $2^3 \times 3 \times 5$ **9.** $2^3 \times 3 \times 7$ **10.** prime **11.** $2 \times 3 \times 13$ **12.** prime **13.** $\frac{2}{7}$ **14.** $\frac{1}{4}$ **15.** $\frac{3}{8}$ **16.** $\frac{13}{17}$ **17.** $\frac{7}{8}$ **18.** $\frac{17}{35}$

19. $\frac{35}{8}$ **20.** $\frac{63}{4}$ **21.** $\frac{37}{7}$ **22.** $\frac{33}{5}$ **23.** $5\frac{5}{8}$ **24.** $4\frac{16}{21}$ **25.** $7\frac{4}{7}$ **26.** $8\frac{2}{9}$ **27.** $3\frac{3}{11}$ **28.** $\frac{117}{8}$ **29.** $4\frac{1}{8}$ **30.** $\frac{20}{77}$ **31.** $\frac{7}{15}$

32. 0 **33.** $\frac{4}{63}$ **34.** $\frac{492}{5}$ or $98\frac{2}{5}$ **35.** $\frac{51}{2}$ or $25\frac{1}{2}$ **36.** $\frac{82}{5}$ or $16\frac{2}{5}$ **37.** 16 **38.** $677\frac{1}{4}$ **39.** $\frac{261}{2}$ or $130\frac{1}{2}$ square feet

40. $\frac{15}{14}$ or $1\frac{1}{14}$ **41.** 6 **42.** 1920 **43.** 1500 **44.** $\frac{1}{2}$ **45.** 8 **46.** 0 **47.** $\frac{46}{33}$ or $1\frac{13}{33}$ **48.** 12 rolls **49.** $\frac{560}{3}$ or $186\frac{2}{3}$ calories

50. 98 **51.** 100 **52.** 90 **53.** $\frac{24}{56}$ **54.** $\frac{33}{72}$ **55.** $\frac{80}{150}$ **56.** $\frac{187}{198}$ **57.** $\frac{2}{7}$ **58.** $\frac{13}{12}$ or $1\frac{1}{12}$ **59.** $\frac{85}{63}$ or $1\frac{22}{63}$ **60.** $\frac{11}{40}$ **61.** $\frac{23}{70}$

62. $\frac{44}{45}$ **63.** $\frac{19}{48}$ **64.** $\frac{61}{75}$ **65.** $5\frac{1}{4}$ **66.** $\frac{49}{9}$ or $5\frac{4}{9}$ **67.** $8\frac{2}{3}$ **68.** $22\frac{3}{7}$ **69.** $\frac{49}{8}$ or $6\frac{1}{8}$ **70.** $\frac{279}{80}$ or $3\frac{39}{80}$ **71.** $\frac{9}{10}$ **72.** $\frac{3}{10}$

73. $8\frac{29}{40}$ miles **74.** $283\frac{1}{12}$ miles **75.** $1\frac{2}{3}$ cups sugar; $2\frac{1}{8}$ cups flour **76.** $206\frac{1}{8}$ miles **77.** 15 lengths **78.** $9\frac{5}{8}$ liters

79. $227\frac{1}{2}$ minutes or 3 hours and $47\frac{1}{2}$ min. **80.** $133 **81.** $577\frac{1}{2}$ **82.** $1\frac{1}{16}$ inch **83.** $242 **84. (a)** 25 miles per gallon **(b)** $58\frac{22}{25}$

85. $\frac{3}{7}$ **86.** $\frac{68}{75}$ **87.** $1\frac{5}{12}$ **88.** $\frac{24}{77}$ **89.** $\frac{17}{6}$ or $2\frac{5}{6}$ **90.** $\frac{64}{343}$ **91.** $\frac{15}{4}$ or $3\frac{3}{4}$ **92.** 99 **93.** 48

How Am I Doing? Chapter 2 Test **1.** $\frac{3}{5}$ (obj. 2.1.1) **2.** $\frac{311}{388}$ (obj. 2.1.3) **3.** $\frac{3}{7}$ (obj. 2.2.2) **4.** $\frac{3}{14}$ (obj. 2.2.2) **5.** $\frac{9}{2}$ (obj. 2.2.2)

6. $\frac{34}{5}$ (obj. 2.3.1) **7.** $10\frac{5}{14}$ (obj. 2.3.2) **8.** 12 (obj. 2.4.2) **9.** $\frac{14}{45}$ (obj. 2.4.1) **10.** 14 (obj. 2.4.3) **11.** $\frac{77}{40}$ or $1\frac{37}{40}$ (obj. 2.5.1)

12. $\frac{39}{62}$ (obj. 2.5.1) **13.** $\frac{90}{13}$ or $6\frac{12}{13}$ (obj. 2.5.3) **14.** $\frac{12}{7}$ or $1\frac{5}{7}$ (obj. 2.5.3) **15.** 36 (obj. 2.6.2) **16.** 48 (obj. 2.6.2) **17.** 24 (obj. 2.6.2)

18. $\frac{30}{72}$ (obj. 2.6.3) **19.** $\frac{13}{36}$ (obj. 2.7.2) **20.** $\frac{11}{20}$ (obj. 2.7.2) **21.** $\frac{25}{28}$ (obj. 2.7.2) **22.** $14\frac{6}{35}$ (obj. 2.8.1) **23.** $4\frac{13}{14}$ (obj. 2.8.2)

24. $\frac{1}{48}$ (obj. 2.8.3) **25.** $\frac{7}{6}$ or $1\frac{1}{6}$ (obj. 2.8.3) **26.** 154 square feet (obj. 2.9.1) **27.** 8 packages (obj. 2.9.1) **28.** $\frac{7}{10}$ mile (obj. 2.9.1)

29. $14\frac{1}{24}$ miles (obj. 2.9.1) **30.** (obj. 2.9.1) **(a)** 40 oranges **(b)** $9\frac{3}{5}$ **31.** (obj. 2.9.1) **(a)** 77 candles **(b)** $1\frac{1}{4}$ **(c)** $827\frac{3}{4}$

Cumulative Test for Chapters 1–2 **1.** eighty-four million, three hundred sixty-one thousand, two hundred eight **2.** 560 **3.** 719,220 **4.** 2075 **5.** 17,216 **6.** 4788 **7.** 840,000 **8.** 4658 **9.** 308 R 11 **10.** 49 **11.** 6,037,000 **12.** 50 **13.** $237 **14.** $306

15. $\frac{83}{112}$ were women; $\frac{29}{112}$ were men **16.** $\frac{7}{13}$ **17.** $\frac{75}{4}$ **18.** $14\frac{2}{7}$ **19.** $\frac{49}{3}$ or $16\frac{1}{3}$ **20.** $\frac{12}{35}$ **21.** 40 **22.** $\frac{61}{54}$ or $1\frac{7}{54}$

23. $\frac{71}{8}$ or $8\frac{7}{8}$ **24.** $\frac{113}{15}$ or $7\frac{8}{15}$ **25.** $\frac{1}{2}$ **26.** $2\frac{3}{4}$ pounds **27.** $24\frac{3}{5}$ miles per gallon **28.** $\frac{35}{8}$ or $4\frac{3}{8}$ cups; $7\frac{5}{8}$ cups

29. 60,000,000 miles **30.** $160

Chapter 3 3.1 Exercises **1.** A decimal fraction is a fraction whose denominator is a power of 10. $\frac{23}{100}$ and $\frac{563}{1000}$ are decimal fractions.
3. hundred-thousandths **5.** fifty-seven hundredths **7.** three and eight tenths **9.** seven and thirteen thousandths **11.** twenty-eight and thirty-seven ten-thousandths **13.** one hundred twenty-four and $\frac{20}{100}$ dollars

15. one thousand, two hundred thirty-six and $\frac{8}{100}$ dollars **17.** twelve thousand fifteen and $\frac{45}{100}$ dollars **19.** 0.7 **21.** 0.96 **23.** 0.481 **25.** 0.006114 **27.** 0.7 **29.** 0.76 **31.** 0.01 **33.** 0.053 **35.** 0.2403 **37.** 10.9 **39.** 84.13 **41.** 3.529 **43.** 235.0104

45. $\frac{1}{50}$ **47.** $3\frac{3}{5}$ **49.** $7\frac{41}{100}$ **51.** $12\frac{5}{8}$ **53.** $7\frac{123}{2000}$ **55.** $8\frac{27}{2500}$ **57.** $235\frac{627}{5000}$ **59.** $\frac{1}{80}$ **61. (a)** $\frac{153}{500}$ **(b)** $\frac{269}{1000}$ **63.** $\frac{1}{250,000}$

65. 525 **66.** 938 **67.** 56,800 **68.** 8,069,000

Quick Quiz 3.1 **1.** five and three hundred sixty-seven thousandths **2.** 0.0523 **3.** $12\frac{29}{50}$ **4.** See Student Solutions Manual

3.2 Exercises **1.** > **3.** = **5.** < **7.** > **9.** < **11.** > **13.** < **15.** > **17.** = **19.** > **21.** 12.6, 12.65, 12.8 **23.** 0.007, 0.0071, 0.05 **25.** 8.31, 8.39, 8.4, 8.41 **27.** 26.003, 26.033, 26.034, 26.04 **29.** 18.006, 18.060, 18.065, 18.066, 18.606 **31.** 6.9 **33.** 29.0 **35.** 578.1 **37.** 2176.8 **39.** 26.03 **41.** 37.00 **43.** 156.17 **45.** 2786.71 **47.** 7.816 **49.** 0.0595 **51.** 12.01578 **53.** 136 **55.** $788 **57.** $15,021 **59.** $96.34 **61.** $5783.72 **63.** 0.599; 0.481 **65.** 365.24 **67.** 0.0059, 0.006, 0.0519, $\frac{6}{100}$, 0.0601, 0.0612, 0.062, $\frac{6}{10}$, 0.61

69. You should consider only one digit to the right of the decimal place that you wish to round to. 86.23498 is closer to 86.23 than to 86.24.

71. $12\frac{1}{8}$ **72.** $10\frac{9}{20}$ **73.** 692 miles **74.** $31,800

Quick Quiz 3.2 **1.** 4.056, 4.559, 4.56, 4.6   **2.** 27.18   **3.** 155.525   **4.** See Student Solutions Manual

3.3 Exercises **1.** 76.8   **3.** 593.9   **5.** 296.2   **7.** 12.76   **9.** 36.7287   **11.** 67.42   **13.** 235.78   **15.** 1112.16   **17.** 21.04 ft.
19. 8.6 pounds   **21.** $78.12   **23.** 47,054.9   **25.** $1411.97   **27.** 3.5   **29.** 25.93   **31.** 49.78   **33.** 508.313   **35.** 135.43
37. 4.6465   **39.** 6.737   **41.** 1189.07   **43.** 1.4635   **45.** 176.581   **47.** 41.59   **49.** 5.2363   **51.** 73.225   **53.** 7.5152 pounds
55. $36,947.16   **57.** $45.30   **59.** 11.64 centimeters   **61.** 2.95 liters   **63.** 0.0061 milligram; yes   **65.** $6.2 billion; $6,200,000,000
67. $162.1 billion; $162,100,000,000   **69.** $8.40; yes; $8.34; very close: the estimate was off by 6¢.   **71.** $x = 8.4$ **73.** $x = 43.7$
75. $x = 2.109$ **77.** 20,288   **78.** 48,793   **79.** 25   **80.** 400

Quick Quiz 3.3 **1.** 72.981   **2.** 2.1817   **3.** 55.675   **4.** See Student Solutions Manual

3.4 Exercises **1.** Each factor has two decimal places. You add the number of decimal places to get 4 decimal places. Multiply 67×8 to get 536.
Place the decimal point 4 places to the left to obtain the result, 0.0536   **3.** When you multiply a number by 100, move the decimal point two places
to the right. The answer is 0.78.   **5.** 0.12   **7.** 0.06   **9.** 0.00288   **11.** 54.24   **13.** 0.000516   **15.** 0.6582   **17.** 2738.4
19. 0.017304   **21.** 768.1517   **23.** 8460   **25.** 53.926   **27.** 6.5237   **29.** $9324   **31.** $494   **33.** 297.6 square feet   **35.** $664.20
37. 514.8 miles   **39.** 28.6   **41.** 5212.5   **43.** 22,615   **45.** 56,098.2   **47.** 1,756,144   **49.** 816,320   **51.** 593.2 centimeters
53. 3281 feet   **55.** $618.00   **57.** $62,279.00   **59.** To multiply by numbers such as 0.1, 0.01, 0.001, and 0.0001, count the number of decimal
places in this first number. Then, in the other number, move the decimal point to the left from its present position the same number of decimal places
as were in the first number.   **61.** 204   **62.** 201   **63.** 127 R 3   **64.** 451 R 100   **65.** 16.6 million or 16,600,000
66. 57.3 million or 57,300,000   **67.** 28.9 million or 28,900,000   **68.** 15.8 million or 15,800,000

Quick Quiz 3.4 **1.** 0.0304   **2.** 3.2768   **3.** 51,620   **4.** See Student Solutions Manual

How Am I Doing? Sections 3.1–3.4 **1.** forty-seven and eight hundred thirteen thousandths (obj. 3.1.1)   **2.** 0.0567 (obj. 3.1.2)
3. $4\frac{9}{100}$ (obj. 3.1.3)   **4.** $\frac{21}{40}$ (obj. 3.1.3)   **5.** 1.59, 1.6, 1.601, 1.61 (obj. 3.2.2)   **6.** 123.5 (obj. 3.2.3)   **7.** 8.0654 (obj. 3.2.3)   **8.** 17.99 (obj. 3.2.3)
9. 19.45 (obj. 3.3.1)   **10.** 27.191 (obj. 3.3.1)   **11.** 10.59 (obj. 3.3.2)   **12.** 7.671 (obj. 3.3.2)   **13.** 0.3501 (obj. 3.4.1)   **14.** 4780.5 (obj. 3.4.2)
15. 37.96 (obj. 3.4.2)   **16.** 7.85 (obj. 3.4.1)   **17.** 6.874 (obj. 3.4.1)   **18.** 0.00000312 (obj. 3.4.1)

3.5 Exercises **1.** 2.1   **3.** 17.83   **5.** 10.52   **7.** 136.5   **9.** 5.412   **11.** 53   **13.** 18   **15.** 130   **17.** 5.3   **19.** 1.2   **21.** 49.3
23. 94.21   **25.** 13.56   **27.** 0.21   **29.** 0.081   **31.** 91.264   **33.** 123   **35.** 213   **37.** $82.73   **39.** approximately 27.3 miles per gallon
41. 24 bouquets   **43.** 182 guests   **45.** 23 snowboards. The error was in putting one less snowboard in the box than was required.
47. $n = 32.2$ **49.** $n = 975$ **51.** $n = 44$ **53.** 41   **54.** $\frac{127}{40}$ or $3\frac{7}{40}$ **55.** $\frac{15}{16}$ **56.** $\frac{91}{12}$ or $7\frac{7}{12}$ **57.** 15
58. $25.21 billion or $25,210,000,000   **59.** $1.22 billion or $1,220,000,000   **60.** about 3.6 times   **61.** about 14.8 times

Quick Quiz 3.5 **1.** 0.658   **2.** 3.258   **3.** 6.58   **4.** See Student Solutions Manual

3.6 Exercises **1.** same quantity   **3.** The digits 8942 repeat.   **5.** 0.25   **7.** 0.8   **9.** 0.125   **11.** 0.35   **13.** 0.62   **15.** 2.25
17. 2.875   **19.** 5.1875   **21.** $0.\overline{6}$ **23.** $0.\overline{45}$ **25.** $3.58\overline{3}$ **27.** $4.\overline{2}$ **29.** 0.308   **31.** 0.905   **33.** 0.146   **35.** 2.036   **37.** 0.404
39. 0.944   **41.** 3.143   **43.** 3.474   **45.** <   **47.** >   **49.** 0.28   **51.** 0.19 inch   **53.** yes; it is 0.025 inch too wide.   **55.** $2.\overline{3}$
57. $3.\overline{36}$ **59.** 0   **61.** 21.414   **63.** 0.0072   **65.** 0.325   **67.** 28.6   **69.** 20.836   **71.** 0.586930   **73. (a)** 0.16   **(b)** $0.144\overline{949}$
(c) b is a repeating decimal and a is a nonrepeating decimal   **75.** $5\frac{3}{4}$ feet deep   **76.** $32\frac{3}{10}$ feet

Quick Quiz 3.6 **1.** 3.5625   **2.** 0.29   **3.** 6.1   **4.** See Student Solutions Manual

3.7 Exercises **1.** 700,000,000   **3.** 30,000   **5.** 8000   **7.** 15   **9.** $20,000   **11.** 2982 kroner   **13.** 2748.27 square feet   **15.** 96 molds
17. 11.59 meters   **19.** 24 servings   **21.** $1263.09   **23.** $510   **25.** 2.763 million or 2,763,000 square kilometers   **27.** $17,319; $5819
29. yes; by 0.149 milligram per liter   **31.** 137 minutes   **33.** 22.9 quadrillion Btu   **35.** approximately 47.8 quadrillion Btu; 47,800,000,000,000,000 Btu
37. $\frac{19}{21}$ **38.** $\frac{7}{38}$ **39.** $\frac{1}{10}$ **40.** 8

Quick Quiz 3.7 **1.** 1.02 inches   **2.** 23.2 miles per gallon   **3.** $9918; $1918   **4.** See Student Solutions Manual

Putting Your Skills to Work **1.** SHELL   **2.** SHELL   **3.** ARCO   **4.** ARCO   **5.** 3.75 gallons   **6.** Answers may vary
7. Answers may vary   **8.** Answers may vary

Chapter 3 Review Problems **1.** thirteen and six hundred seventy-two thousandths   **2.** eighty-four hundred-thousandths   **3.** 0.7
4. 0.81   **5.** 1.523   **6.** 0.0079   **7.** $\frac{17}{100}$ **8.** $\frac{9}{250}$ **9.** $34\frac{6}{25}$ **10.** $1\frac{1}{4000}$ **11.** =   **12.** >   **13.** <   **14.** >
15. 0.901, 0.918, 0.98, 0.981   **16.** 5.2, 5.26, 5.59, 5.6, 5.62   **17.** 0.409, 0.419, 0.49, 0.491   **18.** 2.3, 2.302, 2.36, 2.362   **19.** 0.6   **20.** 19.21
21. 9.8522   **22.** $156   **23.** 77.6   **24.** 152.81   **25.** 14.582   **26.** 113.872   **27.** 0.003136   **28.** 887.81   **29.** 405.6   **30.** 2398.02
31. 0.613   **32.** 123,540   **33.** $8.73   **34.** 0.00258   **35.** 36.8   **36.** 232.9   **37.** 574.4   **38.** 0.059   **39.** $0.91\overline{6}$ **40.** 0.85   **41.** $1.8\overline{3}$
42. 1.1875   **43.** 0.786   **44.** 0.345   **45.** 2.294   **46.** 3.391   **47.** 19.546   **48.** 172.32   **49.** 3.538   **50.** 23.13   **51.** 439.19
52. 64.3   **53.** 4.459   **54.** 0.904   **55.** 20.004   **56.** 1.25   **57.** 112 people   **58.** 24.8 miles per gallon   **59.** $2170.30
60. ABC company   **61.** no; by 0.0005 milligram per liter   **62.** 15.75 inches   **63. (a)** 55.8 feet   **(b)** 175.68 square feet
64. 1396.75 square feet   **65.** 6.1 miles   **66.** 259.9 feet   **67.** $12,750.00; $12,255.00; they should change to the new loan   **68.** $241.00
69. $230.00   **70.** $11.37   **71.** $31.67   **72.** $43.23   **73.** $39.33

How Am I Doing? Chapter 3 Test

1. twelve and forty-three thousandths (obj. 3.1.1)   **2.** 0.3977 (obj. 3.1.2)   **3.** $7\frac{3}{20}$ (obj. 3.1.3)

4. $\frac{261}{1000}$ (obj. 3.1.3)   **5.** 2.19, 2.9, 2.907, 2.91 (obj. 3.2.2)   **6.** 78.66 (obj. 3.2.3)   **7.** 0.0342 (obj. 3.2.3)   **8.** 99.698 (obj. 3.3.1)
9. 37.53 (obj. 3.3.1)   **10.** 0.0979 (obj. 3.3.2)   **11.** 71.155 (obj. 3.3.2)   **12.** 0.5817 (obj. 3.4.1)   **13.** 2189 (obj. 3.4.2)   **14.** 0.1285 (obj. 3.5.2)
15. 47 (obj. 3.5.2)   **16.** $1.\overline{2}$ (obj. 3.6.1)   **17.** 0.875 (obj. 3.6.1)   **18.** 1.487 (obj. 3.6.2)   **19.** 6.1952 (obj. 3.6.2)   **20.** $26.95 (obj. 3.7.2)
21. 18.8 miles per gallon (obj. 3.7.2)   **22.** 3.43 centimeters (obj. 3.7.2)   **23.** $390.55 (obj. 3.7.2)

Cumulative Test for Chapters 1–3

1. thirty-eight million, fifty-six thousand, nine hundred fifty-four   **2.** 479,587   **3.** 54,480
4. 39,463   **5.** 316   **6.** 16   **7.** $\frac{2}{5}$ **8.** $8\frac{7}{24}$ **9.** $\frac{9}{35}$ **10.** $\frac{17}{20}$ **11.** 16   **12.** $\frac{33}{10}$ or $3\frac{3}{10}$ **13.** 24,000,000,000   **14.** 0.039
15. 2.01, 2.1, 2.11, 2.12, 20.1   **16.** 26.080   **17.** 21.946   **18.** 13.118   **19.** 1.435   **20.** 182.3   **21.** 1.058   **22.** 0.8125   **23.** 13.597
24. (a) 110.25 square feet **(b)** 42 feet   **25.** $195.57   **26.** 60 months

Chapter 4

4.1 Exercises

1. ratio   **3.** 5 to 8   **5.** $\frac{1}{3}$ **7.** $\frac{7}{6}$ **9.** $\frac{2}{3}$ **11.** $\frac{11}{6}$ **13.** $\frac{5}{6}$ **15.** $\frac{2}{3}$ **17.** $\frac{8}{5}$ **19.** $\frac{2}{3}$ **21.** $\frac{3}{2}$
23. $\frac{15}{19}$ **25.** $\frac{13}{1}$ **27.** $\frac{10}{17}$ **29.** $\frac{165}{285} = \frac{11}{19}$ **31.** $\frac{35}{165} = \frac{7}{33}$ **33.** $\frac{205}{1225} = \frac{41}{245}$ **35.** $\frac{450}{205} = \frac{90}{41}$ **37.** $\frac{1}{16}$ **39.** $\frac{\$7}{2\text{ pairs of socks}}$
41. $\frac{\$85}{6\text{ bushes}}$ **43.** $\frac{\$19}{2\text{ CDs}}$ **45.** $\frac{410\text{ revolutions}}{1\text{ mile}}$ or 410 revolutions/mile   **47.** $\frac{\$27,500}{1\text{ employee}}$ or $27,500/employee   **49.** $15/hour
51. 16 miles/gallon   **53.** 70 people/sq mi   **55.** 70 books/library   **57.** 66 mi/hr   **59.** 19 patients/doctor   **61.** 5 eggs/chicken
63. $30/share   **65.** $4.50 profit per puppet   **67. (a)** $0.08/oz small box; $0.07/oz large box **(b)** 1¢ per ounce **(c)** The consumer saves $0.48.
69. (a) 13 moose **(b)** 12 moose **(c)** North slope   **71. (a)** $40.95 **(b)** $52.80 **(c)** $11.85   **73.** increased by Mach 0.2
75. $2\frac{5}{8}$ **76.** 5   **77.** $\frac{5}{24}$ **78.** $1\frac{1}{48}$ **79.** $12.25/sq. yard   **80.** $24,150; $16,800

Quick Quiz 4.1

1. $\frac{3}{5}$ **2.** $\frac{340\text{ square feet}}{11\text{ pounds}}$ **3.** 27.18 trees/acre   **4.** See Student Solutions Manual

4.2 Exercises

1. equal   **3.** $\frac{6}{8} = \frac{3}{4}$ **5.** $\frac{20}{36} = \frac{5}{9}$ **7.** $\frac{220}{11} = \frac{400}{20}$ **9.** $\frac{4\frac{1}{3}}{13} = \frac{5\frac{2}{3}}{17}$ **11.** $\frac{6.5}{14} = \frac{13}{28}$ **13.** $\frac{3\text{ inches}}{40\text{ miles}} = \frac{27\text{ inches}}{360\text{ miles}}$
15. $\frac{\$40}{12\text{ cars}} = \frac{\$60}{18\text{ cars}}$ **17.** $\frac{3\text{ hours}}{\$525} = \frac{7\text{ hours}}{\$1225}$ **19.** $\frac{3\text{ teaching assistants}}{40\text{ children}} = \frac{21\text{ teaching assistants}}{280\text{ children}}$ **21.** $\frac{4800\text{ people}}{3\text{ restaurants}} = \frac{11,200\text{ people}}{7\text{ restaurants}}$
23. It is a proportion.   **25.** It is not a proportion.   **27.** It is not a proportion.   **29.** It is a proportion.   **31.** It is a proportion.
33. It is not a proportion.   **35.** It is a proportion.   **37.** It is not a proportion.   **39.** It is a proportion.   **41.** It is a proportion.
43. It is a proportion.   **45.** It is not a proportion.   **47.** no   **49. (a)** no **(b)** The van traveled at a faster rate.   **51.** yes
53. (a) yes **(b)** yes **(c)** the equality test for fractions   **54.** 23.1405   **55.** 17.9968   **56.** 402.408   **57.** 25.8   **58.** $12\frac{3}{8}$ miles

Quick Quiz 4.2

1. $\frac{8}{18} = \frac{28}{63}$ **2.** $\frac{13}{32} = \frac{3\frac{1}{4}}{8}$ **3.** It is not a proportion   **4.** See Student Solutions Manual

How Am I Doing? Sections 4.1–4.2

1. $\frac{13}{18}$ (obj. 4.1.1)   **2.** $\frac{1}{5}$ (obj. 4.1.1)   **3.** $\frac{9}{2}$ (obj. 4.1.1)   **4.** $\frac{9}{11}$ (obj. 4.1.1)
5. (a) $\frac{7}{24}$ **(b)** $\frac{11}{120}$ (obj. 4.1.2)   **6.** $\frac{3\text{ flight attendants}}{100\text{ passengers}}$ (obj. 4.1.2)   **7.** $\frac{31\text{ gallons}}{42\text{ square feet}}$ (obj. 4.1.2)   **8.** 16.25 miles per hour (obj. 4.1.2)
9. $29 per CD player (obj. 4.1.2)   **10.** 160 cookies per pound of cookie dough (obj. 4.1.2)   **11.** $\frac{13}{40} = \frac{39}{120}$ (obj. 4.2.1)   **12.** $\frac{116}{148} = \frac{29}{37}$ (obj. 4.2.1)
13. $\frac{33\text{ nautical miles}}{2\text{ hours}} = \frac{49.5\text{ nautical miles}}{3\text{ hours}}$ (obj. 4.2.1)   **14.** $\frac{3000\text{ shoes}}{\$370} = \frac{7500\text{ shoes}}{\$925}$ (obj. 4.2.1)   **15.** It is a proportion. (obj. 4.2.2)
16. It is not a proportion. (obj. 4.2.2)   **17.** It is not a proportion. (obj. 4.2.2)   **18.** It is a proportion. (obj. 4.2.2)   **19.** It is a proportion. (obj. 4.2.2)
20. It is a proportion. (obj. 4.2.2)

4.3 Exercises

1. Divide each side of the equation by the number a. Calculate $\frac{b}{a}$. The value of n is $\frac{b}{a}$.   **3.** $n = 9$ **5.** $n = 5.6$ **7.** $n = 20$
9. $n = 8$ **11.** $n = 49\frac{1}{2}$ **13.** $n = 15$ **15.** $n = 16$ **17.** $n = 7.5$ **19.** $n = 5$ **21.** $n = 75$ **23.** $n = 22.5$ **25.** $n = 192$
27. $n = 31.5$ **29.** $n = 18$ **31.** $n = 8$ **33.** $n = 162$ **35.** $n \approx 30.9$ **37.** $n \approx 30.1$ **39.** $n \approx 5.5$ **41.** $n = 48$ **43.** $n = 1.25$
45. $n \approx 3.03$ **47.** $n = 3.75$ **49.** $n = 87.36$ **51.** $n = 80$ **53.** $n = 4\frac{7}{8}$ **55.** 11 inches   **57.** $n = 3\frac{5}{8}$ **59.** $n = 10\frac{8}{9}$ **60.** 76
61. 47   **62.** five hundred sixty-three thousandths   **63.** 0.0034   **64.** $1560   **65.** 56 games

Quick Quiz 4.3 1. $n = 1.8$ **2.** $n = 2$ **3.** $n \approx 11.3$ **4.** See Student Solutions Manual

4.4 Exercises
1. He should continue with people on the top of the fraction. That would be 60 people he observed on Saturday night. He does not know the number of dogs, so this would be n. The proportion would be:

$$\frac{12 \text{ people}}{5 \text{ dogs}} = \frac{60 \text{ people}}{n \text{ dogs}}.$$

3. 161 cars   **5.** 3 cups   **7.** $7\frac{1}{2}$ kilometers   **9.** 1404 Hong Kong dollars   **11.** 197.6 feet   **13.** 217 miles   **15.** $12\frac{3}{4}$ cups

17. 102 free throws   **19.** 18.75 gallons   **21.** 40 hawks   **23.** $3570   **25.** 270 chips   **27.** 1 cup of water and $\frac{3}{8}$ cup of milk

29. 2 cups of water and 1 cup of milk   **31.** Albert Pujols, approximately $285,714 for each home run; Alfonso Soriano, approximately $217,391 for each home run   **33.** Ray Allen, approximately $49,145 for each three-point shot; Gilbert Arenas, approximately $51,457 for each three-point shot

35. 56,200   **36.** 196,380,000   **37.** 56.1   **38.** 2.7490   **39. (a)** $\frac{19}{20}$ of a square foot **(b)** 1425 square feet

Quick Quiz 4.4 1. 240 pounds   **2.** 29.09 miles   **3.** 44 free throws   **4.** See Student Solutions Manual

Putting Your Skills to Work 1. Either A or B   **2.** B   **3.** B   **4.** C   **5.** D   **6.** D   **7.** $84.99   **8.** $94.99   **9.** $114.99
10. Answers may vary

Chapter 4 Review Problems
1. $\frac{11}{5}$ **2.** $\frac{5}{3}$ **3.** $\frac{4}{5}$ **4.** $\frac{10}{19}$ **5.** $\frac{28}{51}$ **6.** $\frac{1}{3}$ **7.** $\frac{20}{59}$ **8.** $\frac{14}{25}$ **9.** $\frac{2}{5}$ **10.** $\frac{1}{8}$ **11.** $\frac{7}{32}$

12. $\frac{\$25}{2 \text{ people}}$ **13.** $\frac{4 \text{ revolutions}}{11 \text{ minutes}}$ **14.** $\frac{5 \text{ heartbeats}}{4 \text{ seconds}}$ **15.** $\frac{4 \text{ cups}}{9 \text{ cakes}}$ **16.** $17/share   **17.** $112/credit-hour   **18.** $13.50/square yard

19. $12.50/DVD   **20. (a)** $0.74 **(b)** $0.58 **(c)** $0.16   **21. (a)** $0.22 **(b)** $0.25 **(c)** $0.03   **22.** $\frac{12}{48} = \frac{7}{28}$

23. $\dfrac{1\frac{1}{2}}{5} = \dfrac{4}{13\frac{1}{3}}$ **24.** $\dfrac{7.5}{45} = \dfrac{22.5}{135}$ **25.** $\dfrac{3 \text{ buses}}{138 \text{ passengers}} = \dfrac{5 \text{ buses}}{230 \text{ passengers}}$ **26.** $\dfrac{15 \text{ pounds}}{\$4.50} = \dfrac{27 \text{ pounds}}{\$8.10}$ **27.** It is not a proportion.

28. It is a proportion.   **29.** It is a proportion.   **30.** It is a proportion.   **31.** It is not a proportion.   **32.** It is not a proportion.

33. It is a proportion.   **34.** It is not a proportion.   **35.** $n = 18$ **36.** $n = 7\frac{3}{5}$ or 7.6   **37.** $n = 22.1$ or $22\frac{1}{10}$ **38.** $n = 17$ **39.** $n = 33$

40. $n = 42$ **41.** $n = 7$ **42.** $n = 24$ **43.** $n = 19$ **44.** $n = 5\frac{3}{5}$ or 5.6   **45.** $n \approx 5.0$ **46.** $n \approx 6.8$ **47.** $n = 19$ **48.** $n = 15$
49. $n \approx 7.4$ **50.** $n \approx 5.9$ **51.** $n = 12$ **52.** $n = 550$ **53.** 15 gallons   **54.** 1691 employees   **55.** 2016 francs   **56.** 156.25 Swiss francs
57. 600 miles   **58.** 40 rebounds   **59.** 120 feet   **60. (a)** 7.65 gallons **(b)** $32.13   **61.** 5.71 centimeters tall   **62.** 7.5 grams
63. 477 pavers   **64.** 1680 students   **65.** 9 gallons   **66.** 834 liters   **67.** approximately 113.93 feet   **68.** approximately 27.38 minutes
69. 86 goals   **70.** 552 calories   **71.** 13,680 people   **72.** 195 trips

How Am I Doing? Chapter 4 Test
1. $\frac{9}{26}$ (obj. 4.1.1)   **2.** $\frac{14}{37}$ (obj. 4.1.1)   **3.** $\dfrac{98 \text{ miles}}{3 \text{ gallons}}$ (obj. 4.1.2)   **4.** $\dfrac{140 \text{ square feet}}{3 \text{ pounds}}$ (obj. 4.1.2)
5. 3.8 tons/day (obj. 4.1.2)   **6.** $8.28/hour (obj. 4.1.2)   **7.** 245.45 feet/pole (obj. 4.1.2)   **8.** $85.21/share (obj. 4.1.2)

9. $\frac{17}{29} = \frac{51}{87}$ (obj. 4.2.1)   **10.** $\dfrac{2\frac{1}{2}}{10} = \dfrac{6}{24}$ (obj. 4.2.1)   **11.** $\dfrac{490 \text{ miles}}{21 \text{ gallons}} = \dfrac{280 \text{ miles}}{12 \text{ gallons}}$ (obj. 4.2.1)   **12.** $\dfrac{3 \text{ hours}}{180 \text{ miles}} = \dfrac{5 \text{ hours}}{300 \text{ miles}}$ (obj. 4.2.1)
13. It is not a proportion. (obj. 4.2.2)   **14.** It is a proportion. (obj. 4.2.2)   **15.** It is a proportion. (obj. 4.2.2)   **16.** It is not a proportion. (obj. 4.2.2)
17. $n = 16$ (obj. 4.3.2)   **18.** $n = 22.5$ (obj. 4.3.2)   **19.** $n = 19$ (obj. 4.3.2)   **20.** $n = 29.4$ (obj. 4.3.2)   **21.** $n = 120$ (obj. 4.3.2)
22. $n = 70.4$ (obj. 4.3.2)   **23.** $n = 120$ (obj. 4.3.2)   **24.** $n = 52$ (obj. 4.3.2)   **25.** 6 eggs (obj. 4.4.1)   **26.** 80.95 pounds (obj. 4.4.1)
27. 19 miles (obj. 4.4.1)   **28.** $360 (obj. 4.4.1)   **29.** 136.6 miles (obj. 4.4.1)   **30.** 696.67 kilometers (obj. 4.4.1)   **31.** 88 free throws (obj. 4.4.1)
32. 32 hits (obj. 4.4.1)

Cumulative Test for Chapters 1–4
1. twenty-six million, five hundred ninety-seven thousand, eighty-nine   **2.** 68   **3.** $\frac{11}{32}$ **4.** $2\frac{7}{12}$

5. 65   **6.** 8.2584   **7.** 2.754   **8.** $\frac{9}{35}$ **9.** 16,145.5   **10.** 56.9   **11.** 326.278   **12.** 3.68   **13.** 0.15625   **14.** It is a proportion.

15. It is a proportion.   **16.** $n = 3$ **17.** $n = 2$ **18.** $n \approx 43.9$ **19.** $n = 9$ **20.** $n \approx 3.4$ **21.** $n = 21$ **22.** 8.33 inches
23. $139.50   **24.** 5 pounds   **25.** 214.5 gallons

Chapter 5
5.1 Exercises
1. hundred   **3.** two; left; Drop   **5.** 59%   **7.** 4%   **9.** 80%   **11.** 245%   **13.** 12.5%   **15.** 0.07%
17. 13%   **19.** 9%   **21.** 0.51   **23.** 0.07   **25.** 0.2   **27.** 0.436   **29.** 0.0003   **31.** 0.0072   **33.** 0.0125   **35.** 2.75   **37.** 74%
39. 50%   **41.** 8%   **43.** 56.3%   **45.** 0.2%   **47.** 0.57%   **49.** 135%   **51.** 516%   **53.** 27%   **55.** 20%   **57.** 94%   **59.** 231%
61. 10%   **63.** 8.9%   **65.** 0.62   **67.** 1.38   **69.** 0.003   **71.** 0.75   **73.** 52%   **75.** 1.15   **77.** 0.006   **79.** 0.165; 0.27

81. $36\% = 36 \text{ percent} = 36 \text{ "per one hundred"} = 36 \times \frac{1}{100} = \frac{36}{100} = 0.36$. The rule is using the fact that 36% means 36 per one hundred.

83. (a) 15.62 **(b)** $\frac{1562}{100}$ **(c)** $\frac{781}{50}$ **85.** $\frac{14}{25}$ **86.** $\frac{39}{50}$ **87.** 0.6875   **88.** 0.875   **89.** 5336 vases

Quick Quiz 5.1 1. 0.7%   **2.** 4.5%   **3.** 0.0125   **4.** See Student Solutions Manual

5.2 Exercises 1. Write the number in front of the percent symbol as the numerator of a fraction. Write the number 100 as the denominator of the fraction. Reduce the fraction if possible.   **3.** $\frac{3}{50}$ **5.** $\frac{33}{100}$ **7.** $\frac{11}{20}$ **9.** $\frac{3}{4}$ **11.** $\frac{1}{5}$ **13.** $\frac{19}{200}$ **15.** $\frac{9}{40}$ **17.** $\frac{81}{125}$ **19.** $\frac{57}{80}$

21. $1\frac{17}{25}$ **23.** $3\frac{2}{5}$ **25.** 12   **27.** $\frac{29}{800}$ **29.** $\frac{1}{8}$ **31.** $\frac{11}{125}$ **33.** $\frac{263}{1000}$ **35.** $\frac{7}{250}$ **37.** 75%   **39.** 70%   **41.** 35%   **43.** 72%

45. 27.5%   **47.** 360%   **49.** 250%   **51.** 412.5%   **53.** 33.33%   **55.** 41.67%   **57.** 425%   **59.** 52%   **61.** 2.5%   **63.** 5.95%

65. $37\frac{1}{2}$%   **67.** $7\frac{1}{2}$%   **69.** $26\frac{2}{3}$%   **71.** $22\frac{2}{9}$%

	Fraction	Decimal	Percent
73.	$\frac{11}{12}$	0.9167	91.67%
75.	$\frac{14}{25}$	0.56	56%
77.	$\frac{1}{200}$	0.005	0.5%
79.	$\frac{5}{9}$	0.5556	55.56%
81.	$\frac{1}{32}$	0.0313	$3\frac{1}{8}$%

83. $\frac{463}{1600}$ **85.** 15.375%   **87.** $n = 5.625$ **88.** $n = 4$ **89.** 549,165 documents   **90.** 4500 square feet

Quick Quiz 5.2 1. $\frac{9}{20}$ **2.** $\frac{19}{250}$ **3.** 92%   **4.** See Student Solutions Manual

5.3A Exercises 1. What is 20% of $300?   **3.** 20 baskets out of 25 shots is what percent?
5. This is "a percent problem when we do not know the base."
Translated into an equation: $108 = 18\% \times n$
$$108 = 0.18n$$
$$\frac{108}{0.18} = \frac{0.18n}{0.18}$$
$$600 = n$$

7. $n = 5\% \times 90$ **9.** $30\% \times n = 5$ **11.** $17 = n \times 85$ **13.** 28   **15.** 56   **17.** $51   **19.** 1300   **21.** 1300   **23.** $150   **25.** 84%
27. 11%   **29.** 65%   **31.** 31   **33.** 85   **35.** 12%   **37.** 3.28   **39.** 64%   **41.** 75   **43.** 0.8%   **45.** 18.9   **47.** 80%   **49.** 60.66%
51. 663 students   **53.** 40 years   **55.** $57.60   **57.** 2.448   **58.** 4.1492   **59.** 2834   **60.** 2.36

Quick Quiz 5.3A 1. 127.68   **2.** 9000   **3.** 17%   **4.** See Student Solutions Manual

5.3B Exercises

	p	b	a
1.	75	660	495
3.	22	60	a
5.	49	b	2450
7.	p	50	30

9. 28   **11.** 84   **13.** 56   **15.** 80   **17.** 80   **19.** 600,000   **21.** 20   **23.** 20   **25.** 22   **27.** 40   **29.** 16.4%   **31.** 3.64   **33.** 25%

35. 170   **37.** $960   **39.** 15%   **41.** 18 gallons   **43.** $2280   **45.** 33.5%   **47.** 17.2%   **49.** $1\frac{31}{45}$ **50.** $\frac{1}{26}$ **51.** $4\frac{1}{5}$ **52.** $1\frac{13}{15}$

Quick Quiz 5.3B 1. 1.53   **2.** 120   **3.** 22%   **4.** See Student Solutions Manual

How Am I Doing? Sections 5.1–5.3 1. 17% (obj. 5.1.3)   **2.** 38.7% (obj. 5.1.3)   **3.** 795% (obj. 5.1.3)   **4.** 518% (obj. 5.1.3)
5. 0.6% (obj. 5.1.3)   **6.** 0.04% (obj. 5.1.3)   **7.** 17% (obj. 5.1.1)   **8.** 89% (obj. 5.1.1)   **9.** 13.4% (obj. 5.1.1)   **10.** 19.8% (obj. 5.1.1)
11. $6\frac{1}{2}$% (obj. 5.1.1)   **12.** $1\frac{3}{8}$% (obj. 5.1.1)   **13.** 80% (obj. 5.2.2)   **14.** 50% (obj. 5.2.2)   **15.** 260% (obj. 5.2.2)   **16.** 106.25% (obj. 5.2.2)
17. 71.43% (obj. 5.2.2)   **18.** 28.57% (obj. 5.2.2)   **19.** 75% (obj. 5.2.2)   **20.** 25% (obj. 5.2.2)   **21.** 440% (obj. 5.2.2)   **22.** 275% (obj. 5.2.2)
23. 0.33% (obj. 5.2.2)   **24.** 0.25% (obj. 5.2.2)   **25.** $\frac{11}{50}$ (obj. 5.2.1)   **26.** $\frac{53}{100}$ (obj. 5.2.1)   **27.** $\frac{3}{2}$ or $1\frac{1}{2}$ (obj. 5.2.1)   **28.** $\frac{8}{5}$ or $1\frac{3}{5}$ (obj. 5.2.1)
29. $\frac{19}{300}$ (obj. 5.2.1)   **30.** $\frac{1}{32}$ (obj. 5.2.1)   **31.** $\frac{41}{80}$ (obj. 5.2.1)   **32.** $\frac{7}{16}$ (obj. 5.2.1)   **33.** 42 (obj. 5.3.2)   **34.** 24 (obj. 5.3.2)
35. 94.44% (obj. 5.3.2)   **36.** 44.74% (obj. 5.3.2)   **37.** 3000 (obj. 5.3.2)   **38.** 885 (obj. 5.3.2)

5.4 Exercises **1.** 180,000 pencils **3.** $45 **5.** 20.57% **7.** $3.90 **9.** $550 **11.** 30% **13.** $9,600,000 **15.** 2.48%
17. 216 babies **19.** $761.90 **21.** $150,000 **23.** 2200 pounds **25.** $12,210,000 for personnel, food and decorations; $20,790,000 for
security, facility rental, and all other expenses **27.** $123.50 **29. (a)** $1320 **(b)** $7480 **31.** 1,698,000 **32.** 2,452,400 **33.** 1.63
34. 0.800 **35.** 0.0556 **36.** 0.0792

Quick Quiz 5.4 **1. (a)** $166.88 **(b)** $429.12 **2.** 64.4% **3.** 15,000 people **4.** See Student Solutions Manual

5.5 Exercises **1.** $3400 **3.** $4140 **5.** 20% **7.** 36.6% **9.** $140 **11.** $7.50 **13.** $1040 **15.** 0.6% **17.** $1,600,000
19. $39.75 **21.** 39 boxes **23.** 53.51% **25.** 80% **27. (a)** $85.10 **(b)** $3785.10 **29. (a)** $6.96 **(b)** $122.96 **31.** $10,830

33. (a) $27,920 **(b)** $321,080 **35.** 96.9% **37.** $849.01 **39.** 2 **40.** 120 **41.** $\frac{13}{18}$ **42.** 3.64

Quick Quiz 5.5 **1.** $26,000 **2.** 71.875% **3.** $299 **4.** See Student Solutions Manual

Putting Your Skills to Work **1.** $25,094.08 **2.** $14,970.16 **3.** $25,970.16 **4.** Answers will vary **5.** Answers will vary
6. Answers will vary.

Chapter 5 Review Problems **1.** 62% **2.** 43% **3.** 37.2% **4.** 52.9% **5.** 220% **6.** 180% **7.** 252% **8.** 437%

9. 103.6% **10.** 105.2% **11.** 0.6% **12.** 0.2% **13.** 62.5% **14.** 37.5% **15.** $4\frac{1}{12}$% **16.** $3\frac{5}{12}$% **17.** 317% **18.** 225%

19. 76% **20.** 52% **21.** 55% **22.** 22.5% **23.** 58.33% **24.** 93.33% **25.** 225% **26.** 375% **27.** 277.78% **28.** 555.56%
29. 190% **30.** 250% **31.** 0.38% **32.** 0.63% **33.** 0.32 **34.** 0.68 **35.** 0.1575 **36.** 0.1235 **37.** 2.36 **38.** 1.77

39. 0.32125 **40.** 0.26375 **41.** $\frac{18}{25}$ **42.** $\frac{23}{25}$ **43.** $\frac{7}{4}$ **44.** $\frac{13}{5}$ **45.** $\frac{41}{250}$ **46.** $\frac{61}{200}$ **47.** $\frac{5}{16}$ **48.** $\frac{7}{16}$ **49.** $\frac{1}{1250}$ **50.** $\frac{1}{2500}$

	Fraction	Decimal	Percent
51.		0.6	60%
52.		0.7	70%
53.	$\frac{3}{8}$	0.375	
54.	$\frac{9}{16}$	0.5625	
55.	$\frac{1}{125}$		0.8%
56.	$\frac{9}{20}$		45%

57. 17 **58.** 23 **59.** 60 **60.** 40 **61.** 38.46% **62.** 38.89% **63.** 97.2 **64.** 99.2 **65.** 160 **66.** 140 **67.** 20% **68.** 10%
69. 49 students **70.** 96 trucks **71.** $11,200 **72.** $80,200 **73.** 60% **74.** 7.5% **75.** $183.50 **76.** $756 **77.** $1800 **78.** 4%
79. 6% **80.** $1200 **81. (a)** $362.50 **(b)** $1087.50 **82. (a)** $255 **(b)** $1870 **83.** 16.31% **84.** 38% **85. (a)** $3360 **(b)** $20,640
86. (a) $330 **(b)** $1320 **87. (a)** $60 **(b)** $720

How Am I Doing? Chapter 5 Test **1.** 57% (obj. 5.1.3) **2.** 1% (obj. 5.1.3) **3.** 0.8% (obj. 5.1.3) **4.** 1280% (obj. 5.1.3)

5. 356% (obj. 5.1.3) **6.** 71% (obj. 5.1.1) **7.** 1.8% (obj. 5.1.1) **8.** $3\frac{1}{7}$% (obj. 5.1.1) **9.** 47.5% (obj. 5.2.2) **10.** 75% (obj. 5.2.2)

11. 300% (obj. 5.2.2) **12.** 175% (obj. 5.2.2) **13.** 8.25% (obj. 5.2.3) **14.** 302.4% (obj. 5.2.3) **15.** $1\frac{13}{25}$ (obj. 5.2.3) **16.** $\frac{31}{400}$ (obj. 5.2.3)

17. 20 (obj. 5.3.2) **18.** 130 (obj. 5.3.2) **19.** 55.56% (obj. 5.3.2) **20.** 200 (obj. 5.3.2) **21.** 5000 (obj. 5.3.2) **22.** 46% (obj. 5.3.2)
23. 699.6 (obj. 5.3.2) **24.** 20% (obj. 5.3.2) **25.** $6092 (obj. 5.5.1) **26. (a)** $150.81 (obj. 5.4.3) **(b)** $306.19 **27.** 89.29% (obj. 5.4.1)
28. 23.24% (obj. 5.4.1) **29.** 12,000 registered voters (obj. 5.4.1) **30. (a)** $240 **(b)** $960 (obj. 5.5.3)

Cumulative Test for Chapters 1–5 **1.** 2241 **2.** 8444 **3.** 5292 **4.** 89 **5.** $\frac{67}{12}$ or $5\frac{7}{12}$ **6.** $2\frac{7}{10}$ **7.** $\frac{15}{2}$ or $7\frac{1}{2}$ **8.** 3

9. 77.183 **10.** 34.118 **11.** 8.848 **12.** 0.368 **13.** 4 tiles/square foot **14.** yes **15.** $n = 24$ **16.** 673 faculty members **17.** 35.5%
18. 46.8% **19.** 198% **20.** 3.75% **21.** 2.43 **22.** 0.0675 **23.** 17.76% **24.** 114.58 **25.** 160 **26.** 190 **27.** $544
28. 232 vehicles **29.** 11.31% **30.** $612

Practice Final Examination **1.** eighty-two thousand, three hundred sixty-seven **2.** 30,333 **3.** 173 **4.** 34,103 **5.** 4212

6. 217,745 **7.** 158 **8.** 606 **9.** 116 **10.** 32 mi/gal **11.** $\frac{7}{15}$ **12.** $\frac{42}{11}$ **13.** $\frac{33}{20}$ or $1\frac{13}{20}$ **14.** $\frac{89}{15}$ or $5\frac{14}{15}$ **15.** $\frac{31}{14}$ or $2\frac{3}{14}$

16. 4 **17.** $\frac{14}{5}$ or $2\frac{4}{5}$ **18.** $\frac{22}{13}$ or $1\frac{9}{13}$ **19.** $6\frac{17}{20}$ mi **20.** 5 packages **21.** 0.719 **22.** $\frac{43}{50}$ **23.** > **24.** 506.38 **25.** 21.77

26. 0.757 **27.** 0.492 **28.** 3.69 **29.** 0.8125 **30.** 0.7056 **31.** $\frac{1400 \text{ students}}{43 \text{ faculty}}$ **32.** no **33.** $n \approx 9.4$ **34.** $n \approx 7.7$ **35.** $n = 15$

36. $n = 9$ **37.** $3333.33 **38.** 9.75 in. **39.** $294.12 **40.** 1.6 lb **41.** 0.63% **42.** 21.25% **43.** 1.64 **44.** 17.33% **45.** 302.4
46. 250 **47.** 4284 **48.** $10,856 **49.** 4500 students **50.** 34.3%

Appendix A.1 Balancing a Checking Account Exercises 1. $555.12   3. $2912.65   5. $153.44; Yes, Justin can pay his

car insurance.   7. The account balances.   9. $949.88   11. Jeremy's account balances.

Appendix A.2 Determining the Best Deal When Purchasing a Vehicle Exercises 1. $1259.94   3. $379.98

5. $4245   7. (a) $1244.95; $497.98   (b) $27,741.93   9. (a) $3135.93; $895.98   (b) $50,930.91   11. (a) $7499.85   (b) $42,499.15
13. Dealership 1   15. (a) Dealership 3; $25,678.93   (b) Dealership 1; $28,839   (c) The most expensive purchase price does not guarantee the
most expensive total cost. Many factors need to be considered to determine the best deal.

Appendix C Scientific Calculators Exercises 1. 11,747   3. 352,554   5. 1.294   7. 254

9. 8934.7376; 123.45 + 45.9876 + 8765.3   11. 16.81; $\frac{34}{8}$ + 12.56   13. 10.03   15. 289.387   17. 378,224   19. 690,284   21. 1068.678

23. 1.58688   25. 1383.2   27. 0.7634   29. 16.03   31. $12,562.34   33. 85,163.7   35. $103,608.65   37. 15,782.038   39. 2.500344
41. 897   43. 22.5   45. 84.372   47. 4.435   49. 101.5673   51. 0.24685   53. 30.7572   55. -13.87849 57. -13.0384
59. 3.3174×10^{18} 61. $4.630985915 \times 10^{11}$ 63. -0.5517241379 65. 15.90974544   67. 1.378368601   69. 11.981752   71. 8.090444982
73. 8.325724507   75. 1.08120   77. 1.14773

Essentials of Basic College Mathematics Glossary

Addends (1.2) When two or more numbers are added, the numbers being added are called addends. In the problem $3 + 4 = 7$, the numbers 3 and 4 are both addends.

Amount of a percent equation (5.3A) The product we obtain when we multiply a percent times a number. In the equation $75 = 50\% \times 150$, the amount is 75.

Associative property of addition (1.2) The property that tells us that when three numbers are added, it does not matter which two numbers are added first. An example of the associative property is $5 + (1 + 2) = (5 + 1) + 2$. Whether we add $1 + 2$ first and then add 5 to that, or add $5 + 1$ first and then add that result to 2, we will obtain the same result.

Associative property of multiplication (1.4) The property that tells us that when we multiply three numbers, it does not matter which two numbers we group together first to multiply; the result will be the same. An example of the associative property of multiplication follows: $2 \times (5 \times 3) = (2 \times 5) \times 3$.

Base (1.6) The number that is to be repeatedly multiplied in exponent form. When we write $16 = 2^4$, the number 2 is the base.

Base of a percent equation (5.3A) The quantity we take a percent of. In the equation $8 = 20\% \times 400$, the base is 400.

Billion (1.1) The number 1,000,000,000.

Borrowing (1.3) The renaming of a number in order to facilitate subtraction. When we subtract $42 - 28$, we rename 42 as 3 tens plus 12. This represents 3 tens and 12 ones. This renaming is called borrowing.

Building fraction property (2.6) For whole numbers a, b, and c, where neither b nor c equals zero,

$$\frac{a}{b} = \frac{a}{b} \times 1 = \frac{a}{b} \times \frac{c}{c} = \frac{a \times c}{b \times c}.$$

Building up a fraction (2.6) To make one fraction into an equivalent fraction by making the denominator and numerator larger numbers. For example, the fraction $\frac{3}{4}$ can be built up to the fraction $\frac{30}{40}$.

Caret (3.5) A symbol $\wedge$ used to indicate the new location of a decimal point when performing division of decimal fractions.

Commission (5.5) The amount of money a salesperson is paid that is a percentage of the value of the sales made by that salesperson. The commission is obtained by multiplying the commission rate times the value of the sales. If a salesman sells \$120,000 of insurance and his commission rate is 0.5%, then his commission is $0.5\% \times 120{,}000 = \$600.00$.

Common denominator (2.7) Two fractions have a common denominator if the same number appears in the denominator of each fraction. $\frac{3}{7}$ and $\frac{1}{7}$ have a common denominator of 7.

Commutative property of addition (1.2) The property that tells us that the order in which two numbers are added does not change the sum. An example of the commutative property of addition is $3 + 6 = 6 + 3$.

Commutative property of multiplication (1.4) The property that tells us that the order in which two numbers are multiplied does not change the value of the answer. An example of the commutative property of multiplication is $7 \times 3 = 3 \times 7$.

Composite number (2.2) A composite number is a whole number greater than 1 that can be divided by whole numbers other than itself. The number 6 is a composite number since it can be divided exactly by 2 and 3 (as well as by 1 and 6).

Cross-multiplying (4.3) If you have a proportion such as $\dfrac{n}{5} = \dfrac{12}{15}$, then to cross-multiply, you form products to obtain $n \times 15 = 5 \times 12$.

Debit (1.2) A debit in banking is the removing of money from an account. If you had a savings account and took \$300 out of it on Wednesday, we would say that you had a debit of \$300 from your account. Often a bank will add a service charge to your account and use the word *debit* to mean that it has removed money from your account to cover the charge.

Decimal fraction (3.1) A fraction whose denominator is a power of 10.

Decimal places (3.4) The number of digits to the right of the decimal point in a decimal fraction. The number 1.234 has three decimal places, while the number 0.129845 has six decimal places. A whole number such as 42 is considered to have zero decimal places.

Decimal point (3.1) The period that is used when writing a decimal fraction. In the number 5.346, the period between the 5 and the 3 is the decimal point. It separates the whole number from the fractional part that is less than 1.

Decimal system (1.1) Our number system is called the decimal system or base 10 system because the value of numbers written in our system is based on tens and ones.

Denominator (2.1) The number on the bottom of a fraction. In the fraction $\frac{2}{9}$ the denominator is 9.

Deposit (1.2) A deposit in banking is the placing of money in an account. If you had a checking account and on Tuesday you placed \$124 into that account, we would say that you made a deposit of \$124.

Difference (1.3) The result of performing a subtraction. In the problem $9 - 2 = 7$ the number 7 is the difference.

Digits (1.1) The symbols 0, 1, 2, 3, 4, 5, 6, 7, 8, and 9 are called digits.

Discount (5.4) The amount of reduction in a price. The discount is a product of the discount rate times the list price. If the list price of a television is $430.00 and it has a discount rate of 35%, then the amount of discount is 35% × $430.00 = $150.50. The price would be reduced by $150.50.

Distributive property of multiplication over addition (1.4) The property illustrated by the following: $5 \times (4 + 3) = (5 \times 4) + (5 \times 3)$. In general, for any numbers a, b, and c, it is true that $a(b + c) = a \times b + a \times c$.

Dividend (1.5) The number that is being divided by another. In the problem $14 \div 7 = 2$, the number 14 is the dividend.

Divisor (1.5) The number that you divide into another number. In the problem $30 \div 5 = 6$, the number 5 is the divisor.

Earned run average (4.4) A ratio formed by finding the number of runs a pitcher would give up in a nine-inning game. If a pitcher has an earned run average of 2, it means that, on the average, he gives up two runs for every nine innings he pitches.

Equal fractions (2.2) Fractions that represent the same number. The fractions $\frac{3}{4}$ and $\frac{6}{8}$ are equal fractions.

Equality test of fractions (2.2) Two fractions $\frac{a}{b}$ and $\frac{c}{d}$ are equal if the product $a \times d = b \times c$. In this case, a, b, c, and d are whole numbers and b and $d \neq 0$.

Equivalent fractions (2.2) Two fractions that are equal.

Expanded notation for a number (1.1) A number is written in expanded notation if it is written as a sum of hundreds, tens, ones, etc. The expanded notation for 763 is $700 + 60 + 3$.

Exponent (1.6) The number that indicates the number of times a factor occurs. When we write $8 = 2^3$, the number 3 is the exponent.

Factors (1.4) Each of the numbers that are multiplied. In the problem $8 \times 9 = 72$, the numbers 8 and 9 are factors.

Fundamental theorem of arithmetic (2.2) Every composite number has a unique product of prime numbers.

Improper fraction (2.3) A fraction in which the numerator is greater than or equal to the denominator. The fractions $\frac{34}{29}$, $\frac{8}{7}$, and $\frac{6}{6}$ are all improper fractions.

Inequality symbol (3.2) The symbol that is used to indicate whether a number is greater than another number or less than another number. Since 5 is greater than 3, we would write this with a "greater than" symbol as follows: $5 > 3$. The statement "7 is less than 12" would be written as follows: $7 < 12$.

Interest (5.4) The money that is paid for the use of money. If you deposit money in a bank, the bank uses that money and pays you interest. If you borrow money, you pay the bank interest for the use of that money. Simple interest is determined by the formula $I = P \times R \times T$. Compound interest is usually determined by a table, a calculator, or a computer.

Invert a fraction (2.5) To invert a fraction is to interchange the numerator and the denominator. If we invert $\frac{5}{9}$, we obtain the fraction $\frac{9}{5}$. To invert a fraction is sometimes referred to as *to take the reciprocal of a fraction*.

Irreducible (2.2) A fraction that cannot be reduced (simplified) is called irreducible.

Least common denominator (LCD) (2.6) The least common denominator (LCD) of two or more fractions is the smallest number that can be divided without remainder by each fraction's denominator. The LCD of $\frac{1}{3}$ and $\frac{1}{4}$ is 12. The LCD of $\frac{5}{6}$ and $\frac{4}{15}$ is 30.

Million (1.1) The number 1,000,000.

Minuend (1.3) The number being subtracted from in a subtraction problem. In the problem $8 - 5 = 3$, the number 8 is the minuend.

Mixed number (2.3) A number created by the sum of a whole number greater than 1 and a proper fraction. The numbers $4\frac{5}{6}$ and $1\frac{1}{8}$ are both mixed numbers. Mixed numbers are sometimes referred to as *mixed fractions*.

Multiplicand (1.4) The first factor in a multiplication problem. In the problem $7 \times 2 = 14$, the number 7 is the multiplicand.

Multiplier (1.4) The second factor in a multiplication problem. In the problem $6 \times 3 = 18$, the number 3 is the multiplier.

Number line (1.7) A line on which numbers are placed in order from smallest to largest.

Numerator (2.1) The number on the top of a fraction. In the fraction $\frac{3}{7}$ the numerator is 3.

Odometer (1.8) A device on an automobile that displays how many miles the car has been driven since it was first put into operation.

Order of operations (1.6) An agreed-upon procedure to do a problem with several arithmetic operations in the proper order.

Overtime (2.9) The pay earned by a person if he or she works more than a certain number of hours per week. In most jobs that pay by the hour, a person will earn $1\frac{1}{2}$ times as much per hour for every hour beyond 40 hours worked in one workweek. For example, Carlos earns $6.00 per hour for the first 40 hours in a week and overtime for each additional hour. He would earn $9.00 per hour for all hours he worked in that week beyond 40 hours.

Parentheses (1.4) One of several symbols used in mathematics to indicate multiplication. For example, (3)(5) means 3 multiplied by 5. Parentheses are also used as a grouping symbol.

Percent (5.1) The word *percent* means per one hundred. For example, 14 percent means $\frac{14}{100}$.

Percent of decrease (5.5) The percent that something decreases is determined by dividing the amount of decrease by the original amount. If a tape deck sold for $300 and its price was decreased by $60, the percent of decrease would be $\frac{60}{300} = 0.20 = 20\%$.

Percent of increase (5.5) The percent that something increases is determined by dividing the amount of increase by the original amount. If the population of a town was 5000 people and the population increased by 500 people, the percent of increase would be $\frac{500}{5000} = 0.10 = 10\%$.

Percent proportion (5.3B) The percent proportion is the equation $\frac{a}{b} = \frac{p}{100}$ where a is the amount, b is the base, and p is the percent number.

Percent symbol (5.1) A symbol that is used to indicate percent. To indicate 23 percent, we write 23%.

Placeholder (1.1) The use of a digit to indicate a place. Zero is a placeholder in our number system. It holds a position and shows that there is no other digit in that place.

Place-value system (1.1) Our number system is called a place-value system because the placement of the digits tells the value of the number. If we use the digits 5 and 4 to write the number 54, the result is different than if we placed them in opposite order and wrote 45.

Power of 10 (1.4) Whole numbers that begin with 1 and end in one or more zeros are called powers of 10. The numbers 10, 100, 1000, etc., are all powers of 10.

Prime factors (2.2) Factors that are prime numbers. If we write 15 as a product of prime factors, we have $15 = 5 \times 3$.

Prime number (2.2) A prime number is a whole number greater than 1 that can only be divided by 1 and itself. The first fifteen prime numbers are 2, 3, 5, 7, 11, 13, 17, 19, 23, 29, 31, 37, 41, 43, and 47. The list of prime numbers goes on forever.

Principal (5.4) The amount of money deposited or borrowed on which interest is computed. In the simple interest formula $I = P \times R \times T$, the P stands for the principal. (The other letters are I = interest, R = interest rate, and T = amount of time.)

Product (1.4) The answer in a multiplication problem. In the problem $3 \times 4 = 12$ the number 12 is the product.

Proper fraction (2.3) A fraction in which the numerator is less than the denominator. The fractions $\frac{3}{4}$ and $\frac{15}{16}$ are proper fractions.

Proportion (4.2) A statement that two ratios or two rates are equal. The statement $\frac{3}{4} = \frac{15}{20}$ is a proportion. The statement $\frac{5}{7} = \frac{7}{9}$ is false, and is therefore not a proportion.

Quadrillion (1.1) The number 1,000,000,000,000,000.

Quotient (1.5) The answer after performing a division problem. In the problem $60 \div 6 = 10$ the number 10 is the quotient.

Rate (4.1) A rate compares two quantities that have different units. Examples of rates are $5.00 an hour and 13 pounds for every 2 inches. In fraction form, these two rates would be written as $\frac{\$5.00}{1 \text{ hour}}$ and $\frac{13 \text{ pounds}}{2 \text{ inches}}$.

Ratio (4.1) A ratio is a comparison of two quantities that have the same units. To compare 2 to 3, we can express the ratio in three ways: the ratio of 2 to 3; 2 : 3; or the fraction $\frac{2}{3}$.

Ratio in simplest form (4.1) A ratio is in simplest form when the two numbers do not have a common factor.

Reduced fraction (2.2) A fraction for which the numerator and denominator have no common factor other than 1. The fraction $\frac{5}{7}$ is a reduced fraction. The fraction $\frac{15}{21}$ is not a reduced fraction because both numerator and denominator have a common factor of 3.

Remainder (1.5) When two numbers do not divide exactly, a part is left over. This part is called the remainder. For example, $13 \div 2 = 6$ with 1 left over; the 1 is the remainder.

Repeating decimals (3.6) Decimals that have a digit or a group of digits that repeat. The decimals 0.33333333333 . . . and 1.234234234234 . . . are repeating decimals. The pattern of repeating continues forever. Repeating decimals can be written in a form with a bar over the repeating digit(s). Thus the preceding decimals could be written as $0.\overline{3}$ and $1.\overline{234}$.

Rounding (1.7) The process of writing a number in an approximate form for convenience. The number 9756 rounded to the nearest hundred is 9800.

Sales tax (5.4) The amount of tax on a purchase. The sales tax for any item is a product of the sales tax rate times the purchase price. If an item is purchased for $12.00 and the sales tax rate is 5%, the sales tax is $5\% \times 12.00 = \$0.60$.

Simple interest (5.4) The interest determined by the formula $I = P \times R \times T$ where I = the interest obtained, P = the principal or the amount borrowed or invested, R = the interest rate (usually on an annual basis), and T = the number of time periods (usually years).

Standard notation for a number (1.1) A number written in ordinary terms. For example, $70 + 2$ in standard notation is 72.

Subtrahend (1.3) The number being subtracted. In the problem $7 - 1 = 6$, the number 1 is the subtrahend.

Sum (1.2)   The result of an addition of two or more numbers. In the problem $7 + 3 + 5 = 15$, the number 15 is the sum.

Terminating decimals (3.6)   Every fraction can be written as a decimal. If the division process of dividing denominator into numerator ends with a remainder of zero, the decimal is a terminating decimal. Decimals such as 1.28, 0.007856, and 5.123 are terminating decimals.

Trillion (1.1)   The number 1,000,000,000,000.

Whole numbers (1.1)   The whole numbers are the set of numbers 0, 1, 2, 3, 4, 5, 6, 7, 8, 9, 10, 11, 12, ... The set goes on forever. There is no largest whole number.

Word names for whole numbers (1.1)   The notation for a number in which each digit is expressed by a word. To write 389 with a word name, we would write three hundred eighty-nine.

Zero (1.1)   The smallest whole number. It is normally written 0.

Subject Index

Applications Index

Photo Credits

CHAPTER 1 **CO** Comstock Complete **p. 79** Jean Miele/Corbis/Stock Market **p. 84** Rob & Sas/Corbis/Stock Market **p. 92** Ariel Skelley/Corbis/Stock Market **p. 94** Comstock Complete

CHAPTER 2 **CO** Purestock/Superstock Royalty Free **p. 127** John Paul Endress/Corbis/Stock Market **p. 175** Photolibrary.com **p. 177** Bill Stanton/ImageState Media Partners Limited **p. 182** Lim ChewHow/Shutterstock

CHAPTER 3 **CO** EyeWire Collection/Getty Images-Photodisc **p. 202** © Dale C. Spartas/CORBIS **p. 245** A. Ramey/Woodfin Camp & Associates, Inc. **p. 247** Getty Images-Stockbyte **p. 248** Catherine Ursillo/Photo Researchers, Inc. **p. 249** David R. Frazier Photolibrary/Photo Researchers, Inc. **p. 251** Javier Larrea/Pixtal/Superstock Royalty Free

CHAPTER 4 **CO** Kim Sayer © Dorling Kindersley **p. 271** Lester Lefkowitz/Corbis/Stock Market **p. 286** Peter Saloutos/Corbis/Bettman **p. 288** Photos.com **p. 293** Comstock Complete

CHAPTER 5 **CO** Hans Peter Merton/Robert Harding World Imagery **p. 323** Getty Images **p. 330** SuperStock, Inc. **p. 337** Juice Images/Art Life Images **p. 350** Comstock Complete

METRIC SYSTEM MEASUREMENTS

Length

1 kilometer	(km)	=	1000 meters
1 hectometer	(hm)	=	100 meters
1 dekameter	(dam)	=	10 meters
1 meter	(m)	=	1 meter
1 decimeter	(dm)	=	0.1 meter
1 centimeter	(cm)	=	0.01 meter
1 millimeter	(mm)	=	0.001 meter

Weight

1 metric ton	(t)	=	1,000,000 grams
1 kilogram	(kg)	=	1000 grams
1 hectogram	(hg)	=	100 grams
1 dekagram	(dag)	=	10 grams
1 gram	(g)	=	1 gram
1 decigram	(dg)	=	0.1 gram
1 centigram	(cg)	=	0.01 gram
1 milligram	(mg)	=	0.001 gram

Volume

1 kiloliter	(kL)	=	1000 liters
1 hectoliter	(hL)	=	100 liters
1 dekaliter	(daL)	=	10 liters
1 liter	(L)	=	1 liter
1 deciliter	(dL)	=	0.1 liter
1 centiliter	(cL)	=	0.01 liter
1 milliliter	(mL)	=	0.001 liter

Temperature: Celsius Scale

100°C = Boiling point of water

−273.15°C = Absolute zero: coldest possible temperature

0°C = Freezing point of water

37°C = Normal human body temperature

AMERICAN SYSTEM MEASUREMENTS

Length

1 mile	(mi)	=	1760 yards (yd)
1 mile	(mi)	=	5280 feet (ft)
1 yard	(yd)	=	3 feet (ft)
1 foot	(ft)	=	12 inches (in.)

Volume

1 gallon	(gal)	=	4 quarts (qt)
1 quart	(qt)	=	2 pints (pt)
1 pint	(pt)	=	2 cups (c)

Weight

1 ton	(T)	=	2000 pounds (lb)
1 pound	(lb)	=	16 ounces (oz)

APPROXIMATE EQUIVALENT MEASURES FOR CONVERSION OF UNITS

	American to Metric	Metric to American
Units of Length	1 mile = 1.61 kilometers 1 yard = 0.914 meter 1 foot = 0.305 meter 1 inch = 2.54 centimeters	1 kilometer = 0.62 mile 1 meter = 3.28 feet 1 meter = 1.09 yards 1 centimeter = 0.394 inch
Units of Volume	1 gallon = 3.79 liters 1 quart = 0.946 liter	1 liter = 0.264 gallon 1 liter = 1.06 quarts
Units of Weight	1 pound = 0.454 kilogram 1 ounce = 28.35 grams	1 kilogram = 2.2 pounds 1 gram = 0.0353 ounce